Life Sciences Society

COMPUTATIONAL SYSTEMS BIOINFORMATICS

CONFERENCE PROCEEDINGS – Volume 7

CSB2008

Life Sciences Society

COMPUTATIONAL SYSTEMS BIOINFORMATICS

Stanford University, USA
26–29 August 2008

Editors

Peter Markstein
in silico Labs, LLC, USA

Ying Xu
University of Georgia, USA

ICP Imperial College Press

Published by

Imperial College Press
57 Shelton Street
Covent Garden
London WC2H 9HE

Distributed by

World Scientific Publishing Co. Pte. Ltd.
5 Toh Tuck Link, Singapore 596224
USA office: 27 Warren Street, Suite 401-402, Hackensack, NJ 07601
UK office: 57 Shelton Street, Covent Garden, London WC2H 9HE

British Library Cataloguing-in-Publication Data
A catalogue record for this book is available from the British Library.

COMPUTATIONAL SYSTEMS BIOINFORMATICS
Proceedings of the CSB 2008 Conference — Vol. 7

ISBN-13 978-1-84816-263-1
ISBN-10 1-84816-263-4

Printed in Singapore by Mainland Press Pte Ltd

PREFACE

Looking over the conference papers and the list of invited speakers of the past last seven years, one can say that the Computational Systems Bioinformatics (CSB) group together with the Life Sciences Society have accomplished their goals: CSB and LSS have forged a stronger collaboration between interdisciplinary researchers from different fields: biology, mathematics, engineering, computer science and medicine.

CSB papers describe computational techniques to help bring better understanding of biological processes. We hope that as a result, these techniques will bring significant advances in the understanding of some of the most devastating diseases such as cancer, Alzheimer's and HIV. The hope is that as the increase in understanding continues to advance then so will the computational methods to guide patient therapies.

As an example, Professor Matthew Scott's Keynote talk at CSB2008 is intended to inspire computational scientists to undertake investigations in the area of signal transduction. In future CSB conferences, we look forward to presenting papers that show how computation directly leads to cures of human diseases.

Join LSS and bring your skills to help obtain faster cures.

Vicky Markstein, CSB2008 General Chair

COMMITTEES

Steering Committee

Phil Bourne - University of California, San Diego
Eric Davidson - California Institute of Technology
Steven Salzberg - The Institute for Genomic Research
John Wooley - University of California San Diego, San Diego Supercomputer Center

Organizing Committee

Pat Blauvelt – LSS Membership Director
Ed Buckingham – **Co-Chair**, LSS VP Conferences
Karen Hauge – Palo Alto Medical Foundation, Local Arangements
Kass Goldfein - Finance Consultant
Sami Khuri – San Jose State University, Poster Chair
Ann Loraine – University of North Carolina at Charlotte, CSB Publication Chair
Fenglou Mao – University of Georgia, On-Line Registration and Refereeing Website
Peter Markstein – in silico Labs, Program Co-Chair
Vicky Markstein - Life Sciences Society, **Co-Chair**, LSS President
Josh Stuart – University of California Santa Cruz, Tutorial Chair
Jean Tsukamoto - Graphics Design
Bill Wang - Sun Microsystems Inc, LSS Information Technology Director
Ying Xu – University of Georgia, Program Co-Chair

Program Committee

Tatsuya Akutsu - Kyoto University
Joel Bader – Johns Hopkins University
Jeremy Buhler – Washington University
Jake Chen - Indiana University
Amar Das - Stanford University
Roderic Guigo – IMIM, Barcelona
Wen-Lian Hsu – Academia Sinica
Lydia Kavraki – Rice University
Jing Li – Case Western Reserve
Stefano Lonardi – University of California Riverside
Ann Loraine - University of North Carolina, Charlotte
Bin Ma – University of Western Ontario
Peter Markstein – *in silico* Labs, **Co-chair**
Satoru Miyano - University of Tokyo
Sean Mooney - Indiana University
Chad Myers – University o Minnesota
Isidore Rigoutsos - IBM TJ Watson Research Center
Hershel M. Safer – Weizmann Institute of Science

Mona Singh – Princeton University
Victor Solovyev, Royal Holloway, University of London
David States - University of Michigan
Limsoon Wong – University of Singapore
Dong Xu – University of Missouri
Ying Xu - University of Georgia, **Co-chair**
Xianghong Jasmine Zhou - University of Southern California

Assistant to the Program Co-Chairs

Joan Yantko – University of Georgia

Poster Committee

Sami Khuri – San Jose State University, **Chair**
Lee Kozar – Stanford University
Helen Moore – Genentech
Sue Rhee – Carnegie Institution for Science

Tutorial Committee

Josh Stuart – UC Santa Cruz, **Chair**
Ting Wang – UC Santa Cruz, **Co-Chair**
Jing Zhu – UC Santa Cruz, **Co-Chair**

Stanford Faculty Sponsor

Russ Altman, MD, PhD – Chairman of the Department of BioEngineering

REFEREES

Tatsuya Akutsu

Joel Bader
Gregory Baramidze
Serdar Bozdag
Jeremy Buhler

C. Q. Chang
Huiling Chen
Jake Chen
Yong Chen
Jianlin Cheng
Wayne Cheng-Wei Cheng
Hon Nian Chua
Juan Cui

Hong Jie Dai
Phuongan Dam
Amar Das
Thomas Derrien
Dong Difeng

Stephen Erickson

Anthony Fodor
Richard A. Friedman

Greg Gonye
Roderic Guigo

Elena Harris
Nurit Haspel
Morihiro Hayashida
Allison Heath
Wen-Lian Hsu

Seiya Imoto

Jeesun Jung

Yuki Kato
Lydia Kavraki
Mehmet Koyoturk

Vincent Lacroix
Jing Li
Wenyuan Li
Xiaoli Li
Yunlong Liu
Stefano Lonardi
Ann Loraine

Bin Ma
Peter Markstein
Brett McKinney
Michael Mehan
Samy Meroueh
Satoru Miyano
Mark Moll
Sean Mooney
Chad Myers

Juan Nunez-Iglesias

Victor Olman

Daniel E. Platt
Natasa Przulj

Predrag Radivojac
Isidore Rigoutsos
Ajay K Royyuru

Hershel M. Safer
Mona Singh
Victor Solovyev
David States
Emily Su

Thao Tran

Vladimir Vacic

Hongwei Wu
Xiaogang Wu
Yonghui Wu

Dong Xu
Min Xu
Ying Xu

Rui Yamaguchi

Fengfeng Zhou
Xianghong Jasmine Zhou
Matthias Zytnicki

CONTENTS

Computational Genomics

Computational Methods

Computational Systems Bioinformatics 2008

Genomics

AN ORFOME ASSEMBLY APPROACH TO METAGENOMICS SEQUENCES ANALYSIS

Yuzhen Ye* and Haixu Tang
School of Informatics, Indiana University
Bloomington, Indiana 47408, USA
**Email: yye@indiana.edu*

Metagenomics is an emerging methodology for the direct genomic analysis of a mixed community of uncultured microorganisms. The current analyses of metagenomics data largely rely on the computational tools originally designed for microbial genomics projects. The challenge of assembling metagenomic sequences arises mainly from the short reads and the high species complexity of the community. Alternatively, individual (short) reads will be searched directly against databases of known genes (or proteins) to identify homologous sequences. The latter approach may have low sensitivity and specificity in identifying homologous sequences, which may further bias the subsequent diversity analysis. In this paper, we present a novel approach to metagenomic data analysis, called Metagenomic ORFome Assembly (MetaORFA). The whole computational framework consists of three steps. Each read from a metagenomics project will first be annotated with putative open reading frames (ORFs) that likely encode proteins. Next, the predicted ORFs are assembled into a collection of peptides using an EULER assembly method. Finally, the assembled peptides (i.e., ORFome) are used for database searching of homologs and subsequent diversity analysis. We applied MetaORFA approach to several metagenomics datasets with low coverage short reads. The results show that MetaORFA can produce long peptides even when the sequence coverage of reads is extremely low. Hence, the ORFome assembly significantly increased the sensitivity of homology searching, and may potentially improve the diversity analysis of the metagenomic data. This improvement is especially useful for the metagenomic projects when the genome assembly does not work because of the low sequence coverage.

Keywords: Metagenomics, ORFome, ORFome assembly, Function annotation.

1. INTRODUCTION

Owning to the rapid advancement of the ultra-high throughput DNA sequencing technologies [1], the genomic studies of microorganisms in environmental samples have recently shifted from the focused sequencing of 16sRNA sequences [2] to the shotgun sequencing of the whole DNAs in the sample. This new methodology, now called *metagenomics* or *environmental genomics*, has opened a door for biologists to assess the unknown world of the uncultured microorganisms that are believed to be the majority in any environmental sample. The early attempts of this kind can be traced back to a report published in 2002, in which extremely high diversity of uncultured marine viral communities were revealed through genome sequencing [3]. However, the most important progress in shotgun metagenomics happened in 2004 [4–7], when two research groups published results from their large-scale environmental sequencing projects. The first project studied the sample from the Sargasso Sea, and revealed $\sim$ 2000 distinct species of microorganisms, including 148 types of bacteria that have never been observed before [8]. In the second project, a handful of genomes of bacteria and archaea that had previously resisted attempts to culture them were revealed based on the analysis of the sample from the acid mine drainage [9]. Since then, many more metagenomics projects have been conducted, involving broadened applications from ecology and environmental sciences to chemical industry [10] and human health, e.g., the human gut microbiome projects [11, 12].

The rapid growth of metagenomic data has posed great challenges to the computational analysis [13, 14]. Some metagenomics projects applied directly the data analysis pipeline that includes the whole genome assemblers [15–18] and gene finding programs [19]—originally designed for the conventional Whole Genome Shotgun (WGS) sequencing projects—with only some small parameter modifications [8, 9, 12, 20]. However, it is unclear how accurate these existing tools for fragment assembly and genome annotation are when applied to metagenomic

*Corresponding author.

data. Mavromatis and colleagues have conducted a valuable benchmarking experiment to evaluate the performance of conventional genome assembly and annotation pipeline on simulated metagenomic data [21]. In this experiment, sequencing reads were randomly collected from 113 assembled genomes that are mixed at various complexities. Afterwards, the quality of the results from each processing step (i.e., assembly, gene prediction, and phylogenetic binning) was assessed separately by comparison to the corresponding genomes used in the simulation. This experiment delivered an encouraging message that the number of errors made at each step overall is not high, and some errors (e.g., the chimeric contigs) would not be propagated into the subsequent steps (e.g., binning). Nevertheless, we argue that this experiment may not completely reflect the challenge of metagenomic data analysis, especially the difference between metagenomic data and the data from conventional genome sequencing. Conventional genome projects deal with only one or sometimes a few individual genomes from the same species that are isolated prior to sequencing, whereas metagenomics attempts to analyze simultaneously a huge amount of genomes not only from hundreds of different microorganisms, but also from many individuals of each organism. As a result, even the reads from the same species might be quite different from each other since they might be sampled from different individuals' genomes. Furthermore, those microbial species may exist in the sample at a wide range of abundances. Hence, typically, only a few dominant species can receive good sequence coverage for their genomes, whereas the sequence coverage for the remaining species is low.

More and more metagenomic projects have applied Next-Generation Sequencing (NGS) technologies that produce massive but shorter reads (e.g., $\sim$ 200 bps for 454 pyrosequencing machines) than those from the Sanger sequencing methods. Therefore, many metagenomic sequencing projects that acquired a merely small number of short sequencing reads often skipped the step of fragment assembly, and directly used the short reads for downstream analysis [3, 22, 23]. For instance, short reads can be used to search against protein database using TBLASTX to identify homologous proteins, in which an arbitrary E-value (e.g., $\leq 1e-5$) was chosen as a cutoff [22]. This direct search approach, however, often missed many homologous genes (or proteins) [24], and resulted in a very low false positive rate [a] but high false negative rate. This drawback may bias the further analysis of species diversity (i.e., how many different species are present in the sample) and functional coverage (i.e., how many functional categories of proteins are present in the sample).

In this paper, we present a novel *ORFome assembly* approach to assembling metagenomic sequencing reads. Different from the conventional genome analysis pipeline that first assembles sequencing reads into contigs (or scaffolds) and then predicts protein coding regions within the contigs, our method first identifies putative protein coding regions (i.e., open reading frames, or ORFs) within unassembled reads, and then focuses on the assembly of only these sequences (i.e., *ORFome*). The ORFome assembly approach has several advantages. First, it significantly simplifies the task of fragment assembly that is often complicated by the repetitive sequences present mainly in non-coding regions [25]. Meanwhile, we argue that ORFome assembly does not lose much useful information by neglecting the non-coding sequences due to several reasons: (1) the set of proteins (or the ORFome that encodes them) carry the most important information for the downstream analysis; (2) the microbial genomes are often very compact and protein coding regions comprise a major fraction of them; and (3) microbial proteins are mainly encoded by continuous non-split open reading frames (ORFs), thus the prediction of coding sequences prior to assembly is relatively straightforward. Second, from ORFome assembly, complete proteins (or long peptides) may be derived, thus higher sensitivity and specificity can be achieved in the step of database searching for homologs [24]. Furthermore, most single nucleotide polymorphisms are synonymous mutations that do not change the encoding amino acids so that ORFome assembly does not even feel them. So by working on the peptide sequences (translated from sequencing reads *in silico*)

[a]For example, the MEGAN analysis based on the direct BLAST search method has achieved a 0 false positive rate [23]!

instead of the raw DNA sequences, the ORFome assembly alleviates the assembly difficulty caused by the differences among individual genomes at polymorphic sites. We used four marine viral metagenomic datasets of short reads, acquired using 454 sequencing technique, to test our ORFome assembly method—no genome assemblies are available for these metagenomic datasets because the reads are extremely short and the sequence coverage is low.

2. METHODS

The computational framework of ORFome assembly consists of three steps (Fig. 1 (e-f)): (1) each read is assessed individually and the putative open reading frames (ORFs) that likely encode proteins are annotated; (2) the annotated ORFs are assembled into a collection of peptides using a modified EULER assembly method [26]; and (3) the assembled peptides are used for the database searching of homologs.

A major difference between the ORFome assembly approach and the conventional whole genome assembly is that the former approach conducts gene annotation before assembly, whereas the latter approach conducts gene annotation after assembly. Conventional fragment assembly algorithms are mostly based on the analysis of overlap graph, in which the reads are represented by vertices and the overlaps between reads are represented by edges [27]. The presence of repeats in the genomes often induce many spurious edges in the overlap graph, which is a major challenge in fragment assembly. There are two additional aspects in the metagenomic data that make fragment assembly even more challenging. First, metagenomics projects often apply NGS technique, and produce shorter reads ($\sim$ 200 bps) than Sanger sequencing methods (500-1000 bps). As a result, many short repeats (with lengths between 200 bps and 500 bps) may increase the complexity of the overlap graph, and cause many more mis-assemblies [28]. Second, unlike the conventional genome shotgun sequencing, which handles a single species, metagenomics sequencing reads are collected from a large amount of different genomes. Hence, we anticipate these reads should be assembled into not one but many sequences that may even share high similarity on multiple regions. Therefore, the straightforward application of conventional fragment assemblers may encounter difficulties. In contrast, the ORFome assembly approach attempts to assemble only the most important portions of the target genomes, i.e., the protein coding regions, which can highly reduce the complexity of the overlap graph and thus improve the assembly quality.

It is worth pointing out the idea of ORFome assembly can be viewed as an extension of the repeat masking approach used in whole genome assembly of large eukaryotic (including human) genomes. To avoid the complication induced by the many interspersed repeat copies present in most eukaryotic genomes, Celera Assembler first masked out putative repeats in the unassembled reads, and then focused on the assembly of the remaining reads from non-repetitive regions [29, 30]. The resulting overlap graph, which consists of a number of connected components each representing reads from continuous non-repetitive regions, is much simpler and easy to be analyzed. Similarly, the ORFome assembly approach divides the complex overlap graph into a number of components each representing reads from a single gene or several highly similar genes from the same family.

We applied the ORFome assembly approach to several metagenomics datasets from Ocean samples with low coverage and short reads [22]. The results show that MetaORFA can produce long peptides even when the sequence coverage of reads is extremely low. Hence, further analysis of assembled peptides significantly increased the sensitivity for subsequent homology searching, and may potentially improve the diversity analysis of the metagenomic data.

2.1. ORFome Assembly Algorithm

We implemented a tool called MetaORFA in C/C++ under linux platforms for the ORFome assembly. MetaORFA consists of two programs. One program takes as input a set of reads and predicts a number of putative ORFs; and the other program takes as input the set of putative ORFs, and reports a set of peptides corresponding to the assembled ORFs. Prior to be supplied to MetaORFA, the original reads were first processed by MDUST (a popular tool for autonomous masking from TIGR, which implements the DUST algorithm [31]) to mask out low-

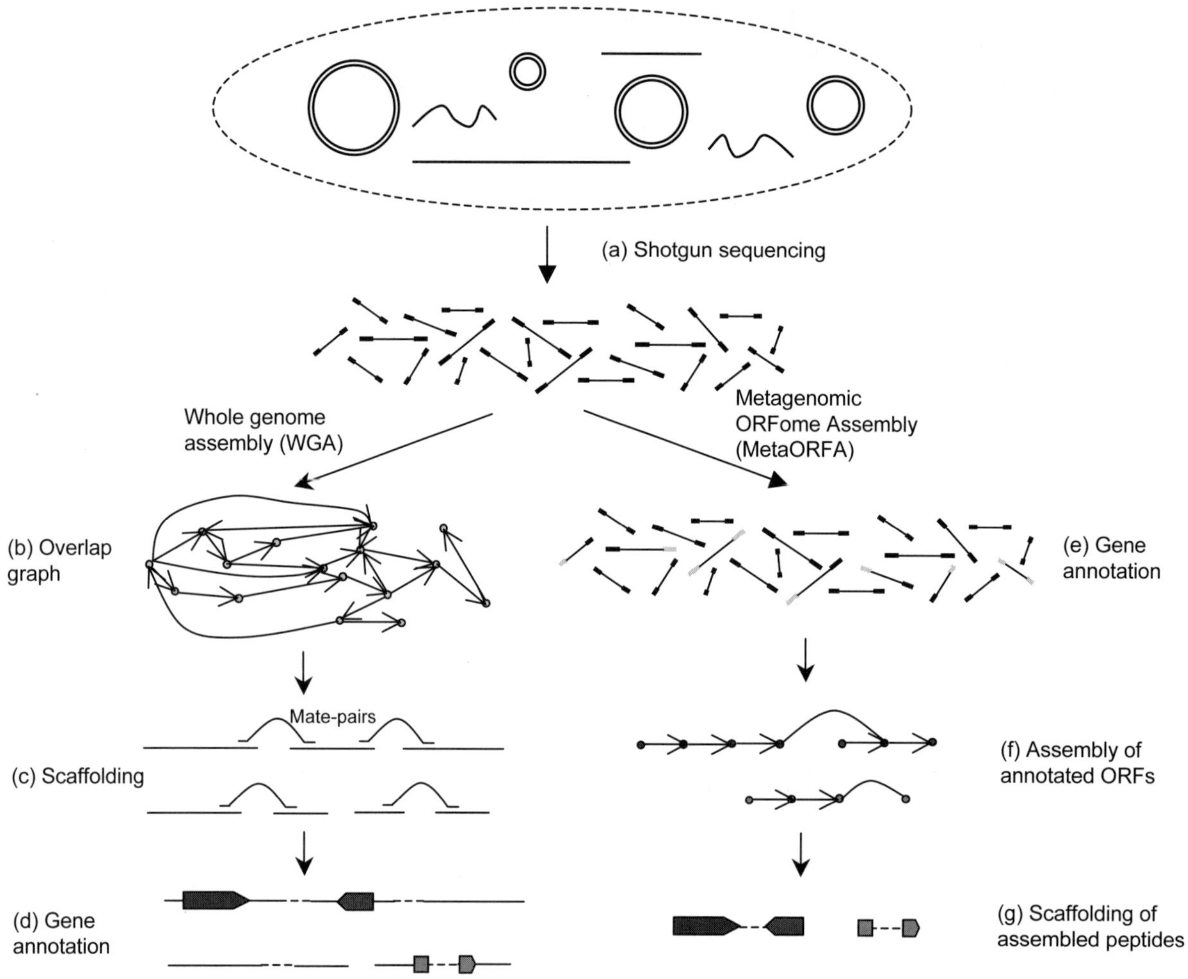

Fig. 1. A schematic comparison of the ORFome assembly approach with the whole genome assembly (WGA) pipeline for the metagenomic sequence analysis. Both approaches attempt to characterize the protein coding genes in the shotgun sequencing reads from the metagenomic analysis of an environmental sample containing a number of different microorganisms (the reads are shown as double-barreled, as currently several NGS techniques are capable of generating such data; however, some early metagnomics projects, including the datasets used in this paper, did not produce double-barreled sequencing reads, and thus the scaffolding step is not feasible) (a). The whole genome assembly (WGS) pipeline (b-d) first assembles the reads into contigs and scaffolds, and then annotates the genes in the assembled sequences. In comparison, ORFome assembly approach (e-g) first applies gene finding in the unassembled reads, and then assembles only those annotated (partial) ORFs into peptides. These peptides may be further connected to form scaffolds if there are mate-pairs available from double-barreled sequencing (g).

complexity regions, and then processed by Tandem Repeat Finder (TRF V4.0) [32] to mask out short tandem repeats.

In this preliminary study, we adopted a very simple method for ORF prediction. For each read (and its reverse complement), a region from the beginning (i.e., position 1, 2, or 3, depending on the frame) or a start codon to the end of the read or a stop codon is considered as a potential ORF. Only ORFs with more than a threshold K (default $K = 30$) codons were reported. These ORFs will be then transformed into peptide sequences, and subsequently assembled using a modified EULER algorithm originally designed for DNA fragment assembly [26]. In this process, we first build a de Bruijn graph using all k-mers (default $k = 10$) in the putative peptides from previous step, and then apply the equivalent transformations as described in Ref. 26 to resolve short repeats among peptides. Unlike many other genome assemblers that assemble reads into linear contigs, EULER aims at constructing from the reads a *repeat graph* that represents not only the unique regions but also the repeat structures [33]. Although we anticipate there are not many repeats in the coding sequences, the similar parts of homologous proteins from the same family may act like repeats during the ORFome assembly. EULER assembly method can generate a compact graph structure representing the architecture of domain combinations, including domain recurrences and shuffling[34]. We note further analysis of the ORFome assembly results, as described below, has not fully taken advantage of the information embodied in the repeat graph. Ideally, one can adopt a network matching approach to identify a path in the repeat graph representing a peptide sequence that is most similar to a protein in databases of known proteins. Nevertheless, our analysis has demonstrated that even the simple analysis of individual assembled peptides (corresponding to the edges in the repeat graph) revealed more proteins in the sample.

2.2. Functional Coverage Assessment

The ORFome, i.e., the set of assembled peptides, is ready for further computational analysis with different purposes, e.g., searching against database for homologous sequences, or mapping to biological pathways to study metabolic diversity [35]. Here we show that we can improve the functional coverage of metagenomics sequences by using assembled peptides instead of unassembled reads. There are various ways to estimate functional coverage of a sample. In this study we used PANTHER (Protein ANalysis THrough Evolutionary Relationships) protein family classification [36] for such assessment. The comparison of the functional coverage between different ORFomes is then straightforward, and we can simply count the number of families (subfamilies) found in each assembled ORFome and calculate their differences.

In the PANTHER classification system, proteins are classified into families and subfamilies of shared function by experts. Families and subfamilies are presented as Hidden Markov Models (HMMs). We downloaded the PANTHER HMM library Version 6.1 (release date December 17, 2007) from ftp://ftp.pantherdb.org, which contains 5547 protein family HMMs, divided into 24,582 functionally distinct protein subfamily HMMs. We also downloaded the HMM searching tool (pantherScore.pl, version 1.02), which utilized fast BLAST search prior to the more sensitive but time-consuming HMM matching procedure to speed up the process. The query protein sequence will first be blasted against the consensus sequences of each PANTHER HMMs, and then based on the results, some heuristics are applied to determine which HMMs (i.e., protein families or subfamilies) that the query should be compared with using hmmsearch from the hmmer package (http://hmmer.janelia.org).

2.3. Metagenomic Sequences Datasets

We tested our algorithm on four datasets each containing metagenomics sequences of a major oceanic region community (the four regions are Sargasso Sea, Coast of British Columbia, Gulf of Mexico, and Arctic Ocean) (referred to as Ocean Virus datasets) [22]. The reads were acquired by 454 sequencing machine, and they are typically very short. All the metagenomic sequences were downloaded from CAMERA website (http://camera.calit2.net/) [37].

3. RESULTS

We applied our ORFome assembly tool MetaORFA to assemble the four Ocean Virus datasets. The assembly of a dataset took about from several minutes to half an hour for the four datasets we used here (on a linux machine with Intel(R) Core(TM)2 CPU@ 2.40GHz). The unassembled reads and assembled peptides were searched against Integrated Microbial Genomics (IMG) database [38] using BLASTP to identify known homologous proteins in pre-sequenced microbial genomes. To show the improvement of functional coverage after the ORFome assembly, we also searched both sets of sequences against PANTHER families and subfamilies. Below we first report the basic statistics of the assembled peptides as compared to the unassembled reads, and then show the annotation of the ORFs by BLAST search and PANTHER family annotation.

Table 1. Statistics of the ORFs for Ocean Virus datasets

Sample		Num	Min	Max	Ave	Num60
Arctic Ocean	Reads	688590	35	370	99	-
	UA-ORF	1015432	30	58	33	0
	A-Pep	368278	30	175	37	12428
Sargasso Sea	Reads	399343	36	282	104	-
	UA-ORF	345411	30	49	33	0
	A-Pep	214330	30	162	34	1530
Coast of British	Reads	16456	37	254	102	
Columbia	UA-ORF	426666	30	61	33	1
	A-Pep	304106	30	196	34	2559
Gulf of Mexico	Reads	771849	38	246	95	-
	UA-ORF	467085	30	54	33	0
	A-Pep	206111	30	157	34	1703

Num, Min, Max and Ave represent the total number, the minimum, maximum, and average length of the reads (in nucleotides), unassembled ORFs (UA-ORF, in amino acid residues) and assembled peptides (A-Pep, in amino acid residues), respectively. Num60 represents the total number of unassembled ORFs and assembled peptides of length ≥ 60. We note that the minimum length of unassembled ORFs is 30 because we used a default cutoff 30 to detect ORFs in original short reads, which are used as inputs for MetaORFA.

3.1. Assembled Peptides from the ORFome Assembly

Table 1 shows the statistics of the reads, unassembled putative ORFs and assembled peptides for the four Ocean Viruses datasets. For all four datasets, the ORFome assembly successfully produced long peptides (≥ 60) that are not present in the unassembled reads. However, the number and the length of long peptides are different from one dataset to another. For example, the ORFome assembly produced the largest number (12,428) of long peptides with longest average length (37 aa) in the Arctic Ocean dataset, even though comparable number of sequencing reads were acquired in each of these four datasets. This may indicate either the diversity of the microorganisms in Arctic Ocean sample is lower than the diversity in the other samples, or the microorganism genomes in this sample are more compact than the genomes in the other samples.

We use the second longest peptide assembled from the Gulf of Mexico dataset as an example to illustrate the advantages of the ORFome assembly. [b] Fig. 2 shows that 12 putative ORFs detected from different short reads were assembled into a long peptide (144 aa) by the ORFome assembly, which shows strong similarity across the entire peptide with an annotated protein in IMG database.

3.2. Homology Search of Assembled Peptides

One of the commonly used analysis of metagenomic data is the searching of the unassembled reads against databases of known microbial proteins in an attempt to use the identified homologous proteins to assess the function and species diversity in the sample [39, 23]. In this type of analysis, a quite high cutoff is often chosen for the BLAST E-values (i.e., less significant) because the query sequences (i.e., reads) are quite short. As a result, there may be many false hits included in the final list of homologous proteins, which can mislead the diversity analysis. Comparing with this straightforward approach, we anticipate the homology search using the assembled peptides from the ORFome assembly can achieve higher sensitivity and result in more hits with higher significance (i.e., lower E-values).

We compared the results of homology searches using assembled peptides with the results using unassembled reads. The four Ocean Virus datasets were tested separately against IMG database. As reported in Ref. 22 [c], only few reads hit proteins in the database. We emphasize that the assembled

[b]The longest peptide has 157 amino acids, which only has very weak similarity to the protein sequences collected in IMG; it is hard to identify this peptide based on the similarity search result.

[c]We note that a direct comparison is not feasible since different databases were used for homology searching in these two studies.

(a)

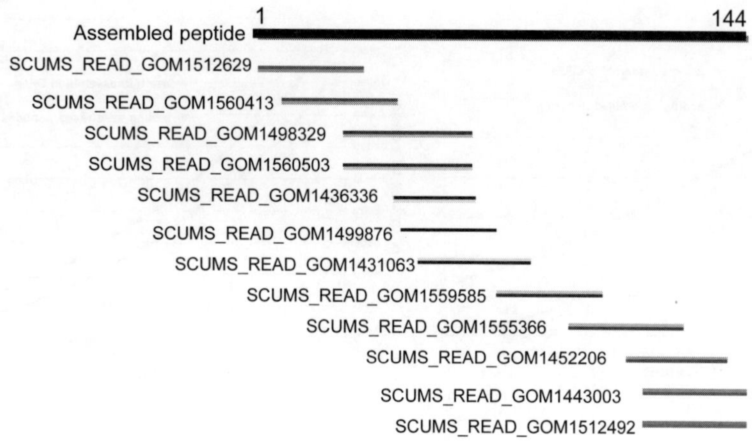

(b)

```
Query  1    ALEHGAGYTPYGDHFIMQCGMEVVLADGEVVRTGQGALAGSKHWQVTKHAAGPQFDGMFT  60
            ALEHGAGYTPYGDHF+MQCGMEVVLADG+VVRTGQGA+ GS+HWQ TKHAAGP FDGMFT
Sbjct  160  ALEHGAGYTPYGDHFVMQCGMEVVLADGQVVRTGQGAIEGSQHWQSTKHAAGPHFDGMFT  219

Query  61   QSNFGVVTKMGIWLMPEPPGYKPFMITYEREEDLEAIFEITRPLKVNQIIPNAAVAVDLL  120
            QSNFG+VTKMGIWLMPEPPGYKPFMITYEREEDL AIF+  +PLKVNQ+IPNAAVAVDLL
Sbjct  220  QSNFGIVTKMGIWLMPEPPGYKPFMITYEREEDLAAIFDAVKPLKVNQVIPNAAVAVDLL  279

Query  121  WEASAKTTRRHYFDGKGP  138
            WE SAKTTRRHYFDGKGP
Sbjct  280  WEVSAKTTRRHYFDGKGP  297
```

Fig. 2. A long peptide with 144 aa (contig11088, highlighted in bold line) assembled from 12 putative ORFs (represented as thin lines below the contig) in the Gulf of Mexico dataset shows strong similarity with proteins in IMG database with known function (a). (b) shows the BLAST alignment between the peptide and the flavoprotein subunit p-cresol methylhydroxylase from *Novosphingobium aromaticivorans* in IMG database.

peptides increase the number of significant hits (i.e., E-value $\leq 1e-5$) in all four datasets, from 26% in the Gulf of Mexico dataset (i.e., 2489 read hits were added to 9517 read hits received from the searching using unassembled reads) to 43% in the Arctic Ocean dataset (26,578 read hits were added to 61,578 original read hits). Fig. 3 shows the detailed comparison of the added number of read hits when various E-value cutoffs were applied. For all four datasets, a nearly constant number of read hits can be added by using assembled peptides. In comparison, a majority of read hits from the similarity searching using unassembled reads received high E-values. For instance, there are only 11,144 read hits in the Arctic Ocean dataset with E-values ≤ 1e-10, whereas 30,151 additional read hits (i.e., 270% more!) can be added from the similarity searching using assembled peptides. Among four datasets, the improvement of read hits is most significant in the Arctic Ocean dataset, which is consistent with the result that this dataset also achieved the best assembly performance.

3.3. Novel Assignments of Functional Categories by Assembled Peptides

We further assessed the performance of the ORFome assembly in improving the function annotation on

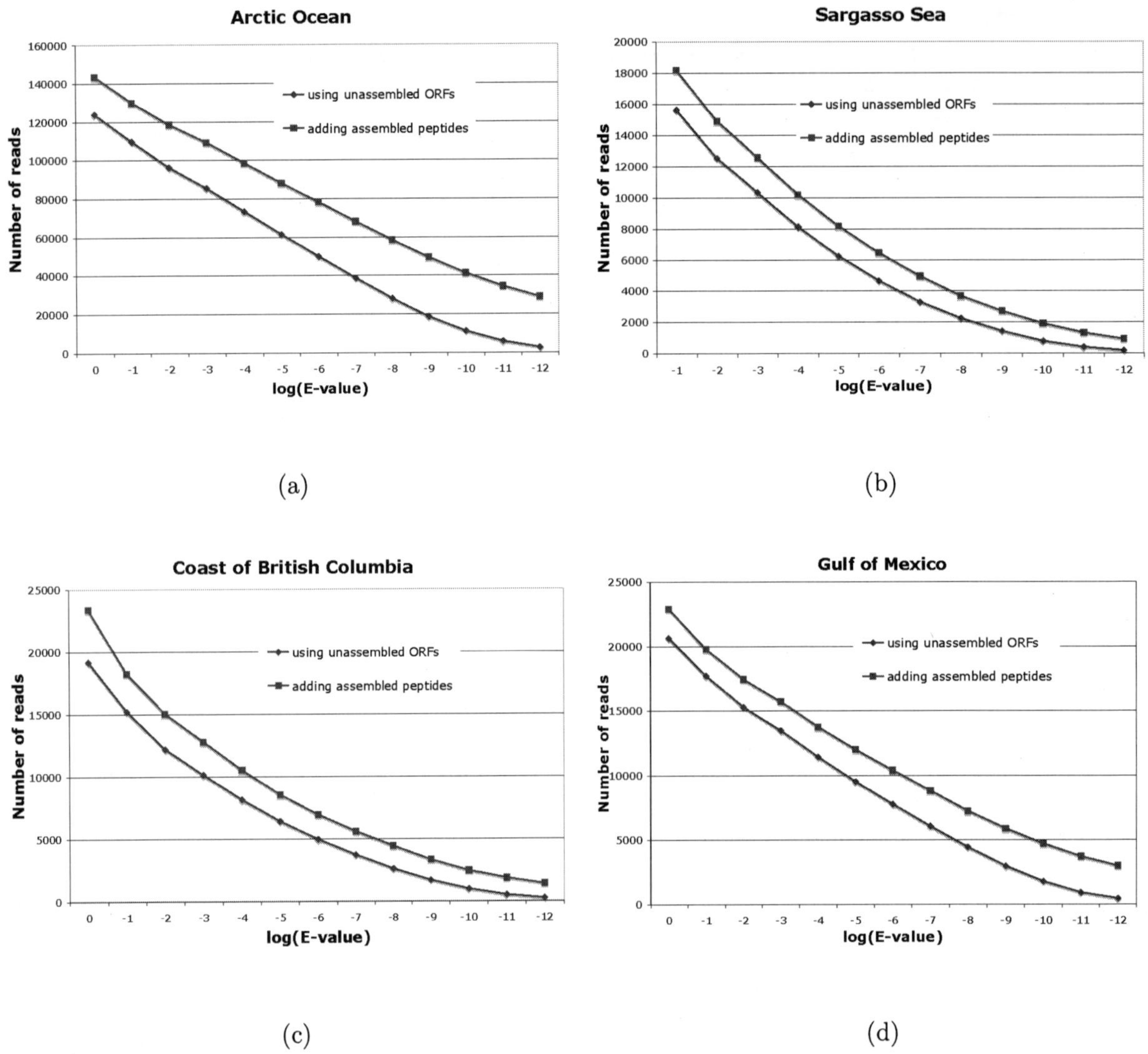

Fig. 3. Detailed comparison of the total number of read hits in IMG database using unassembled and the total number of read hits including those read hits belonging to the assembled peptides at different BLAST E-value cutoffs. The deviation between the two lines indicates the gain of read hits by using assembled peptides from the ORFome assembly.

the Ocean Virus datasets. Table 2 summarizes the statistics of the number of matched families in PANTHER database for all four datasets. Both the number of hits from the searching of unassembled reads as well as the additional number of hits from the searching of assembled peptide are listed. Although the additional numbers of families detected by using assembled peptides are relatively low for all datasets, there are still some new protein families (or novel protein functions) that can be annotated when assembled peptides were used. This suggests that we may be able to improve the protein function annotation using assembled peptides.

In the Gulf of Mexico dataset, the assembled peptides hit additional 20 PANTHER protein families. One of them is ATP synthase mitochondrial F1 complex assembly factor 2 (Panther family ID PTHR21013). The peptide corresponding to this hit is assembled from two ORFs from different reads (see Fig. 4).

SCUMS_READ_GOM1547922: 1-99

RTPQRAPLIVASAALAETIAAEW***QDQGDTVDPA***

RTPQRAPLIVASAALAETIAAEW***QDQGDTVDPA***AMPITGLTNAAIDLALPDPLGFAE

QDQGDTVDPAAMPITGLTNAAIDLALPDPLGFAE

SCUMS_READ_GOM1406881: 2-103

Fig. 4. An example of assembled peptide in the Gulf of Mexico dataset, which is assembled from two reads. The sequence of the assembled ORF is shown in the box; the overlap of the two putative ORFs from two reads are highlighted by italic and bold fonts. This assembled peptide hits the protein family of ATP synthase mitochondrial F1 complex assembly factor 2 in PANTHER database.

Table 2. Summary of the family annotation of assembled peptides versus unassembled reads for the four ocean virus datasets

Sample	Family	Add-on	Example
Arctic Ocean	590	33	PTHR22748
Sargasso Sea	265	4	PTHR11527
Coast of British Columbia	352	7	PTHR10566
Gulf of Mexico	413	21	PTHR17630

The Family column lists the total number of protein families that are found from unassembled reads. The "Add-on" column lists the additional panther protein families that are detected by using assembled peptides. The last column gives a few examples of the additional protein families (or functions) that are annotated based the assembled peptides only: PTHR22748, AP endonuclease (E-value = 5.4e-12); PTHR11527 (subfamily SF15), heat shock protein 16 (E-value = 1.5e-07); PTHR10566 (subfamily SF7), ubiquinone biosynthesis protein AARF (E.coli)/ABC (Yeast)-related (E-value = 7.3e-11); PTHR17630 (subfamily SF20), carboxymethylenebutenolidase (Evalue = 4.7e-08) .

4. DISCUSSION

One of the main issues in whole genome assembly is the chimeric contigs that are resulted from mis-assemblies. Tremendous finishing efforts have to be invested in order to identify and correct these errors. This issue is expected to be more serious in metagenomics data analysis because of the higher complexity of metagenomics sample and the introduction of short reads. Although it remains unclear whether the mis-assemblies will dramatically influence the conclusion on the principal aims of metagenomics, such as the assessment of species diversity in the sample, many metagenomics project avoided assembling sequencing reads, and analyzed the original reads directly. The ORFome assembly provides a simple solution to conduct a small-scale but accurate assembly of protein coding regions that can improve the sensitivity of homology search. In this study, although we showed the homology searching was improved after the ORFome assembly, we have not systematically evaluated the influence of these improvements on the diversity analysis. We will apply the ORFome assembly approach to more datasets with various sequence coverage and sample complexities (i.e., the approximate number of species and the range of abundances among these species). Our intention is to estimate the sequencing efforts that are required to get a good assessment of species diversity for samples with different complexities.

There are several ways to further improve the ORFome assembly algorithm described here. First, the current method for predicting putative ORFs in sequencing reads can be improved by incorporating additional features of gene coding sequences (e.g., the codon usages) and utilizing sophisticated probabilistic models. Second, a few parameters (e.g., the length cutoff of putative ORFs) used in the ORFome assembly should be optimized. This indicates that there is still room for the further improvement of the ORFome assembly approach by selecting more appropriate parameters. Finally, as we mentioned in the METHODS section, the advantages of the

ORFome assembly have not been fully taken in the downstream data analysis in this study. The EULER method used here can assemble putative ORFs into a repeat graph, in addition to the peptides represented by edges in the graph. Therefore, one can adopt a network matching approach as used in Ref. 40 to achieve a more sensible database searching.

Finally we point out that the basic method we adopted for ORF prediction may generate some spurious peptides, and some of the assembled ORFs may be not real proteins. Those spurious peptides may not cause serious problems in applications such as the homology search based annotations as used in this paper. However, we should not neglect them in other types of applications, such as comparison of the number of protein clusters (families) among different metagenomic datasets.

5. CONCLUSION

We present a novel ORFome assembly approach to metagenomics data analysis. The application of this method on four metagenomics datasets achieved promising results. Even with low coverage short reads from these datasets, our method has assembled many long peptides, which can hit on annotated proteins in the database that are not detectable otherwise. The ORFome assembly provides a useful tool to retrieve rich information from metagenomic sequencing reads, and it shows potential to facilitate an accurate assessment of the species and functional diversity in metagenomics.

Acknowledgement

This research was supported by the Indiana META-Cyt Initiative of Indiana University, funded in part through a major grant from the Lilly Endowment, Inc, and School of Informatics. The authors thank the University Information Technology Services team in Indiana University for their help with high-performance computing (for BLAST search).

References

1. Mardis E. Anticipating the 1,000 dollar genome. *Genome Biol*, 2006; **7**:112.
2. Lane D, Pace B, Olsen G, et al. Rapid determination of 16S ribosomal RNA sequences for phylogenetic analyses. *Proc Natl Acad Sci USA*, 1985; **82**:6955–6959.
3. Breitbart M, Salamon P, Andresen B, et al. Genomic analysis of uncultured marine viral communities. *Proc Natl Acad Sci USA*, 2002; **99**:14250–14255.
4. Galperin M. Metagenomics: from acid mine to shining sea. *Environ Microbiol*, 2004; **6**:543–545.
5. Eyers L, George I, Schuler L, et al. Environmental genomics: exploring the unmined richness of microbes to degrade xenobiotics. *Appl Microbiol Biotechnol*, 2004; **66**:123–130.
6. Streit W, Schmitz R. Metagenomics–the key to the uncultured microbes. *Curr Opin Microbiol*, 2004; **7**:492–498.
7. Riesenfeld C, Schloss P, Handelsman J. Metagenomics: genomic analysis of microbial communities. *Annu Rev Genet*, 2004; **38**:525–552.
8. Venter J, Remington K, Heidelberg J, et al. Environmental genome shotgun sequencing of the Sargasso Sea. *Science*, 2004; **304**:66–74.
9. Tyson G, Chapman J, Hugenholtz P, et al. Community structure and metabolism through reconstruction of microbial genomes from the environment. *Nature*, 2004; **428**:37–43.
10. Lorenz P, Eck J. Metagenomics and industrial applications. *Nat Rev Microbiol*, 2005; **3**:510–516.
11. Turnbaugh P, Ley R, Mahowald M, et al. An obesity-associated gut microbiome with increased capacity for energy harvest. *Nature*, 2006; **444**:1027–1031.
12. Gill S, Pop M, Deboy R, et al. Metagenomic analysis of the human distal gut microbiome. *Science*, 2006; **312**:1355–1359.
13. Chen K, Pachter L. Bioinformatics for whole-genome shotgun sequencing of microbial communities. *PLoS Comput Biol*, 2005; **1**:106–112.
14. Foerstner K, von Mering C, Bork P. Comparative analysis of environmental sequences: potential and challenges. *Philos Trans R Soc Lond B Biol Sci*, 2006; **361**:519–523.
15. Batzoglou S, Jaffe D, Stanley K, et al. ARACHNE: a whole-genome shotgun assembler. *Genome Res*, 2002; **12**:177–189.
16. Jaffe D, Butler J, Gnerre S, et al. Whole-genome sequence assembly for mammalian genomes: Arachne 2. *Genome Res*, 2003; **13**:91–96.
17. Huson D, Reinert K, Kravitz S, et al. Design of a compartmentalized shotgun assembler for the human genome. *Bioinformatics*, 2001; **17 Suppl 1**:S132–139.
18. Aparicio S, Chapman J, Stupka E, et al. Whole-genome shotgun assembly and analysis of the genome of Fugu rubripes. *Science*, 2002; **297**:1301–1310.
19. Azad R, Borodovsky M. Probabilistic methods of identifying genes in prokaryotic genomes: connections to the HMM theory. *Brief Bioinformatics*, 2004; **5**:118–130.
20. Yooseph S, Sutton G, Rusch D, et al. The Sorcerer II Global Ocean Sampling expedition: expanding the universe of protein families. *PLoS Biol*, 2007; **5**:e16.

21. Mavromatis K, Ivanova N, Barry K, et al. Use of simulated data sets to evaluate the fidelity of metagenomic processing methods. *Nat Methods*, 2007; **4**:495–500.
22. Angly F, Felts B, Breitbart M, et al. The marine viromes of four oceanic regions. *PLoS Biol*, 2006; **4**:e368.
23. Huson D, Auch A, Qi J, et al. MEGAN analysis of metagenomic data. *Genome Res*, 2007; **17**:377–386.
24. Wommack K, Bhavsar J, Ravel J. Metagenomics: read length matters. *Appl Environ Microbiol*, 2008; **74**:1453–1463.
25. Tang H. Genome assembly, rearrangement, and repeats. *Chem Rev*, 2007; **107**:3391–3406.
26. Pevzner P, Tang H, Waterman M. An Eulerian path approach to DNA fragment assembly. *Proc Natl Acad Sci USA*, 2001; **98**:9748–9753.
27. Myers E. Toward simplifying and accurately formulating fragment assembly. *J Comput Biol*, 1995; **2**:275–290.
28. Chaisson M, Pevzner P, Tang H. Fragment assembly with short reads. *Bioinformatics*, 2004; **20**:2067–2074.
29. Myers E, Sutton G, Delcher A, et al. A whole-genome assembly of Drosophila. *Science*, 2000; **287**:2196–2204.
30. Venter J, Adams M, Myers E, et al. The sequence of the human genome. *Science*, 2001; **291**:1304–1351.
31. Morgulis A, Gertz E, Scher A, et al. A fast and symmetric DUST implementation to mask low-complexity DNA sequences. *J Comput Biol*, 2006; **13**:1028–1040.
32. Benson G. Tandem repeats finder: a program to analyze DNA sequences. *Nucleic Acids Res*, 1999; **27**:573–580.
33. Pevzner PA, Tang H, Tesler G. De novo repeat classification and fragment assembly. *Genome Res*, 2004; **14**(9):1786–1796.
34. Raphael B, Zhi D, Tang H, et al. A novel method for multiple alignment of sequences with repeated and shuffled elements. *Genome Res*, 2004; **14**:2336–2346.
35. Rodriguez-Brito B, Rohwer F, Edwards R. An application of statistics to comparative metagenomics. *BMC Bioinformatics*, 2006; **7**:162.
36. Thomas P, Campbell M, Kejariwal A, et al. PANTHER: a library of protein families and subfamilies indexed by function. *Genome Res*, 2003; **13**:2129–2141.
37. Seshadri R, Kravitz S, Smarr L, et al. CAMERA: a community resource for metagenomics. *PLoS Biol*, 2007; **5**:e75.
38. Markowitz V, Szeto E, Palaniappan K, et al. The integrated microbial genomes (IMG) system in 2007: data content and analysis tool extensions. *Nucleic Acids Res*, 2007; .
39. Edwards R, Rohwer F. Viral metagenomics. *Nat Rev Microbiol*, 2005; **3**:504–510.
40. Ye Y, Jaroszewski L, Li W, et al. A segment alignment approach to protein comparison. *Bioinformatics*, 2003; **19**:742–749.

A PROBABILISTIC CODING BASED QUANTUM GENETIC ALGORITHM FOR MULTIPLE SEQUENCE ALIGNMENT

Hongwei Huo*, Qiaoluan Xie, and Xubang Shen
School of Computer Science and Technology, Xidian University
Xi'an, Shaanxi 710071, P.R.China
**Email: hwhuo@mail.xidian.edu.cn*

Vojislav Stojkovic
Computer Science Department, Morgan State University
Baltimore, Maryland 21251, USA
Email: vojislav.stojkovic@morgan.edu

This paper presents an original Quantum Genetic algorithm for Multiple sequence ALIGNment (QGMALIGN) that combines a genetic algorithm and a quantum algorithm. A quantum probabilistic coding is designed for representing the multiple sequence alignment. A quantum rotation gate as a mutation operator is used to guide the quantum state evolution. Six genetic operators are designed on the coding basis to improve the solution during the evolutionary process. The features of implicit parallelism and state superposition in quantum mechanics and the global search capability of the genetic algorithm are exploited to get efficient computation. A set of well known test cases from BAliBASE2.0 is used as reference to evaluate the efficiency of the QGMALIGN optimization. The QGMALIGN results have been compared with the most popular methods (CLUSTALX, SAGA, DIALIGN, SB_PIMA, and QGMALIGN) results. The QGMALIGN results show that QGMALIGN performs well on the presenting biological data. The addition of genetic operators to the quantum algorithm lowers the cost of overall running time.

1. INTRODUCTION

Multiple Sequence Alignment (MSA) is one of the most challenging tasks in bioinformatics. Most of the MSA methods are based on the dynamic programming approach. The dynamic programming approach requires time proportional to the product of the lengths of sequences which makes it computationally difficult. In the general case, the theoretical sound and biologically motivated scoring methods are not straightforward connected. Usually, it is hard to efficiently align more than a few sequences. For larger instances, a variety of heuristics strategies have been developed. In general, two basic classes of MSA methods have been proposed: progressive alignment and iterative alignment[1].

Progressive alignment methods use dynamic programming to build MSA. The best known software system based on progressive alignment method is maybe CLUSTALW[2]. Other well-known MSA systems based on progressive alignment method are MULTALIGN[3], T-COFFEE[4], MAFFT[5], MUSCLE[6], Align-m[7], and PROBCONS[8]. Mostly, they target proteins or short DNA sequences. The main advantages of progressive alignment methods are speed and simplicity. The main disadvantage of progressive alignment methods is that mistakes in the initial alignments of the most closely related sequences are propagated to the multiple alignments.

Iterative alignment methods depend on algorithm that produces an alignment and refines it through a serious of iterations until no more improvement can be made. Iterative alignment methods can be deterministic or stochastic. The deterministic iterative strategies involve extracting sequences one by one from a multiple alignment and realigning them to the remaining sequences. Stochastic iterative alignment methods include Hidden Markov Model (HMM) training, simulated annealing[9] and evolutionary computation[10]. The main advantage of stochastic iterative alignment methods is a good separation between the optimization process and evaluation criteria. The main disadvantages of stochastic iterative alignment methods are local optima, slow convergent speed, and lacking a specific termination condition.

* Corresponding author.

In the last twenty years a growing interest in quantum computation and quantum information is due to the possibility to efficiently solve hard problems for conventional computer science paradigms. Quantum algorithms exploit the laws of quantum mechanics. The quantum computation can dramatically improve performance for solving problems like factoring and search in an unstructured database. Genetic algorithms are stochastic search algorithms based on the principles of natural selection and natural genetics. They work on a set of chromosomes, called population that evolves by means of crossover and mutation towards a maximum of the fitness function. Genetic algorithms are efficient and flexible algorithms.

Han-Kim[11] proposed the possibility to integrate the quantum and genetic algorithms. Huo and Stojkovic[12] presented Quantum-inspired Evolutionary Algorithms (QEA) with a quantum representation. By adapting a qubit chromosome representation, a quantum population is generated. Classical population is generated by performing measurements on the quantum population. The best elements are searched in the classical population and used to update the quantum population. Experiments are carried out on the knapsack problem.

Now we go one step further. We redesigned QEA to solve the multiple sequence alignment problem. This paper presents a Quantum Genetic algorithm for Multiple sequence ALIGNment (QGMALIGN). It exploits the expression power of quantum mechanics in the coding and shows how to take advantage of quantum phenomena to efficiently speed up classical computation. A new probabilistic coding method for the MSA representation is given. A quantum rotation gate as a mutation operator is used to guide the quantum state evolution of the population. Six genetic operators are designed on the basis of the coding to help to improve the solutions during the evolutionary process. The features of implicit parallelism and state superposition in quantum mechanics and the global search capability of the genetic algorithm are exploited to perform efficient computation. The COFFEE (Consistency based Objective Function For alignmEnt Evaluation)[13] function is used to measure individual fitness. To demonstrate QGMALIGN's effectiveness, a set of well known test cases from BAliBASE2.0 is used as reference to evaluate the efficiency of the optimization for QGMALIGN. The QGMALIGN results have been compared with the most popular methods (CLUSTALX, SAGA, DIALIGN, SB_PIMA, and QGMALIGN) results. The QGMALIGN results show that QGMALIGN performs well on the presenting biological data.

2. CODING AND FITNESS EVALUATION

2.1. Quantum probabilistic coding

The basic information unit of quantum computation is the qubit. A qubit is a two-level quantum system and can be considered a superposition of two independent basis states $|0\rangle$ and $|1\rangle$, denoted by:

$$|\psi\rangle = \alpha|0\rangle + \beta|1\rangle. \tag{1}$$

where α and β are complex number such that $|\alpha|^2 + |\beta|^2 = 1$.

A two-level classical system can be only in one of the basis states $|0\rangle$ or $|1\rangle$. α and β are probability amplitudes associated with the $|0\rangle$ state and the $|1\rangle$ state, respectively. If we want to transfer information from the quantum system to a classical 0-1 system, we have to perform measurement of the quantum state, whose result is probabilistic: we get the state $|0\rangle$ with probability $|\alpha|^2$ and the state $|1\rangle$ with probability $|\beta|^2$. There is no way to know exactly both values. We cannot clone the unknown state $|\psi\rangle$ as stated by the No cloning theorem.

The evolution of a quantum system is described by a special linear operator, unitary operator U_f, which operates on qubits.

$$U_f|\psi\rangle = U_f[\alpha|0\rangle + \beta|1\rangle] = \alpha U_f|0\rangle + \beta U_f|1\rangle$$

An important consequence of the linearity of quantum operators is that the evolution of a two-level quantum system is the linear combination of the evolution of the basis states $|0\rangle$ and $|1\rangle$. It is possible to compute $f(x)$ for many different values of x simultaneously in a single application of U_f.

A system of m-qubits can represent 2^m different states simultaneously. The observing quantum state collapses to a single state among these states.

A qubit individual in a quantum genetic algorithm is defined as follows:

$$q = \left[\begin{array}{c|c|c|c} \alpha_1 & \alpha_2 & \cdots & \alpha_m \\ \beta_1 & \beta_2 & \cdots & \beta_m \end{array}\right]. \tag{2}$$

where $|\alpha_i|^2 + |\beta_i|^2 = 1$, $i = 1, 2, \ldots, m$. The quantum coding is inspired by the features of quantum mechanics. During the evolution process of a quantum system, we need to compute $|\alpha_i|^2$ to obtain the probability matrix of the quantum system and then to transform it to the corresponding binary matrix by performing the quantum observation. The quantum variation has an indirect effect on the qubit state by changing the values of α_i at the expense of some extra space for storing probability matrix. It is disadvantageous for solving complex problems. The new quantum probabilistic coding is proposed for representing the multiple sequence alignment. This way of coding shields the underlying information of complex α and β. The genetic operators can perform directly on the probabilistic matrix while the feature of superposition from quantum mechanics is preserved.

Assume that $Q(t) = \{q^t_1, q^t_2, \ldots, q^t_n\}$ is a population of the generation t, where n is the number of chromosomes in the population. The chromosome q^t_j is defined as

$$q^t_j = [p^t_{j1}| \quad p^t_{j2}| \quad p^t_{j3}| \quad \cdots \quad |p^t_{jm}]. \tag{3}$$

where $p^t_{ji} = |\beta^t_{ji}|^2$, p^t_{ji} is the probability of the letter being observed with value one at that position, p^t_{ji} is the length of the chromosome. When $p^t_{ji} = 1/2$, there are 2^m underlying different linear superposition states occurring with the same probability. The probabilistic coding that substitutes the form of (2) simplifies the encoding and saves the running time of the algorithms while maintaining the quantum properties.

2.2. Mapping the coding to the solution to MSA

The MSA problem can be formulated mathematically as follows: Given n sequences $S = \{S_1, S_2, \ldots, S_n\}$ defined over the finite alphabet Σ, where $n \geq 2$. S_{ij} where $1 \leq i \leq n$, $1 \leq j \leq l_i$ is a character of the alphabet Σ, where l_i is the length of S_i. A potential alignment is the set $S' = \{S'_1, S'_2, \ldots, S'_n\}$, satisfying the following conditions: (i) The sequence S'_i is the extension of S_i and is defined over the alphabet $\Sigma' = \Sigma \cup \{-\}$. '-' denotes a gap. The deletion of gaps from S'_i gives S_i; (ii) S'_i and S'_j have the same length; (iii) An objective function is a reference to biological significance that evaluates the quality of alignments.

An alignment for MSA can be obtained by measuring the quantum probabilistic matrix. The system collapses to a superposition of states that have the observed fitness. The measurement operation stems from quantum observation on a quantum computer. The difference is that the quantum observation on a quantum state can be performed many times without destroying all other configurations as it is not done in pure quantum systems. The quantum observation allows us to extract one state from the superposition of quantum probabilistic representation, having value of one with probability p_{ji} and zero with probability $1-p_{ji}$. The result of this operation is a binary matrix (BM, see Fig. 1). '1'means that there is a letter at the position of the original sequence. '0' means a gap. The number of '1s' in a row has to be the length of the sequence. The result must be repaired to fit the length of the sequence. Fig. 2 shows the alignment obtained from the binary matrix.

$$\begin{bmatrix} (0.85| & 0.34| & 0.85| & 0.95) \\ (0.77| & 0.50| & 0.87| & 0.45) \\ (0.85| & 0.42| & 0.90| & 0.65) \end{bmatrix} \rightarrow \begin{pmatrix} 1 & 0 & 1 & 1 \\ 1 & 0 & 1 & 0 \\ 1 & 1 & 1 & 1 \end{pmatrix}$$

Fig. 1. Measurement operation.

$$\begin{pmatrix} 1 & 0 & 1 & 1 \\ 1 & 0 & 1 & 0 \\ 1 & 1 & 1 & 1 \end{pmatrix} \rightarrow \begin{pmatrix} A & \_ & F & R \\ A & \_ & F & \_ \\ A & E & F & R \end{pmatrix}$$

Fig. 2. From binary matrix to an alignment.

2.3. Objective function

Objective function is used to measure the quality of MSA, which provides the basis for selection mechanism of the algorithm. Ideally, what score is better, then the multiple alignment is more biologically relevant. In this paper, we have used the COFFEE function as fitness criterion. Firstly we have a set of pairwise reference alignments (library), which includes $n*(n-1)/2$ pairwise alignments. The COFFEE function evaluates the consistency between the current multiple alignment and the pairwise alignments contained in the library. It can be formalized as follows:

$$COST(A) = \left[\sum_{i=1}^{N-1} \sum_{j=i+1}^{N} W_{ij} \times SCORE(A_{ij}) \right] \Big/ \left[\sum_{i=1}^{N-1} \sum_{j=i+1}^{N} W_{ij} \times LEN(A_{ij}) \right] \tag{4}$$

where N is the number of sequences to be aligned. A_{ij} is the pairwise projection (obtained from the multiple alignment) of sequences S_i and S_j, $LEN(A_{ij})$ is the length of this alignment, $SCORE(A_{ij})$ is the number of aligned pairs of residues that are shared between A_{ij} and the corresponding pairwise alignment in the library, and W_{ij} is the weight associated with the pairwise alignment.

3. THE OPTIMIZATION MECHANISM OF QUANTUM GENETIC ALGORITHM

3.1. Quantum mutation

The mutation operator in standard genetic algorithms is performed randomly. Individual variation of the evolutionary process with random disturbances can slow the convergent process. Quantum evolutionary processes are unitary transformations: rotations of complex space. Repeated application of a quantum transform may rotate the state closer and closer to the desired state. The basic result for quantum evolutionary process is that an unitary matrix can be represented by a finite set of universal gates. The quantum state evolution is guided by adding the optimal individual information to the variation so as to increase the probability of some quantum states to observe the better alignments and improve the convergence for the algorithm. The quantum rotation gate is the quantum unitary transformation U, defined as follows:

$$U(\delta\theta) = \begin{bmatrix} \cos(\delta\theta) & -\sin(\delta\theta) \\ \sin(\delta\theta) & \cos(\delta\theta) \end{bmatrix}$$

the angles $\delta\theta$ can be found in table 1. The rotation gate is used to update the quantum state.

$$\begin{bmatrix} \alpha_i' \\ \beta_i' \end{bmatrix} = U(\delta\theta)\begin{bmatrix} \alpha_i \\ \beta_i \end{bmatrix} = \begin{bmatrix} \cos(\delta\theta) & -\sin(\delta\theta) \\ \sin(\delta\theta) & \cos(\delta\theta) \end{bmatrix}\begin{bmatrix} \alpha_i \\ \beta_i \end{bmatrix}. \quad (5)$$

The quantum rotation gate is implemented by rotating the complex space. In Fig. 3, $|\alpha|^2$ gives the probability that the qubit will be found in the $|0\rangle$ state and $|\beta|^2$ gives the probability that the qubit will be found in the $|1\rangle$ state. Counterclockwise rotation in the first and third quadrants will increase the probability amplitude $|\beta|^2$, while in the second and fourth quadrants will increase the probability amplitude $|\alpha|^2$.

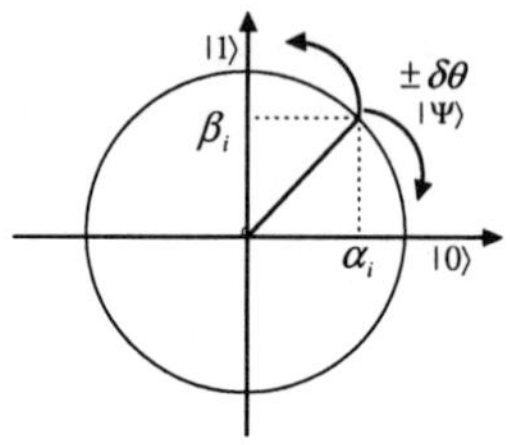

Fig. 3. Quantum observation: a projection on the basis.

According to the quantum probability coding, expression (5) can be simplified as follows:

$$p_i' = \cos^2(\delta\theta)p_i + \sin^2(\delta\theta)(1-p_i) + 2\cos(\delta\theta)\sin(\delta\theta)\sqrt{p_i(1-p_i)}\,. \quad (6)$$

Eq. (6) hides the influence of the sign of α_i and β_i on p_i. The unitary transformation makes p_i to take real values between 0 and 1. Only angles in the first quadrant can be took into account, as shown in Fig. 2.

The setting of rotation angle $\delta\theta$ is through experimentation and refer to the results in reference 11. The settings of $\delta\theta$ are application dependent. Many factors have an influence on the selection of the rotation angles, including the numbers of iterations associated with the characteristics of the sequences, the diversity of the population and convergent rate. Following the experimentation on the multiple sequence alignment problem, a lookup table for the choice of $\delta\theta$ is shown in the table 1. The values of the fitness for the best chromosome in the third column in the table have all values false because when the genetic operators perform on the chromosomes - the best one has been optimal in the current population.

Table 1. Lookup table of the rotation angle $\delta\theta$.

x_i	$best_i$	$f(x) \le f(best)$	$\delta\theta$
0	0	false	-0.005π
0	1	false	0.025π
1	0	false	-0.025π
1	1	false	0.005π

3.2. Genetic operators

The quantum mutation operator can bring good diversity of population. However, for the complexity of MSA, it is more probably for the evolutionary process to trap into the local optimum. Therefore, several genetic operators are designed to avoid local optimum inspired by SAGA[10], which enhanced the capabilities to find the global optimal solutions.

3.2.1. *Local adjustment mutation operators*

To improve the convergence - the better evolutionary strategies are needed. Inserting a gap to the left or to the right of the same position in each of the selected sequences often generate a better arrangement. An operator is designed to move blocks of residues or gaps inside an alignment. Two local adjustment operators are designed: the ResidueBlockShuffle operator and the GapBlockShuffle operator.

ResidueBlockShuffle: Move a full block without gaps to the right or to the left one position. A gap is inserted into the left or the right to that position. The block of randomly chosen length is chosen at a random position. Fig. 4(a) outlines this mechanism.

GapBlockShuffle: Split the block horizontally with the probability 15% and move one of the sub-blocks to the left or to the right. Move a full block of gaps with the probability 85% to the right or left until it merges with the next block of gaps, as Fig. 4(b) indicates.

$$\begin{bmatrix} 1.00 & 0.98 & 0.01 & 0.01 & 0.98 & 0.98 \\ 0.98 & 0.02 & 0.01 & 0.01 & 0.99 & 0.99 \\ 0.01 & 0.99 & 0.99 & 0.99 & 0.99 & 0.99 \\ 0.99 & 0.98 & 0.01 & 0.98 & 0.99 & 0.01 \end{bmatrix}$$

(a) ResidueBlockShuffle

$$\begin{bmatrix} 1.00 & 0.98 & 0.01 & 0.01 & 0.98 & 0.98 \\ 0.98 & 0.02 & 0.01 & 0.01 & 0.99 & 0.99 \\ 0.01 & 0.99 & 0.99 & 0.99 & 0.99 & 0.99 \\ 0.99 & 0.98 & 0.01 & 0.98 & 0.99 & 0.01 \end{bmatrix}$$

(b) GapBlockShuffle

Fig. 4. Local adjustment mutation operator examples.

3.2.2. *Global mutation operators*

BlockMove: Find a block with gaps randomly in an alignment, with width between two and length of the sequence and exchange position of the block with the position of a non-gap block. A special treatment for the gap-column. Fig. 5 shows how the BlockMove operator works. The length of the migration block is generated at random. The new location is taken from the nearby position including non-gaps with a large probability and randomly generated. Migrates to the neighbor with a large probability. The operator has a good effect on avoiding the local optimum.

$$\begin{bmatrix} 1.00 & 0.98 & 0.01 & 0.01 & 0.98 & 0.98 \\ 0.98 & 0.02 & 0.01 & 0.01 & 0.99 & 0.99 \\ 0.01 & 0.99 & 0.99 & 0.99 & 0.99 & 0.99 \\ 0.99 & 0.98 & 0.01 & 0.98 & 0.99 & 0.01 \end{bmatrix}$$

Fig. 5. Global mutation operator example: BlockMove.

ConsistencyShuffle: To make full use of the information from pairwise alignment library to perform the corresponding positions of adjustment and alignment, the ConsistencyShuffle operator, inspired by PHGA-COFFE, is designed to adjust the relative position of the residues. The process is as follows: Find a non-gap location of a sequence randomly in the multiple alignment, such as the positions with box in Fig. 6(a); Find the relative positions at which the selected sequence is aligned in the pairwise alignments library and record them in an array; Adjust gaps in the alignment so that the letters of the site for the multiple alignment coincide with the corresponding ones in pairwise alignment library, see Fig. 6(b).

$$\begin{bmatrix} 1.00 & 0.98 & 0.01 & 0.01 & 0.98 & 0.98 \\ 0.98 & 0.02 & 0.01 & 0.01 & 0.99 & 0.99 \\ 0.01 & 0.99 & 0.99 & 0.99 & 0.99 & 0.99 \\ 0.99 & 0.98 & 0.01 & 0.98 & 0.99 & 0.01 \end{bmatrix}$$

(a) Finding the position in the pairwise library

$$\begin{bmatrix} 1.00 & 0.98 & 0.01 & 0.01 & 0.98 & 0.98 \\ 0.02 & 0.00 & 0.01 & 0.05 & 0.98 & 0.99 \\ 0.99 & 0.99 & 0.99 & 0.99 & 0.99 & 0.01 \\ 0.99 & 0.98 & 0.01 & 0.01 & 0.98 & 0.09 \end{bmatrix}$$

(b) Adjustment

Fig. 6. ConsistencyShuffle.

The Crossover operators merge two different alignments with a higher quality into a new one. QGMALIGN implemented two different types of crossover: SingleCrossover and UniformCrossover. The former may be very disruptive. To avoid this drawback, the UniformCrossover operator is designed to promote multiple exchanges between two parents in a more subtle manner. Exchanges are promoted between zones of homology. In QGMALIGN, check whether or not the two chromosomes can do UniformCrossover, otherwise do SingleCrossover.

SingleCrossover: The X-shaped crossover is performed at the point where the perfected matched column belongs to, as shown in Fig. 7. After the crossover, the two new alignments maybe don't satisfy the constraints to the length of the sequence. The new chromosomes and the original chromosomes have different number of gaps. So we have to adjust the new chromosomes. We change p_{ij} with 1- p_{ij} in the shadowed area at random until the requirement for the number of gaps is met.

UniformCrossover: Find the crossover position in the two selected alignments, respectively. Children are produced by swapping blocks between the two parents where each block is randomly chosen between two positions. The shadowed blocks (see Fig. 8) are different areas between the two new alignments, coming from the two parents. During the process, the gaps are adjusting at random and the strategies are the same as the ones used in SingleCrossover. The choice of crossover points must satisfy the constraints: (i) The distance between the crossover positions is at least ten; (ii) At least one of the points is not available in another alignment.

3.2.3. *The Selection operator*

The Selection operator chooses the good alignments with a probability based on their fitness measured by OF(Objective Funcation). The selection operator makes sure that the good alignments survive and an optimal alignment can be found. It acts the same roles as the process of migration in evolutionary algorithms. The

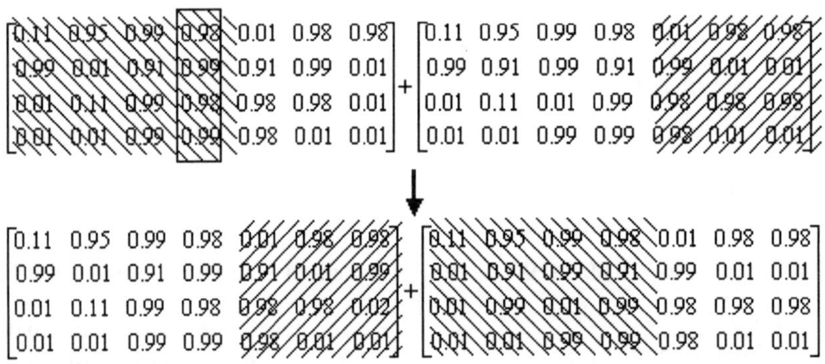

Fig. 7. SingleCrossover operator.

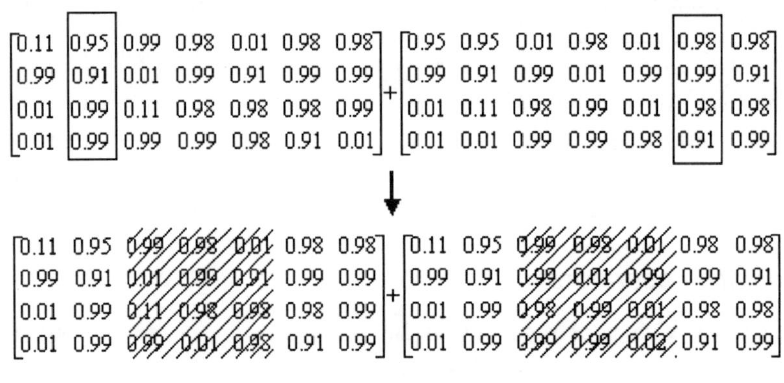

Fig. 8. UniformCrossover operator.

selection mechanisms in QGMALIGN are: typically 30% of the new generation is directly from the previous generation with the fittest alignments and the remaining 70% of the chromosomes in the new generation are created by roulette wheel selection.

3.3. Building the pairwise alignment library

In the QGMALIGN, the Needleman-Wunsch algorithm is used to build the pairwise alignment library and $n*(n-1)/2$ pairwise alignments are obtained. The BLOSUM matrices are chosen as the substitute matrix for protein sequences. The BLOSUM series ranges from BLOSUM30 to BLOSUM90, which one to choose depends on the distance between the two sequences, that is, the similarity of the two sequences. The penalty function is defined as follows:

$$penalty(gaps) = GOP + NG*GEP \tag{7}$$

where *GOP* (Gap Open Penalty) is a penalty for opening a new gap, *GEP* (Gap Extension Penalty) is a penalty for extending the length of an existing gap, and *NG* is the length of the gaps after the extension.

4. ALGORITHM

To perform multiple sequence alignment, the MSA method QGMALIGN is presenetd. QGMALIGN is derived from applying genetic algorithm in quantum computation. It uses a m-qubit representation variation of the form (3). For each representation, a binary matrix is defined, where each entry is selected using the corresponding qubit probability, $|\alpha_i|^2$ or $|\beta_i|^2$. It follows that if $|\alpha_i|^2$ or $|\beta_i|^2$ approaches to 1 or 0, then the qubit chromosome converges to a single state and the diversity given by the superposition of states disappears gradually.

The quantum-inspired computing algorithm QGMALIGN can be summarized in four steps:

(i) Initialize the population $Q(t) = \{q^t_1, q^t_2, \ldots, q^t_n\}$ of n-qubit chromosomes, where

$$q^t_j = [p^t_{j1} | \; p^t_{j2} | \; p^t_{j3} | \; \ldots \; | p^t_{jm}] \; , j = 1, \ldots, n;$$

(ii) Apply Hadamard gate to chromosome of the population and generates a superposition of all 2^n possible states;

(iii) A sequence of rotate gate and genetic operators to evolve the population;

(iv) Quantum measurements and evaluation.

4.1. QGMALIGN algorithm

The QGMALIGN can be presented as a pseudocode as follows:

Algorithm QGMALIGN

1. Build pairwise library.
2. Initial population QM of 10 chromosomes.
3. Measurement from QM to BMs.
4. Evaluate the solutions of BMs and save the best one to Best_BM.
5. **while** (not termination-condition) **do**
    6. Apply global mutation with a probability.
    7. Apply local mutation with a probability.
    8. Apply the quantum mutation according to the best solution Best_BM.
    9. Measurement from QM to BMs.
    10. **for** each BM_i **do** evaluate the corresponding alignment
    11. **if** (fitness of Best_BM < fitness of BMbest)
        12. Best_BM = BMbest
    13. Elite selection.

The procedure QGMALIGN works as follows. Line 1 uses the Needleman-Wunsch algorithm to build the pairwise alignment library with $n*(n-1)/2$ pairwise alignments. Line 2 initializes the population QM to 10 chromosomes. Line 3 extracts one state from the superposition of quantum probabilistic representation, having value of one with probability p_{ji} and zero with probability $1-p_{ji}$. The result of this operation is a binary matrix. '1' means that there is a letter at the position of the original sequence. '0'means a gap. Line 4 uses the COFFEE function to evaluate the alignment and saves the current best alignment. Lines 2-13 refine an alignment through a serious of optimization mechanisms. A quantum rotation gate as a mutation operator is used to guide the quantum state evolution. Six genetic operators are designed on the coding basis to improve the solution during the evolutionary process. The procedure ternimates when the current best alignment is not improved after 1000 times iterations.

5. THE EXPERIMENTAL RESULTS AND ANALYSIS

5.1. The experimental results

The parameters in QGMALIGN have been set as follows: *GOP* = 5, *GEP* = 0.1, the size of population is 10, and *Tmax* = 30000. The probabilities for various operators are given in table 2.

The experimental database comes from benchmark BAliBASE2.0. SPS (Sum-of-Pairs Score), is used to evaluate the final alignment. Comparisons (see tables 3~7) of the experimental results have been made with the the most popular methods (CLUSTALX, SAGA, DIALIGN, SB_PIMA, and QGMALIGN). Experimental results show that QGMALIGN performs well.

Table 2. The probabilities for five genetic operators.

Name	Probability
ResidueBlockShuffle	0.36
GapBlockShuffle	0.36
BlockMove	0.25
ConsistencyShuffle	0.65
Crossover	0.15

Table 3. SPS of Ref1.

dataset	identity	Seq_noxlength	CLUSTAL X	SAGA	DIALIGN	SB_PIMA	QGMALIGN
1idy	14%	5x65	0.705	0.342	0.018	0.145	0.450
1r69	13%	4x80	0.481	0.550	0.406	0.681	0.467
2trx	17%	4x95	0.754	0.801	0.728	0.451	0.515
1havA	15%	5x200	0.446	0.411	0.130	0.300	0.200
2hsdA	19%	4x260	0.691	0.771	0.679	0.470	0.313
Kinase	20%	5x280	0.736	0.862	0.764	0.733	0.345
1lvl	19%	4x450	0.632	0.619	0.699	0.559	0.223
1hfh	31%	5x125	0.917	0.945	0.410	0.868	0.687
1hpi	33%	4x75	0.861	0.916	0.785	0.909	0.762
1pfc	28%	5x110	0.988	0.994	0.894	0.927	0.808
451c	27%	5x85	0.719	0.662	0.729	0.541	0.554
1aym3	32%	4x235	0.969	0.955	0.962	0.976	0.720
1pii	32%	4x255	0.864	0.896	0.890	0.832	0.575
1pkm	34%	4x440	0.921	0.955	0.927	0.907	0.717
1csp	51%	5x70	0.993	0.993	0.980	1.000	0.921
1dox	46%	4x105	0.919	0.879	0.859	0.868	0.835
1fmb	49%	4x105	0.981	0.979	0.959	0.952	0.901
1plc	46%	5x95	0.958	0.931	0.931	0.904	0.931
2fxb	51%	5x65	0.945	0.951	0.945	0.945	0.946
9rnt	57%	5x105	0.974	0.965	0.864	0.970	0.978
1led	43%	4x250	0.946	0.923	0.516	0.987	0.765
1ppn	46%	5x230	0.989	0.983	0.648	0.962	0.910
1thm	49%	4x280	0.961	0.956	0.946	0.971	0.809
5ptp	43%	5x250	0.966	0.940	0.888	0.966	0.694
1gtr	42%	5x430	0.986	0.995	0.961	0.960	0.755
1rthA	42%	5x540	0.977	0.960	0.958	0.962	0.786

Table 4. SPS of Ref2.

dataset	identity	Seq_no×length	CLUSTALX	SAGA	DIALIGN	SB_PIMA	QGMALIGN
1idy	28%	19×65	0.515	0.548	F	F	0.920
1ubi	32%	15×100	0.482	0.492	F	0.129	0.763
1aboA	28%	15×85	0.650	0.489	0.384	0.391	0.573
1csy	29%	19×100	0.154	0.154	F	F	0.684
1r69	26%	20×80	0.675	0.475	0.675	0.675	0.738
1tvxA	34%	16×70	0.552	0.448	F	0.241	0.832
1tgxA	35%	19×80	0.727	0.773	0.630	0.678	0.697
2trx	34%	19×95	0.870	0.870	0.734	0.850	0.883
1sbp	23%	16×280	0.231	0.217	0.374	0.043	0.364
2hsdA	28%	20×250	0.484	0.498	0.262	0.039	0.601
1ajsA	35%	18×390	0.324	0.311	F	F	0.612
1pamA	35%	18×500	0.761	0.623	0.576	0.393	0.572
2myr	32%	17×490	0.904	0.825	0.840	0.727	0.736
4enl	48%	17×450	0.375	0.739	0.122	0.096	0.655

Table 5. SPS of Ref3.

dataset	identity	Seq_no×length	CLUSTALX	SAGA	DIALIGN	SB_PIMA	QGMALIGN
1idy	20%	27×70	0.273	0.364	F	F	0.468
1r69	19%	23×85	0.524	0.524	0.524	F	0.247
1ubi	20%	22×105	0.146	0.585	F	F	0.321
1pamA	34%	19×530	0.678	0.579	0.683	0.546	0.526
1ped	32%	21×425	0.627	0.646	0.641	0.450	0.372
1wit	22%	19×110	0.565	0.484	0.500	0.645	0.548
2myr	24%	21×540	0.538	0.494	0.272	0.278	0.547
4enl	41%	19×480	0.547	0.672	0.050	0.393	0.394

Table 6. SPS of Ref4.

dataset	identity	Seq_no×length	CLUSTAL X	SAGA	DIALIGN	SB_PIMA	QGMALIGN
1csp	32%	6×700	F	F	0.889	F	0.304
1vln	43%	14×230	0.879	0.606	0.545	0.636	0.372
1ckaA	26%	10×820	F	0.375	1.000	1.000	0.120
1mfa	18%	8×480	1.000	0.385	1.000	0.846	0.345
1ycc	36%	9×210	0.485	0.485	0.727	0.970	0.436
2abk	30%	7×520	F	F	1.000	0.471	0.126

Table 7. SPS of Ref5.

dataset	identity	Seq_noxlength	CLUSTAL X	SAGA	DIALIGN	SB_PIMA	QGMALIGN
S51	21%	15×335	0.938	0.831	0.646	0.338	0.363
S52	29%	5×340	1.000	1.000	1.000	0.515	0.789
1eft	19%	8×310	F	F	0.579	F	0.088
1pysA	25%	10×320	0.429	0.429	0.762	0.190	0.176
1qpg	35%	5×510	1.000	0.521	1.000	1.000	0.525
1thm2	38%	7×240	0.774	0.774	1.000	0.194	0.546
Kinase1	26%	5×380	0.806	0.484	0.806	0.677	0.346

CLUSTAL X is a greedy based progressive alignment method. When there are more sequences to be aligned, the major problem with the methods is that mistakes in the initial alignments of the most closely related sequences are propagated to the multiple alignments. The approach doesn't work well on ref4. The DIALIGN program constructs multiple alignment iteratively using the results from segment-to-segment comparisons. It works well on ref4 and ref5, but not very good for ref1 to ref3. SAGA uses twenty-two different genetic operators and each operator has a probability of being chosen - that is to be dynamically optimized during the run. QGMALIGN performs better on ref2 than the other listed methods. In addition, QGMALIGN can compete with CLUSTAL X and SAGA on ref3 and ref4. Experimental results showed that QGMALIGN obtained a better alignment with advantages on global optimization when there are more sequences to be aligned and the length of sequence is nearly 400.

5.2. Comparisons and analysis

To study the effects of the various genetic operators on the alignment, the comparisons of test results of the use of quantum mutation operator and adding genetic operators in it have been made. (See Table 8). The experimental results show that the genetic optimization operators are essential to obtain the better alignment. They can improve the alignment with a lower cost, because the program performs iterations from the 279 per 10 seconds before adding the genetic operators to 282 per 10 seconds after adding the genetic operators on the average.

The quantum rotation angle mutation operator guides the evolutionary process using a single optimal information. Although the optimal solutions of information constantly changes, if the information varies a little, it is easier for the process to fall in the local optimal solution, especially for the difficult multiple sequence alignment problem. The problem is not the unimodal extreme optimization and also the solutions of the problem are not unique. With genetic operators, the QGMALIGN algorithm guides the population towards the optimal solution, while maintaining the diversity of the population during the process of iterations. The added genetic operators improve the efficiency of the algorithm.

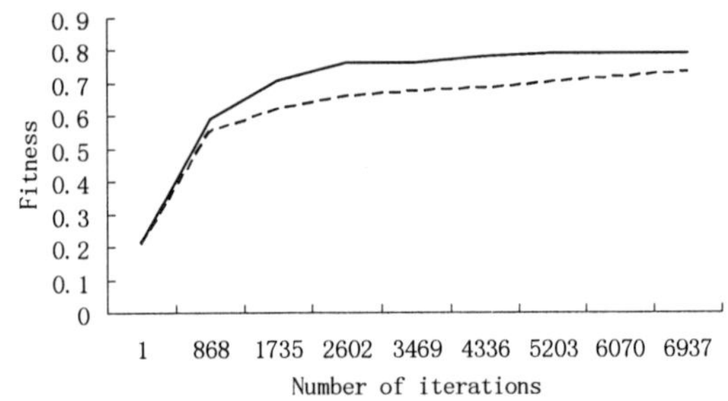

(a) the convergent rate for 9rnt

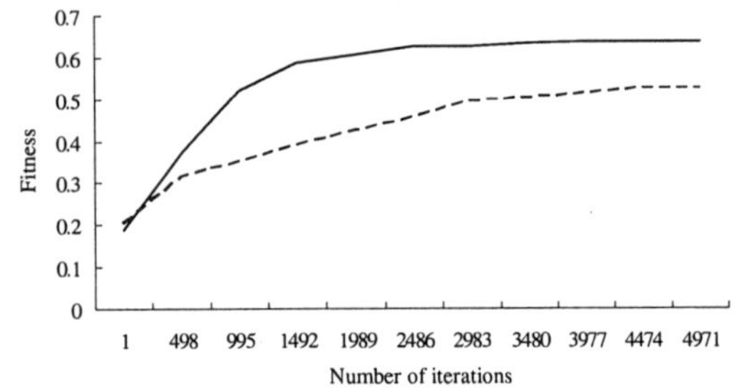

(b) the convergent rate for 1aho

Fig. 9. The convergent rate of the algorithm in two cases.

The dashed line represents the convergent rate of the algorithm with only quantum mutation operator in it.

The real line represents the convergent rate of the algorithm with six genetic operators in it.

The results in Fig. 9 show that the quantum algorithm with six genetic operators performs better on the data 1aho and 9rnt than the pure quantum mutation algorithm.

The algorithm with genetic operators converges faster to the better solution and the quality of the alignment is improved significantly.

6. CONCLUSION

This paper presents the Quantum Genetic algorithm for Multiple sequence ALIGNment - QGMALIGN. The QGMALIGN results show that QGMALIGN performs better on ref2 than the most popular methods (CLUSTALX, SAGA, DIALIGN, SB_PIMA, and QGMALIGN). Also, QGMALIGN can compete with CLUSTAL X and SAGA on ref3 and ref4. If there are a lot of sequences to be aligned and the lengths of sequences are near to 400, then QGMALIGN obtaines the better alignment with advantages on global optimization. The added genetic operators produced a lower cost running time.

Table 8. The experimental results of the algorithm with different operators in it.

sequences	Seq_no×length	All operators		Quantum mutation	
		SPS	#Iterations/time	SPS	#Iterations/time
1plc_ref1	5×100	0.931	4061/28s	0.874	6535/35s
1csy_ref2	19×102	0.684	23479/677s	0.465	27134/789s
1idy_ref3	27×72	0.468	7044/228s	0.320	18633/653s
1pysA_ref4	4×820	0.207	16459/598s	0.103	30000/1223s
1pysA_ref5	10×340	0.176	30000/1375s	0.068	30000/1308s
Average	—	0.493	16209/581.2s	0.366	22460/795.6s

Acknowledgments

This paper was supported by the National Natural Science Foundation of China, Grant No.69601003 and the National Young Natural Science Foundation of China, Grant No.60705004.

References

1. Serafim B. The many faces of sequence alignment. *Briefings in Bioinformatics* 2005; 6(1): 6-22.
2. Thompson JD, Higgins DG, Gibson TJ. CLUSTAL W: Improving the sensitivity of progressive multiple sequence alignment through sequence weighting, position-specific gap penalties and weight matrix choice. *Nucleic Acids Res* 1994; 22(22): 4673–4680.
3. Barton, G. J. and Sternberg, M. J. E. A strategy for the rapid multiple alignment of protein sequences. *J. Mol. Biol.* 1987; 198: 327–337.
4. Notredame, C., Higgins, D. G. and Heringa, J. T-Coffee: A novel method for fast and accurate multiple sequence alignment. *J. Mol. Biol.* 2000; 302: 205– 217.
5. K., Kuma, K. and Miyata, T. MAFFT: A novel method for rapid multiple sequence alignment based on fast Fourier transform. *Nucleic Acids Res.* 2002; 30(14): 3059–3066.
6. Edgar, R. C. MUSCLE: Multiple sequence alignment with high accuracy and high throughput. *Nucleic Acids Res.* 2004; 32(5): 1792–1797.
7. Van Walle, I., Lasters, I. and Wyns, L. Align-m – a new algorithm for multiple alignment of highly divergent sequences. *Bioinformatics* 2004; 20(9):1428– 1435.
8. Do, C. B., Brudno, M. and Batzoglou, S. ProbCons: Probabilistic consistency-¬based multiple alignment of amino acid sequences. *Genome Research* 2005; 15:330-340.

9. Alexander V. Lukashin, Jacob Engelbrecht, and SØren Brunak. Multiple alignment using simulated annealing: branch point definition in human mRNA splicing. *Nucleic Acids Res.* 1992; 20(10):2511–2516.
10. Alexander V Lukashin, Notredame C, Higgins D G. SAGA: Sequence alignment by genetic algorithm. *Nucleic Acids Res.* 1996; 24(8):1515–1524.
11. K H Han, J -H Kim. Quantum-inspired Evolutionary Algorithm for a Class of Combinatorial Optimization. *IEEE Trans. Evolutionary Computation* 2002; 6(6): 580-593.
12. Hongwei Huo and Vojislav Stojkovic. Applications of Quantum Computing in Bioinformatics. *The 6th annual international conference on computational systems bioinformatics CSB*2007, *Tutorial program PM*2, San Diego, California, August 13-17, 2007.
13. Notredame C, Holm L, Higgins D G. COFFEE: an objective function for multiple sequence alignments. *Bioinformatics* 1998; 14(5): 407-422.

SCALABLE COMPUTATION OF KINSHIP AND IDENTITY COEFFICIENTS ON LARGE PEDIGREES

En Cheng*, Brendan Elliott, and Z. Meral Ozsoyoglu

Electrical Engineering and Computer Science Department,

Case Western Reserve University, 10900 Euclid Avenue, Cleveland, OH, 44106, USA

**Email: exc92@case.edu*

With the rapidly expanding field of medical genetics and genetic counseling, genealogy information is becoming increasingly abundant. An important computation on pedigree data is the calculation of identity coefficients, which provide a complete description of the degree of relatedness of a pair of individuals. The areas of application of identity coefficients are numerous and diverse, from genetic counseling to disease tracking, and thus, the computation of identity coefficients merits special attention. However, the computation of identity coefficients is not done directly, but rather as the final step after computing a set of generalized kinship coefficients. In this paper, we first propose a novel Path-Counting Formula for calculating generalized kinship coefficients, which is motivated by Wright's path-counting method for computing the inbreeding coefficient for an individual. We then present an efficient and scalable scheme for calculating generalized kinship coefficients on large pedigrees using NodeCodes, a special encoding scheme for expediting the evaluation of queries on pedigree graph structures. We also perform experiments for evaluating the efficiency of our method, and compare it with the performance of the traditional recursive algorithm for three individuals. Experimental results demonstrate that the resulting scheme is more scalable and efficient than the traditional recursive methods for computing generalized kinship coefficients.

1. INTRODUCTION

In human genetics, pedigree diagrams are utilized to trace the inheritance of a specific trait, abnormality, or disease, calculate genetic risk ratios, identify individuals at risk, and facilitate genetic counseling. A sample pedigree diagram is shown in Figure 1a. Pedigrees are hierarchical hereditary structures and are typically represented as directed acyclic graphs. More specifically, a pedigree can be defined as "a simplified diagram of a family's genealogy that shows family members' relationships to each other and how a specific trait, abnormality, or disease has been inherited"[7]. Generally speaking, genetic counseling is the process by which patients or relatives, at risk of an inherited trait or disease, are advised of the consequences and nature of the trait or disease, the probability of developing or transmitting it, and the options open to them in management and family planning in order to prevent, avoid or ameliorate it. In order to calculate genetic risk ratios and identify individuals at risk, we need a measure of the degree of relatedness of two or more individuals. It is worthwhile to mention that calculating genetic risk ratios allows mainstream epidemiologists to leverage genetics for the study of diseases. In addition to the study of qualitative diseases, many developments in quantitative genetics also require knowledge of the probability that a pair of relatives have specified genotypes. Calculation of correlations between relatives

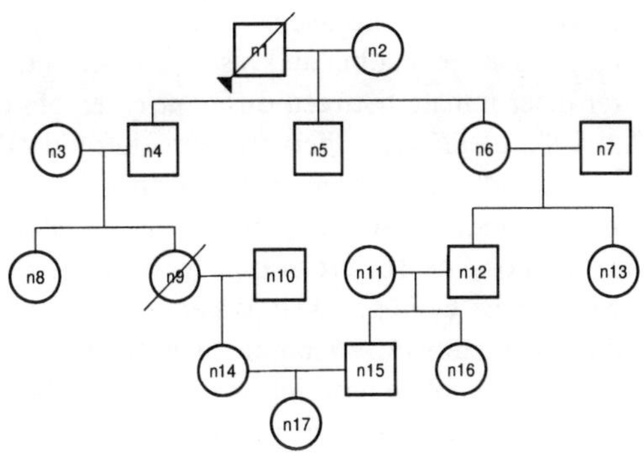

Fig. 1a. Small pedigree diagram

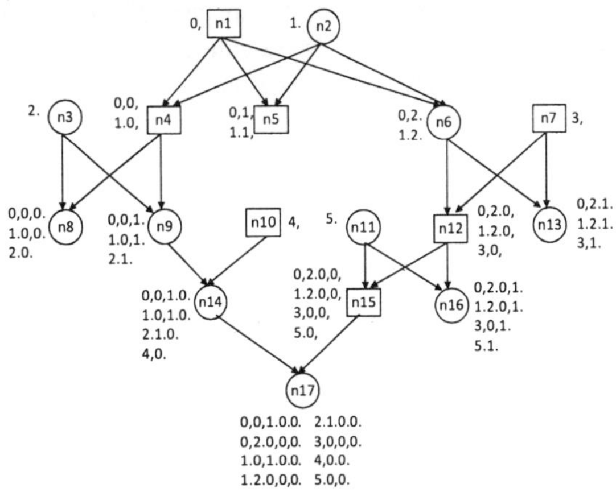

Fig. 1b. Pedigree as a graph with NodeCodes

* Corresponding author.

forms the foundation of classical biometrical analyses of quantitative traits such as height, weight, and cholesterol level[10]. In summary, making full use of genealogy information by measuring the degree of relatedness of a pair of individuals is a significant and practical issue in modern genetics.

Note that all measures of relatedness are based on the concept of identical by descent. Two genes are *identical by decent* (IBD) if one is a physical copy of the other or if they are both physical copies of the same ancestral gene. This concept is primarily due to Cotterman[3] and Malecot[14] and has been successfully applied to many problems in population genetics. The simplest measure of relationship between two individuals *a* and *b* is their kinship coefficient Φ_{ab}. The *kinship coefficient* Φ_{ab} *is the probability that a gene selected randomly from* a *and a gene selected randomly from the same autosomal locus of* b *are IBD.* While useful in many applications, the kinship coefficient does not completely summarize the genetic relation between two individuals. For instance, siblings and parent-offspring pairs share a common kinship coefficient of ¼. To better discriminate between different types of "pairs of relatives", *identity coefficients* were introduced by Gillois[6], Harris[8], and Jacquard[11]. Considering four genes of two individuals on a fixed autosomal locus, there are 15 possible IBD relations due to the fact that identity may exist *within* as well as *between* individuals. A notable characteristic of identity coefficients is that they provide a complete description of the probability of identity by descent between single loci of two individuals. Hence, this unique feature of identity coefficients has resulted in their application in a diverse range of fields. This includes the calculation of risk ratios for qualitative disease, the analysis of quantitative traits, genetic counseling in medicine, and wider studies of the genetic structure of populations.

A recursive algorithm for the calculation of identity coefficients proposed by Karigl[12] has been known for some time. This method requires that one calculate a set of generalized kinship coefficients, from which one can obtain the identity coefficients via a linear transformation. Although this recursive approach works well for small to moderate-size pedigrees, it can take impractical amounts of time when applied to very large pedigrees, particularly when coefficients are desired for many pairs of individuals. As data collection and storage technology are becoming more readily available at a lower cost, the size and variety of usable pedigree data has been increasing at a high rate. There are already large, heavily used pedigree data collections such as the Utah Population Database[15] with 1.6 million genealogy records. Thus, there is an urgent need for scalable techniques for efficiently calculating identity coefficients on large pedigrees due to both increasing volume of available pedigree data, and increasing use of pedigree data analysis in medical genetics for hereditary diseases.

In this paper, we propose a novel path-counting formula for the calculation of generalized kinship coefficients, motivated by Wright's path-counting formula for the computation of inbreeding coefficients. It has been previously shown that inbreeding coefficient queries can be efficiently evaluated using Wright's path-counting formula in conjunction with the NodeCodes encoding scheme[4]. Thus, once we have defined the path-counting formula, we can utilize NodeCodes and develop an efficient and scalable scheme for calculating the generalized kinship coefficients on very large pedigrees. We also present experimental results evaluating the performance of our strategy for calculating generalized kinship coefficients.

The main contributions of our work are as follows:

I. A novel path-counting formula for the calculation of generalized kinship coefficients.

II. An efficient and scalable scheme for calculating the generalized kinship coefficients and identity coefficients on large pedigrees using NodeCodes.

III. Experimental results demonstrating significant performance gains for calculating the generalized kinship coefficients for three individuals versus the traditional recursive algorithms.

2. RELATED WORK

There are two main approaches for computing kinship coefficients: a path-counting approach and an iterative approach[1]. The path-counting approach requires the detection of common ancestors and the summation of their contributions to the kinship coefficient. The iterative approach does not require the identification of paths through pedigrees. It begins with an initial group of individuals, and proceeds through the pedigree, computing successively the kinship between individuals who are descended from the initial population. The path-counting method has minimal storage requirements, but with some penalty in terms of computing time. The iterative approach is feasible to compute kinship coefficients for many individuals only if the kinship matrix is relatively sparse.

Among previous studies concerning the computation of identity coefficients, Karigl presented a description of identity coefficients and generalized kinship coefficients and proposed a technique that calculates the identity coefficient via a series of recursive calls to first calculate the generalized kinship coefficients and then a linear transformation is applied[12]. The generalized kinship coefficients include kinship coefficients for two, three, four, and two pairs of individuals. The basic problem is that each generalized kinship coefficient requires a separate recursion through the pedigree, which can be very time-consuming if the pedigree is very deep. Thus, the recursive algorithm can be infeasible when applied to very large pedigrees, particularly when coefficients are desired for many pairs of individuals.

Wright's formula[17], for computing the inbreeding coefficient of an individual is a typical example of path-counting formula. Utilizing an encoding scheme called NodeCodes in conjunction with Wright's formula, an efficient method for computing inbreeding coefficient is proposed by Elliott[4]. This paper was motivated by the question that whether we can extend the benefit of utilizing encoding schemes in calculation of the inbreeding coefficient to the computation of generalized kinship coefficients for more than 2 individuals..

3. BACKGROUND

This section describes condensed identity coefficients, generalized kinship coefficients, and path-counting formulas for standard kinship coefficient in more detail.

3.1. Condensed Identity Coefficients

If we consider four genes of two individuals on a fixed autosomal locus, then the 15 possible states can be reduced to 9 condensed identity states if we ignore the distinction between maternally and paternally derived genes. The states range from state 1 in which all four genes are IBD to state 9 in which none of the four genes are IBD. The probabilities associated with each condensed identity state, Δ_1 to Δ_9, are called *condensed identity coefficients*. The 15 states and their respective condensed identity coefficients are shown in Figure 2a.

The condensed identity coefficients can be computed from the generalized kinship coefficients (Φ_{ab} , Φ_{abc} , Φ_{abcd} , and $\Phi_{ab,cd}$) using the linear transformation shown in Figure 2b. Hence, we focus on the computation of generalized kinship coefficients.

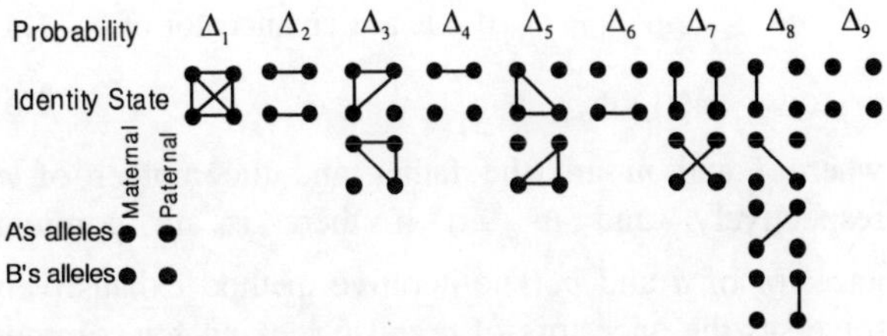

Fig. 2a. The 15 possible identity states for individuals *A* and *B*, grouped by their 9 condensed states. Lines indicate alleles that are IBD.

$$\begin{bmatrix} 1 & 1 & 1 & 1 & 1 & 1 & 1 & 1 & 1 \\ 2 & 2 & 2 & 2 & 1 & 1 & 1 & 1 & 1 \\ 2 & 2 & 1 & 1 & 2 & 2 & 1 & 1 & 1 \\ 4 & 0 & 2 & 0 & 2 & 0 & 2 & 1 & 0 \\ 8 & 0 & 4 & 0 & 2 & 0 & 2 & 1 & 0 \\ 8 & 0 & 2 & 0 & 4 & 0 & 2 & 1 & 0 \\ 16 & 0 & 4 & 0 & 4 & 0 & 2 & 1 & 0 \\ 4 & 4 & 2 & 2 & 2 & 2 & 1 & 1 & 1 \\ 16 & 0 & 4 & 0 & 4 & 0 & 4 & 1 & 0 \end{bmatrix} \begin{bmatrix} \Delta_1 \\ \Delta_2 \\ \Delta_3 \\ \Delta_4 \\ \Delta_5 \\ \Delta_6 \\ \Delta_7 \\ \Delta_8 \\ \Delta_9 \end{bmatrix} = \begin{bmatrix} 1 \\ 2\Phi_{aa} \\ 2\Phi_{bb} \\ 4\Phi_{ab} \\ 8\Phi_{aab} \\ 8\Phi_{abb} \\ 16\Phi_{aabb} \\ 4\Phi_{aa,bb} \\ 16\Phi_{ab,ab} \end{bmatrix}$$

Fig. 2b. Linear transformation to calculate identity coefficients

3.2. Condensed Identity Coefficients

In addition to the kinship coefficients Φ_{ab} for two individuals, there is a set of generalized kinship coefficients for three, four, and two pairs of individuals, which are denoted as Φ_{abc} , Φ_{abcd} , and $\Phi_{ab,cd}$, respectively. Φ_{abc} (or Φ_{abcd}) is the probability that three (or four) randomly chosen genes, one from each individual, are IBD. $\Phi_{ab,cd}$ is the probability that a random gene from *a* is IBD with a random gene from *b* and that a random gene from *c* is IBD with a random gene from *d*.

Recursive equations for generalized kinship coefficients Φ_{abc} , Φ_{abcd} , and $\Phi_{ab,cd}$ are proposed by Karigl[12]. For example, the generalized kinship coefficient for three individuals, Φ_{abc} , is expressed as follows.

$\Phi_{abc} = \frac{1}{2}(\Phi_{fbc} + \Phi_{mbc})$ if *a* is not an ancestor of *b* or *c* (1.1)

$\Phi_{aab} = \frac{1}{2}(\Phi_{ab} + \Phi_{fmb})$ if *a* is not an ancestor of *b* (1.2)

$$\Phi_{aaa} = \frac{1}{4}(1 + 3\Phi_{fm}) \qquad (1.3)$$

where *f* and *m* are the father and the mother of *a*, respectively, and $\Phi_{abc} = 0$ if there is no common ancestor of *a*, *b*, and *c*.

3.3. Path-Counting Formula

The approaches for computing the kinship coefficient Φ_{ab} are the iterative approach[12] and the path-counting approach[17]. The recursive formulas for Φ_{ab} used in the iterative approach[12] are:.

$$\Phi_{ab} = \frac{1}{2}(\Phi_{fb} + \Phi_{mb}) \text{ if } a \text{ is not an ancestor of } b \quad (1.4)$$

$$\Phi_{aa} = \frac{1}{2}(1 + \Phi_{fm}) \quad (1.5)$$

where f and m are the father and the mother of a, respectively, and $\Phi_{ab} = 0$ if there is no common ancestor of a and b. The iterative method exhaustively traverses the ancestors of a and b looking for common ancestors; when it finds them, it also recursively calculates each ancestor's inbreeding.

The path-counting approach is Wright's formula[17]:

$$\Phi_{ab} = \sum_{A}\sum_{P}(\frac{1}{2})^{L_{P(A,a)}+L_{P(A,b)}+1}[1+INC(A)] \quad (1.6)$$

where A is a common ancestor of a and b, $L_{P(A,a)}$ is the length of a path from A to a, $L_{P(A,b)}$ is the length of a path from A to b, and $INC(A) = \Phi_{fm}$ is the inbreeding coefficient of A. Paths from a to A to b that do not pass through the same individual more than once are identified and the probability of a gene being IBD is based on the number and length of these paths, modified by the common ancestor's own inbreeding.

4. PATH-COUNTING FORMULAS FOR GENERALIZED KINSHIP COEFFICIENTS

The recursive equations for generalized kinship coefficients were described in section 3.2. To make the computation of identity coefficients feasible for large pedigrees, we propose a set of path-counting formulas for generalized kinship coefficients. In this work, we will focus on showing how to generalize the path-counting formula for calculating the generalized kinship coefficient for three individuals (Φ_{abc}).

4.1. Terminology and Definitions

The following terminology and definitions for *path level* concepts will be utilized in presenting our path-counting formula for Φ_{abc}.

Triple-common ancestor: Given three individuals a, b and c, if A is a common ancestor of the three individuals, then we call A a *triple-common ancestor* of a, b and c.

Double-common ancestor: Given three individuals a, b and c, if D is a common ancestor of two of the three individuals, but it's not the common ancestor of the 3rd individual, then we say that D is a *double-common ancestor* of a, b and c.

P(A,a) denotes the set of all possible paths from A to a, where the paths can only traverse edges in the direction of parent to child such that $P(A,a) \neq \emptyset$ if and only if A is an ancestor of a. P_{Aa} denotes a particular path from A to a, where $P_{Aa} \in P(A,a)$. Let $I(P_{Aa})$ be the set of individuals on P_{Aa}.

Path-Triple denoted as $<P_{Aa}, P_{Ab}, P_{Ac}>$, where $P_{Aa} \in P(A,a)$, $P_{Ab} \in P(A,b)$, $P_{Ac} \in P(A,c)$.

Shared individual(s): The set of *shared individual(s) between two paths* P_{Aa} and P_{Ab}, denoted as $S_2(A, P_{Aa}, P_{Ab}) = I(P_{Aa}) \cap I(P_{Ab}) - \{A\}$, is non-empty if both P_{Aa} and P_{Ab} pass through a common set of individuals (excluding A). Likewise, the set of *shared individual(s) among three paths* P_{Aa}, P_{Ab}, and P_{Ac} is denoted as $S_3(A, P_{Aa}, P_{Ab}, P_{Ac}) = I(P_{Aa}) \cap I(P_{Ab}) \cap I(P_{Ac}) - \{A\}$.

Crossover & Overlap individual(s): If $s \in S_2(A, P_{Aa}, P_{Ab})$ (e.g. a double-common ancestor), we call s a *crossover individual* with respect to P_{Aa} and P_{Ab} if the two paths pass through different parents of s (i.e. one path passes through the mother and one passes through the father). On the other hand, if P_{Aa} and P_{Ab} pass through *same* parent of s, then we call s an *overlap individual* with respect to P_{Aa} and P_{Ab}.

Overlap Path: If s is an overlap individual with respect to P_{Aa} and P_{Ab}, then both P_{Aa} and P_{Ab} pass through the same parent-child edge (i.e. both mother or both father) and this edge is called an *overlap edge*. If this parent of s, denoted by p, is also an overlap individual on both paths, then there is an overlap edge regarding p as well. These two overlap edges are *consecutive* with respect to P_{Aa} and P_{Ab}. All consecutive edges constitute a path and this path is called an *overlap path*. If p is not an overlap individual, then s is simply a *crossover individual* and there is no overlap path. However, if the overlap path extends all the way to the triple common ancestor A, we instead call it a *root overlap path*. The *length of a path-triple* $<P_{Aa}, P_{Ab}, P_{Ac}>$ is denoted as $L_{<P_{Aa}, P_{Ab}, P_{Ac}>}$. Computing the length of a path-triple is given in the next section.

4.2. Path-Counting Formula for Φ_{abc}

Given a path-triple, we use the logic in Figure 3 to decide if a path-triple is counted toward the kinship value or rejected and the traversal through this diagram determines which *case* the path-triple belongs to. Identifying the case for a path-triple involves processing crossover, overlap, and shared individuals among three paths.

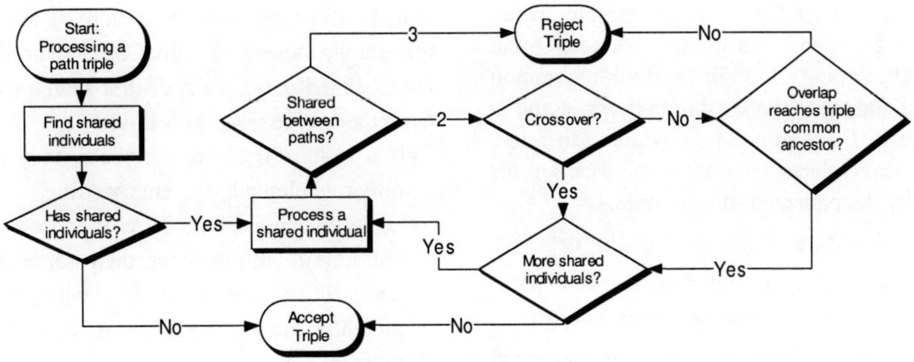

Fig. 3. Processing a path-triple

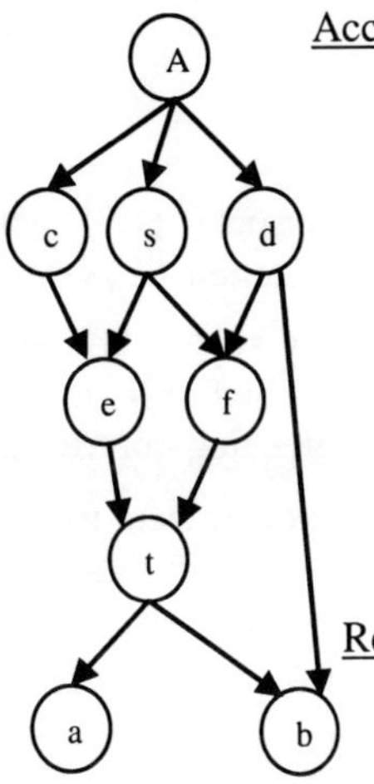

Accept Cases 1-4

$$Case\,1:<P_{Aa},P_{Ab},P_{Ac}>=\begin{cases}A\to s\to e\to t\to a\\ A\to d\to b\\ A\to c\end{cases}$$

$$Case\,2:<P_{Aa},P_{Ab},P_{Ac}>=\begin{cases}A\to s\to e\to t\to a\\ A\to d\to f\to t\to b\\ A\to c\end{cases}$$

where t is a crossover individual.

$$Case\,3:<P_{Aa},P_{Ab},P_{Ac}>=\begin{cases}A\to s\to e\to t\to a\\ A\to s\to e\to t\to b\\ A\to c\end{cases}$$

where t is an overlap individual and the overlap path is a root overlap path.

$$Case\,4:<P_{Aa},P_{Ab},P_{Ac}>=\begin{cases}A\to s\to e\to t\to a\\ A\to s\to f\to t\to b\\ A\to c\end{cases}$$

where t is a crossover individual; s is an overlap individual and the overlap path is a root overlap path.

Reject Cases 5-6

$$Case\,5:<P_{Aa},P_{Ab},P_{Ac}>=\begin{cases}A\to c\to e\to t\to a\\ A\to c\to e\to t\to b\\ A\to c\end{cases}$$

where c is a shared indiviudal among three paths.

$$Case\,6:<P_{Aa},P_{Ab},P_{Ac}>=\begin{cases}A\to c\to e\to t\to a\\ A\to s\to e\to t\to b\\ A\to c\end{cases}$$

where t is is an overlap individual and the overlap path is not a root overlap path.

Fig. 4. Six cases with respect to a path-triple

$$\Phi_{abc}=\sum_{A}\left(\sum_{\substack{<P_{Aa},P_{Ab},P_{Ac}>\in Case\,1\ or\\ <P_{Aa},P_{Ab},P_{Ac}>\in Case\,2}}(\tfrac{1}{2})^{L_{<P_{Aa},P_{Ab},P_{Ac}>}+2}[1+3*INC(A)]+\sum_{\substack{<P_{Aa},P_{Ab},P_{Ac}>\in Case\,3\ or\\ <P_{Aa},P_{Ab},P_{Ac}>\in Case\,4}}(\tfrac{1}{2})^{L_{<P_{Aa},P_{Ab},P_{Ac}>}+2}[1+INC(A)]\right)\quad(1.7)$$

According to Figure 3, we categorize all possible cases regarding a path-triple to 6 cases, and an example for each case is shown in Figure 4. Four of them are *accept cases* (1-4), in which case, they will contribute to the computation of Φ_{abc}. The other two cases are *reject cases* (5-6), and the path-triple does not contribute to the compuation of Φ_{abc}. A detailed description follows.

Case 1: $S_3(A,a,b,c)=\varnothing$ and no shared individual between any two of the three paths.

Case 2: only crossover(s) exist between any two of the three paths.

Case 3: only overlap(s) exist between any two of the three paths, but the overlap path is a root overlap path.

Case 4: both crossover(s) and overlap(s) exist between any two of the three paths, but the overlap path is a root overlap path.

Case 5: $S_3(A,a,b,c)\neq\varnothing$.

Case 6: overlap exists between any two of the three paths, but the overlap path is not a root overlap path.

Now, we can formally introduce a path-counting formula for Φ_{abc} (1.7) where A is a triple-common ancestor of a, b and c, and $INC(A)$ is the inbreeding coefficient of A.

Intuitively, case 1 and case 2 are simple triple-common ancestor paths to A (as in eq. 1.3), case 3 and case 4 are paths going through a double-common ancestor D which reduce to the kinship between A and D plus the distance to D (as in eq. 1.5), while case 5 and case 6 are the equivalents to traditional overlap for calculating Φ_{ab} by the path counting formula.

To utilize the equation (1.7) for computing Φ_{abc}, we need a method to calculate the length of a path-triple $L_{<P_{Aa},P_{Ab},P_{Ac}>}$. Let $L_{P_{Aa}}$ denote the total number of parent-child edges in P_{Aa}. Then $L_{<P_{Aa},P_{Ab},P_{Ac}>}$ is computed as follows.

$$L_{<P_{Aa},P_{Ab},P_{Ac}>} = \begin{cases} L_{P_{Aa}} + L_{P_{Ab}} + L_{P_{Ac}} & \textit{for case } 1 \,\&\, 2 \\ L_{P_{Aa}} + L_{P_{Ab}} + L_{P_{Ac}} - L_{P_{As}} & \textit{for case } 3 \,\&\, 4 \end{cases} \tag{1.8}$$

where s is an overlap individual and the overlap path is a root overlap path.

The path-counting formulas for Φ_{abcd} and $\Phi_{ab,cd}$ can be formulated using the approach given above for Φ_{abc}. For the rest of this paper, we focus on the computation of the generalized kinship coefficient for three individuals. The generalized kinship coefficients can be then directly utilized for the computation of identity coefficients.

5. CALCULATING Φ_{abc} USING NODECODES

In this section, we present an efficient and scalable NodeCodes-based scheme for our path-counting formula, motivated by the effectiveness of NodeCodes in conjunction with Wright's formula for inbreeding coefficient[4].

5.1. NodeCodes

NodeCodes is a graph encoding scheme originally proposed for encoding single source directed graphs[2,16], which was later adapted to encode pedigree data[5]. Pedigree data is represented by a directed acyclic graph, where the nodes represent individuals and directed edges represent parent-child relationships. Using NodeCodes, each node is assigned labels which are sequences of integers and delimiters. The integers represent the sibling order, and the delimiters denote the generations as well as indicating the gender of the node. We use ".", ",", and ";" to denote female, male or unknown respectively.

First the *progenitors* (nodes with in-degree 0) are labeled (we may consider adding a virtual root r and making all progenitors children of r). For each node u in the graph, the set of NodeCodes of u, denoted NC(u), are assigned using a depth-first-search traversal starting from the source node as follows:

- If u is the virtual root node r, then NC(u) contains only one element, the empty string.
- Let u be a node with NC(u), and v_0, v_1, ... v_k be u's children in sibling order, then for each x in NC(u), a code $xi*$ is added to NC(v_i), where $0 \le i \le k$, and $*$ indicates the gender of the individual represented by node v_i.

An example of NodeCodes is shown in Figure 1b using the pedigree from Figure 1a converted to a graph of parent-child edges.

5.2. Calculating Φ_{ab} and $INC(A)$

According to our path-counting formula (1.7), the calculation of Φ_{abc} requires the computation of $INC(A)$ as a final step. In our work, we utilize the efficient NodeCodes-based method described by Elliott[4] to compute $INC(A) = \Phi_{fm}$. Note that, inbreeding coefficient of an individual is actually the kinship coefficient for the individuals' parents. As a result, the method for computing inbreeding coefficient described by Elliott[4] can be utilized to calculate Φ_{ab} in general.

5.3. Calculating Φ_{abc}

The basic idea of the path counting formula for Φ_{abc} is to identify the common ancestors of a, b and c and sum their contributions to Φ_{abc}. Note that, the NodeCodes of an individual i effectively capture all ancestors that pass genes to i. Thus, given the NodeCodes of three individuals a, b, and c, we can identify all triple-common ancestors of a, b, and c via longest common prefix matching and each NodeCode from a, b, and c containing the shared prefix represents a path to the shared individual. We process each triple-common ancestor at path-level to form path-triples by taking the cross products of the sets of prefix-matched NodeCodes from a, b, and c to obtain all path-triples to be processed for that common ancestor. For each path-triple, we identify crossover, overlap, and shared individuals among three paths, and then utilize the logic described in Figure 3 to decide the triple's case and thus how it should contribute to the sum according to equation (1.7). This process is repeated for each such shared NodeCode prefix which is a *Longest Common Prefix* (LCP) for matching (which will be defined shortly) to obtain the final sum as the value for Φ_{abc}. The general outline for calculating Φ_{abc} using NodeCodes is presented in algorithm *Generalized-Kinship-Coefficient-* Φ_{abc}.

Algorithm Generalized-Kinship-Coefficient- Φ_{abc}

Input: NodeCodes NC(a), NC(b), and NC(c)
Output: Φ_{abc}
1. Initialize $\Phi_{abc} = 0$.
2. Identify a set of triple-common ancestors of a, b and c.
3. For each common ancestor A
   a. Find a set of $\langle P_{Aa}, P_{Ab}, P_{Ac}\rangle$.
   b. For each $\langle P_{Aa}, P_{Ab}, P_{Ac}\rangle$
      - ***Process-Path-Triple*** ($\langle P_{Aa}, P_{Ab}, P_{Ac}\rangle$).
      - If $\langle P_{Aa}, P_{Ab}, P_{Ac}\rangle \in$ *Case 1* or *Case 2*,
        then $\text{var} = (\tfrac{1}{2})^{L_{\langle P_{Aa}, P_{Ab}, P_{Ac}\rangle}+2}[1+3*INC(A)]$.
      - If $\langle P_{Aa}, P_{Ab}, P_{Ac}\rangle \in$ *Case 3* or *Case 4*,
        then $\text{var} = (\tfrac{1}{2})^{L_{\langle P_{Aa}, P_{Ab}, P_{Ac}\rangle}+2}[1+INC(A)]$.
      - Otherwise, $\text{var} = 0$.
      - $\Phi_{abc} = \Phi_{abc} + \text{var}$.
4. Return Φ_{abc}.

Algorithm Process-Path-Triple

Input: $\langle P_{Aa}, P_{Ab}, P_{Ac}\rangle$
Output: the case that $\langle P_{Aa}, P_{Ab}, P_{Ac}\rangle$ fits in
1. Initialize *crossover=false*, *overlap=false*.
2. Identify a set of shared individuals between any two of the three paths, and among all three paths.
3. If no shared indiviudal,
   then return $\langle P_{Aa}, P_{Ab}, P_{Ac}\rangle \in$ *Case 1*.
4. For each shared individual s_i
   - If s_i is shared among all three paths,
     then return $\langle P_{Aa}, P_{Ab}, P_{Ac}\rangle \in$ *Case 5*.
   - If s_i is a crossover individual, then *crossover=true*.
   - Else, check if the overlap path is a root overlap path.
     - If it is a root overlap path, then *overlap=true*.
     - Otherwise, return $\langle P_{Aa}, P_{Ab}, P_{Ac}\rangle \in$ *Case 6*.
5. If *crossover=true && overlap=false*,
   then return $\langle P_{Aa}, P_{Ab}, P_{Ac}\rangle \in$ *Case 2*.
6. If *crossover=false && overlap=true*,
   then return $\langle P_{Aa}, P_{Ab}, P_{Ac}\rangle \in$ *Case 3*.
7. If *crossover=true && overlap=true*,
   then return $\langle P_{Aa}, P_{Ab}, P_{Ac}\rangle \in$ *Case 4*.

In this algorithm, step 2 and step 3.a are based on finding the LCP for matching and then find the unique set of shared individuals by treating the prefixes as NodeCode and retrieving individual identifiers by the NodeCodes to eliminate duplicates. Step 3.b calls the algorithm *Process-Path-Triple*, which implements the logic presented in Figure 3, to return path-triple's case. In this procedure, we identify crossover, overlap individuals, and root overlap paths, which are the critial steps for processing a path-triple. We will explain them in detail.

Longest Common Prefix (LCP) for matching: Let X, Y, and Z be (sub)sets of the NodeCodes for a, b, and c. Then p is the longest common prefix for matching X, Y, and Z, if there is no p' where p is a prefix of p', and p' is a common prefix of **all** x_i in X, **all** y_i in Y, and **all** z_i in Z.

Identifying triple-common ancestors: We use the notation p=LCP(X,Y,Z) to denote that p is the LCP for matching sets X, Y, and Z. Given NodeCodes NC(a), NC(b), and NC(c), identifying triple-common ancestors requires matching NC(a), NC(b), and NC(c) having the longest common prefix for matching sets.

Identifying path-triples: Let A be a triple-common ancestor of a, b, and c, p_i, $1 \leq i \leq k$, be the NodeCodes of A such that p_i=LCP(X_{pi},Y_{pi},Z_{pi}) for some nonempty subsets X_{pi}, Y_{pi}, and Z_{pi} of NC(a), NC(b), and NC(c), respectively. Let p be any one of such p_i's. Then, the set of path-triple from A to a, b, and c can be represented as PT(A, p)={(x,y,z)| p=LCP(X_p, Y_p ,Z_p) and $x \in X_p$, $y \in Y_p$, and $z \in Z_p$ }.

Identifying crossover and overlap individuals: If s is a shared individual between two paths P_{Aa} and P_{Ab}, then there must be a NodeCode $n_{Aa} \in NC(s)$ that is proper prefix of P_{Aa} and a NodeCode $n_{Ab} \in NC(s)$ that is proper prefix of P_{Ab}. We call s a crossover individual with respect to P_{Aa} and P_{Ab} if n_{Aa} and n_{Ab} pass through different parents of s (i.e. one code passes through the mother and one passes through the father, identified by gender delimiters). However, if n_{Aa} and n_{Ab} pass through *same* parent of s, then s is an overlap individual with respect to P_{Aa} and P_{Ab}.

Identifying the root overlap path: If s is an overlap individual with respect to P_{Aa} and P_{Ab}, then there must be a NodeCode $n_{Aa} \in NC(s)$ that is proper prefix of P_{Aa} and a NodeCode $n_{Ab} \in NC(s)$ that is proper prefix of P_{Ab}. We identify an overlap path with respect to s as a root overlap path if n_{Aa} is equal to n_{Ab}; otherwise, it is not a root overlap path.

5.4. Computing Φ_{aab} and Φ_{aaa}

When calculating the condensed identity coefficients, we also need to directly calculate Φ_{aab} and Φ_{aaa}. However, these cases can be transformed and reduced to Φ_{abc} and Φ_{ab}, respectively, which can directly be computed according to (1.7) and Wright's formula (1.6).

For Φ_{aab}, assume a has two virtual children x and y, and we first compute Φ_{xyb}. According to the recursive formula (1.1), we get $\Phi_{xyb} = (\frac{1}{2})^2 * \Phi_{aab}$, which can be rewritten as $\Phi_{aab} = 4.0 * \Phi_{xyb}$. To evaluate this using NodeCodes, we can artificially construct the NodeCodes for x and y based on the NodeCodes for a. With $NC(x)$ and $NC(y)$, we can apply the formula (1.7) to compute Φ_{xyb}.

For Φ_{aaa}, we evaluate it by substituting equation (1.3). Again, finding the inbreeding of a is done using the NodeCodes-based method proposed by Elliott[4]. Thus, we can now fully compute the generalized kinship coefficient for two or three individuals.

6. EXPERIMENTS

In this section, we show the efficiency of our path-counting method using NodeCodes for Φ_{abc} by making comparisons with the performance of a recursive method proposed by Karigl[12]. We examine the performance of Φ_{abc} using data from the Cleveland Clinic's Familial Polyposis Registry[9] and synthetic pedigrees[4]. Results for Φ_{ab} are equivalent to finding the inbreeding coefficient as in Elliott's work[4], where experiments showed speed improvements of 3-9 times.

6.1. Experimental Setup

The Cleveland Clinic's Familial (CCF) Polyposis Registry[9] database contains pedigrees of 750 families and 11,350 patient histories recorded in the past twenty-five years at CCF. We performed experiments on this dataset using 654 pedigrees containing 8,345 individuals, with the largest pedigree consisting of 118 individuals spanning 8 generations. In order to test scalability of our method, we used twelve synthetic pedigrees[4] ranged from 77 individuals spanning 3 generations for the smallest to 195,197 individuals spanning 19 generations for the largest. The data is stored in a SQLServer database.

We compared the execution time required to calculate Φ_{abc} by the recursive method described by Karigl[12] and the path-counting method using NodeCodes. We analyzed the effects of pedigree size (# individuals), the depth of individuals in the pedigree (the longest path between the individual and a progenitor), and the kinship coefficient value.

6.2. Experimental Results

In the first experiment, 500 random triples were selected from each of our 12 synthetic pedigrees. For each triple, the query was run on cold cache starting with no memoization data to show how the cost of calculating kinship increases with pedigree size for the recursive algorithm and the path-counting method using NodeCodes. We refer to the recursive method as *KinshipIter* and we refer to the path-counting method using NodeCodes as *KinshipNC*.

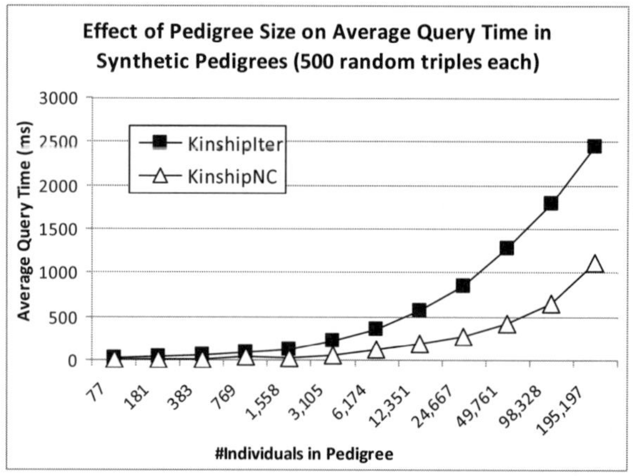

Fig. 5. Effect of pedigree size on average query time in synthetic pedigrees

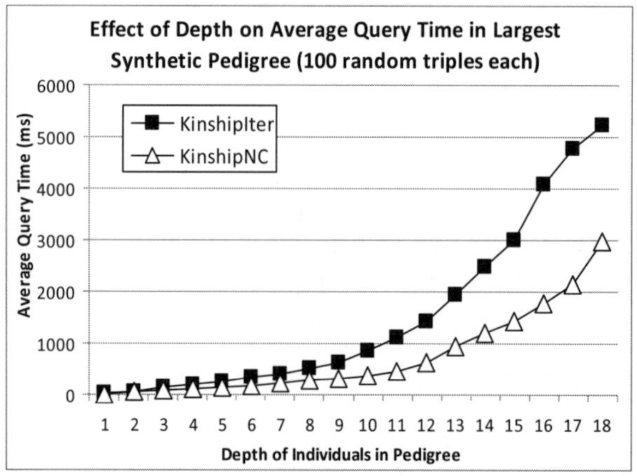

Fig. 6. Effect of depth on average query time in largest synthetic pedigree

Figure 5 shows the average time per query for each pedigree. As can be seen, the average time per query grew increasingly larger for *KinshipIter* method compared to *KinshipNC* as the pedigree size increased, from a comparable amount of time on the small pedigrees (<800 individuals) to 2.2-3.1 times slower per query than *KinshipNC* on the larger pedigrees (>1200 individuals).

In our next experiment, we examined the effect of the depth of the individual in the pedigree (number of steps in the longest NodeCode) on the query time. For each depth, we generated 100 random triples from the largest synthetic pedigree. Figure 6 shows how the average time per query grows as the individual's depth increases. We can see that *KinshipNC* scales better than *KinshipIter*, 1.7-2.3 times faster than *KinshipIter* for large pedigrees. The reason for this is that *KinshipNC* can skip intermediate generations and can jump straight to the common ancestors.

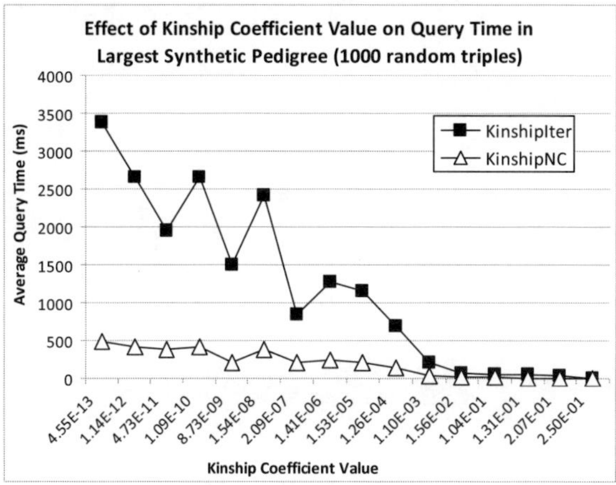

Fig. 7. Effect of kinship coefficient value on average query time in largest synthetic pedigree

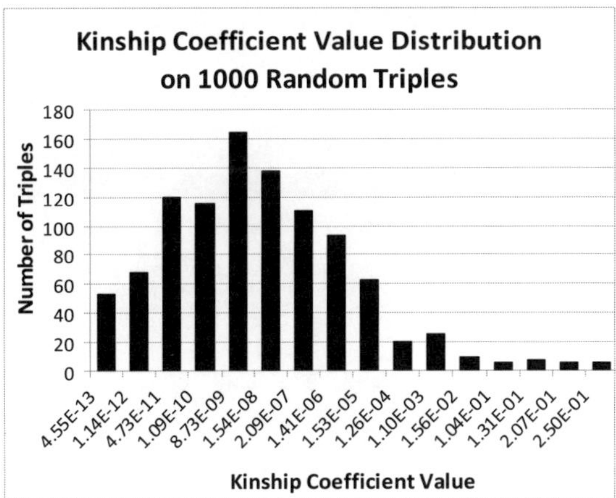

Fig. 8. Kinship coefficient value distribution for triples in Fig. 7

Next, we generated 1,000 random triples from the largest synthetic pedigree and investigated the effect of the kinship coefficient value on query time. For the kinship coefficient value, we expected that individuals with larger kinship coefficients would be more inbred and have more NodeCodes, causing *KinshipNC* to suffer slightly. Figure 7 shows the average query time for each distinct kinship coefficient value, and we can see that for most values of the kinship coefficient, *KinshipNC* outperformed *KinshipIter* by 4.9-7.3 times. As expected, for a few of the highest kinship coefficient values, we see slightly less improvement with respect to the performance of *KinshipNC* over that of *KinshipIter* (only 2.5 times faster). Figure 8 shows the distribution of the kinship coefficient value for the triples used in Figure 7. It clearly shows that the low-range values account for most of the triples, and for those values, *KinshipNC* outperforms *KinshipIter.*

Finally, we compared the results from our experiment on all the real pedigrees, which are all relatively small in comparison; the results are shown in Table 1. We randomly picked 43,862 triples on the real pedigrees. According to the ratio in the table, we can tell the *KinshipNC* is around 8.90 times faster than *KinshipIter*.

Table 1. Performance results on real data

	KinshipIter	KinshipNC	Ratio
Average Time Elapsed (ms)	29.17	3.28	8.90
Average SQL Queries Run	26.10	3.15	8.30

7. CONCLUSION

We have proposed a path-counting formula (PCF) for generalized kinship coefficient by generalizing Wright's path-counting method for three individuals. Based on our PCF, we presented an efficient and scalable method using NodeCodes for the computation of generalized kinship coefficient. We also implemented and tested our method using both real and synthetic data of various sizes to test scalability. Experimental results show that the use of NodeCodes for PCF achieves 2.2-8.9 times faster performance for computing generalized kinship coefficient on pedigree data, especially for real pedigrees as well as synthetic pedigrees of sizes between 800 and 200,000. Our future work includes (i) generalizing PCF for remaining generalized kinship coefficients, (ii) developing a scalable method for calculating identity coefficients utilizing the PCF and an encoding of paths such as NodeCodes.

Acknowledgement

We would like to thank Elena Manilich, Dr. James Church, and the Cleveland Clinic's Familial Polyposis Registry[9] for kindly allowing us to use their data for this study. This work is partially supported by the US National Science Foundation grants DBI-0218061, ITR-0312200 and CNS-0551603.

References

1. Boyce AJ. Computation of inbreeding and kinship coefficients on extended pedigrees, *Journal of Heredity* 1983; **74**:400-404.
2. Bozkaya T, Balkir N, Lee T. Efficient Evaluation of Path Algebra Expressions. *CWRU Tech. Report*, 1997.
3. Cotterman CW. A calculus for statistico-genetics. Unpublished Ph.D thesis, Ohio State University, Columbus, Ohio. Reprinted in Ballonoff, P. (Ed.). *Genetics and Social Structure*, Dowden, Hutchinson & Ross, Stroudsburg, P.A., 1974.
4. Elliott B, Akgul SF, Mayes S, Ozsoyoglu ZM. Efficient Evaluation of Inbreeding Queries on Pedigree Data. In *Proceedings of SSDBM* 2007; **9**: 3-12.
5. Elliott B, Akgul SF, Ozsoyoglu ZM, Manilich E. A Framework for Querying Pedigree Data. In *Proceedings of SSDBM* 2006; **18**:71-80.
6. Gillois M. La relation d'identité en génétique. *Ann Inst Henri Poincare B* **2** :1-94
7. Glosary of Genetic Terms, National Human Genome Research Institute http://www.genome.gov/glossary.cfm?key=pedigree
8. Harris DL. Genotypic covariances between inbred relatives. *Genetics* **50**: 1319-1348.
9. http://www.clevelandclinic.org/registries/
10. Jacquard A. The Genetic Structure of Populations. Springer-Verlag, New York, 1974.
11. Jacquard A. Logique du calcul des coefficients d'identite entre deux individuals. *Population* (Paris), 1966 ; **21**: 751-776.
12. Karigl G. A recursive algorithm for the calculation of identity coefficients. *Ann Hum Genet* 1981; 45:299–305.
13. Lange K. Mathematical and Statistical Methods for Genetic Analysis. Springer-Verlag, NY. 2002.
14. Malecot G. Les mathématique de l'hérédité, Masson, Pairs. Translated edition: The Mathematics of Heredity, Freeman, San Francisco, 1969.
15. Pedigree and Population Resource: Utah Population Database. http://www.hci.utah.edu/groups/ppr/
16. Sheng L, Ozsoyoglu ZM, Ozsoyoglu G. A Graph Query Language and Its Query Processing. In *Proceedings of ICDE Conference*, 1999.
17. Wright S. Coefficients of Inbreeding and Relationship. *The American Naturalist*, Vol. 56, No. 645, 1922.

VOTING ALGORITHMS FOR THE MOTIF FINDING PROBLEM

Xiaowen Liu and Bin Ma
Department of Computer Science, University of Western Ontario
London, ON, Canada, N6A 5B7
Email: liuxiaowencs@gmail.com, bma@uwo.ca

Lusheng Wang*
Department of Computer Science, City University of Hong Kong
Kowloon, Hong Kong
Email: lwang@cs.cityu.edu.hk

Finding motifs in many sequences is an important problem in computational biology, especially in identification of regulatory motifs in DNA sequences. Let c be a motif sequence. Given a set of sequences, each is planted with a mutated version of c at an unknown position, the motif finding problem is to find these planted motifs and the original c. In this paper, we study the VM model of the planted motif problem, which is proposed by Pevzner and Sze [1]. We give a simple Selecting One Voting algorithm and a more powerful Selecting k Voting algorithm. When the length of motif and the number of input sequences are large enough, we prove that the two algorithms can find the unknown motif consensus with high probability. In the proof, we show why a large number of input sequences is so important for finding motifs, which is believed by most researchers. Experimental results on simulated data also support the claim. Selecting k Voting algorithm is powerful, but computational intensive. To speed up the algorithm, we propose a progressive filtering algorithm, which improves the running time significantly and has good accuracy in finding motifs. Our experimental results show that Selecting k Voting algorithm with progressive filtering performs very well in practice and it outperforms some best known algorithms.
Availability: The software is available upon request.

1. INTRODUCTION

The motif finding problem in molecular biology is to find similar regions common to each sequence in a given set of DNA, RNA, or protein sequences. This problem has many applications, such as locating binding sites and finding conserved regions in unaligned sequences, genetic drug target identification and designing genetic probes.

Since DNA bases and protein residues are subject to mutations, motifs with similar functions in different sequences are not identical. From an algorithmic point of view, the motif finding problem can be considered as the consensus pattern problem. Different variants of this problem are NP-hard [2, 3] and several polynomial time algorithm schemes have been proposed [3–5].

In practice, the motif finding problem has been intensely studied. Many methods and software have been proposed [6–20]. Basically, there are two different approaches for the motif finding problem [21]. The first approach uses the pattern-driven method. For DNA and RNA sequences, all 4^L possible patterns are tried to find the best motif consensus [14, 18], where L is the length of motif. When the length of motif is large, the running time of pattern-driven algorithm is formidable. The other approach uses the sample-driven method. Sample-driven algorithms use all substrings of length L in input sequences (all L-mers) as the set of patterns [1, 3, 9]. These algorithms start from L-mers in input sequences, then use heuristic search to find the motif consensus. Due to mutations, the sample-driven algorithms may miss some good starting patterns and fail to find the real motif consensus. Based on this observation, the extended sample-driven approach is developed. This approach is a hybrid of the pattern-driven method and the sample-driven method. Both L-mers in input sequences and close neighbors of the L-mers are used as the patterns [15, 21].

To test and evaluate different methods, Pevzner and Sze proposed a planted motif model [1]. In the

*Corresponding author.

planted model, the input is a set of random samples, each sample is planted with a motif with mutations (errors). The planted motif problem is to find the planted motifs in the samples and the motif consensus. If we find the correct motif consensus, the planted motifs can be found easily. Therefore, we will focus on finding the motif consensus in this study. There are two different mutation models. The first model is the FM model, where each sequence contains one instance of an (L, D) motif, i.e. each instance of length L contains D mutated positions, where the D positions are randomly selected. The second model is the VM model, where each sequence contains exactly one instance, and each position of the instance is mutated independently with probability p.

This first model has been studied and tested with many algorithms [21–23]. In this paper, we mainly study the second model. In the experiments, our algorithms are tested on both the FM and VM models.

The main contributions of this paper are the following. First, we give a simple Selecting One Voting algorithm and a more powerful Selecting k Voting algorithm. We prove that, when the length of motif and the number of input sequences are large, the two algorithms can find the unknown motif consensus with high probability. Second, most researchers believe that a large number of sequences containing mutated motifs can help us find motifs. When the number of input sequences increases, the probabilities that motifs can be found will increase. In the proof, we show that the number of input sequences can help our algorithms find motifs. Our experiments on simulated data also support the relationship between the number of input sequences and the performance of our algorithms. Select k Voting algorithm is powerful, but the time complexity of the algorithm is too high to be practical for real problems. Finally, we propose a progressive filtering method to speed up Selecting k Voting algorithm. With the filtering method, the time complexity of Selecting k Voting algorithm is improved from $O(Lm^{k+1}n)$ to $O(\alpha Lm(k^2+n))$, where n is the number of input sequences, m is the length of each sequence and α is an input parameter. Our experimental results show that Selecting k Voting algorithm with progressive filtering performs very well in practice and it outperforms some best known algorithms.

2. PROBABILITY MODELS

In this paper, we consider DNA sequences and use a fixed alphabet $\Sigma = \{A, C, G, T\}$. For a string $s \in \Sigma^m$, the i-th letter of s is denoted by $s[i]$. Thus, $s = s[1]s[2]\ldots s[m]$. A string s is called a *uniform random DNA sequence* if for each letter $s[i]$ in s, $Pr(s[i] = A) = Pr(s[i] = C) = Pr(s[i] = G) = Pr(s[i] = T) = \frac{1}{4}$. Let s_1 and s_2 be two strings of the same length. The Hamming distance between s_1 and s_2 is denoted by $d(s_1, s_2)$. For a string $t \in \Sigma^L$ and a string $s \in \Sigma^m$, where $L < m$, the distance between t and s is the minimum distance between t and any L-mer in s, denoted by $d(t, s)$. For a set of strings $S = \{s_1, s_2, \ldots, s_n\}$ and a string s of the same length m, if each letter $s[i]$ in s is a majority letter in $\{s_1[i], s_2[i], \ldots, s_n[i]\}$, s is called a *consensus string* of S.

In the VM probability model, n input strings with planted motifs are generated as follows. We first generate n uniform random DNA sequences $\overline{s_1}, \overline{s_2}, \ldots, \overline{s_n}$, each of length m. Suppose a uniform random DNA sequence $c \in \Sigma^L$ is the original motif consensus. Based on c, we generate n motifs $c_1, c_2, \ldots, c_n \in \Sigma^L$ with mutations (errors) by changing each letter in c independently with probability $p = \frac{3}{4} - \epsilon$. That is, for each letter $c_i[j]$ in c_i, $Pr(c_i[j] = c[j]) = \frac{1}{4} + \epsilon$, and for $u \in \Sigma \backslash \{c[j]\}$, $Pr(c_i[j] = u) = \frac{1}{4} - \frac{\epsilon}{3}$. Then, for each string $\overline{s_i}$, we randomly select a position h in $\{1, 2, \ldots, m - L + 1\}$ and replace $s[h]s[h+1]\ldots s[h+L-1]$ with c_i to get s_i. We say that c_i is planted to s_i. In this way, we get a new set of strings $s_1, s_2, \ldots, s_n$, which is a set of random strings with planted motifs. From now on, we will consider the string set $S = \{s_1, s_2, \ldots, s_n\}$ as the input sequences. For each string s_i, the set of all L-mers of s_i is denoted by P_i.

Since the FM probability model is used in our experiments, we also give the definition of the FM model. In the (L, D) FM model, with n uniform random DNA sequences and a motif consensus $c \in \Sigma^L$, the method of generating n mutated motifs is different from the VM model. In motif consensus c, we randomly select exact D positions and each of the D letters is changed to any of other three letters to get a mutated motif c_i. For each of the D positions, $Pr(c_i[j] = c[j]) = 0$, and for $u \in \Sigma\backslash\{c[j]\}$, $Pr(c_i[j] = u) = \frac{1}{3}$. Finally, the n mutated motifs are planted to the n uniform random DNA sequences to get a set of n input sequences.

With the VM probability model, we give the definition of the planted motif problem.

Definition 2.1. Given a set $S = \{s_1, s_2, \ldots, s_n\}$ of strings each of length m, generated as described in the VM probability model, and an integer L, the *planted motif problem* is to find the unknown motif consensus c.

In some cases, we want to find the closest substrings in the input sequences. Then, we have another similar problem.

Definition 2.2. Given a set $S = \{s_1, s_2, \ldots, s_n\}$ of strings each of length m, generated as described in the VM probability model and an integer L, the *planted closest substring problem* is to find a length L substring t_i for each string $s_i \in S$ and a consensus string t such that $\sum_{1\le i\le n} d(t, t_i)$ is minimized.

Algorithm 1

Input A sequence set $S = \{s_1, s_2, \ldots, s_n\} \subset \Sigma^m$, a starting pattern $t \in \Sigma^L$.

Output A motif consensus with length L.

1. For each sequence s_i, find an L-mer $t_i \in P_i$ such that $d(t, t_i) = d(t, s_i)$.
2. Output a consensus string of $\{t_1, t_2, \ldots, t_n\}$.

Fig. 1. The voting algorithm.

3. ALGORITHMS

In this section, we give several algorithms for the planted motif problem.

3.1. Voting Algorithm

For the planted motif problem, our algorithms have two parts. The first part is to find a starting pattern (L-mer). The second part is to use the starting pattern to find the motif consensus and planted motifs. Here, we first give a simple voting algorithm for finding motif consensus from a given starting pattern t. In details, our algorithm has two steps: (1) in each sequence s_i, find an L-mer $t_i \in P_i$ with the minimum distance to t; (2) find a consensus string of $t_1, t_2, \ldots, t_n$ (Fig. 1).

In practice, we can use the resulting consensus string of the voting operation as a new starting pattern, and do voting operation repeatedly until there is no further improvement. Compared with other pattern refinement methods, such as Gibbs sampling [7] and EM [8, 9] methods, the voting algorithm uses a consensus string instead of a profile to represent the pattern and uses a simple voting operation to do pattern refinement. The main advantage of the voting algorithm is its high speed. Since the pattern is represented by a string, the voting algorithm converges faster than the Gibbs sampling and EM methods. In addition, the voting algorithm avoids the time consuming computation of likelihoods in Gibbs sampling and EM methods. So, the voting algorithm is much faster than the Gibbs sampling and EM approaches. With the fast speed, we can try much more starting patterns to find a good motif. Our experimental results also show that the voting algorithm is powerful for finding motifs.

Algorithm 2

Input A sequence set $S = \{s_1, s_2, \ldots, s_n\} \subset \Sigma^m$ and integers L and r.

Output A motif consensus with length L.

1. Repeat Steps 2-5 r times.
2. Randomly select a sequence $s_i \in S$ which has not been selected in previous rounds.
3. **For** each L-mer $t \in P_i$, **do**
4. Use the voting algorithm to find a consensus t^* from starting pattern t.
5. Add t^* to candidate motif consensus set C.
6. Output motif consensus c_A such that $\sum_{1\le i\le n} d(c_A, s_i)$ is the minimum in all candidates in C.

Fig. 2. Selecting One Voting algorithm.

3.2. Selecting One Voting Algorithm

The performance of the voting algorithm depends on the qualities of starting patterns. So we need to find good starting patterns for it. Our method follows the sample-driven approach. We use L-mers in input sequences as the set of patterns. We randomly select an input sequence $s_i \in S$ and find the planted motif c_i by enumerating all L-mers in s_i. Then the motif c_i is used as a starting pattern of the voting algorithm. The above procedure is repeated r times and the best motif consensus is output, where r is an input parameter. The algorithm is called Selecting One Voting algorithm (Fig. 2).

We can prove that, when L and n are large enough, Algorithm 2 can correctly find the motif consensus with high probability. In the following analysis, we use an important lemma about Chernoff-Hoeffding bound [24].

Lemma 3.1. *Let $X_1, X_2, \ldots, X_n$ be n independent random binary (0 or 1) variables, where X_i takes on the value of 1 with probability p_i, $0 < p_i < 1$. Let $X = \sum_{i=1}^{n} X_i$ and $\mu = E[X]$. Then for any $0 < \lambda < 1$,*

(1) $\mathbf{Pr}(X \geq \mu + \lambda n) \leq e^{-2\lambda^2 n}$,
(2) $\mathbf{Pr}(X \leq \mu - \lambda n) \leq e^{-2\lambda^2 n}$.

From the VM probability model, each planted motif has $(\frac{3}{4} - \epsilon)L$ mutations on average. When r is large, we repeat the voting operation many times. Then, the probability that we can find a planted motif with no more than $(\frac{3}{4} - \epsilon)L$ mutations is high.

Lemma 3.2. *The probability that one planted motif t with $d(c,t) \leq (\frac{3}{4} - \epsilon)L$ is selected in Step 3 of Algorithm 2 is at least $1 - \left(\frac{3}{4}\right)^r$.*

Proof. Based on the VM probabilistic model, the distance between c and a mutated motif c_i fits the binomial distribution $B(L, \frac{3}{4} - \epsilon)$. Therefore, for each c_i, $Pr(d(c_i, c) > \frac{3}{4} - \epsilon) \leq \frac{3}{4}$ (The inequality can be proved by enumerating all possible L's). In Algorithm 2, we randomly select r planted motifs as starting patterns. Since the r motifs are independently generated, the probability that each selected motif c_i has $d(c_i, c) > (\frac{3}{4} - \epsilon)L$ is no more than $\left(\frac{3}{4}\right)^r$. Therefore, the lemma is proved. □

When the length of motif is large enough, with high probability, two planted motifs have smaller distance than two random L-mers. Based on this observation, we have the following lemma.

Lemma 3.3. *When $d(c,t) \leq (\frac{3}{4} - \epsilon)L$ and $L > \frac{9}{8\epsilon^4} \log \frac{3m}{\epsilon}$, for each sequence $s_i \in S$, the probability that c_i is selected in Step 1 of the voting algorithm is no less than $1 - \frac{1}{3}\epsilon$.*

Proof. First, we consider the planted motif c_i in s_i. Let $X_1, X_2, \ldots, X_L$ be the random variables such that $X_j = 1$, if $c_i[j] = t[j]$; $X_j = 0$, otherwise. From the assumption, we have $d(c,t) \leq (\frac{4}{3} - \epsilon)L$. From the generation method, for each letter $c_i[j]$, $Pr(c_i[j] = c[j]) = \frac{1}{4} + \epsilon$. Therefore, $Pr(X_j = 1) = \frac{1}{4} + \frac{4}{3}\epsilon^2$ and $Pr(X_j = 0) = \frac{3}{4} - \frac{4}{3}\epsilon^2$. Let $X = \sum_{1 \leq j \leq L} X_j$. Then $E(X) = (\frac{1}{4} + \frac{4}{3}\epsilon^2)L$. By Lemma 3.1,

$$Pr\left(X \leq \left(\frac{1}{4} + \frac{2}{3}\epsilon^2\right)L\right) \leq Pr\left(X \leq E(X) - \frac{2}{3}\epsilon^2 L\right) \leq e^{-\frac{8}{9}\epsilon^4 L}.$$

When $L > \frac{9}{8\epsilon^4} \log \frac{3m}{\epsilon}$, the probability is

$$Pr\left(X \leq \left(\frac{1}{4} + \frac{2}{3}\epsilon^2\right)L\right) \leq \frac{\epsilon}{3m}. \quad (1)$$

Second, we consider an L-mer $t' \in P_i \backslash \{c_i\}$. Let $Y_1, Y_2, \ldots, Y_L$ be the random variables such that $Y_j = 1$, if $t'[j] = c[j]$; $Y_j = 0$, otherwise. For each letter $t'[j]$, $Pr(t'[j] = A) = Pr(t'[j] = C) = Pr(t'[j] = G) = Pr(t'[j] = T) = \frac{1}{4}$. Therefore, $Pr(Y_j = 1) = \frac{1}{4}$ and $Pr(Y_j = 0) = \frac{3}{4}$. Let $Y = \sum_{1 \leq j \leq L} Y_j$. We have $E(Y) = \frac{1}{4}L$. By Lemma 3.1,

$$Pr\left(Y \geq \left(\frac{1}{4} + \frac{2}{3}\epsilon^2\right)L\right) \leq Pr\left(Y \geq E(Y) + \frac{2}{3}\epsilon^2 L\right) \leq e^{-\frac{8}{9}\epsilon^4 L}.$$

When $L > \frac{9}{8\epsilon^4} \log \frac{3m}{\epsilon}$, the probability is

$$Pr\left(Y \geq \left(\frac{1}{4} + \frac{2}{3}\epsilon^2\right)L\right) \leq \frac{\epsilon}{3m}.$$

Consider all the L-mers in $P_i \backslash \{c_i\}$, the probability that there is an L-mer $t' \in P_i \backslash \{c_i\}$ such that $d(t', c) \geq (\frac{1}{4} + \frac{2}{3}\epsilon^2)L$ is no more than $\frac{m-L}{3m}\epsilon$.

Together with (1), when $d(c,t) \leq (\frac{3}{4} - \epsilon)L$ and $L > \frac{9}{8\epsilon^4} \log \frac{3m}{\epsilon}$, the probability that c_i is selected in Step 1 of the voting algorithm is no less than $1 - \frac{1}{3}\epsilon$. □

When the planted motifs can be found with high probability, the voting algorithm can find the motif consensus with high probability.

Lemma 3.4. *Suppose $d(c,t) \leq (\frac{3}{4} - \epsilon)L$ and each planted motif c_i is selected in Step 1 of the voting algorithm with probability no less than $1 - \frac{1}{3}\epsilon$. When $\frac{n}{\log n} \geq \frac{9}{2\epsilon^2}$, the probability that $t^* = c$ is no less than $1 - \frac{4L}{n}$.*

Proof. Consider a position j , $1 \leq j \leq L$, such that $t[j] \neq c[j]$. Let $X_1, X_2, \ldots, X_n$ be the random variables such that $X_i = 1$, if $t_i[j] = c[j]$; $X_i = 0$, otherwise. For all motifs $c_1, c_2, \ldots, c_n$, the expected number of motifs c_i such that $c_i[j] = c[j]$ is $(\frac{1}{4} + \epsilon)n$. Let M^+ and M^- be the sets of planted motifs selected and not selected into $\{t_1, t_2, \ldots, t_n\}$, respectively. The probability that a planted motif is not selected in the voting algorithm is no more than $\frac{1}{3}\epsilon$. Therefore, the expectation of $|M^-|$ is no more than $\frac{1}{3}\epsilon n$. In the worse case, each motif in M^- has the same letter with c at position j. Then, the expected number of t_i's such that $t_i \in M^+$ and $t_i[j] = c[j]$ is no less than $(\frac{1}{4} + \epsilon)n - \frac{\epsilon}{3}n = (\frac{1}{4} + \frac{2}{3}\epsilon)n$. Let $X = \sum_{1 \leq i \leq n} X_i$. We have $E(X) \geq (\frac{1}{4} + \frac{2}{3}\epsilon)n$. By Lemma 3.1,

$$Pr\left(X \leq \frac{n}{4} + \frac{1}{3}\epsilon\right) \leq Pr\left(X \leq E(X) - \frac{1}{3}\epsilon n\right) \leq e^{-\frac{2}{9}\epsilon^2 n}.$$

When $\frac{n}{\log n} \geq \frac{9}{2\epsilon^2}$, the probability is

$$Pr\left(X \leq \frac{n}{4} + \frac{1}{3}\epsilon\right) \leq \frac{1}{n}.$$

For a letter $u \in \Sigma \backslash \{c[j]\}$, let $Y_1, Y_2, \ldots, Y_n$ be the random variables such that $Y_i = 1$, if $t_i[j] = u$; $Y_i = 0$, otherwise. For all motifs $c_1, c_2, \ldots, c_n$, the expected number of c_i's with $c_i[j] = u$ is $(\frac{1}{4} - \frac{1}{3}\epsilon)n$. Note that the expected number of t_i's not being a planted motif is no more than $\frac{1}{3}\epsilon n$. In the worse case, each t_i not being a planted motif has letter u at position j. Then, the expected number of t_i's with $t_i[j] = u$ is no more than $(\frac{1}{4} - \frac{\epsilon}{3})n + \frac{\epsilon}{3}n = \frac{1}{4}n$. Let $Y = \sum_{1 \leq i \leq n} Y_i$. It follows that $E(Y) \leq \frac{1}{4}n$. By Lemma 3.1,

$$Pr\left(Y \geq \frac{n}{4} + \frac{1}{3}\epsilon n\right) \leq Pr\left(Y \geq E(Y) + \frac{1}{3}\epsilon n\right) \leq e^{-\frac{2}{9}\epsilon^2 n}.$$

When $\frac{n}{\log n} \geq \frac{9}{2\epsilon^2}$, the probability that $t^*[j] = u$ is no more than n^{-1}. There are three possible letters in $\Sigma \backslash \{c[j]\}$. The probability that none of them has $\frac{n}{4} + \frac{1}{3}\epsilon n$ letters in column j is at least $1 - \frac{3}{n}$. Therefore, the probability that $c[j]$ is the majority letter at column j is at least $1 - \frac{4}{n}$.

Then, we consider a position j' such that $t[j'] = c[j']$. Similar to the previous case, we can prove that when $\frac{n}{\log n} \geq \frac{9}{2\epsilon^2}$, the probability that $t^*[j'] = c[j']$ is at least $1 - \frac{4}{n}$.

Consider all the L positions in c, the probability that $t^* = c$ is no less than $1 - \frac{4L}{n}$. □

From the Lemmas 3.2, 3.3 and 3.4, we get the following theorem.

Theorem 3.1. *When $\sum_{1 \leq i \leq n} d(c, s_i)$ is the minimum in all strings in Σ^L, $L > \frac{9}{8\epsilon^4} \log \frac{3m}{\epsilon}$, and $\frac{n}{\log n} \geq \frac{9}{2\epsilon^2}$, the probability that Algorithm 2 can find the motif consensus c is no less than $1 - \left(\frac{3}{4}\right)^r - \frac{4L}{n}$.*

Theorem 3.1 shows that m, L and n are all important factors of Selecting One Voting algorithm. The increase of n will increase the probability of finding the motif consensus. Many researchers believe that a large number of sequences containing mutated motifs can help us find motifs. This proof shows why motifs can be found more easily in many similar sequences than in a few sequences. That is the reason why multiple alignment of many similar sequences is important for finding useful biological information.

Suppose $\sum_{1 \leq i \leq n} d(c, s_i)$ is the minimum in all strings in Σ^L. When the length of motif and the mutation rate can not guarantee that Algorithm 2 outputs the motif consensus with high probability, we can prove the algorithm is a good approximation algorithm for the planted closest substring problem. For output c_A of Algorithm 2,

$$E\left(\sum_{1 \leq i \leq n} d(c_A, s_i)\right) \leq E\left(\sum_{1 \leq i \leq n} d(c_1, s_i)\right) = \left(\frac{3}{4} - \frac{4}{3}\epsilon^2\right) nL.$$

For the optimal solution c,

$$E\left(\sum_{1\le i\le n} d(c,s_i)\right) = \left(\frac{3}{4}-\epsilon\right) nL.$$

Therefore, the expected approximation ratio of Selecting One Voting algorithm is $1+\frac{4}{3}\epsilon$.

In Selecting One Voting algorithm, we try $r(m-L+1)$ different starting patterns. For each starting pattern, the voting operation takes $O(Lmn)$ time. So the time complexity of the whole algorithm is $O(rLm^2n)$.

Algorithm 3

Input A sequence set $S=\{s_1,s_2,\ldots,s_n\}\subset \Sigma^m$ and integers L, r and k.

Output A motif consensus with length L.

1. **Repeat** Steps 2-6 r times.
2. Randomly select k sequences $s_{x_1}, s_{x_2}, \ldots, s_{x_k}$ from S which have not been selected in previous rounds.
3. **For** each L-mer set $\{a_1,a_2,\ldots,a_k\}$ where $a_1\in P_{x_1}, a_2\in P_{x_2},\ldots,a_k\in P_{x_k}$, **do**
4. Find a consensus string t of $a_1,a_2,\ldots,a_k$. (If there are several consensus strings, randomly select one).
5. Use the voting algorithm to find a consensus t^* from starting pattern t.
6. Add t^* to candidate motif consensus set C.
7. Output motif consensus c_B such that $\sum_{1\le i\le n} d(c_B,s_i)$ is the minimum in all candidates in C.

Fig. 3. Selecting k Voting algorithm.

3.3. Selecting k Voting Algorithm

Selecting One Voting algorithm only uses L-mers in input strings as starting patterns of voting. When motifs are short and mutation rate is high, Selecting One Voting algorithm may miss some good patterns and fail to find the motif consensus. To get more good patterns, one simple idea is to use a consensus string of several planted motifs as a starting pattern of voting. Intuitively, when we know k planted motifs, the consensus string of the k mutated motifs will be more similar to the unknown motif consensus than one mutated motif. Based on this observation, we give a new powerful Selecting k Voting algorithm. In details, we first randomly select k sequences $s_{x_1}, s_{x_2}, \ldots, s_{x_k}$ from S and select one L-mer in each of the k sequence to get k L-mers. By enumerating all L-mers in $s_{x_1}, s_{x_2}, \ldots, s_{x_k}$, the k planted motifs can be selected. Then, we find a consensus string of $c_{x_1}, c_{x_2}, \ldots, c_{x_k}$. When there are several consensus strings, we randomly select one. In this way, we can get a consensus string of the k planted motifs and use this consensus string as a starting pattern of voting. Similar to Algorithm 2, the above procedure is repeated r times. The algorithm is shown in Fig. 3.

We can show that Selecting k Voting algorithm is more powerful than the simple Selecting One Voting algorithm. Suppose k planted motifs are selected as $a_1,a_2,\ldots,a_k$. We consider one column in the multiple alignment of the k planted motifs. From the VM probabilistic model, the numbers of occurrences of the letters in Σ fit the multinomial distribution. Suppose $u\in\Sigma$ is selected as the majority letter, the number of occurrences of u is x_1, and the numbers of occurrences of the other three letters are x_2,x_3,x_4, respectively. Obviously, x_1,x_2,x_3 and x_4 are non-negative integers and $\sum_{i=1}^4 x_i = k$. In addition, x_1 is a maximum number in x_1,x_2,x_3,x_4. From the observation, we define a set $Q=\{(x_1,x_2,x_3,x_4)|x_1,x_2,x_3,x_4\in Z^* \ \&\ \sum_{i=1}^4 x_i = k \ \&\ x_1=\max_{i=1}^4 x_i\}$, where Z^* is the set of non-negative integers. Set Q contains all possible values of (x_1,x_2,x_3,x_4) such that u can be selected as the majority letter. Sometimes, more than one letter has the maximum number of occurrences. So, set Q can be divided into four disjoint subsets Q_1,Q_2,Q_3 and Q_4. If $(x_1,x_2,x_3,x_4)\in Q_i$, then there are i letters with the maximum number of occurrences. Based on the property of multinomial distribution, we have the following observation.

Observation 3.1. When $\{a_1,a_2,\ldots,a_k\}\subseteq\{c_1,c_2,\ldots,c_n\}$, for each letter $t[j]$ in the consensus string t, the probability that $t[j]=c[j]$ is

$$q=\sum_{i=1}^{4}\sum_{(x_1,x_2,x_3,x_4)\in Q_i}\frac{1}{i}\frac{k!}{x_1!x_2!x_3!x_4!}\left(\frac{1}{4}+\epsilon\right)^{x_1}\left(\frac{1}{4}-\frac{1}{3}\epsilon\right)^{x_2+x_3+x_4}$$

Let $\frac{1}{4}+\delta=q$. When $k\ge 1$, we have $\delta\ge\epsilon$. For example, suppose the error rate of the VM model is $p = 0.27$ and $\epsilon = \frac{3}{4} - p = 0.48$. (Notice that $4/15 \approx 0.27$ and the error rate 0.27 is corresponding to the $(15, 4)$ FM model in the Motif Challenge Problem.) If $k = 3$, we have $\delta = \frac{5}{4}\epsilon + \frac{2}{3}\epsilon^2 - \frac{4}{3}\epsilon^3 \approx 0.60$. And the error rate of the consensus string of three planted motifs is $1 - q \approx 0.15$. The change from 0.27 to 0.15 can increase the accuracy of the voting algorithm significantly. Our experimental results also support this point. In addition, we can use Chernoff-Hoeffding bound to prove that when $k > \frac{9}{2}\epsilon^2$, $\delta > \frac{1}{4}$.

For Selecting k Voting algorithm, we can show similar result to Theorem 3.1. That is, when L and n are large enough, Selecting k Voting algorithm can find the motif consensus with high probability. The difference is that the condition $L > \frac{9}{8\epsilon^4} \log \frac{3m}{\epsilon}$ is changed to $L > \frac{9}{8\epsilon^2\delta^2} \log \frac{3m}{\epsilon}$, where $\delta \geq \epsilon$. Similar to Lemma 3.2, we have

Lemma 3.5. *The probability that one consensus string t with $d(c,t) \leq (\frac{3}{4} - \delta)L$ is selected in Algorithm 3 is at least $1 - \left(\frac{3}{4}\right)^r$.*

Suppose we find a consensus string t with $d(c,t) \leq (\frac{3}{4} - \delta)L$. For a letter $c_i[j]$ in c_i, the probability that $c_i[j] = t[j]$ is no less than $\frac{1}{4} + \frac{4}{3}\epsilon\delta$. Similar to Lemma 3.3, we have the following lemma.

Lemma 3.6. *When $d(c,t) \leq (\frac{3}{4} - \delta)L$ and $L > \frac{9}{8\epsilon^2\delta^2} \log \frac{3m}{\epsilon}$, for each string $s_i \in S$, the probability that c_i is selected in Step 1 of the voting algorithm is no less than $1 - \frac{1}{3}\epsilon$.*

When the consensus string of k motifs is used, the length of L can be reduced from $L > \frac{9}{8\epsilon^4} \log \frac{3m}{\epsilon}$ to $L > \frac{9}{8\epsilon^2\delta^2} \log \frac{3m}{\epsilon}$ where $\delta > \epsilon$. From this point of view, Selecting k Voting algorithm is more powerful than Selecting One Voting algorithm.

From Lemmas 3.4, 3.5 and 3.6, we get the following theorem.

Theorem 3.2. *When $\sum_{1 \leq i \leq n} d(c, s_i)$ is the minimum in all strings in Σ^L, $L > \frac{9}{8\epsilon^2\delta^2} \log \frac{3m}{\epsilon}$ and $\frac{n}{\log n} \geq \frac{9}{2\epsilon^2}$, the probability that Selecting k Voting algorithm can find the motif consensus c is no less than $1 - \left(\frac{3}{4}\right)^r - \frac{4L}{n}$.*

Suppose $\sum_{1 \leq i \leq n} d(c, s_i)$ is the minimum in all strings in Σ^L. When the mutate rate is so high that Selecting k Voting algorithm can not find the motif consensus, we can prove Selecting k Voting algorithm is a good approximation algorithm for the planted closest substring problem. For output c_B of Algorithm 3,

$$E\left(\sum_{1 \leq i \leq n} d(c_B, s_i)\right) \leq \left(\frac{3}{4} - \frac{4}{3}\epsilon\delta\right) nL.$$

For the optimal solution c,

$$E\left(\sum_{1 \leq i \leq n} d(c, s_i)\right) = \left(\frac{4}{3} - \epsilon\right) nL.$$

Therefore, the expected approximation ratio of Selecting k Voting algorithm is $1 + (1 - \frac{4}{3}\delta)\frac{4\epsilon}{3-4\epsilon}$. When k is large enough, δ is approximate to $\frac{3}{4}$, and the ratio is approximate to 1.

We try $r(m-L+1)^k$ different L-mer sets in Step 3 of Algorithm 3. For each consensus string, the voting operation takes $O(Lmn)$ time. So the time complexity of the whole algorithm is $O(rLm^{k+1}n)$.

In practice, the length of input sequences is from several hundred to thousands. The time complexity of Selecting k Voting algorithm is too high to be practical. Therefore, we will introduce a progressive method to speed up Selecting k Voting algorithm in the next subsection.

3.4. Progressive Filtering Algorithm

In Selecting k Voting algorithm, we need to enumerate all possible L-mer sets of the selected k sequences. Then, we need to do voting operations for $r(m - L + 1)^k$ times. To speed up the algorithm, we can filter out random L-mers to decrease the number of voting operations. If the number of voting operations is decreased, the time complexity of the algorithm will be improved.

Consider two sequences s_i and $s_j \in S$. From the VM probabilistic model, $d(c_i, c_j)$ is a relatively small value. Intuitively, the distance $d(c_i, c_j)$ tends to be less than the distance between two random L-mers. The property of the pairwise distance gives us the inspiration of designing a progressive filtering algorithm.

In the Motif Challenge Problem proposed by Pevzner and Sze [1], the distance between a pair of planted motifs is often not the shortest in all L-mer pairs, because there are $O(m^2)$ random L-mer pairs. Many real data also has the same property. Although the distance between a pair of planted motifs may be not the shortest in all L-mer pairs, the distance is smaller than the distances of a large portion of two random L-mers. With the above analysis, we design a progressive filtering algorithm. In the selected k sequences $\{s_{x_1}, s_{x_2}, \ldots, s_{x_k}\} \subset S$ in Selecting k Voting algorithm, we first consider two sequences s_{x_1} and s_{x_2}. In all pairs of L-mers (t_1, t_2), where $t_1 \in P_{x_1}$ and $t_2 \in P_{x_2}$, we keep the best α pairs of L-mers based on $d(t_1, t_2)$ and delete other pairs, where α is an input parameter. The set of the α pairs of L-mers is represented by $\mathcal{S}_2$. In practice, we can set $\alpha \approx m^{1.5}$. The reason is that if $\alpha \approx m^{1.5}$, the planted motif c_{x_1} is contained in $m^{0.5}$ pairs on average. As a result, the probability that (c_{x_1}, c_{x_2}) is not in the $m^{1.5}$ pairs is small.

Then, we consider the third sequence s_{x_3}. For each $t_3 \in P_{x_3}$ and each pair $(t_1, t_2) \in \mathcal{S}_2$, we compute the sum of pair distance $d(t_1, t_2) + d(t_1, t_3) + d(t_2, t_3)$. Base on this value, we also keep the best α triples in set $\mathcal{S}_3$. Similarly, we do the same operation for $s_{x_4}, \ldots, s_{x_k}$. Finally, we get a set $\mathcal{S}_k$ containing α k-tuples and use the k-tuples to get consensus strings and do voting operations. The algorithm is shown in Fig. 4.

Algorithm 4

Input A set of k sequences $s_{x_1}, s_{x_2}, \ldots, s_{x_k} \in S \subset \Sigma^m$, and integers L and α.
Output A set $\mathcal{S}_k$ with α k-tuples.

1. Find set $\mathcal{S}_2$ of the best α pairs of L-mers $(t_1, t_2) \in P_{x_1} \times P_{x_2}$ based on $d(t_1, t_2)$.
2. **For** $i = 3$ to k
3. In $\mathcal{S}_{i-1} \times P_{x_i}$, find set $\mathcal{S}_i$ of the best α i-tuples $(t_1, t_2, \ldots, t_i)$ based on $\sum_{t', t'' \in \{t_1, t_2, \ldots, t_i\}} d(t', t'')$.
4. Output set $\mathcal{S}_k$.

Fig. 4. Progressive filtering algorithm.

The time complexity of the progressive filtering algorithm is $O(\alpha k^2 Lm)$. If we combine the progressive filtering algorithm and Selecting k Voting algorithm, the time complexity of the new algorithm is $O(\alpha r Lm(k^2 + n))$, which is much better than the original Selecting k Voting algorithm. When $\alpha = m^{1.5}$, the time complexity is $O(rLm^{2.5}(k^2 + n))$.

We note that the special case of the progressive filtering algorithm for $k = n$ can be used directly to find motifs. When $k = n$, the progressive filtering algorithm can output α different n-tuples. Then we can find a consensus string from the L-mers in each n-tuple and output the best consensus string.

3.5. Motif Refinement

In practice, we can use some heuristic methods to improve the accuracy of the voting algorithm. Here we introduce two methods.

First, after we get a consensus string t^* from a voting operation, we do not directly output t^*. We can use the resulting string t^* as a new starting pattern and do voting operation repeatedly until there is no further improvement.

Second, we can do local search based on a candidate motif consensus. For a candidate motif consensus t^*, we can change one letter in t^* to get a new motif consensus t^{**}. There are totally $L(|\Sigma| - 1)$ ways to change t^*. From the $L(|\Sigma| - 1)$ ways, we select the best way to change t^* based on the score function. The local search can be repeated until there is no further improvement.

There are some techniques in speeding up the local search in implementation. Let a be the selected L-mer with the minimum distance to t^* in s_i, and b another L-mer in s_i. Note that $d(t^{**}, b) \geq d(t^*, b) - 1$ and $d(t^{**}, a) \leq d(t^*, a) + 1$. If $d(t^*, b) \geq d(t^*, a) + 2$, then $d(t^{**}, b) \geq d(t^*, a) + 1 \geq d(t^{**}, a)$. Therefore, when we search an L-mer with the minimum distance to a new motif consensus t^{**}, it is not necessary to compare all L-mers in s_i with t^{**}. We only consider L-mers with distance no more than $d(t^*, s_i) + 1$ to t^*. This technique can increase the speed of local search dramatically. Another technique is to use the bit representation of L-mers. In this way, the distance between L-mers can be computed with bit operations, which are much faster than counting different letters. This is also an advantage of the local search method compared with the EM method, which needs to compute the likelihood of each L-mer and can not use bit operation strategy to speed up.

4. EXPERIMENTS

We implemented the algorithms in Java. The software is available upon request. To get more starting patterns, we used selection with replacement instead of selection without replacement in Step 2 of Algorithm 2 and Algorithm 3 in the implementation. We tested the algorithms on a PC with an AMD 2.0G CPU and 1.0G Memory. In the experiments, we generated several sets of simulated data. Basically, we followed the settings of the Motif Challenge Problem proposed by Pevzner and Sze [1].

First, we tested the algorithms on simulated data of VM model with a small mutation rate. The parameters were $m = 600$, $L = 15$ and $\epsilon = 0.55$. That is, the mutation rate was $p = \frac{3}{4} - 0.55 = 0.2$. To discover the relationship between n and the accuracy rates of the algorithms, we set $n = 3, 5, 10, 20, 40, 100$ and generated six groups of data respectively. In each group of data, we generated 1000 instances. Selecting One Voting algorithm and Selecting $k = 3$ Voting algorithm were tested with parameter $r = 20$. For an instance with $n = 20$, the running time of Selecting One Voting algorithm is 20.7 seconds. The time complexity of Selecting k Voting is too high to finish the tests in reasonable time. In the tests of Selecting k Voting algorithm, instead of using all L-mer sets of selected k sequences, we only used L-mer sets containing planted motifs to do voting operations. The performance of this method is similar to that of Selecting k Voting algorithm. So we use the results of this method for reference. The results are reported in Table 1.

Table 1. The percentages of correct outputs of Selecting One Voting algorithm and Selecting $k = 3$ Voting algorithm with $m = 600$, $L = 15$, $\epsilon = 0.55$ on the VM model.

	Selecting One	Selecting $k = 3$
$n = 3$	8.0	23.4
$n = 5$	34.1	52.7
$n = 10$	86.3	92.6
$n = 20$	99.7	99.9
$n = 40$	100	100
$n = 100$	100	100

Table 1 shows that when the error rate is not high, both algorithms can find the planted motifs in most cases. The results also show that the number of input sequences is an important factor of the accuracy rates of the algorithms. When n increases, the accuracy rates of the algorithms increase, which is consistent with the results in the proof in Section 3. This fact explains why motifs can be found more easily in many similar sequences than in a few sequences.

Second, we increased the error rate from 0.2 to 0.27 ($\epsilon = 0.48$). Notice that $4/15 \approx 0.27$ and the error rate 0.27 is corresponding to the $(15, 4)$ FM model in the Motif Challenge Problem. Similar to the previous tests, we set $m = 600$, $L = 15$ and $\epsilon = 0.48$, and generated six groups of simulated data each containing 1000 instances. The results are reported in Table 2.

Table 2. The percentages of correct outputs of Selecting One Voting algorithm, Selecting $k = 3$ Voting algorithm and Selecting $k = 3$ Voting algorithm with progressive filtering, with $m = 600$, $L = 15$, $\epsilon = 0.48$ on the VM model.

	Selecting One	Selecting $k = 3$	Selecting $k = 3$ with Progressive Filtering
$n = 3$	1.1	8.4	1
$n = 5$	6.5	21.7	5
$n = 10$	38.7	59.9	38
$n = 20$	89.6	94.9	88
$n = 40$	99.9	100	99
$n = 100$	100	100	100

When the error rate increases, the accuracy rates of Selecting One Voting algorithm decrease much faster than those of Selecting k Voting algorithm. The tests show that Selecting k Voting algorithm is more powerful than Selecting One Voting algorithm.

Although Selecting k Voting algorithm is powerful, the time complexity of the algorithm is too high. To speed up Selecting k Voting algorithm, we proposed a progressive filtering algorithm. To evaluate the progressive filtering algorithm, we selected 100 instances from each group of previous simulated data with error rate 0.27, and tested Selecting $k = 3$ Voting algorithm with progressive filtering on the instances. The parameters were set to $r = 20$ and $\alpha = 20000$. For an instance with $n = 20$, the running time of Selected k Voting algorithm with progressive filtering is 725 seconds. The results are also reported in Table 2. The experimental results show that Selecting k Voting algorithm with progressive filtering has good accuracy rates, while the running time is improved significantly compared with the original Selecting k Voting algorithm.

Since Selecting One Voting and Selecting $k = 3$ Voting algorithm with progressive filtering have good performance and short running time, we tested their performance on difficult cases. We compared the algorithms with the well known random projection algorithm [23]. We followed the test method of Table 1 in Ref. 23 and tested on several difficult FM models where $n = 20$, $m = 600$ and $(L, D) = (12,3), (14,4), (16,5), (18,6)$ and $(19,6)$. For each model, 100 instances were generated. For Selecting One Voting, we selected all 20 sequences to do voting operations. For Selecting $k = 3$ Voting algorithm with progressive filtering, we set $r = 100$ and $\alpha = 20000$.

In the difficult FM models, some random L-mers in input sequences may have small distances to the motif consensus c and only a part of planted motifs can be found. We did tests on the difficult FM models and found that, in some extreme cases, the motif consensus c does not have the optimal score function, and another length L string c' with $d(c, c') = 1$ has the optimal score function $\sum_{1 \le i \le n} d(c', s_i) < \sum_{1 \le i \le n} d(c, s_i)$. In this case, even if our algorithm can find the motif consensus c, the motif consensus will not be output as the optimal solution.

In the experiments of FM model, we followed the objective function used in Refs. 1 and 23. We assumed that D was known and counted the number of mutated motifs with distance no more than D to the motif consensus as the score function. Although it is not practical to use a fixed D in real problems, the function is used for comparison.

The accuracy rates of the random projection algorithm are from Table 1 in Ref. 23. The details are listed in Table 3. Since each input instance contains only 20 sequences, Selecting One Voting algorithm can only select 20 planted motifs as the starting patterns. Moreover, the planted motifs have many errors. Therefore, the accuracy rates of Selecting One Voting algorithm are not good for some difficult models such as $(16,5)$ and $(18,6)$ models. For Selecting $k = 3$ Voting algorithm with progressive filtering, there are many possible starting patterns and the consensus strings contain less errors. Therefore, it performs well on the difficult models and outperforms the best known random projection algorithm. From the experimental results, we can conclude that, although the ideas of our algorithms are simple, our algorithms are effective and powerful in finding planted motifs.

Table 3. The percentages of correct outputs of the random projection algorithm, Selecting One Voting algorithm and Selecting $k = 3$ Voting algorithm with progressive filtering (The results of PROJECTION are from Table 1 in Ref. 23).

L	D	PROJECTION	Selecting One	Selecting $k = 3$ with progressive filtering
12	3	96	97	100
14	4	86	91	100
16	5	77	53	100
18	6	82	36	99
19	6	98	90	100

5. CONCLUSION

In this paper, we studied the planted motif problem. We proposed Selecting One Voting algorithm and Selecting k Voting algorithm for finding planted motifs. We formally proved the common belief that a large number of input sequences can help us find motifs. To speed up Selecting k Voting algorithm, we also gave a progressive filtering algorithm. The experimental results validated the relationship between the number of input sequences and the accuracy rates of our algorithms, and showed that Selecting k algorithm with progressive filtering is powerful in finding planted motifs in difficult planted motif models.

References

1. Pevzner PA, Sze S-H. Combinatorial approaches to finding subtle signals in DNA sequence. In *Proceedings of the Eighth International Conference on Intelligent Systems for Molecular Biology (ISMB)* 2000: 269–278.
2. Lanctot JK, Li M, Ma B, Wang S, Zhang L. Distinguishing string selection problems. In *Proceedings of the Tenth Annual ACM-SIAM Symposium on Discrete Algorithms (SODA)* 1999: 633–642.
3. Li M, Ma B, Wang L. Finding similar regions in many sequences. *Journal of Computer and System Sciences* 2002; **65**: 73–96.

4. Andoni A, Indyk P, Patrascu M. On the optimality of the dimensionality reduction method. In *Proceedings of the 47th Annual IEEE Symposium on Foundations of Computer Science (FOCS)* 2006: 449–458.
5. Ma B, Sun X. More efficient algorithms for closest string and substring problems. In *Proceedings of the 12th Annual International Conference on Research in Computational Molecular Biology (RECOMB)* 2008: 396-406.
6. Hertz GZ, Stormo GD. Identifying DNA and protein patterns with statistically significant alignments of multiple sequences. *Bioinformatics* 1999; **15**(7-8): 563–577.
7. Lawrence CE, Altschul SF, Boguski MS, Liu JS, Neuwald AF, Wootton JC. Detecting subtle sequence signals: a Gibbs sampling strategy for multiple alignment. *Science* 1993; **262**(5131): 208–214.
8. Bailey TL, Elkan C. Fitting a mixture model by expectation maximization to discover motifs in biopolymers. In *Proceedings of the Second International Conference on Intelligent Systems for Molecular Biology (ISMB)* 1994: 28–36.
9. Bailey TL, Elkan C. The value of prior knowledge in discovering motifs with MEME. In *Proceedings of the Third International Conference on Intelligent Systems and Molecular Biology (ISMB)* 1995: 21–29.
10. Bailey TL, Elkan C. Unsupervised learning of multiple motifs in biopolymers using expectation maximization. *Machine Learning* 1995; **21**: 51–83.
11. Rocke E, Tompa M. An algorithm for finding novel gapped motifs in DNA sequences. In *Proceedings of the Second Annual International Conference on Research in Computational Molecular Biology (RECOMB)* 1998: 228–233.
12. Eskin E, Pevzner PA. Finding composite regulatory patterns in DNA sequences. *Bioinformatics* 2002; **18 Suppl 1**: S354–S363.
13. Blanchette M, Schwikowski B, Tompa M. An exact algorithm to identify motifs in orthologous sequences from multiple species. In *Proceedings of the Eighth International Conference on Intelligent Systems for Molecular Biology (ISMB)* 2000: 37–45.
14. Brazma A, Jonassen I, Vilo J, Ukkonen E. Predicting gene regulatory elements *in silico* on a genomic scale. *Genome Research* 1998; **8**(11): 1202–1215.
15. Price A, Ramabhadran S, Pevzner PA. Finding subtle motifs by branching from sample strings. *Bioinformatics* 2003; **19 Suppl. 2**: ii149–ii155.
16. Sinha S, Tompa M. Discovery of novel transcription factor binding sites by statistical overrepresentation. *Nucleic Acids Research* 2002; **30**(24): 5549–5560.
17. Prakash A, Blanchette M, Sinha S, Tompa M. Motif discovery in heterogeneous sequence data. *Pacific Symposium on Biocomputing* 2004: 348–359.
18. Tompa M. An exact method for finding short motifs in sequences, with application to the ribosome binding site problem. In *Proceedings of the Seventh International Conference on Intelligent Systems for Molecular Biology (ISMB)* 1999: 262–271.
19. Tompa M, Li N, Bailey TL, Church GM, Moor BD, Eskin E, Favorov AV, Frith MC, Fu Y, Kent WJ, Makeev VJ, Mironov AA, Noble WS, Pavesi G, Pesole G, Régnier M, Simonis N, Sinha S, Thijs G, van Helden J, Vandenbogaert M, Weng Z, Workman C, Ye C, Zhu Z. Assessing computational tools for the discovery of transcription factor binding sites. *Nature Biotechnology* 2005; **23**(1): 137–144.
20. van Helden J, André B, Collado-Vides J. Extracting regulatory sites from the upstream region of yeast genes by computational analysis of oligonucleotide frequencies. *Journal of Molecular Biology* 1998; **281**(5): 827–842.
21. Keich U, Pevzner PA. Finding motifs in the twilight zone. *Bioinformatics* 2002; **18**(10): 1374–1381.
22. Keich U, Pevzner, PA. Subtle motifs: defining the limits of motif finding algorithms. *Bioinformatics* 2002; **18**(10): 1382–1390.
23. Buhler J, Tompa M. Finding motifs using random projections. *Journal of Computational Biology* 2002; **9**(2): 225–242.
24. Hoeffding W. Probability inequalities for sums of bounded random variables. *Journal of the American Statistical Association* 1963; **58**(301): 13–30.

Computational Systems Bioinformatics 2008

Proteomics

A MAX-FLOW BASED APPROACH TO THE IDENTIFICATION OF PROTEIN COMPLEXES USING PROTEIN INTERACTION AND MICROARRAY DATA

Jianxing Feng*

Department of Computer Science and Technology, Tsinghua University, Beijing 100084, China.

**Email: fengjx06@mails.tsinghua.edu.cn*

Rui Jiang

MOE Key Laboratory of Bioinformatics, Bioinformatics Division TNLIST/Department of Automation, Tsinghua University, Beijing 100084, China.

Email: ruijiang@tsinghua.edu.cn

Tao Jiang

Department of Computer Science, University of California, Riverside, CA 92521.

Email: jiang@cs.ucr.edu

The emergence of high-throughput technologies leads to abundant protein-protein interaction (PPI) data and microarray gene expression profiles, and provides a great opportunity for the identification of novel protein complexes using computational methods. Although it has been demonstrated in the literature that methods using protein-protein interaction data alone can successfully predict a large number of protein complexes, the incorporation of gene expression profiles could help refine the putative complexes and hence improve the accuracy of the computational methods.

By combining protein-protein interaction data and microarray gene expression profiles, we propose a novel *Graph Fragmentation Algorithm* (GFA) for protein complex identification. Adapted from a classical max-flow algorithm for finding the (weighted) densest subgraphs, GFA first finds large (weighted) dense subgraphs in a protein-protein interaction network and then breaks each such subgraph into fragments iteratively by weighting its nodes appropriately in terms of their corresponding log fold changes in the microarray data, until the fragment subgraphs are sufficiently small. Our extensive tests on three widely used protein-protein interaction datasets and comparisons with the latest methods for protein complex identification demonstrate the superior performance of our method in terms of accuracy, efficiency, and capability in predicting novel protein complexes. Given the high specificity (or precision) that our method has achieved, we conjecture that our prediction results imply more than 200 novel protein complexes.

1. INTRODUCTION

With the advances in modern biophysics and biochemistry, it has been widely accepted that the rise of complicated biological functions is largely due to the cooperative effects of multiple genes and/or gene products. This understanding leads to the emergence of high-throughput technologies for identifying interactions between biological molecules and results in the prosperity of interactomics in the post genomics and proteomics era. For example, with the use of yeast two-hybrid assays [1–3] and pull-down mass spectrometry experiments [4, 5], genome-wide protein-protein interactions (PPIs) have been identified and encoded into global PPI networks for the model species *Saccharomyces cerevisiae* (*i.e.* baker's yeast) [6–8]. With the improvement of instruments and increase in the throughput, these technologies have also been applied to identify interactions of human proteins, providing an increasing understanding of the global human PPI network [9]. Parallel to the boom of high-throughput identification of PPIs, genome-wide microarray experiments regarding the expression of genes and their products across a number of different conditions have also been conducted and resulted in publicly available databases such as the *gene expression omnibus* [10].

*To whom correspondence should be addressed.

As a major form of the cooperative effects of two or more proteins, protein complexes play important roles in the formation of complicated biological functions such as the transcription of DNA, the translation of mRNA, and many others. Traditionally, protein complexes are identified using experimental techniques such as the X-ray crystallography and the nuclear magnetic resonance (NMR), or computational methods such as protein-protein docking. These methods, though successful, can hardly meet the requirement of identifying all protein complexes in known organisms, due to the large number of proteins, the cost of biological experiments, and the limited availability of protein structure information. On the other hand, since a protein complex is composed of a group of two or more proteins that are associated by stable protein-protein interactions, computational methods that can make use of abundant data given by the above high-throughput technologies have been demonstrating increasing successes [11–15].

Many studies use PPI data alone for the purpose of identifying protein complexes or biologically functional modules. These methods assume that densely connected components in PPI networks are likely to form functional modules and hence are likely to be protein complexes [16]. With this assumption, the methods generally use the density of interactions as a main criterion and identify protein complexes by finding dense regions in PPI networks. To mention a few, Bader and Hoque proposed a clustering algorithm called MCODE that isolates dense regions in a PPI network by weighting each vertex according to the topological properties of its neighborhood [11]. Andreopoulos *et al.* presented a layered clustering algorithm that groups proteins by the similarity of their direct neighborhoods [17]. Spirin and Mirny applied three methods (*i.e.* clique enumeration, super paramagnetic clustering, and Monte Carlo simulation) to the MIPS PPI network for yeast [7] and produced about 100 dense subgraphs that were predicted to be protein complexes [12]. Their results were found to be superior to many others in terms of accuracy. Pei and Zhang introduced the use of a subgraph quality measure as well as a "seed-refine" algorithm to search for possible subgraphs in a PPI network [13]. King *et al* gave a clustering algorithm based on restricted neighborhood search to partition a PPI network into clusters using some cost function [18]. Bu *et al.* introduced a spectral method derived from graph theory to uncover hidden topological structures that consist of biologically relevant functional groups [19]. Li *et al.* found maximal dense regions by merging local cliques according to their affinity [14]. In a subsequent work, Li *et al.* devised an algorithm, called DECAFF, to address two major issues in current high-throughout PPI data, namely, incompleteness and high data noise [15].

Another group of methods combine PPI data and microarray gene expression profiles for the purpose of identifying protein complexes. These methods regard PPIs as static descriptions of the potential collaborative effects between proteins and treat gene expression profiles as dynamic information of genes under various conditions. Since proteins of a complex usually work together to complete certain biological functions, and there exists a simple mapping between genes and their products, the combination of PPI and microarray gene expression data can clearly help the discovery of protein complexes or functional modules [20, 21]. Moreover, such a combination is also often used in the search for regulatory modules and signalling circuits [22]. As an example, Guo *et al.* identified condition-responsive subnetworks in a PPI network by weighting its edges based on gene expression profiles [23].

Besides these methods, there exist some other methods that aim at identifying protein complexes by using comparative interactomics. For example, Sharan *et al.* identified protein complexes by a comparative analysis of the PPI networks from yeast and bacteria [24]. Hirsh and Sharan developed a probabilistic model for protein complexes that are conserved across two species and applied it to yeast and fly [25]. These methods based on comparative analysis require the availability of quality PPI networks from multiple species and can only identify protein complexes conserved in multiple species.

Despite differences in the approach and use of data, most of the computational methods mentioned above follow a bottom-up local search strategy. For example, Li *et al.* first finds small dense subgraphs (or components) in a PPI network and then merges these components gradually to form protein complex-like subgraphs [15]. Pei and Zhang greedily expands some carefully selected seed subgraphs until a given criterion is met [13]. Because a local search strategy does not return the optimal solutions in general, the above bottom-up methods are not guaranteed to find the densest subgraphs in the input PPI network and therefore may miss many important protein complexes that are truely dense.

To overcome this drawback, we present a top-down method to identify protein complexes that explicitly utilizes the density information in PPI networks as well as microarray gene expression profiles. This work combines the classic maximum network-flow based *Densest Subgraph Algorithm* (DSA) [26] to find the densest subgraphs with a novel application of microarray data. Our algorithm, named the *Graph Fragmentation Algorithm* (GFA), first finds dense subgraphs in a PPI network (many of which could potentially be large), and breaks each of them into fragments iteratively by weighting its nodes appropriately in terms their corresponding log fold changes in the microarray data, until the fragment subgraphs are sufficiently small. In order to test the performance of our method, we apply GFA to three widely used yeast PPI networks (*i.e.* the MIPS, DIP and BioGRID PPI networks) and compare our predictions with the known protein complexes in the MIPS database as well as with those of the latest methods for protein complex identification (that are not based on comparative analysis) [12, 15]. The test results clearly demonstrate the superior performance of our method in terms of accuracy, efficiency, and capability in predicting novel protein complexes. For example, GFA could be tuned to achieve sensitivity 73% and specificity 85% simultaneously on the DIP PPI network. Our method also provides a ranking of the predicted complexes, taking advantage of the multiple conditions (or samples) in the microarray expression data. Putative complexes with higher ranks are believed to have a larger likelihood to be true protein complexes. Moreover, our predictions result in more than 200 highly ranked dense subgraphs that share no common proteins with the known complexes in MIPS and are thus likely to be novel protein complexes.

For the convenience of presentation, some of the figures and tables are omitted in the main text and given in the appendix.

2. MATERIALS AND METHODS

2.1. Data sources

Three PPI datasets concerning *Saccharomyces cerevisiae* are used. The first one is the MIPS protein-protein interaction network dataset [7], which is believed to contain the most credible PPI data and will simply be denoted as MIPS-PPI. The second one is the DIP protein-protein interaction network dataset [6], denoted as DIP-PPI. The third one is BioGRID protein-protein interaction dataset [8], which is the most comprehensive one and will be denoted as BioGRID-PPI. Because a PPI network is treated as an undirected simple graph, at most one edge will be kept between any pair of proteins. The numbers of nodes (or edges) in the MIPS, DIP and BioGRID PPI networks are 4,554 (or 12,319), 4,932 (or 17,201) and 5,201 (or 71,044), respectively.

We retrieved 58 sets of microarray gene expression data concerning yeast from the GEO database [10]. The expression levels have been log transformed, and the microarray data contain a total of 716 samples (or conditions). Since the genes expressed in each sample are different, and they could also be different from the genes contained in a PPI network, we will use a sample of the microarray data on a PPI network if it covers at least 90% of the genes in the network. This criterion results in 477, 571 and 623 samples that can be applied to the MIPS, DIP and BioGRID PPI networks, respectively.

As in the previous studies [11, 12, 14, 15], the MIPS complex database [7] is used as the benchmark (*i.e.* the truth) to evaluate the protein complexes predicted by our method. This database contains protein complexes manually verified and those identified by high throughput experiments. We denote the set of complexes verified manually as MIPS-MAN and the set of all protein complexes in the database as MIPS-ALL. Furthermore, our GFA algorithm only outputs connected subgraphs, but many complexes in MIPS-ALL are not connected in the above PPI networks. To evaluate our results in a more reasonable way, we decompose each MIPS complex into connected components according to the PPI network under study. We will use MIPS-MAN-COMP and MIPS-ALL-COMP to denote the sets of connected complex components obtained from MIPS-

MAN and MIPS-ALL, respectively. Finally, since GFA does not output subgraphs consisting of a single node or edge (because they are trivial), all complexes or complex components with sizes 1 or 2 are removed from MIPS-MAN-COMP and MIPS-ALL-COMP. Note that the contents of MIPS-MAN-COMP and MIPS-ALL-COMP depend on the underlying PPI network used. In Table 1, we summarize sizes of the benchmark sets with respect to each PPI network.

Table 1. Sizes of the benchmark sets of protein complexes with respect to each PPI network.

Benchmark	MIPS-PPI	DIP-PPI	BioGRID-PPI
MIPS-MAN-COMP	100	114	134
MIPS-ALL-COMP	272	759	804

2.2. An outline of GFA

A PPI network is considered as an undirected simple graph, where nodes represent proteins and edges denote interactions between two nodes. To find dense subgraphs, various computational methods have been proposed (see the introduction section). Nevertheless, these methods are mostly based on local search strategies and can hardly find the densest subgraphs in a given PPI network.

A widely used definition of the density for a subgraph is $\delta = 2 \cdot |E|/(|V| \cdot (|V|-1))$ [11, 12], where E and V denote the sets of edges and nodes in the subgraph, respectively. An alternative definition is $\delta = |E|/|V|$. In general, the former definition favors small subgraphs (see Spirin and Mirny [12]), while the latter favors large subgraphs. However, both definitions are sensitive to the size of a subgraph. In fact, when the first definition is applied, we have to add a lower bound on $|V|$ to make the result interesting. Considering this, we use the latter definition of density in this work, since there exists an elegant algorithm to find the densest subgraph under this definition. Besides, our experimental results also demonstrate that this definition of density works well in finding protein complexes.

Theoretically, the problem of finding a subgraph with the greatest density in a graph under the first definition is much harder than that under the second one. The problem under the first definition is basically equivalent to finding the largest clique in a graph, a classical NP-hard problem in theoretical computer science [27]. However, there is an elegant and fast algorithm to solve the problem under the second density definition. This algorithm, simply denoted as DSA (*i.e.* the *Densest Subgraph Algorithm*), finds the densest subgraph in a graph by iteratively solving a series of maximum flow problems and has the time complexity of $O(|E| \cdot |V| \cdot \log(|V|^2/|E|))$ [26]. Although DSA can be iterated to find many dense subgraphs in a PPI network, this approach (alone) will likely not work well in terms of finding protein complex-like subgraphs, since it tends to find large dense subgraphs while protein complexes are usually small (*i.e.* containing no more than 20 proteins). Nevertheless, DSA will form the core ingredient of our GFA algorithm for finding protein complexes. GFA actually uses a generalized version of the second density definition: $\delta = |E|/w(V)$, where we assume that the nodes in the graph are weighted (*e.g.* using the log fold changes in some sample of microarray data) and $w(V)$ denotes the total weight of the nodes in the subgraph. The DSA algorithm mentioned above also works for this generalized definition.

GFA consists of two phases: (1) identify candidate subgraphs from the input PPI network using a single sample of gene expression data, and (2) combine candidate subgraphs from multiple samples to form a ranked list of predicted protein complexes. The basic idea behind the first phase is to iterate DSA to obtain (large) dense subgraphs and then break each large dense subgraph into fragment subgraphs by weighting its nodes appropriately using the log fold changes of the nodes in the sample. This phase is executed on each sample separately. In the second phase, we detect and remove redundant (or overlapping) subgraphs found using different samples and rank the subgraphs according to the times that they are found in all samples. The worst case time complexity of GFA is largely determined by the time complexity of phase 1, which is $O(|E| \cdot |V|^2 \cdot \log(|V|^2/|E|) \cdot \mathit{MaxIter} \cdot \mathit{SampleSize})$, where *MaxIter* is a parameter defined below and *SampleSize* is the number of samples of the microarray data used in the computation.

2.3. Identification of candidate subgraphs

Again, the idea is to break each large dense subgraph found by DSA into smaller ones by weighting its nodes appropriately using gene expression data. Recall that the gene expression data contains hundreds of samples. In this phase, we look at one sample at a time. The log fold change of the expression value of gene A in the sample is denoted as $expr(A)$. At the beginning, the nodes in the input PPI network with degree 1 are removed. This will reduce the size of the network and will not affect our final result much because a dense subgraph is not expected to contain nodes with degree 1. Then we weight every node uniformly as 1 and run DSA to find the densest subgraph. If the size of the subgraph identified is above a certain threshold (denoted as *MaxSize*), the weight of each node A in the subgraph is multiplied by a factor of $e^{-expr(A)}$ and DSA is applied again to the subgraph. The effect of this multiplication is that the weights of highly differentially expressed genes in the subgraph are reduced. The exponentially factor of $e^{-expr(A)}$ in this adjustment was chosen empirically. Note that, since DSA maximizes the ratio $|E|/w(V)$, it tends now to find a subgraph with nodes bearing small weights. In other words, the above weighting adjustment favors genes that are highly differentially expressed in the sample. As an effect, some nodes with large weights may be removed and the subgraph is fragmented. This step is executed iteratively, until either a given maximum iteration count (denoted as *MaxIter*) is reached or the size of the subgraph is below *MaxSize*.

Once a sufficiently small dense subgraph is found, all the nodes in the subgraph and all the edges adjacent to any one of the nodes in the subgraph are removed from the PPI network. Then, we remove all the nodes with degree 1 in the remaining network and reiterate the above process of using DSA to find the next sufficiently small dense subgraph. The whole process ends when the PPI network exhausts.

Now we discuss the two parameters *MaxSize* and *MaxIter*. *MaxSize* determines the maximum size of a subgraph found by GFA. In principle, it should be set as the largest possible size of an expected protein complex component (see Section 2.1 for the definition of protein complex components) for a given PPI network. For example, in our experiments, for MIPS-PPI, we select 20 as the bound because the maximum size of a protein complex component in MIPS-ALL-COMP does not exceed 20. However, our experiments show that GFA is quite robust with respect to this parameter and it is fine to use a slightly larger *MaxSize*, especially when the microarray data contains many samples, because only the "core" of a subgraph will be found in multiple samples. For example, we also tried to set *MaxSize* as 30 on MIPS-PPI and got almost the same result. The parameter *MaxIter* reflects how strictly we enforce the size bound. A small *MaxIter* may lead to subgraphs with sizes above *MaxSize*. This is useful when there are a few protein complexes that are very dense and much larger than the other protein complexes and we do not want to make *MaxSize* too large. So, the parameters *MaxSize* and *MaxIter* together control the sizes of the output subgraphs. Fortunately, our test results show that the final result of GFA is not very sensitive to either of these parameters.

2.4. Combining candidate subgraphs

The above phase 1 of GFA generates a set of candidate subgraphs for each sample of the microarray data. However, many of these subgraphs are duplicated or similar. We define the *overlap score* of two sets A and B as $overlap(A, B) = 2|A \bigcap B|/(|A| + |B|)$. This step aims to distill promising subgraphs from the candidate subgraphs. More specifically, duplicates and trivial subgraphs with sizes 1 or 2 are removed and similar subgraphs will be merged. However, because of the drastic difference in the densities of the three PPI networks considered in this paper, we have to use two different strategies in this phase. We use a simple strategy for MIPS-PPI and a more general, slightly more complicated strategy for DIP-PPI and BioGRID-PPI. The latter networks are much denser.

2.4.1. *The simple strategy*

Here we simply count the frequency of each candidate subgraph in all samples and rank the subgraphs by their frequencies. A subgraph with a high frequency is expected to be a promising protein complex (or complex component), since it is dense and many of its nodes are highly differentially expressed in multiple samples. After the frequency of each candidate subgraph is calculated, we check if two candidate subgraphs overlap. If the overlap score between two graphs (computed using their vertex sets) is above a certain cutoff (denoted as *MaxOverlap*), they are deemed duplicates and the one with a smaller frequency is simply removed.

Note that, the result of this removal step depends on the order that we process the candidate subgraphs. For example, consider subgraphs A, B, C with sizes a, b, c respectively, with $a > b > c$. A overlaps with B and B overlaps with C, but A and C do not overlap according to the given overlap criterion. If A and B are processed after B and C are processed, only A remains. But if we process A and B first, then both A and C will remain. So, for consistency, we consider the pairs of candidate subgraphs in decreasing order of their overlap. This simple strategy is also applied to the following more general strategy and the combined strategy.

As shown in our experimental results, this simple strategy works very well on MIPS-PPI, mostly due to its sparsity. It also works on the DIP-PPI and BioGRID-PPI, although it appears to be too conservative in dealing with similar candidate subgraphs.

2.4.2. *The more general strategy*

Dense subgraphs in dense PPI networks tend to be larger and we cannot expect that the subgraph corresponding to a real protein complex will be discovered by GFA from many samples exactly, since the samples generally have different expression levels. Thus, the simple strategy is too conservative for this situation. Moreover, when the input PPI network is large (such as BioGRID-PPI), DSA becomes quite slow and we may not want to spend the time to examine every sample of the microarray data. Hence, in this case, we need revise the definition of frequency and introduce a more general strategy to combine results from different samples. Our basic idea here is to combine similar candidate subgraphs. Due to the page limit, this general strategy and a combined method to integrate it with the simple stragegy is omitted in this extended abstract but will be given in the full paper.

3. RESULTS

3.1. Some useful definitions and notations

Before discussing the results, we need introduce several definitions and notations. First, since we will mainly validate our predictions against benchmark protein complexes in MIPS, we define the *effective size* of a predicted protein complex as the number of proteins shared by this predicted complex and the complexes in the relevant benchmark (*i.e.* MIPS-MAN-COMP or MIPS-ALL-COMP). Obviously, we could only hope to validate predicted protein complexes with large effective sizes. We say that a protein complex (component) A in a benchmark set is *identified* by a predicted complex B with some cutoff p if $|A \bigcap B|^2/(|A| \cdot |B|) \geq p$. Since a commonly used value for p in the literature is 0.2 [11, 15], we say that B *matches* A if A is identified by B with the cutoff $p = 0.2$. The following several (shorthand) notations will be convenient to use in tables and figures:

(1) *predicted* (or P for short): The number of predicted protein complexes.
(2) *matched* (or M for short): The number of predicted complexes that match some protein complex component in the relevant benchmark set.
(3) $P_{e \geq n}$: The number of predicted complex with effective sizes at least n.
(4) $P_{e=n}$: The number of predicted complexes with effective sizes exactly n.
(5) *identified(p)* (or $I(p)$ for short): The number of complex components in the relevant benchmark set that have been identified by any one of the predicted complexes with cutoff p. This parameter generally reflects the sensitivity of the prediction. Although the widely used p value is 0.2, we will also consider $p = 0.5$ since it could provide more insight into the prediction result.
(6) *effective specificity* : The number of predicted protein complexes that match complex components in the relevant benchmark set divided by the number of predicted complexes with effective sizes at least 2. In other words, it is equal to $M/P_{e \geq 2}$. Hereafter, the term *specificity* refers to *effective specificity* unless stated otherwise.

Note that, because of overlaps in the predicted results and the benchmark sets, the number of matched predicted complexes may not be the same as the number of the identified complex components in the relevant benchmark. In other words, M may be different from $I(0.2)$. For example, $M = 1$ and $I(0.2) = 2$ means that the result consists of one predicted complex that matches (and perhaps contains) two complex components in the benchmark. On the other hand, $M = 2$ and $I(0.2) = 1$ means that the result consists of two predicted complexes that match (and are perhaps contained in) a single benchmark complex component. In general, let us define the *efficiency* of a prediction as the ratio between $I(p)$ and M. Clearly, with the same $I(p)$ value (*i.e.* the same sensitivity), we would prefer prediction results with a small M since a smaller M would imply a higher efficiency. In our test results, an important property is that the number $P_{e \geq 2}$ is very close to M when the parameter *MinFrequency* is large. Hence, among the protein complexes predicted by GFA, a top ranked protein complex either has a match in the benchmark or has a very small effective size (*i.e.* it is largely disjoint from the benchmark).

3.2. Matching to the benchmark

For succinctness, we give a detailed report of the prediction result on MIPS-PPI and their matches in MIPS-MAN-COMP, and sketch the other results. As mentioned before, on MIPS-PPI, the simple strategy in phase 2 of GFA is applied. MIPS-MAN-COMP contains 100 complex components with respect to MIPS-PPI. The actual output of GFA depends on the parameters *MinFrequency* and *MaxOverlap* involved in phase 2. By choosing different values for these two parameters, we obtain prediction results with different combinations of sensitivity and specificity. In general, a big *MinFrequency* implies a high specificity and a low sensitivity.

Figure 1 shows the number of predicted complexes and their matching benchmark complexes with respect to various combinations of *MinFrequency* and *MaxOverlap*. An interesting observation is the high accordance among $P_{e \geq 2}$, M and $I(0.2)$. The accordance between the former two terms implies (as mentioned before) that a predicted protein complex has either a match in the benchmark or a very small (*i.e.* at most 1) effective size. While the accordance between the latter two terms indicates that GFA is very efficient and the accordance between the 1st and 3rd terms implies that GFA maintains a good (effective) specificity. The comparison between the prediction results for $MaxOverlap = 0.2$ and $MinOverlap = 0.5$ shows that the parameter *MaxOverlap* has little impact when *MinFrequency* is greater than 2. This means that the predicted protein complexes in general do not overlap too much with each other.

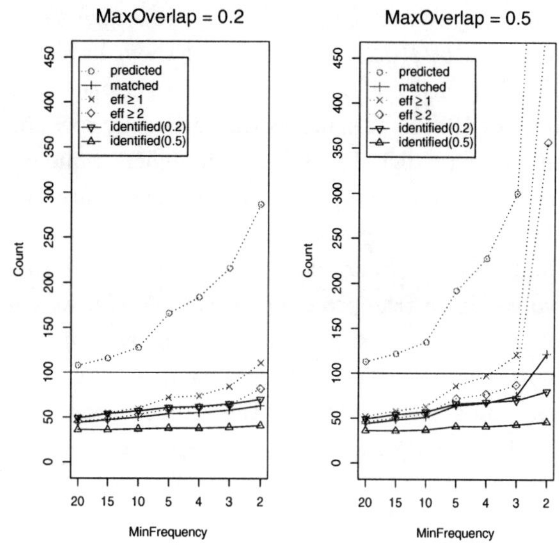

Fig. 1. Protein complexes predicted by GFA on MIPS-PPI and their matches in MIPS-MAN-COMP. Two *MaxOverlap* values, 0.2 (left) and 0.5 (right), are considered. The notation $eff \geq 1$ stands for $P_{e \geq 1}$.

Table 2 gives the detailed results when two extremal values of *MinFrequency* are considered, with *MaxOverlap* being fixed at 0.2. In the first group of results where MIPS-MAN-COMP is used as the benchmark (*i.e.* the more reliable benchmark), when $MinFrequency = 20$, 49 out of the 100 complex components in the benchmark are identified. Although the sensitivity is only 49%, 44 out of the 45 predicted complexes with large effective sizes (*i.e.* at least 2) have matches in the benchmark, which means that the (effective) specificity of this prediction is 97%. Moreover, among the 64 predicted protein complexes that have no matches in the benchmark, 58 of them have zero effective size. In other words, their proteins do not appear in the benchmark at all. We conjecture that these 58 predicted complexes represent novel protein complexes (or at least are involved in novel protein complexes).

On the other hand, if $MinFrequency = 2$, the predicted protein complexes identify 70% of the complex components in the benchmark, but the specificity drops. Among the 82 predicted complexes with large effective sizes, 63 of them match complex components in the benchmark, *i.e.* the specificity is 77%. Comparing the values of $I(0.2)$ and $I(0.5)$, we see that GFA could identify 21 additional complex components in the benchmark using $MinFrequency = 2$ than using $MinFrequency = 20$, as suggested by the values of $I(0.2)$, but only 5 of them have been identified with a high accuracy, as suggested by the values of $I(0.5)$. This means that generally speaking, complexes predicted by GFA with higher frequencies identify complex components in the benchmark more accurately. In other words, a predicted complex with a higher rank is more likely to be (or at least to be involved in) a true protein complex. Again, we conjecture that the 176 predicted complexes that share no proteins with the benchmark complexes represent novel complexes. Note that, the 176 novel complexes actually include the 58 novel complexes mentioned above.

By examining the sets of subgraphs output by GFA with $MinFrequency = 20$ and $MinFrequency = 2$ in detail, we find that the former set could already identify most of the large and dense complex components in the benchmark MIPS-MAN-COMP. 18 out of the 30 complex components in the benchmark missed by the latter (larger) set are trees with at most 6 nodes, and the remaining 12 missing complex components have densities at most 2 in MIPS-PPI. The details of these results are not shown here. It is also clear that GFA achieves very good efficiency in both cases, with the ratio $I(0.2)/M$ being about 1.11.

In the second part of Table 2 where MIPS-ALL-COMP is used as the benchmark, when $MinFrequency = 20$, 57 out of the 61 predicted complexes with large effective sizes have matches in the benchmark. Thus, we still have the same property that a protein complex predicted by GFA with a high frequency has either a match in the benchmark or a very small effective size. The sensitivity and specificity of the prediction are generally a bit worse than those using MIPS-MAN-COMP. The sensitivity is $135/272 = 50\%$ for $MinFrequency = 20$ and $179/272 = 66\%$ for $MinFrequency = 2$ and the specificity is $57/61 = 93\%$ for $MinFrequency = 20$ and $89/127 = 70\%$ for $MinFrequency = 2$. This is perhapss due to the noise in MIPS-ALL.

Table 2. Protein complexes predicted by GFA on MIPS-PPI and their matches in MIPS-MAN-COMP and MIPS-ALL-COMP. MAN and ALL stands for MIPS-MAN-COMP and MIPS-ALL-COMP, respectively. f stands for $MinFrequency$, and $MaxOverlap$ is set to 0.2.

	P	$I(0.2)$	$I(0.5)$	M	$P_{e\geq 2}$	$P_{e=0}$
MAN, f=20	108	49	36	44	45	58
MAN, f=2	287	70	41	63	82	176
ALL, f=20	108	135	82	57	61	43
ALL, f=2	287	179	91	89	127	129

It is interesting to note that only a small fraction of the novel protein complexes conjectured above have matches in MIPS-ALL-COMP (*i.e.* at most $15 = 58 - 43$ for $MinFrequency = 20$ and at most $47 = 176 - 129$ for $MinFrequency = 2$).

The GFA prediction results on DIP-PPI and BioGRID-PPI and their matches in MIPS-MAN-COMP are sketched below. The details are given in the appendix (see Tables 5 and 6, and Figures 2 and 3). On DIP-PPI and BioGRID-PPI, the combined strategy in phase 2 of GFA is used. In both cases, $MinFrequency = 3$ is selected as the smallest frequency threshold instead of $MinFrequency = 2$. This is because the combined strategy in phase 2 introduces noise (*i.e.* spurious subgraphs) as it relaxes the definition of frequency. Such spurious subgraphs typically have very low frequencies and could potentially be eliminated by using a moderate *MinFrequency* threshold.

On DIP-PPI, the parameter *MaxOverlap* is set as 0.2 as before. With $MinFrequency = 20$, GFA predicts 116 protein complexes. 51 of them are conjectured to be novel based on their (zero) effective sizes, using MIP-MAN-COMP as the benchmark. The sensitivity and specificity are 50% and 91%, respectively. With $MinFrequency = 3$, GFA predicts 318 protein complexes, and 204 of them are conjectured to be novel. The sensitivity and specificity are 73% and 85%, respectively. Unlike on the MIPS or DIP PPI networks, the parameter *MaxOverlap* has a significant impact on the prediction results for BioGRID-PPI, since the network is much denser. We will take $MaxOverlap = 0.5$ as an example to show the results in this paper. With $MinFrequency = 20$, GFA predicts 166 protein complexes and 111 of them are conjectured to be novel. The sensitivity and specificity are 31% and 83%, respectively, still using MIPS-MAN-COMP as the benchmark. With $MinFrequency = 3$, GFA predicts 870 protein complexes and 529 of them are conjectured to be novel. The sensitivity and specificity in this case are 48% and 63%, respectively.

We note that in all of the above tests, GFA achieves the best sensitivity (of 73%) with a decent specificity (of 85%) on DIP-PPI, whereas its accuracy deteriorates significantly on BioGRID-PPI. This does not surprise us because although MIPS-PPI is supposed to be the most reliable one among the three PPI networks, it may also miss many true edges (interactions). In other words, it may be too conservative. These missing edges, some of which may exist in DIP-PPI, could provide useful density information in the computation of GFA. On the other hand. BioGRID-PPI may contain many false interactions that could mislead GFA. The prediction efficiency of GFA remains good on both DIP-PPI and BioGRID-PPI.

3.3. Comparison with the previous methods

In this section, we compare the performance of GFA with those of two existing methods for identifying protein complexes from PPI networks that are proposed or surveyed in Spirin and Mirny [12] and Li *et al.* [15]. We will not consider methods based on comparative analysis of PPI networks in this comparison, since we are mostly interested in the interplays between PPI data and microarray gene expression data in the current study and the issue of how gene expression profiles could help analysis of PPI networks. Because the previous methods all predict complexes that are connected in the input PPI network and contain at least three proteins, MIPS-MAN-COMP will be used as the benchmark for a fair comparison.

Table 3. Comparison of GFA and Spirin and Mirny [12] on MIPS-PPI. The row $MinFrequency = 58$ shows the result of GFA when *MinFrequency* is set as 58.

	P	$I(0.2)$	$I(0.5)$	M	$P_{e\geq 2}$	$P_{e=0}$
V. Spirin	76	39	28	46	51	21
$MinFrequency = 58$	77	39	30	35	35	40

The first comparison is with the result reported in Spirin and Mirny [12], which is a bit old but still among the most accurate protein complex predictions. After removing duplicates from the protein complexes predicted in this reference, we obtain 76 subgraphs. To match this number, we set $MinFrequency = 58$ so that the number of subgraphs output by GFA is close to 76. Table 3 summarizes the performance of both prediction results. Both results identify almost the same number of complex components in the benchmark (so the same sensitivity) with the cutoff $p = 0.2$ or $p = 0.5$. But the values of the parameter M show that our result is more efficient, since the 35 matched predicted complexes in our result achieve the same sensitivity as that achieved by the 46 matched predicted complexes in Spirin and Mirny [12]. More importantly, because of this efficiency, our result suggests 19 more novel complexes that are completely disjoint from the proteins in the benchmark complexes, as shown in the $P_{e=0}$ column.

The comparison should be taken with a grain of salt because Spirin and Mirny [12] used an older PPI data from MIPS, which is no longer available. Note that, a more recent MIPS-PPI may not give their method a better result. Nonetheless, our algorithm GFA is much simpler than theirs, especially when the simple strategy is used in phase 2, because their result is a combination of the outputs of three totally different algorithms.

The second comparison is with DECAFF, an algorithm proposed by Li *et al.* [15] recently. Since they gave a detailed comparison between DECAFF and many existing methods for protein complex identification in the literature, including MCODE [11], LCMA [14], and an algorithm proposed by Altaf-Ul-Amin *et al.* [28], and demonstrated the superiority of DECAFF over these methods, we will only compare GFA with DECAFF in this paper.

DECAFF uses the same MIPS PPI data as that used by GFA and predicts 1220 complexes. The first group of results in Table 4 shows the matching of the 1,220 complexes to the benchmark complexes. For comparison, the matching of the 287 complexes predicted by GFA with $MinFrequency = 2$ is listed here too. As we can see, the GFA prediction result contains less than 1/4 of the complexes predicted by DECAFF while only losing 3% sensitivity. This comparison also suggests that the complexes produced by DECAFF overlap with each other a lot. For a more informative comparison, we remove overlapped putative complexes as described in Section 2.4.1. Since the removal depends on the cutoff *MaxOverlap*, we consider two cutoff values here: 0.5 and 0.2.

The second and third groups of results in Table 4 compare the predictions of GFA and DECAFF after the removal. In each case, the *MinFrequency* parameter in GFA is selected so that the number of predicted complexes by GFA is close to that by DECAFF. The comparison shows that GFA outperforms DECAFF in terms of sensitivity ($I/100$), specificity ($M/P_{e\geq2}$) and efficiency (I/M). Moreover, GFA is able to find more novel protein complexes, as shown in the $P_{e=0}$ column.

Table 4. Comparison of GFA and DECAFF on MIPS-PPI. o and f stand for $MaxOverlap$ and $MinFrequency$, respectively.

	P	$I(0.2)$	$I(0.5)$	M	$P_{e\geq2}$	$P_{e=0}$
DECAFF	1,220	73	48	505	757	280
o=0.2, f=2	287	70	41	63	82	176
o=0.5, DECAFF	242	61	25	64	109	87
o=0.5, f=4	228	68	41	67	77	131
o=0.2, DECAFF	111	43	21	41	55	44
o=0.2, f=18	111	53	36	46	47	58

We also compare our results on BioGRID-PPI with that generated by DECAFF on the same PPI network as reported in Li *et al.* [15]. A comparison of the 2,840 predicted complexes predicted by DECAFF and the benchmark complexes is given in the first row of Table 7 in the appendix. Although this prediction has a high (perfect) sensitivity and decent specificity, it has a very low efficiency as the 118 complex components in the benchmark are identified by a large number (*i.e.* 1,141) of the predicted complexes. In other words, the predicted complexes overlap a lot with each other. For a more informative comparison, we again remove overlapped putative complexes using the method described in Section 2.4.1, with $MaxOverlap = 0.5$ or $MaxOverlap = 0.2$. The second and third groups of results in Table 7 compare the predictions of GFA and DECAFF after the removal. In each case, $MinFrequency$ is selected so that the number of predicted complexes by GFA is close to that by DECAFF. The table shows that GFA outperforms DECAFF significantly in terms of specificity ($M/P_{e\geq2}$), efficiency (I/M), and the ability to predict novel protein complexes ($P_{e=0}$). It is only outperformed by DECAFF in sensitivity when $p = 0.2$. In fact, it achieves a better sensitivity than DECAFF when $p = 0.5$, although the sensitivities of both methods are all pretty low.

3.4. The effects of microarray data and parameters in phase 1

The experiments on the three PPI datasets show that the number of samples combined in GFA has a big impact on the final result, but the prediction results of GFA are not very sensitive to the parameters in phase 1. Due to the page limit, a detailed discussion on these effects or non-effects is omitted in this extended abstract but will be given in the full paper.

4. CONCLUSIONS AND DISCUSSION

We have presented a max-flow based algorithm, GFA, to identify complexes from PPI networks by incorporating microarray data. Compared to the previous methods, GFA is actually able to find the densest subgraphs in the input PPI network efficiently, rather than using some local search heuristic. Our experiments on the MIPS, DIP, and BioGRID PPI networks have demonstrated that GFA outperforms the previous methods in terms of specificity, efficiency and ability in predicting novel protein complexes, and it has a comparable sensitivity as those of the previous methods. One of the reasons that GFA was not able to identify some of the benchmark protein complexes is that it removes nodes of degree 1 from the network in every iteration. This step is necessary since it prevents GFA from producing many small spurious complexes. We may have to explore a different strategy in order to improve the sensitivity.

In phase 1 of GFA, multiple rounds of DSA have to be executed in order to find a dense subgraph of a sufficiently small size. This is time consuming. To speed up this step, we can set a small *MaxIter*. We have demonstrated that the final result is not very sensitive to this parameter. An alternative is to assign larger weights to nodes based on expression data in each round.

Our discussion in the previous section shows that the performance of GFA generally improves when more samples are combined. However, the running time of GFA is proportional to the number of samples and could become a concern when the PPI network is large/dense.

Acknowledgements

This work was partly supported by the Natural Science Foundation of China grants 60621062, 60503001, 60528001, and 60575014, the Hi-Tech Research and Development Program of China (863 project) grants 2006AA01Z102 and 2006AA02Z325, the National Basic Research Program of China grant 2004CB518605, NSF grant IIS-0711129, NIH grant LM008991, a startup supporting plan at Tsinghua University, and a Changjiang Visiting Professorship at Tsinghua University.

References

1. Peter Uetz et al. A comprehensive analysis of protein-protein interactions in saccharomyces cerevisiae. *Nature* 2000; 403:623–627.
2. Takashi Ito et al. Toward a protein-protein interaction map of the budding yeast: a. comprehensive system to examine two-hybrid interactions in all possible combinations between the yeast proteins. *Proc. Natl Acad. Sci. USA* 2000; 97(3):1143–1147.
3. Takashi Ito et al. A comprehensive two-hybrid analysis to explore the yeast protein interactome. *Proc. Natl Acad. Sci. USA* 2001; 98(8):4569–4574.
4. Yuen Ho et al. Systematic identification of protein complexes in saccharomyces cerevisiae by mass spectrometry. *Nature* 2002; 415:180–183.
5. Anne-Claude Gavin et al. Functional organization of the yeast proteome by systematic analysis of protein complexes. *Nature* 2002; 415:141–147.
6. Lukasz Salwinski et al. The database of interacting proteins: 2004 update. *Nucleic Acids Research* 2004; 32:D449–D451.
7. U. Güldener et al. CYGD: the comprehensive yeast genome database. *Nucleic Acids Research* 2005; 33:D364–D368.
8. Chris Stark et al. BioGRID: a. general repository for interaction datasets. *Nucleic Acids Research* 2006; 34:D535–D539.
9. Ulrich Stelzl et al. A human protein-protein interaction network: A resource for annotating the proteome. *Cell* 2005; 122(6):957–968.
10. Tanya Barrett et al. NCBI GEO: mining tens of millions of expression profiles–database and tools update. *Nucleic Acids Research* 2007; 35:D760–D765.
11. Gary D. Bader and Christopher W. V. Hogue. An automated method for finding molecular complexes in large protein interaction networks. *BMC Bioinformatics* 2003; 4(2).
12. Victor Spirin and Leonid A. Mirny. Protein complexes and functional modules in molecular networks. *Proc. Natl Acad. Sci. USA* 2003; 100(21):12123–12128.
13. Peng Jun Pei and Ai Dong Zhang. A 'seed-refine' algorithm for detecting protein complexes from protein interaction data. *IEEE Transactions on Nanobioscience* 2007; 6(1):43–50.
14. Xiao-Li Li et al. Interaction graph mining for protein complexes using local clique merging. *Genome Informatics* 2005; 16(2):260–269.
15. Xiao-Li Li et al. Discovering protein complexes in dense reliable neighborhoods of protein interaction networks. *Comput. Syst. Bioinformatics Conf.* 2007; 6:157–168.
16. Amy Hin Yan Tong et al. A combined experimental and computational strategy to dene protein interaction networks for peptide recognition modules. *Science* 2002; 295(5553):321–324.
17. Bill Andreopoulos et al. Clustering by common friends finds locally significant proteins mediating modules. *Bioinformatics* 2007; 23(9):1124–1131.
18. A. D. King et al. Protein complex prediction via cost-based clustering. *Bioinformatics* 2004; 20(17):3013–3020.
19. Dongbo Bu et al. Topological structure analysis of the protein-protein interaction network in budding yeast. *Nucleic Acids Research* 2003; 31(9):2443–2450.
20. Sabine Tornow and H. W. Mewes. Functional modules by relating protein interaction networks and gene expression. *Nucleic Acids Research* 2003; 31(21):6283–6289.
21. Yu Huang et al. Systematic discovery of functional modules and context-specific functional annotation of human genome. *Bioinformatics* 2007; 23(13):i222–i229.
22. Trey Ideker et al. Discovering regulatory and signalling circuits in molecular interaction networks. *Bioinformatics* 2002; 18(90001):S233–S240.
23. Zheng Guo et al. Edge-based scoring and searching method for identifying condition-responsive

protein-protein interaction sub-network. *Bioinformatics* 2007; 23(16):2121–2128.

24. Roded Sharan et al. Identification of protein complexes by comparative analysis of yeast and bacterial protein interaction data. *J. Comput. Biol* 2005; 12(6):835–846.
25. Eitan Hirsh and Roded Sharan. Identification of conserved protein complexes based on a model of protein network evolution. *Bioinformatics* 2007; 23(2):e170–e176.
26. Giorgio Gallo et al. A fast parametric maximum flow algorithm and applications. *SIAM J. Comput* 1989; 18(1):30–55.
27. Michael R. Garey and David S. Johnson. *Computers and intractability : a guide to the theory of NP-completeness.* W. H. Freeman & Co., New York, NY, USA, 1979.
28. M. Altaf-Ul-Amin et al. Development and implementation of an algorithm for detection of protein complexes in large interaction networks. *BMC Bioinformatics* 2006; 7(207).

Appendix: additional figures and tables

Table 5. Protein complexes predicted by GFA on DIP-PPI and their matches in MIPS-MAN-COMP and MIPS-ALL-COMP. MAN and ALL stands for MIPS-MAN-COMP and MIPS-ALL-COMP, respectively. f stands for *MinFrequency*, and *MaxOverlap* is set to 0.2.

	P	$I(0.2)$	$I(0.5)$	M	$P_{e\geq 2}$	$P_{e=0}$
MAN, $f = 20$	116	57	35	49	54	51
MAN, $f = 3$	318	83	46	69	81	204
ALL, $f = 20$	116	171	75	77	97	6
ALL, $f = 3$	318	303	106	160	241	35

Table 6. Protein complexes predicted by GFA on BioGRID-PPI and their matches in MIPS-MAN-COMP. f and o stand for *MinFrequency* and *MaxOverlap*, respectively.

	P	$I(0.2)$	$I(0.5)$	M	$P_{e\geq 2}$	$P_{e=0}$
$o = 0.5, f = 20$	166	42	28	38	46	111
$o = 0.5, f = 3$	870	85	41	108	223	529
$o = 0.2, f = 20$	157	38	25	35	44	106
$o = 0.2, f = 3$	453	73	30	69	103	296

Table 7. Comparison of GFA and DECAFF on BioGRID-PPI. Again, o and f stand for *MaxOverlap* and *MinFrequency*, respectively.

	P	$I(0.2)$	$I(0.5)$	M	$P_{e\geq 2}$	$P_{e=0}$
DECAFF	2,840	118	81	1141	1,871	533
$o = 0.5$, DECAFF	610	101	30	144	264	215
$o = 0.5$, $f = 4$	582	75	40	79	144	369
$o = 0.2$, DECAFF	226	53	15	48	78	113
$o = 0.2$, $f = 10$	221	51	25	46	56	150
$o = 0.2$, $f = 9$	234	52	25	46	58	160

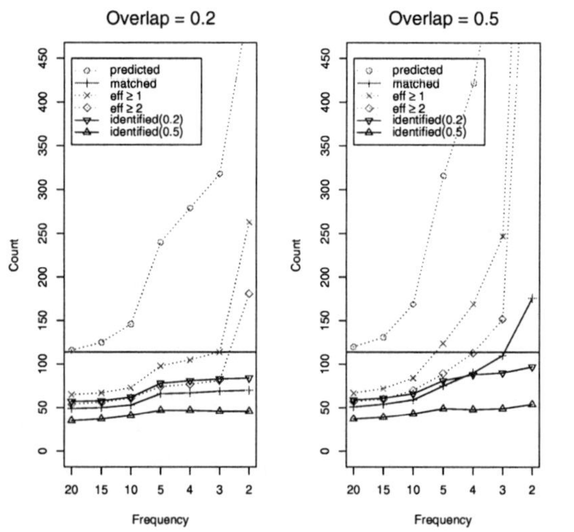

Fig. 2. Protein complexes predicted by GFA on DIP-PPI and their matches in MIPS-MAN-COMP. Two *MaxOverlap* values, 0.2 (left) and 0.5 (right), are considered. Here, *eff* ≥ 1 stands for $P_{e\geq 1}$.

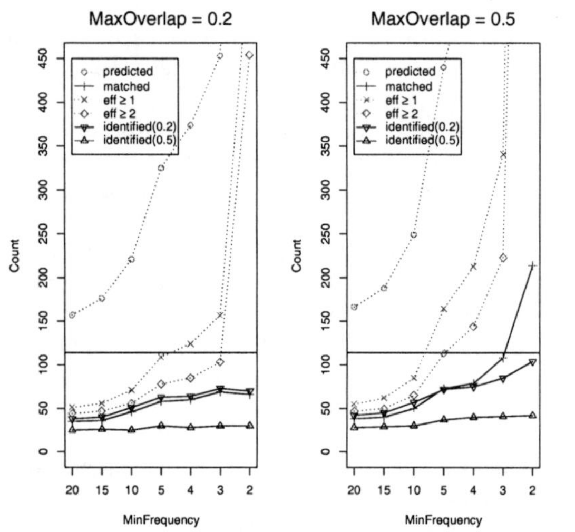

Fig. 3. Protein complexes predicted by GFA on BioGRID and their matches in MIPS-MAN-COMP. Two *MaxOverlap* values are considered: 0.2 (left) and 0.5 (right). Again, *eff* ≥ 1 stands for $P_{e\geq 1}$.

MSDASH: MASS SPECTROMETRY DATABASE AND SEARCH

Zhan Wu*
Department of Computer Science, University of Western Ontario,
London, Ontario N6A 5B8, Canada
** Email: zwu47@csd.uwo.ca*

Gilles Lajoie
Department of Biochemistry, University of Western Ontario,
London, Ontario N6A 5B8, Canada
Email: glajoie@uwo.ca

Bin Ma
Department of Computer Science, University of Western Ontario,
London, Ontario N6A 5B8, Canada
Email: bma@csd.uwo.ca

Along with the wide application of mass spectrometry in proteomics, more and more mass spectrometry data are becoming publicly available. Several public mass spectrometry data repositories have been built on the Internet. However, most of these repositories are devoid of effective searching methods. In this paper we describe a new mass spectrometry data library, and a novel method to efficiently index and search in the library for spectra that are similar to a query spectrum. A public online server have been set up and demonstrated outstanding speed and scalability of our methods.

Together with the mass spectrometry library, our searching method can improve the protein identification confidence by comparing a spectrum with the ones that are already characterized in the database. The searching method can also be used alone to cluster the similar spectra in a mass spectrometry dataset together, in order to to improve the speed and accuracy of the protein identification or quantification.

1. INTRODUCTION

Mass spectrometry has become the standard high-throughput method for protein identification, and more recently, for protein quantification [1, 2]. In a typical protein identification experiment using mass spectrometry, proteins are enzymatically digested to peptides, and the tandem mass (MS/MS) spectra of the peptides are measured using a tandem mass spectrometer. The limitation of the current experiment procedure results in spectra that are difficult to interpret due to poor fragmentation and contaminations from chemical noise.

Many software programs have been developed to identify the sequence of a peptide from its MS/MS spectrum. All these programs more or less depend on a model to predict a theoretical spectrum of a given peptide sequence. By using either a search in a protein database, or by constructing a sequence from scratch, a peptide that gives the best match between the predicted and the experimental spectra is deduced. The approach using a protein database is referred to as database search [3–6], and the construction from scratch is called *de novo* sequencing [7, 8].

The prediction of the theoretical spectrum is a difficult task, partially because the mobile proton model [9] for the peptide fragmentation is not a quantitative model. Limited success was achieved in predicting the theoretical spectrum on a specific mass spectrometer type with a fixed parameter setting [10]. However, in order to do the data analysis in a high-throughput manner, most of the software programs use over-simplified models. Normally these programs expect good y-ion series and/or b-ion series to be observed in order to confidently identify the peptide sequence.

This creates the following situation. Some peptides with certain sequences do not produce good

*Corresponding author.

y-ion or b-ion series and therefore cannot be confidently identified by high-throughput experiments and software. The imperfect spectra are often due to the inherent nature of the peptides. That is, very similar spectra will be produced if the experiment is repeated under similar conditions. In a typical dataset, these imperfect spectra are mixed together with other low-quality spectra that are contaminated by chemical noise. This further complicates the data interpretation. According to our experience and the literature [11], a typical MS/MS dataset can contain as many as 80% of tandem mass spectra that are not characterized by current software.

In spectrometry analysis there have been another approach for spectrum interpretation, which matches the spectrum with a library of confidently characterized spectra (called Annotated Spectrum Library). Such an approach does not need to predict the spectrum from a peptide sequence and can potentially interpret more spectra with higher confidence. However, the huge number of possible peptide sequences and the lack of an efficient matching method make this approach computationally expensive for peptide identification. X! Hunter [12] is the only search engine that adopted this approach in peptide identification. By limiting the search in only the consensus spectra of confidently identified peptides of certain organisms, X! Hunter managed to perform the search relatively efficiently.

In this paper we propose to extend this Annotated Spectrum Library approach a step further. We compare a spectrum with all of the publicly-available spectra, annotated or not, and find the similar spectra. This makes the computation even more expensive, but has the advantages as discussed below.

Two situations may arise when matches are found:

(1) There are one or more previously-characterized spectra that match the current one. The current spectrum can use the previous characterization. This is the same as the Annotated Spectrum Library approach. Note that the previous characterization might have been done under a better experimental condition (such as a simpler protein mixture, more abundant sample, or better instrument).
(2) There are several uncharacterized spectra that match the current one. This implies that these spectra are unlikely random noises and deserve further examination by more optimized experiments or more extensive computation. In addition, because the MS/MS spectra of the same peptide on different instruments may produce slightly different spectra, the combination of these similar but not identical spectra (and their associated information such as organisms and experimental conditions) will reveal more information about the peptide than any single spectrum would. As a result, the chance of successful peptide identification will be greatly increased.

Clearly, this strategy comes with the price of increased computational complexity. A method based on locality sensitive hashing was proposed to speed up the matching [13]. The method first clusters the database spectra into clusters according locality sensitive hashing. Then a query spectrum is compared only with the clusters that are "neighbors" of the query spectrum. This avoided the one-against-all comparison between the query spectrum and the database. The method provides good trade off between the sensitivity and the speed of the search. However, this method is complicated and the 100 times theoretical speed up factor claimed in the paper would be diminished when implemented in a real system.

In this paper we propose a new algorithm for speeding up the spectrum matching in a large MS/MS database, based on a novel "thumbnail" idea. The method is simple and easy to implement. Written in Java, our algorithm achieves an average matching speed of comparing one million pairs of spectra per second on a single CPU. When the precursor ion mass of an MS/MS spectrum is known (which is usually the case in peptide identification), we can use the precursor ion mass to pre-select the database spectra for the matching. Depending on the mass accuracy of the data, this further improves the speed by hundreds to thousands of times, resulting in a final speed of searching one spectrum in 10^8-10^{10} spectra per second. We believe such a speed should be sufficient for most applications nowadays. Our method is also very memory-friendly: the index of each spectrum requires only 8 bytes in the main memory. This drastically reduces the memory usage, because keeping a spectrum in memory will usually require thousands of bytes. The method can also be easily parallelized. All these properties enables the inexpensive implementation of a real system.

In the past several years, more and more mass spectrometry data have been made publicly available. Among many available mass spectrometry data repositories, some popular ones are Open Proteomics Database [14], Peptide Atlas [15–17], and Sashimi Repository [18]. They provide great testing data for the research of new data analysis software. However, none of these data repositories supports efficient searching of similar spectra. This makes the repositories to be no much more than well-organized FTP sites, and the data in them are not fully utilized in the analysis of a newly measured dataset. In this paper we introduce our new public mass spectrometry data library. Equipped with our efficient searching method, the library allows the user to query with an MS/MS spectrum and efficiently retrieve all the similar spectra (together with their annotations if there are any) in the library.

Our efficient searching method can also be used without a spectrum database. The method can be used to cluster similar spectra in one or a few datasets together. This not only speeds up the subsequent data analysis by removing redundancies, but also improves the peptide identification confidence by gathering information from different MS/MS scans together (possibly from repeated experiments under slightly different conditions).

The rest of the paper is organized as follows: Section 2 defines the mass spectrometry terms used in the paper. All definitions are standard and can be skipped by a reader who is familiar with this area. Section 3 introduces our fast searching algorithm. Section 4 introduces the implementation of the public data library. The speed and the sensitivity of our searching method is demonstrated in Section 5.

2. TERMS AND NOTATIONS

An *MS/MS spectrum* of a peptide contains a list of peaks. Each signal *peak* is caused by some fragment ions of the peptide, and can be encoded with two real values: the *m/z value* of the peak represents the mass to charge ratio of the fragment ions, and the *intensity* value of the peak indicating the abundance of the fragment ions. There are different types of fragment ions for a peptide, where the most important ones are the *y-ions* and *b-ions*. The *precursor mass* of a spectrum is the mass of the whole peptide. The mass unit for m/z and precursor mass is *dalton*. Peptides are obtained by digesting proteins using enzymes and the most commonly used enzyme is *trypsin*. The resulting peptides using trypsin are called *tryptic peptides* and typical tryptic peptides range from 500 to 3000 daltons. A mass spectrometer measures m/z and precursor mass with small errors. For this reason mass errors are allowed in spectrum matching. The maximum error allowed for matching two m/z values is called the *mass error tolerance*. Typical mass error tolerance ranges from ± 0.01 dalton to ± 1 dalton depending on the types of the spectrometer.

3. SEARCHING METHOD

The main idea of our searching method is an efficient filtration that efficiently rejects the apparently unmatched spectra and keeps only the possible matches for further examination using more time-consuming but more accurate criteria. This filtering method is a common practice to speed up approximate pattern matching. A good filtration method should reject as many false matches as possible (to maximize the selectivity), while keeping as many true matches as possible (to maximize the sensitivity).

Our searching method consists of the following steps: First, the database spectra are preprocessed and the major peaks of each spectrum are stored in a relational database. Then a "thumbnail" is computed for each spectrum and put in a computer's main memory. For each spectrum, this thumbnail is a 64-bit integer. The filtration is done using this 64-bit integer. Lastly, the spectra passing the filtration will be retrieved from the relational database for examination using a more accurate scoring function, and outputs are generated.

These steps are described in more details in the following subsections.

3.1. Spectrum Preparation

For each spectrum in the library, data preprocessing is needed to prepare the spectrum for the fast matching. First, due to the random measurement error of the instrument, multiple copies of the identical ion can cause a cluster of adjacent peaks with very small difference in their m/z values. These adjacent peaks need to be centroided together to form only one peak. Any standard centroiding method can be used in this step.

After centroiding, each spectrum can still possibly contain hundreds to thousands of peaks. A large portion of these peaks are very weak and should be regarded as noise. Keeping them in the comparison will not only reduce the speed, but also add errors to the scoring function for comparing two spectra. For a typical tryptic peptide of length 15, there are only 28 y-ions and b-ions that are the most useful for peptide identification. Therefore, for the purpose of spectrum comparison, it is safe to only examine the strongest 50 peaks of each centroided spectrum. In our method, the strongest 50 peaks of each centroided spectrum are selected and added into a relational database as a BLOB for fast retrieval. This greatly reduces the spectrum complexity with only negligible loss in the accuracy of the scoring function.

3.2. Thumbnail of a Spectrum and Rapid Filtration

We propose here a novel "thumbnail" idea for fast filtration. Basically, a thumbnail of a spectrum is a bit array where each bit indicates whether the spectrum contains a strong peak at some given mass value. Then the comparing of two spectra can be done in a rapid way by a bitwise-and operation on their thumbnail, and counting the number of 1s in the result.

More precisely, let $[0, K-1] = \{0, 1, \ldots, K-1\}$ and $h : R^+ \to [0, K-1]$ be a hash function that maps the positive numbers to integers between 0 and $K-1$. Let S be a spectrum with peaks at m/z values $x_1, x_2, \ldots, x_m$. We denote $mz(S) = \{x_1, x_2, \ldots, x_m\}$. Then the thumbnail of S is defined as

$$h(S) = \{i \mid \text{there is } x_j \text{ such that } h(x_j) = i\}.$$

In a computer, $h(S)$ can be equivalently represented by a length-K bit array T such that $T[i] = 1$ if and only if $i \in h(S)$. We will sometimes call T as the thumbnail of S too.

Lemma 3.1. *Let h be a hash function. Let S_1 and S_2 be two spectra of length m. Suppose S_2 is a random spectrum independent to S_1 and h, then*

$$Pr(|h(S_1) \cap h(S_2)| > (1+\delta)m^2/K) < \left(\frac{e^\delta}{(1+\delta)^{(1+\delta)}}\right)^{m^2/K}.$$

Proof. Because S_1 has at most m peaks, $|h(S_1)| \leq m$. For any $x \in mz(S_2)$, the probability that $h(x) \in h(S_1)$ is therefore at most $p = m/K$. The expected number of peaks from S_2 that are mapped into $h(S_1)$ is then at most $mp = m^2/K$. By using Chernoff's bound [19] straightforwardly,

$$\begin{aligned} &Pr(|h(S_1) \cap h(S_2)| > (1+\delta)m^2/K) \\ &\leq Pr((1+\delta)m^2/K \text{ peaks of } S_2 \text{ mapped in } S_1) \\ &< \left(\frac{e^\delta}{(1+\delta)^{(1+\delta)}}\right)^{m^2/K} \end{aligned}$$

□

When $\delta \geq 0$, $\frac{e^\delta}{(1+\delta)^{(1+\delta)}}$ is a monotonically decreasing function that approaches 0 rapidly when δ increases. As a result, by selecting proper $t = \lfloor (1+\delta)m^2/K \rfloor$, m, and K, we can let $Pr(|h(S_1) \cap h(S_2)| > t)$ become very small. For example, when $m = 20$, $K = 128$, and $t = 12$, $Pr(|h(S_1) \cap h(S_2)| > t) < 1.74 \times 10^{-4}$ according to Lemma 3.1. We note that the bound given in Lemma 3.1 is not tight and the real probability is much lower than this. This suggests that we can use the size of $h(S_1) \cap h(S_2)$, i.e., the intersection of the thumbnails of the query spectrum and the database spectrum, to filter out the random spectra. In order to be useful, this filter should not reject the correct matches. This is guaranteed by Lemma 3.2.

Lemma 3.2. *Let h be a hash function. Let S_1 and S_2 be two length-m spectra. Suppose S_2 is such that $|mz(S_1) \cap mz(S_2)| = n$, and the hash function h is independent to S_1 and S_2. Then for any $\delta > 0$,*

$$\begin{aligned} &Pr(|h(S_1) \cap h(S_2)| \leq t) \\ &\leq \binom{K}{t} \times (t/K)^n && (1) \\ &\leq \frac{1}{\sqrt{2\pi t}} \times e^t \times (t/K)^{n-t} && (2) \end{aligned}$$

Proof. Denote $X = mz(S_1) \cap mz(S_2)$. Then $|X| = n$ and the number of possible mappings from X to $[0..K-1]$ is K^n.

Clearly, for any $x \in X$, $h(x) \in h(S_1) \cap h(S_2)$. Therefore, if $|h(S_1) \cap h(S_2)| \leq t$, then all the n values in X need to be mapped into a size-t subset of $[0, K-1]$. There are $\binom{K}{t}$ such subsets. For each of them, there are t^n possible ways to map X to it. Hence, the total number of possible mappings that satisfies $|h(S_1) \cap h(S_2)| \leq t$ is upper bounded by $\binom{K}{t} \times t^n$. Consequently,

$$\begin{aligned}
&Pr(|h(S_1) \cap h(S_2)| \leq t) \\
&\leq \binom{K}{t} \times t^n \times K^{-n} \\
&\leq \frac{K^t}{t!} \times (t/K)^n \\
&\leq \frac{K^t}{\sqrt{2\pi t} \times (t/e)^t} \times (t/K)^n \qquad (3)\\
&= \frac{1}{\sqrt{2\pi t}} \times e^t \times (t/K)^{n-t}
\end{aligned}$$

The inequation (3) is true because of Stirling's formula [20]. □

From (2) it is clear that when t is much smaller than n and K larger than n, the probability becomes very small. For example, when $m = 20$, $K = 128$, and $t = 12$ as before and $n = 19$, $Pr(|h(S_1) \cap h(S_2)| \leq t) \leq 1.2 \times 10^{-3}$ according to (1). Again, the bound proved in Lemma 3.2 is not tight and the real probability is much lower than this.

From the above two examples, we can see that by choosing a right threshold t for the size of the intersection thumbnail, Lemma 3.1 guarantees that a random S_2 can be rejected by the threshold with high probability; whereas Lemma 3.2 guarantees that a spectrum S_2 similar to S_1 can pass the threshold with high probability.

In our implementation of this filtering method, we use $m = 20$, $K = 64$, and $t = 12$. Given a query spectrum S_1, a spectrum S_2 passes the filtration if and only if $h(S_1) \cap h(S_2)$ contains more than t elements. By randomly sampling one million spectrum pairs, the probability that a random spectrum can pass the filtration is estimated and shown in Figure 1. In particular, when $t = 12$, this probability is only 0.000166.

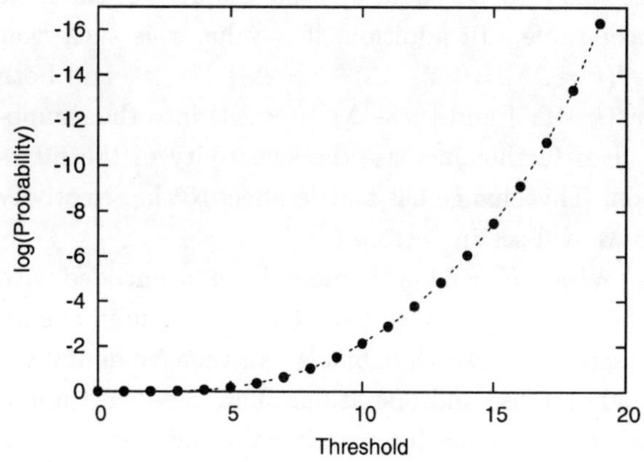

Fig. 1. When $m = 20, K = 64$, the base-10 logarithm of the probability that a random spectrum passes the filtration with threshold t. The x axis is the threshold. The y axis is the probability in logarithmic scale.

For a spectrum S_2 such that $|mz(S_1) \cap mz(S_2)| = n$, the sensitivity of the filtration, i.e., the probability that S_2 can pass the filtration is also estimated by random sampling and given in Figure 2.

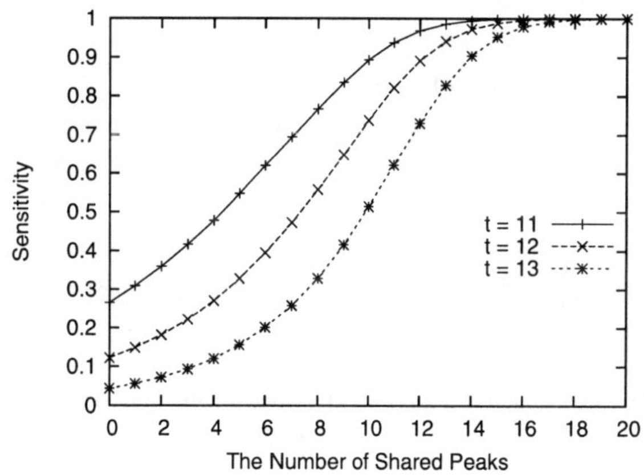

Fig. 2. When $m = 20, K = 64$, the sensitivity of the filtration for a similar spectrum sharing n out of the 20 peaks with the query.

In the discussion above, we assumed h to be a perfect hashing function. However, in reality we must consider the mass error tolerance. That is, two m/z values $x \in mz(S_1)$ and $y \in mz(S_2)$ match if $|x - y| \leq \Delta$ for a predefined Δ. To allow such m/z values to be mapped together by h, we require $h(x) = h'(\lfloor x/(k \times \Delta) \rfloor)$ for a constant k and another hash function $h' : N \to [0, K-1]$. This way, when $|x - y| \leq \Delta$, $\lfloor x/(k \times \Delta) \rfloor$ and $\lfloor y/(k \times \Delta) \rfloor$

are likely to be equal and therefore give the same hash value. In addition, if a value x is such that $\lfloor x/(k \times \Delta) \rfloor = \lfloor (x - \Delta)/(k \times \Delta) \rfloor + 1$, we add both $\lfloor x/(k \times \Delta) \rfloor$ and $\lfloor (x - \Delta)/(k \times \Delta) \rfloor$ into the thumbnail to further increase the sensitivity of the filtration. This change has a little affect to the selectivity as we will see in Section 5.

When $K = 64$, a thumbnail can be encoded with a 64-bit long int type. On a 64-bit computer, the intersection of two thumbnails can then be done by a single bitwise-and operation. The size of a thumbnail can be done by counting the number of 1s in the long integer, which can be done either by some very efficient programs [a] or by a single CPU instruction if such operation is supported by the CPU. In our system, a spectrum is considered to be divided into 64 segments, and the highest 20 segments of each spectrum are used to compute the thumbnail of the spectrum. The thumbnails of all the spectra in the database are pre-computed and loaded into the main memory of the computing servers. This requires $8N$ bytes of main memory if there are N spectra. When a search is performed, the thumbnail of the query spectrum is computed using the same hash function and then compared with each thumbnail using the above mentioned fast operations. Only those passing the filtration are further compared using the more accurate measurement described in the following subsection.

3.3. Spectrum Similarity

Once a spectrum passes the filtration, the strongest 50 peaks of the spectrum stored in the relational database are retrieved and compared against the strongest 50 peaks of the query spectrum. A similarity score is calculated as follows.

Let (x_i, h_i) be a peak with m/z value x_i and intensity h_i. Let $S_1 = \{(x_1, h_1), \ldots, (x_m, h_m)\}$ and $S_2 = \{(x'_1, h'_1), \ldots, (x'_m, h'_m)\}$. We assume the peaks in each spectrum are sorted in ascending order according to the m/z values. The peaks in the two spectra are compared using a merge-sort type of procedure to find all the pairs of peaks such that $|x_{i_k} - x'_{j_k}| \leq \Delta$ for $k = 1, \ldots, l$, where Δ is the mass error tolerance. Then the similarity score of the two spectra can be defined as

$$sc(S_1, S_2) = \frac{\sum_{k=1}^{l} h_{i_k} h'_{j_k}}{\sqrt{\sum_{i=1}^{m} h_i^2}\sqrt{\sum_{j=1}^{m} h'^2_j}}$$

According to our experience, many low-quality spectra in the library often contain one or few very strong peaks. If the above formula is used, then two spectra that share one strong peak may become very similar, despite the fact that their other peaks do not match each other. To reduce this risk, we convert the intensity of each peak to the logarithm of the intensity before the calculation given above.

4. THE DATABASE SYSTEM

We have implemented the search method mentioned above in Java, together with a public spectra database server that allows public users to deposit data to the database and search for similar spectra in the database. The system is called MSDash and available online at http://ala.bin.csd.uwo.ca:8080/msdash.

The system consists a web server, a database sever and ten computing servers. Each sever has a single-core AMD Opteron CPU. The web server runs Apache Tomcat and the database server uses MySQL. As soon as the user submit the query MS data file, the web server will forward the task to the computing servers. After the computing servers finish the matching process, the matched list of mass spectra will be transferred back to the web server and displayed to the user.

Currently, some publicly available data downloaded from the Open Protein Database [14] and the Sashimi data repository [18] have been added to the database as test data. These include about 3.3 million tandem mass spectra. The raw data are stored in mzXML format on the hard drives of the servers. Fifty strongest peaks of each spectrum are stored in the MySQL database. And the twenty strongest peaks of each spectrum are used to generate a thumbnail and all of the thumbnails are loaded in the main memory of the computing servers.

---

[a] See http://infolab.stanford.edu/~manku/bitcount/bitcount.html for some examples.

5. EXPERIMENTS RESULT

Figure 3 shows the average searching time for one spectrum with unknown precursor mass in our spectra database, at different database size using 10 CPUs. Clearly the searching time grows linearly to the size of the database, indicating the excellent scalability of our system. The time does not approach zero when the database size approaches zero. This is because of the overhead due to network communication for query submission and result display. Besides the overhead, the average search speed (indicated by the slope of the line) is approximately 10 million matches per second on 10 CPUs, i.e., 1 million matches per second on one CPU.

In our experiment testing for speed, we assumed that the precursor ion mass of a spectrum is unknown. Therefore, the query spectrum needs to be compared with every spectrum in the database. However, when the precursor ion mass is known, one only needs to compare it with the spectra in the database with similar precursor ion mass. So that the query spectrum only needs to match 10^{-2} to 10^{-4} of the database spectra depending on the precursor ion mass error tolerance. If this precursor filtration option is selected by the user, our system can search one query spectrum in a database with 10^8 to 10^{10} spectra per second on a single CPU.

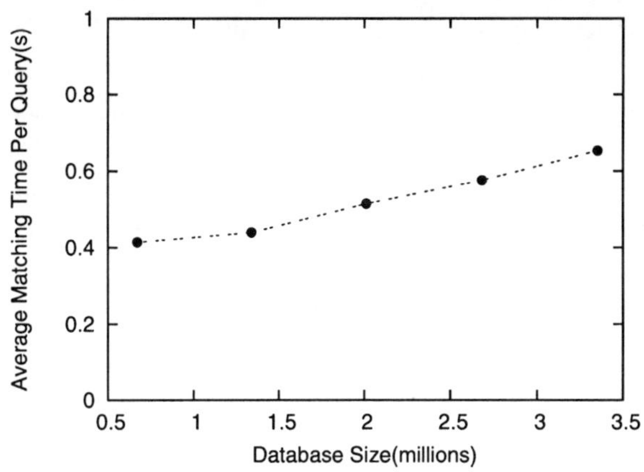

Fig. 3. Average searching time for searching one spectrum with unknown precursor ion mass in MSDash system with varying database size. Ten CPUs are used. The x axis shows the number of spectra in the database, and the y axis shows the searching time per query in seconds.

The thumbnail filtration contributed significantly to the speed of our system. Using our default threshold $t = 12$, Figure 4 shows the percentage of remaining spectra after the filtration. The figure shows that only 0.14% of the spectra in the database can pass the filtration and need further examination by the more time-consuming similarity function defined in Section 3.3 [b]. We note that because the hash function we use in the real system is not a perfect hash function and needs to be modified as described in Section 3.2, this percentage is higher than the estimation with simulation in Figure 1.

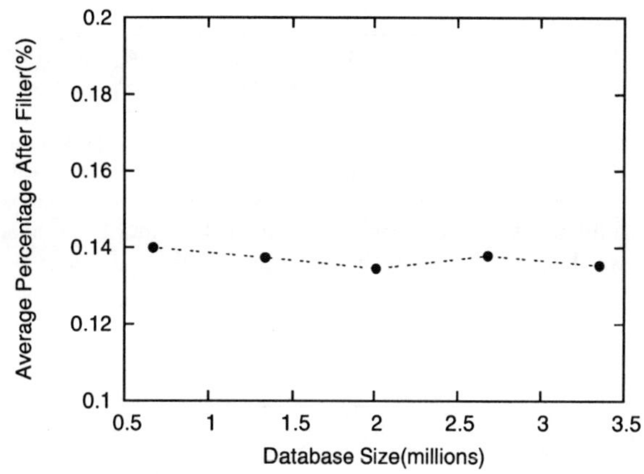

Fig. 4. The percentage of database spectra that pass the fast thumbnail filtration with $t = 12$. The y axis is the percentage and the x axis shows the number of spectra in the database.

Figure 5 shows the average query time for searching real spectra according to different values of the threshold parameter t and the total number of spectra in the database is approximate 3.3 million. From Figure 5, we can find that the average speed per query is improved dramatically when the value of the threshold t is increased. Similarly, Figure 6 shows the average percentage of remaining spectra after the filtration with different threshold t and the same database size. The figure shows that the sensitivity is dropping quickly when the threshold t is increased. Clearly the threshold t can be used to control the trade-off between the speed and the sensitivity of the query. In our system, the default threshold t is

[b]This results in a speed up of 700 times, comparing to the 100 times speed up factor claimed in the paper [13].

12, but the user can change this value. The recommended value for the threshold t is from 11 to 13. In the following part we will examine the sensitivity when $t = 12$.

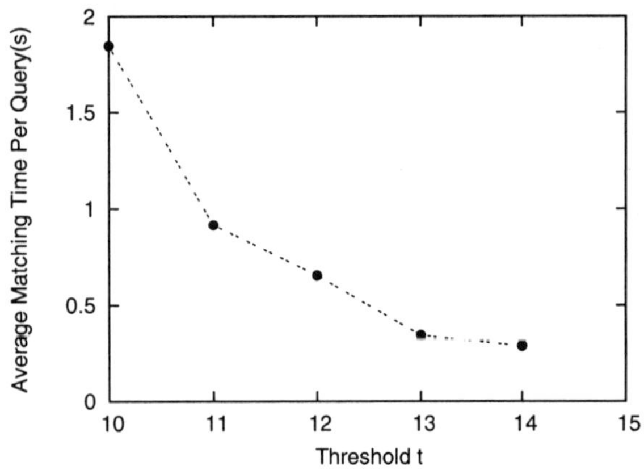

Fig. 5. Average searching time for searching one spectrum with unknown precursor ion mass in MSDash system with different threshold t. Ten CPUs are used. The Database size is 3.3 million. The x axis shows the threshold t, and the y axis shows the searching time per query in seconds.

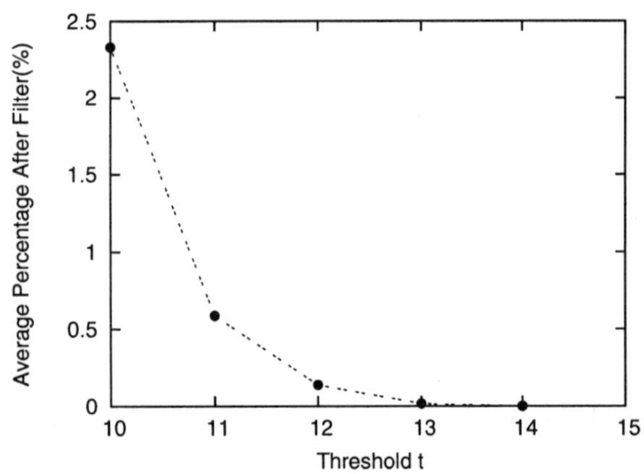

Fig. 6. The percentage of database spectra that pass the fast thumbnail filtration with different values of threshold t. The Database size is 3.3 million. The y axis is the percentage and the x axis shows the threshold t.

To test the sensitivity of our system, we selected 100 spectra from the database and randomly modified each of them 10 times. Then the modified spectra are searched in the database to see the percentage that the original spectra can be retrieved.

The random modifications are applied on both the intensities of peaks and the m/z values of the peaks. The m/z modification with probability p is done as follows: for *each* peak in the query spectrum, with probability p, the m/z value of the peak is replaced with another random value. The intensity modification with error range $\pm x\%$ is done as follows: for *each* peak in the query spectrum, add a uniformly random error between $-x\%$ to $x\%$ to the intensity of the peak. Table 1 shows the sensitivity of our method under different levels of modifications. From the table we can see that our method can find all real matches even if every peak's intensity is modified by up to ±30%, and 5% of the peaks are randomly moved around. Even when 20% of the peaks are randomly moved around, our method still keeps high sensitivity.

Table 1. Sensitivity under different levels of modifications.

	0%	±10%	±20%	±30%
$p = 0$	100%	100%	100%	100%
$p = 0.05$	100%	100%	100%	100%
$p = 0.1$	100%	99.8%	99.9%	99.8%
$p = 0.2$	98.3%	98.6%	98.5%	98.3%
$p = 0.3$	94.8%	93.0%	94.8%	93.6%

6. CONCLUSION

Based on a novel spectrum thumbnail concept, we introduced an efficient spectrum searching method. Comparing to other methods for the similar purpose, our strategy is not only significantly faster, but also much easier to implement. The method achieves a speed of searching one spectrum in one million spectra per second per CPU without knowing the precursor ion mass, or 10^8 to 10^{10} spectra per second per CPU if knowing the precursor ion mass. Our searching method has very high sensitivity. We have also implemented a public online server that allows users to deposit data to our database, and search a spectrum in the database. The server is available at http://ala.bin.csd.uwo.ca:8080/msdash.

References

1. R Aebersold1, M Mann.Mass spectrometry-based proteomics. *Nature Journal* 2003; bf 422:198–20.
2. MA Baldwin. Protein Identification by Mass Spectrometry: Issues to be Considered. *Molecular Cellular Proteomics* 2004; **3**:1–9.

3. DN Perkins, DJC Pappin, DM Creasy, JS Cottrell. Probability-based protein identification by searching sequence databases using mass spectrometry data. *Electrophoresis* 1999; **20(18)**:3551–3567.
4. JK Eng, AL McCormack, JR Yates. An approach to correlate tandem mass spectral data of peptides with amino acid sequences in a protein database. *Journal of The American Society for Mass Spectrometry* 1994; **5**:976–989.
5. RE Moore, MK Young, TD Lee. Qscore: An Algorithm for Evaluating SEQUEST Database Search Results. *Journal of The American Society for Mass Spectrometry* 2003; **13(4)**:378–386.
6. R Craig, RC Beavis. TANDEM: matching proteins with mass spectra. *Bioinformatics* 2004; **20(9)**:1466–1467.
7. B Ma, K Zhang, C Liang. An Effective Algorithm for the Peptide De Novo Sequencing from MS/MS Spectrum. *Journal of Computer and System Sciences* 2005; **70(3)**:418–430.
8. A Frank, P Pevzner. PepNovo: De Novo Peptide Sequencing via Probabilistic Network Modeling. *Analytical Chemistry* 2005; **77(4)**:964–973.
9. VH Wysocki, G Tsaprailis, LL Smith, LA Breci. Mobile and localized protons: a framework for understanding peptide dissociation. *Journal of Mass Spectrometry* 2000; **35(12)**:1399–1406.
10. Z Zhang. Prediction of Low-Energy Collision-Induced Dissociation Spectra of Peptides. *Analytical Chemistry* 2004; **76(14)**:3908–3922.
11. A Keller, S Purvine, A Nesvizhskii, S Stolyar, DR Goodlett, E Kolker. Experimental Protein Mixture for Validating Tandem Mass Spectral Analysis. *OMICS* 2002; **6(2)**:207–212.
12. R Craig, JP Cortens, D Fenyo, RC Beavis. Using Annotated Peptide Mass Spectrum Libraries for Protein Identification. *Journal of Proteome Research* 2006; **5(8)**:1843–1849.
13. D Dutta, T Chen. Speeding up Tandem Mass Spectrometry Database Search: Metric Embeddings and Fast Near Neighbor Search. *Bioinformatics* 2007; **23(5)**:612–618.
14. JT Prince, MW Carlson, R Wang, P Lu, EM Marcotte. The need for a public proteomics repository. *Nature Biotechnoly* 2004; **22(4)**:471–472.
15. F Desiere, EW Deutsch, NL King, AI Nesvizhskii, P Mallick, J Eng, S Chen, J Eddes, SN Loevenich, R Aebersold. The PeptideAtlas Project. *Nucleic Acids Research* 2006; **34**:655–658.
16. EW Deutsch, JK Eng, H Zhang, NL King, AI Nesvizhskii, B Lin, H Lee, EC Yi, R Ossola, R Aebersold. Human Plasma PeptideAtlas. *Proteomics* 2005; **5(13)**:3497–3500.
17. F Desiere, EW Deutsch, AI Nesvizhskii, P Mallick, N King, JK Eng, A. Aderem, R Boyle, E Brunner, S Donohoe, N Fausto, E Hafen, L Hood, MG Katze, K Kennedy, F Kregenow, H Lee, B Lin, D Martin, J Ranish, DJ Rawlings, LE Samelson, Y Shiio, J Watts, B Wollscheid, ME Wright, W Yan, L Yang, E Yi, H Zhang, R Aebersold. Integration of Peptide Sequences Obtained by High-Throughput Mass Spectrometry with the Human Genome. *Genome Biology* 2004; **6**:R9.
18. http://sashimi.sourceforge.net/repository.html
19. H Chernoff. A measure of asymptotic efficiency for tests of a hypothesis based on the sum of observations. *Annals of Mathematical Statistics* 1952; **23**:493–507.
20. W Feller. Stirling's Formula. *An Introduction to Probability Theory and Its Applications* 1968; **1**:50–53.

ESTIMATING SUPPORT FOR PROTEIN-PROTEIN INTERACTION DATA WITH APPLICATIONS TO FUNCTION PREDICTION

Erliang Zeng
Bioinformatics Research Group (BioRG), School of Computing and Information Sciences
Florida International University
Miami, FL 33199, USA
Email: ezeng001@cs.fiu.edu

Chris Ding
Department of Computer Science and Engineering, University of Texas at Arlington
Arlington, TX 76019, USA
Email: CHQDing@uta.edu

Giri Narasimhan*
Bioinformatics Research Group (BioRG), School of Computing and Information Sciences
Florida International University
Miami, FL 33199, USA
Email: giri@cs.fiu.edu

Stephen R. Holbrook†
Computational Research Division, Lawrence Berkeley National Laboratory
Berkeley, CA 94720, USA
Email: srholbrook@lbl.gov

Almost every cellular process requires the interactions of pairs or larger complexes of proteins. High throughput protein-protein interaction (PPI) data have been generated using techniques such as the *yeast two-hybrid systems*, *mass spectrometry method*, and many more. Such data provide us with a new perspective to predict protein functions and to generate protein-protein interaction networks, and many recent algorithms have been developed for this purpose. However, PPI data generated using high throughput techniques contain a large number of false positives. In this paper, we have proposed a novel method to evaluate the support for PPI data based on gene ontology information. If the semantic similarity between genes is computed using gene ontology information and using Resnik's formula, then our results show that we can model the PPI data as a mixture model predicated on the assumption that true protein-protein interactions will have higher support than the false positives in the data. Thus semantic similarity between genes serves as a *metric of support* for PPI data. Taking it one step further, new function prediction approaches are also being proposed with the help of the proposed metric of the support for the PPI data. These new function prediction approaches outperform their conventional counterparts. New evaluation methods are also proposed.

1. INTRODUCTION

Protein-protein interactions (PPI) are essential for cellular activities considering the fact that almost every biological function requires the cooperation of many proteins. Recently, many high-throughput methods have been developed to detect pairwise protein-protein interactions. These methods include the yeast two-hybrid approach, mass spectrometry techniques, genetic interactions, mRNA coexpression, and *in silico* methods[1]. Among them, the yeast two-hybrid approach and mass spectrometry techniques aim to detect physical binding between proteins.

The huge amount of protein-protein interaction data provide us with a means to begin elucidating protein function. Functional annotation of proteins is a fundamental problem in the post-genomic era. To date, a large fraction of the pro-

*Corresponding author.
†Corresponding author.

teins have no assigned functions. Even for one of the most well-studied organisms such as *Saccharomyces cerevisiae*, about a quarter of the proteins remain uncharacterized[2].

There are several functional annotation systems. These annotation systems include COGs (Clusters of Orthologous Groups)[3], Funcat (Functional Catalogue)[4] and GO (Gene Ontology)[5]. GO is the most comprehensive system and is widely used. In this paper, we will focus on functional annotations based on GO terms associated with individual genes and proteins.

A lot of previous work has been done on protein function prediction by using the recently available protein-protein interaction data (see review by Sharan *et al.*[2]). The simplest and most direct method for function prediction determines the function of a protein based on the known function of proteins lying in its neighborhood in the PPI network. Schwikowski *et al.*[6] used the so-called majority-voting technique to predict up to three functions that are frequently found among the annotations of its network neighbors. Hishigaki *et al.*[7] approached this problem by also considering the background level of each function across the whole genome. The χ^2-like score was computed for every predicted function. Hua *et al.*[8] proposed to improve the prediction accuracy by investigating the relation between network topology and functional similarity.

In contrast to the local neighborhood approach, several methods have been proposed to predict functions using global optimization. Vazquez *et al.*[7] and Nabieva *et al.*[9] formulated the function prediction problem as a minimum multiway cut problem and provided an approximation algorithm to this NP-hard problem. Vazquez *et al.*[7] used a simulated annealing approach and Nabieva *et al.*[9] applied a integer programming method. Karaoz *et al.*[10] used a similar approach but handled one annotation label at a time. Several probabilistic models were also proposed for this task such as the *Markov random field model* used by Letovsky *et al.*[11] and Deng *et al.*[12], and a statistical model used by Wu *et al.*[13].

Despite some successful applications of the aforementioned algorithms in functional annotation of uncharacterized proteins, many challenges still remain. One of the big challenges is that PPI data has a high degree of noise[1]. Most methods that generate interaction networks or perform functional prediction do not have a preprocessing step to clean the data or filter out the noise. Although some methods include the reliability of experimental sources as suggested by Nabieva *et al.*[14], the reliability estimations are crude and do not consider the variations in the reliability of instances within the same experimental source. Some approaches were proposed to predict protein-protein interaction based on evidence from multi-source data. The evidence score calculated from multi-source data is a type of reliability measure of the protein-protein interaction data. Such approaches include those developed by Jansen *et al.*[15], Bader *et al.*[16], Zhang *et al.*[17], Ben-Hur *et al.*[18], Lee *et al.*[19], Qi *et al.*[20], and many more. Jansen *et al.*[15] combined multiple sources of data using a Bayes classifier. Bader *et al.*[16] developed statistical methods that assign a confidence score to every interaction. Zhang *et al.*[17] predicted co-complexed protein pairs by constructing a decision tree. Ben-Hur *et al.*[18] used kernel methods for predicting protein-protein interactions. Lee *et al.*[19] developed a probabilistic framework to derive numerical likelihoods for interacting protein pairs. Qi *et al.*[20] used a *Mixture-of-Experts* method to predict the set of interacting proteins. The challenges of integrating multi-source data are manily due to the heterogeneity of the data and the effect of a functionally-biased reference set. Another problem is that most multi-source data are unstructured but often correlated.

Another important shortcoming of most function prediction methods is that they do not take all annotations and their relationships into account. Instead, they have either used arbitrarily chosen functional categories from one level of annotation hierarchy or some arbitrarily chosen so-called informative functional categories based on some *ad hoc* thresholds. Such arbitrarily chosen functional categories only cover a small portions of the whole annotation hierarchy, making the predictions less comprehensive and hard to compare. Predicting functions using the entire annotation system hierarchy is necessary and is a main focus of this paper.

In this paper, we propose a method to address the above two problems. We hypothesize that the distribution of similarity values of pairs of proteins can be modeled as a sum of two log-normal distributions (i.e., a mixture model) representing two populations – one representing pairs of proteins that interact with high support (high confidence), and the other representing pairs that interact with low support (low confidence) (section 2.2). The parameters of the mixture model were then estimated from a large database. This mixture model was then used to differentiate interactions with high confidence from the ones that have low confidence, and was integrated into the function prediction methods. A new evaluation method was also proposed to evaluate the predictions (section 2.4). The new evaluation method captures the similarity between GO terms and reflects the relative hierarchical positions of predicted and true function assignments.

Note that while PPI data involves proteins, GO terms are associated with genes and their products. For the rest of this paper, we will use the terms *genes* and their associated *proteins* interchangeably.

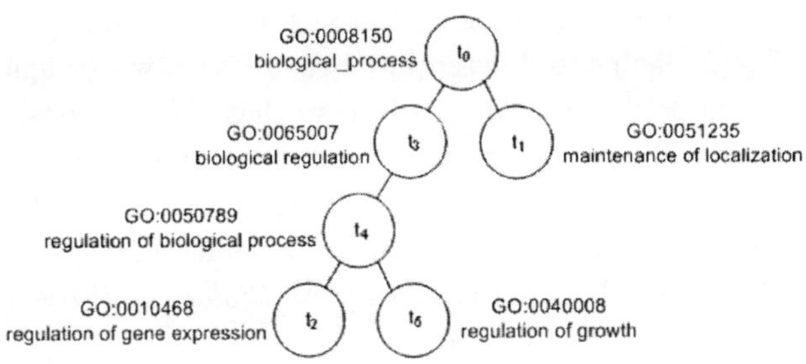

Fig. 1. An example showing the hierarchy of sample GO terms.

2. METHODS

In this section, we first introduce the concepts of similarity between genes calculated based on gene ontology. Next, we investigate inherent properties of some previously known methods used to calculate such similarity. Then a mixture model is introduced to model the distribution of the similarity values between pairs of genes. Next, we present the new function prediction methods using this mixture model. Finally, we present improved evaluation methods for function prediction.

2.1. Similarity between Genes Based on Gene Ontology Data

Suppose that a gene A is associated with the following GO terms $\{t_{a1}, ..., t_{ai}\}$, and that a gene B is associated with the following GO terms $\{t_{b1}, ..., t_{bj}\}$, The similarity between genes A and B based on gene ontology is defined as

$$sim_X(A, B) = \max_{i,j}\{sim_X(t_{ai}, t_{bj})\}. \tag{1}$$

where $sim_X(t_{ai}, t_{bj})$ is the similarity between the GO terms t_{ai} and t_{bj} using method X.

Thus, in order to calculate the similarity between genes, we need to calculate the similarity between individual GO terms, for which many methods have been proposed. Below we discuss the methods proposed by Resnik[21], Jiang and Conrath[22], Lin[23], and Schlicker *et al.*[24]. The methods proposed by Resnik, Jiang and Conrath, and Lin have been used in other domain and was introduced to this area by Lord *et al.*[25].

Resnik:

$$sim_R(t_1, t_2) = \max_{t \in S(t_1,t_2)} \{IC(t)\} \tag{2}$$

Jiang-Conrath:

$$dist_{JC}(t_1, t_2) = \min_{t \in S(t_1,t_2)} \{IC(t_1) + IC(t_2) - 2IC(t)\} \tag{3}$$

Lin:

$$sim_L(t_1, t_2) = \max_{t \in S(t_1,t_2)} \left\{\frac{2IC(t)}{IC(t_1) + IC(t_2)}\right\} \tag{4}$$

Schlicker:

$$sim_S(t_1, t_2) = \max_{t \in S(t_1,t_2)} \left\{\frac{2IC(t)}{IC(t_1) + IC(t_2)}(1 + IC(t))\right\}. \tag{5}$$

Here $IC(t)$ is the information content of term t:

$$IC(t) = -\log(p(t)), \tag{6}$$

where $p(t)$ is defined as $freq(t)/N$, $freq(t)$ is the number of genes associated with term t or with any child term of t in the data set, N is total number of genes in the genome that have at least one GO term associated with them, and $S(t_1, t_2)$ is the set of common subsumers of the terms t_1 and t_2. Note that the Jiang-Conrath proposal uses the complementary concept of distance instead of similarity.

The basic objective of these methods is to capture the specificity of each GO term and to calculate the similarity between GO terms in a way that reflects their positions in the GO hierarchy. However, as discussed below, we argue that the methods of Lin and Jiang-Conrath are not best suited for this purpose. For example, consider the non-root terms t_2 (GO:0010468) and t_3 (GO:0065007) in Figure 1. Then $dist_{JC}(t_2, t_2) = dist_{JC}(t_3, t_3) = 0$, and $sim_L(t_2, t_2) = sim_L(t_3, t_3) = 1$. In other words, the methods of Lin and Jiang-Conrath cannot differentiate between two pairs of genes, one of which is associated with the term t_2 (GO:0010468), and the other with t_3 (GO:0065007) because it ignores the fact that t_2 (GO:0010468, regulation of gene expression) is more specific than t_3 (GO:0065007, biological regulation). In contrast, $sim_R(t_2, t_2) = -\log p(t_2) > sim_R(t_3, t_3) = -\log p(t_3)$, if t_2 is more specific than t_3, thus reflecting the relative positions (and the specificities) of t_2 and t_3 in the GO hierarchy. For example, in *Saccharomyces cerevisiae*, genes *YCR042C* and *YMR227C* encode TFIID subunits. Both are annotated with GO terms GO:0000114 (G1-specific transcription in mitotic cell cycle) and GO:0006367 (transcription initiation from RNA polymerase II promoter). According to the definition, $sim_L(\text{YCR042C}, \text{YMR227C}) = 1$ and $dist_{JC}(\text{YCR042C}, \text{YMR227C}) = 0$. Now consider another pair of genes *YCR046C* and *YOR063W*, both of which encode components of the ribosomal large subunits, however, one is mitochondrial and the other cytosolic. Both are annotated with the GO term GO:0006412 (translation). Again, according to the definition, $sim_L(\text{YCR046C}, \text{YOR063W}) = 1$ and $dist_{JC}(\text{YCR046C}, \text{YOR063W}) = 0$. Thus, we have

$$\begin{aligned} &sim_L(\text{YCR042C}, \text{YMR227C}) \\ &\quad = sim_L(\text{YCR046C}, \text{YOR063W}) = 1, \end{aligned}$$

and

$$\begin{aligned} &dist_{JC}(\text{YCR042C}, \text{YMR227C}) \\ &\quad = dist_{JC}(\text{YCR046C}, \text{YOR063W}) = 0. \end{aligned}$$

But clearly, the annotations of genes YCR042C and YMR227C are much more specific than the annotations of genes YCR046C and YOR063W. So the similarity between genes YCR042C and YMR227C should be greater than the similarity between genes YCR046C and YOR063W. The similarity between genes calculated by the method of Resnik reflects this fact, since

$$\begin{aligned} &sim_R(\text{YCR042C}, \text{YMR227C}) \\ &\quad = -\log p(\text{GO}:0000114) = 9.69 \\ &\quad > sim_R(\text{YCR046C}, \text{YOR063W}) \\ &\quad = -\log p(\text{GO}:0006412) = 4.02. \end{aligned}$$

2.2. Mixture Model and Parameter Estimation

The contents of this entire subsection are among the novel contributions of this paper.

As mentioned earlier, PPI data generated using high throughput techniques contain a large number of false positives[1]. Thus the PPI data set contains two groups, one representing true positives and the other representing false positives. However, differentiating the true and false positives in a large PPI data set is a big challenge due to the lack of good quantitative measures. An *ad hoc* threshold can be used for such measures. Our proposed method avoids such choices. Instead, we propose a mixture model to differentiate the two groups in a large PPI data set. One group contains pairs of interacting proteins that have strong support, the other of pairs of interacting proteins that have weak or unknown support. Here we hypothesize that the similarity between genes based on Gene Ontology using the method of Resnik (see Eq.(2)) helps to differentiate between the two groups in the PPI data. We conjecture that the true positives will have higher gene similarity values than the false positives. A mixture model is used to model the distribution of the similarity values (using the Resnik method for similarity of *Biological Process* GO terms). In particular,

$$p(x) = w_1 p_1(x) + w_2 p_2(x), \tag{7}$$

where $p_1(x)$ is the probability density function for the similarity of pairs of genes for pairs of genes with true interactions in the PPI data, and $p_2(x)$ is the probability density function for the similarity of pairs of genes in the false positives; w_1 and w_2 are the weights for p_1 and p_2, respectively. Given a large data set, p_1, p_2, w_1, and w_2 can be inferred by the maximum likelihood estimation (MLE) method. For our case, we conclude that the similarity of pairs of genes can be modeled as a mixture of two log-normal distributions with probability density functions

$$p_1(x) = \frac{1}{\sqrt{2\pi}\sigma_1 x} exp\left(-\frac{(\log x - \mu_1)^2}{2\sigma_1^2}\right) \tag{8}$$

and

$$p_2(x) = \frac{1}{\sqrt{2\pi}\sigma_2 x} exp\left(-\frac{(\log x - \mu_2)^2}{2\sigma_2^2}\right). \tag{9}$$

After parameter estimation, we can calculate a value s such that for any $x > s$, $p(x \in \text{Group 2}) > p(x \in$

Group 1). This value s is the threshold meant to differentiate the PPI data with high support (Group 2) from those with low support (Group 1). The further away the point is from s, the greater is the confidence. Furthermore, the confidence can be measured by computing the p-value since the parameters of distribution are known.

Thus our mixture model suggests a way of differentiating the true positives from the false positives by only looking at the similarity value of pairs of genes (using the method of Resnik in Eq.(2) for similarity of *Biological Process* GO terms), and by using a threshold value specified by the model (Group 1 contains false positives and Group 2 contains the true positives). Note that no *ad hoc* decision are involved.

2.3. Function Prediction

The major advantage of the method presented above is that the p-values obtained from the mixture model provide us with a metric of support of a reliability measure for the PPI data set. However, the limitation of our technique is that it can only be applied to pairs of genes with annotations. In order to overcome this limitation, it has been suggested that function prediction should be performed first to predict the functional annotation of unannotated genes. As mentioned earlier, many computational approaches have been developed for this task[2]. However, the prediction methods are prone to high false positives. Schwikowski *et al.*[6] proposed the *Majority-Voting* (MV) algorithm for predicting the functions of an unannotated gene u by the following objective function,

$$\alpha_u = \arg\max_{\alpha} \sum_{v \in N(u), \alpha_v \in A(v)} \delta(\alpha_v, \alpha), \tag{10}$$

where $N(u)$ is the set of neighbors of u, $A(v)$ is the set of annotations associated with gene v, α_i is the annotation for gene i, $\delta(x, y)$ is a function that equals 1 if $x = y$, and 0 otherwise. In other words, gene u is annotated with the term α associated with the largest number of its neighbors. The main weakness of this conventional majority voting algorithm is that it weights all its neighbors equally, and is prone to errors because of the high degree of false positives in the PPI data set. Using the metric of support proposed in section 2.2, we propose a modified *"Reliable" Majority-Voting* (RMV) algorithm which assigns a functional annotation to an unannotated gene u based on the following objective function

$$\alpha_u = \arg\max_{\alpha} \sum_{v \in N(u), \alpha_v \in A(v)} w_{v,u} \delta(\alpha_v, \alpha), \tag{11}$$

where $w_{v,u}$ is the reliability of the interaction between genes v and u, that is, $w_{v,u} = sim(A(v), \{\alpha\})$.

Another weakness of the conventional MV algorithm is that it only allows exact matches of annotations and will reject even approximate matches of annotations. Here we propose the *Weighted Reliable Majority-Voting* (WRMV) method, a modification of RMV, with the following objective function

$$\alpha_u = \arg\max_{\alpha} \sum_{v \in N(u)} w_{v,u} \Big(\max_{\alpha_v \in A(v)} sim(\alpha_v, \alpha) \Big), \tag{12}$$

where $sim(x, y)$ is a function that calculates the similarity between the GO terms x and y.

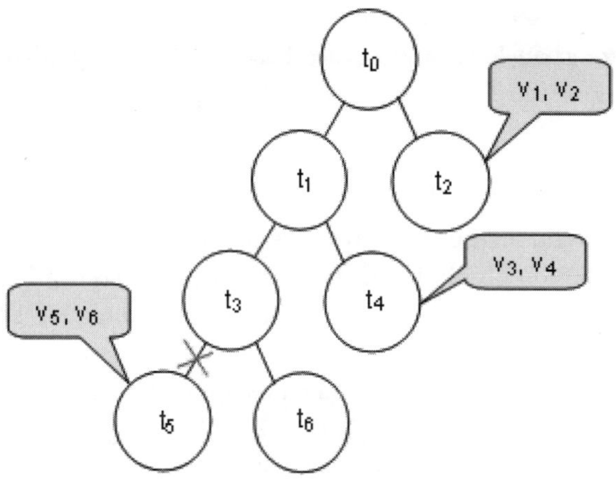

Fig. 2. An example showing the hierarchy of GO terms associated with a set of genes. GO term t_2 is associated with genes v_1 and v_2; GO term t_4 is associated with genes v_3 and v_4; GO term t_5 is associated with genes v_5 and v_6.

Note that the aforementioned algorithms only predict one functional annotation term for an uncharacterized gene. But they can be adapted to predict k functional annotation terms for any uncharacterized gene by picking the k best values of α in each case.

The example in Figure 2 illustrates the necessity of considering both the metric of support for the PPI data and the relationships between GO terms during function prediction. Assume we need to predict functions for a protein u, whose neighbors in the interaction network include proteins v_1, v_2, v_3, v_4, v_5, and v_6. As shown in Figure 2, suppose proteins v_1 and v_2 are annotated with GO term t_2, v_3 and v_4 with GO term t_4, and v_5 and v_6 with GO term t_5. According to the MV algorithm, protein u will be assigned all the GO terms t_2, t_4, and t_5, since each of the three terms has equal votes (2 in this case). However, as can be seen from Figure 2, GO term t_5 is more specific than GO terms t_2 and t_4. So GO term t_5 should be the most favored as an annotation for protein u, assuming that all the PPI data are equally reliable. On the other hand, if the interactions between proteins u and v_5 and v_6 are less reliable (or false positives), then there is less support for associating protein u with term t_5.

Note that the metric of support can also be used to improve other approaches besides the MV algorithm. In this paper, we have employed only local approaches, because as argued by Murali *et al.*[26] methods based on global optimization do not perform better than local approaches based on majority-voting algorithm.

2.4. Evaluating the Function Prediction

Several measures are possible in order to evaluate the function prediction methods proposed in section 2.3. For the traditional cross-validation technique, the simplest method to perform an evaluation is to use *precision* and *recall*, defined as follows:

$$Precision = \frac{\sum_i k_i}{\sum_i m_i}, \quad Recall = \frac{\sum_i k_i}{\sum_i n_i}, \tag{13}$$

where n_i is the number of known functions for the protein i, m_i is the number of predicted functions for the protein i when hiding its known annotations, and k_i is the number of matches between known and predicted functions for protein i. The conventional method to count the number of matches between the annotated and predicted functions only considers the exact overlap between predicted and known functions, ignoring the structure and relationship between functional attributes. Using again the simple example illustrated in Figure 2, assume that the correct function annotation of a protein u is GO term t_4, while term t_1 is the only function predicted for it. Then both recall and precision would be reported to be 0 according to the conventional method. However, it overlooks the fact that GO term t_4 is quite close to the term t_1. Here we introduce a new definition for precision and recall. For a known protein, suppose the known annotated functional terms are $\{t_{o1}, t_{o2}, ..., t_{on}\}$, and the predicted terms are $\{t_{p1}, t_{p2}, ..., t_{pm}\}$. We define the success of the prediction for function t_{oi} as

$$RecallSucess(t_{oi}) = \max_j sim(t_{oi}, t_{pj}),$$

and the success of the predicted function t_{pj} as

$$PrecisionSuccess(t_{pj}) = \max_i sim(t_{oi}, t_{pj}).$$

We define the new precision and recall measures as follows:

$$Precision = \frac{\sum_j PrecisionSuccess(t_{pj})}{\sum_j sim(t_{pj}, t_{pj})}, \tag{14}$$

$$Recall = \frac{\sum_i RecallSucess(t_{o_i})}{\sum_i sim(t_{o_i}, t_{o_i})}. \tag{15}$$

3. EXPERIMENTAL RESULTS

3.1. Data Sets

Function prediction methods based on a protein-protein interaction network can make use of two data sources - the PPI data set and a database of available functional annotations. In this section, we will introduce the two data sources we used in our experiments.

3.1.1. *Gene Ontology*

We used the available functional annotations from the Gene Ontology (GO) database[5]. GO consists of sets of structured vocabularies each organized as a rooted directed acyclic graph (DAG), where every node is associated with a GO term and edges represent either a "IS-A" or a "PART-OF" relationship. Three independent sets of vocabularies are provided: *cellular component*, *molecular function* and *biological process*. Generally, a gene is annotated by one or more GO terms. The terms at the lower levels correspond to more specific function descriptions. If a gene is annotated with a GO term, it is also annotated with the ancestors of that GO term. Thus, the terms at the higher levels have more associated genes. The GO database is constantly being updated; we used version 5.403, and the gene-term associations for *Saccharomyces cerevisiae* from version 1.1344 from SGD.

3.1.2. *Protein-Protein Interaction Data*

Several PPI data sets were used in this paper for our experiments. The first PPI data set was downloaded from the BioGRID database[27]. Henceforth, we will refer to this data set as the *BioGRID* data set. The *confirmation number* for a given pair of proteins is defined as the number of independent confirmations that support that interaction. A pseudo-negative data set was also generated by picking pairs of proteins that were not present in the PPI data set. Thus each pair of proteins in the pseudo-negative data set has a confirmation number of 0. There were 87920 unique interacting pairs in total with confirmation numbers ranging from 0 to 40. This data set is used to estimate the metric of support for pairs of proteins.

Two so-called *gold-standard data sets* (gold-standard positive and gold-standard negative) were used to test the performance of our method. These two gold-standard data sets were hand-crafted by Jansen *et al.*[15]. The gold-standard positives came from the MIPS (Munich Information Center for Protein Sequence) complexes catalog[28] since the proteins in a complex are guaranteed to bind to each other. The number of gold-standard positive pairs used in our experiments was 7727. A gold-standard negative data set is harder to define. Jansen *et al.* created such a list by picking pairs of proteins known to be localized in separate subcellular compartments[15], resulting in a total of 1838501 pairs.

3.2. Results on Using the Mixture Model

The similarity between genes based on the *Biological Process* categorization of the GO hierarchy was calculated using Eq.(1) and Eq.(2). The method was separately applied to the *BioGRID* data set, in which PPI data have non-negative, integral confirmation numbers k. Interacting pairs of proteins from *BioGRID* data set were grouped based on their confirmation number. It is clear that the PPI data set may include a large number of false positives. Thus, the challenge is to differentiate the true interactions from the false ones. We hypothesize that each of these groups generated above contains two subgroups, one representing pairs of proteins that interact with high support, and the other representing pairs that interact with low support. Data sets with larger confirmation numbers are expected to have less noise.

As shown in Figure 3, a histogram of the (logarithm of) similarity measure (using the *Resnik* method for similarity of GO terms) was plotted for pairs of genes within each group (i.e., same degree of confirmation from the PPI data set). In order to visualize the whole histogram, we have arbitrarily chosen $\log(0) = \log(0.01) \approx -4.61$. Based on our earlier assumptions, we conjectured that each of these histograms can be modeled as a mixture of two normal distributions (since the original is a mixture two log-normal distribution), one for the Group 1, and the other for the Group 2.

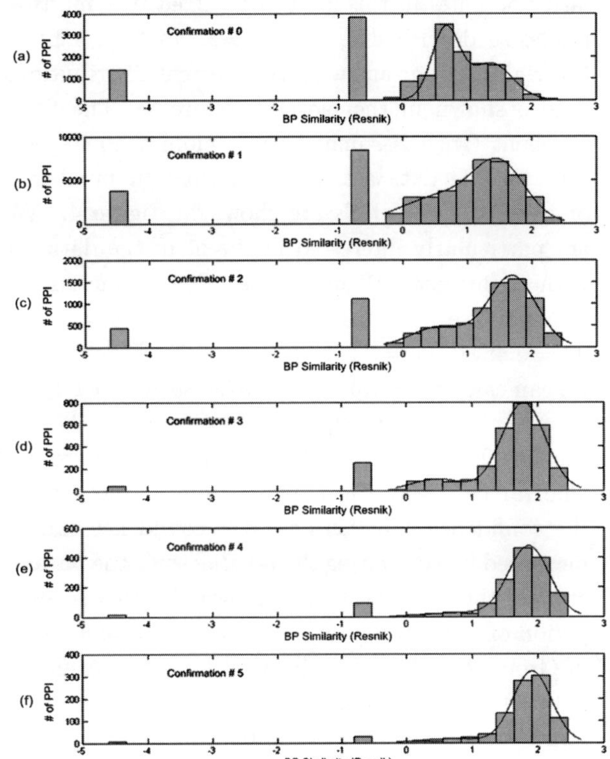

Fig. 3. Distribution of similarity of genes based on the Resnik method using: (a) the entire PPI data set, (b) 1 or more independent confirmations, (c) 2 or more independent confirmations, (d) 3 or more independent confirmations, (e) 4 or more independent confirmations, (f) 5 or more independent confirmations.

All the plots in Figure 3 have three well-separated subgroups. Note that the leftmost subgroup corresponds to those pairs of genes for which at least one has the GO terms associated with the root of the GO hierarchy; the subgroup in the middle corresponds to those pairs of genes at least one of which is associated with a node close to the root of the GO hierarchy. The reason for the existence of these two subgroups is that there are some PPI data sets containing genes with very non-specific functional annotations. As we can see from Figure 3, the larger the confirmation number, the less pronounced are these two groups. Thus, for the two leftmost subgroups, similarity of genes based on GO annotation cannot be used to differentiate signal from noise (Thus function prediction for these genes are necessary and is an important focus of this paper). However, for PPI data containing genes with specific functions (i.e., the rightmost group in the plots of Figure 3), similarity of genes in this group was fitted to a mixture model as described in section 2.2. In fact, a fit of the rightmost group with two normal distributions is also shown in the plots of Figure 3. The fit is excellent (with R-squared value more than 98 percent for the data set with confirmation number 1 or more). The details are shown in Figure 4. We are particularly interested in the fit of the data set with confirmation 1 and above. The estimated parameters are $\mu_1 = 0.3815$, $\sigma_1 = 0.4011$, $\mu_2 = 1.552$, $\sigma_2 = 0.4541$, $w_1 = 0.23$, and $w_2 = 0.77$. From the fit, we can calculate a value $s = 0.9498$ such that for any $x > s$, $p(x \in \text{Group 2}) > p(x \in \text{Group 1})$. This is the threshold meant to differentiate the two groups. The further away the point is from s, the greater the confidence. Furthermore, the confidence can be measured by computing the *p-value* since the parameters of the distribution are known. Further investigation of these two groups reveal that proteins pairs in Group 2 contain proteins that have been well annotated (associating with GO terms that have levels larger or equal to 3). The components of Group 1 are more complicated. It consists of the interactions between two poorly annotated genes, the interactions between a well annotated gene and a poorly annotated gene, and the interactions between two well annotated genes.

The results of further experiments performed on the PPI data sets from the human proteome[27] also displayed similar results (data not shown).

To test the power of our estimation, we applied it to the gold-standard data set. In particular, for each pair of genes in the gold-standard data set, we calculated the similarity between the genes in that pair and compared it to the threshold value $s = 0.9498$. If the similarity was larger than s, we labeled it as Group 2, otherwise, as Group 1. We then calculated the percentage of pairs of proteins in Group 2 and Group 1 in the gold-standard positive and negative data sets.

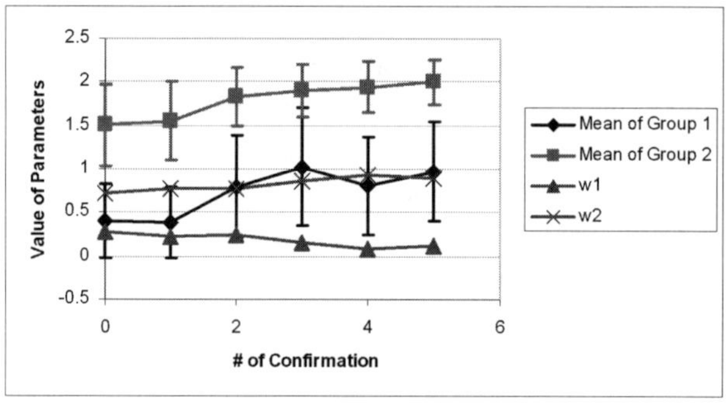

Fig. 4. Parameters for the density function, fitting $p(x) = w_1 p_1(x) + w_2 p_2(x)$ for the metric of support for PPI data with different confirmation numbers. Group 1 corresponds to noise, and Group 2 to signal.

As shown in Table 1, majority of the pairs in the gold-standard positive data (GSPD) set were labeled correctly as Group 2 (99.61%), and most of the pairs in the gold-standard negative data set (GSND) were correctly labeled as Group 1 (83.03%). These high percentage values provide further support for our mixture-model based technique. It is worth pointing out that the GSPD set is clearly more reliable than the GSND set as described in section 3.1.2.

Table 1. Mixture model on gold-standard data set.

	total PPI pairs	subgroup PPI pairs	percentage
GSPD[a]	7727	7696[1]	99.61
GSND[b]	1838501	1526467[2]	83.03

[a] Golden Standard Positive Data set.
[b] Golden Standard Negative Data set.
[1] Number of PPI pairs in Group 2.
[2] Number of PPI pairs in Group 1.

One possible objection to the application in this paper is that the results of the mixture model is an artifact of functional bias in the PPI data set. To address this objection, we applied the mixture model to PPI data after separating out the data from the three main different high-throughput methods, i.e., yeast two-hybrid systems, mass spectrometry, and synthetic lethality experiments. As reported by Mering *et al.*[1], the overlap of PPI data detected by the different methods is small, and each technique produces a unique distribution of interactions with respect to functional categories of interacting proteins. In other words, each method tends to discover different types of interactions. For example, the yeast two-hybrid system largely fails to discover interactions between proteins involved in translation; mass spectrometry method predicts relatively few interactions for proteins involved in transport and sensing.

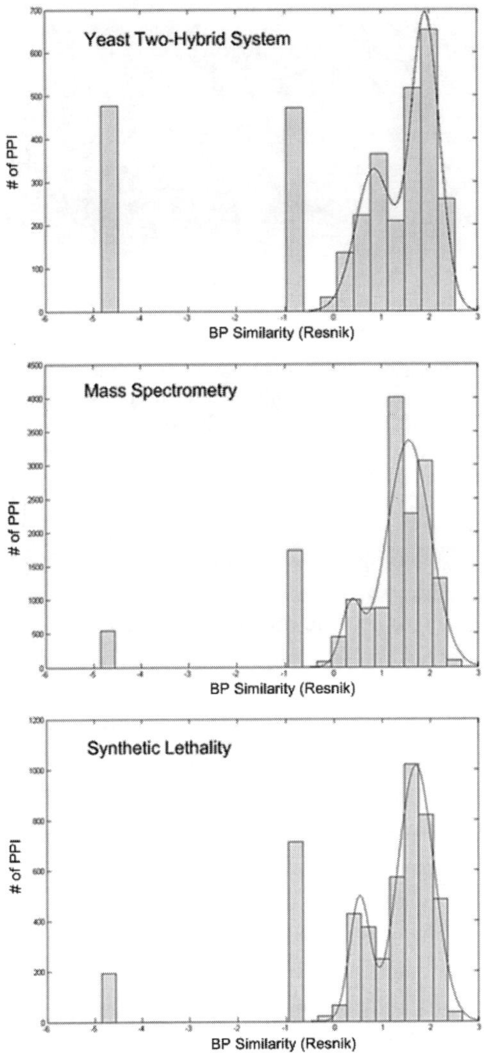

Fig. 5. Distribution of similarity of pairs of genes based on the method by Resnik for PPI data generated by high-throughput methods yeast two-hybrid systems (top), mass spectrometry (middle), and Synthetic Lethality (bottom).

In summary, if the PPI data set has a functional bias, then the PPI data produced by individual methods should have an even greater functional bias, with each one biased toward different functional categories.

Despite the unique functional bias of each method, the mixture model when applied to the PPI data from the individual methods showed the same bimodal mixture distribution (Figure 5), indicating that the mixture model is tolerant to severe functional bias and is therefore very likely to be a the reflection of inherent features of the PPI data.

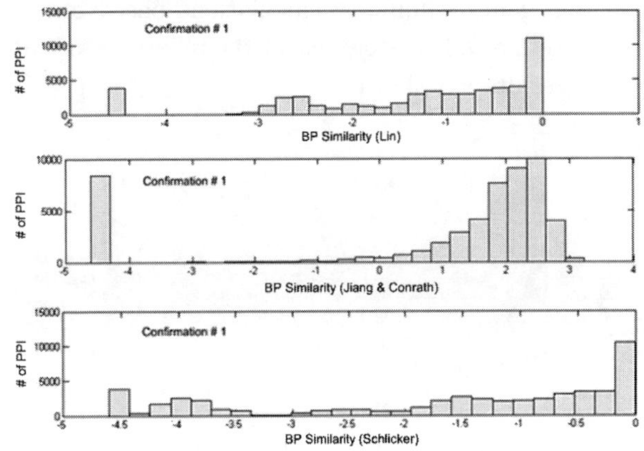

Fig. 6. Distribution of similarity of genes based on method Lin, Jiang-Conrath, and Schlicker for PPI data with confirmation number of 1 and more (Confirmation # 1).

In order to justify our choice of the Resnik similarity measure, we also applied the Lin (Eq.(4)), Jiang-Conrath (Eq.(3)), and Schlicker (Eq.(5)) methods to the PPI data set with confirmation number 1 or more. The results shown in Figure 6 confirms our analysis that the Lin and Jiang-Conrath methods are inappropriate for similarity computation. As shown in Figure 6, the histogram of similarity values between genes calculated by Lin's formula has a peak at the rightmost end. Additionally, the rest of the histogram fails to display a bimodal distribution, which is necessary to flush out the false positives. Furthermore, the peak increases with increasing confirmation number (data not shown). In contrast, the histograms of distance measures between genes calculated by the Jiang-Conrath's method (middle in Figures 6) produces a peak at its leftmost end with a unimodal distribution for the remaining, thus showing that the mixture model is unlikely to produce meaningful results. Schlicker's method was devised to combine Lin's and Resnik's methods. However, its performance was similar to that of Lin's method (see in Figure 6). We also applied these methods to the same PPI data set, but with higher confirmation numbers (data not shown). Since those data sets are likely to have fewer false positives, it is no surprise that the histograms were even less useful for discriminatory purpose.

Finally, we tried our methods on the other two GO categorizations, i.e., *cellular component* and *molecular function*. Since those categorizations are less comprehensive with a large number of unannotated genes, similarity calculations based on the them did not adequately reflect the reliability of PPI data (results not shown).

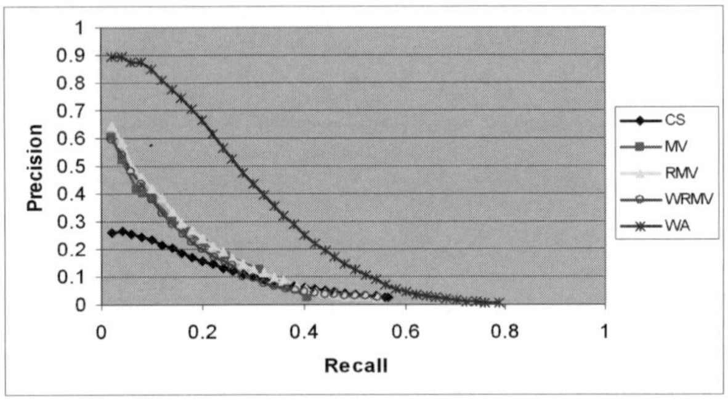

Fig. 7. Precision-recall analysis of five function prediction methods using the conventional evaluation metric as described in Eq.(13) for 1) Chi-Square method (CS), 2) Majority-Voting method (MV), 3) Reliable Majority-Voting method (RMV), 4) Weighted Reliable Majority-Voting (WRMV), and 5) FS-Weighted Averaging method (WA).

3.3. Function Prediction

Five different function prediction approaches based on neighborhood-counting – three introduced in section 2.3, one introduced in section 1, and the last one called FS-Weighted Averaging method (WA) developed by Hua *et al.*[8] – were compared. We note that in our implementation of the WA method, we use the similarity measure given in Eq.(2) from Hua *et al.*[8] to compute the reliability measure, $w_{v,u}$, in Eq.(11) of this paper. The precision and recall for each approach was calculated on the *BioGRID* PPI data set using 5-fold cross validation. First, a conventional evaluation method was employed, which consisted of computing precision and recall as a simple count of the predictions for the gold-standard positive and negative sets. As shown in Figure 7, when conventional evaluation methods were applied to calculate the precision and recall, the FS-Weighted Averaging (WA) method performed the best, and there was no significant difference among the other three methods (MV, RMV, and WRMV). However, when the new evaluation method (as described in Eq.(14) and Eq.(15)) was applied, both WA and WRMV performed well (see Figure 8). Among the three versions of Majority-Voting methods (MV, RMV, and WRMV), Weighted Reliable Majority-Voting method performed the best, and the conventional Majority-Voting method performed the worst.

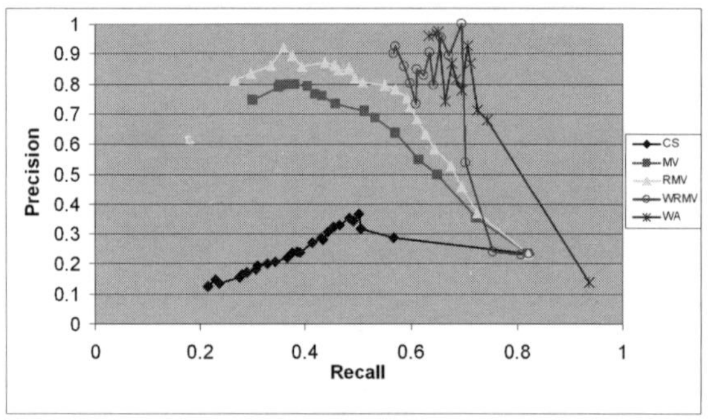

Fig. 8. Precision-recall analysis of five function prediction methods using new evaluation metric as described in Eq.(14) and Eq.(15) for 1) Chi-Square method (CS), 2) Majority-Voting method (MV), 3) Reliable Majority-Voting method (RMV), 4) Weighted Reliable Majority-Voting method (WRMV), and 5) FS-Weighted Averaging method (WA).

In order to see the effectiveness of the new evaluation metric, the precision-recall curves of the three function prediction methods (RMV, WRMV and WA) using new and conventional evaluation metrics are compared by combining the related curves in Figure 7 and Figure 8. As shown in Figure 9, the proposed new evaluation method has two advantages over the conventional one. First, the new evaluation method provides wider precision and recall coverage, that is, at the same precision (recall) value, the recall (precision) calculated by the new method is larger than that calculated by the old one. This is due to the strict definition of conventional precision and recall, while ignoring the fact that some pairs of true and predicted annotations are very similar to each other. Second, the new evaluation method has more power to measure the performance of function prediction methods. For example, the precision-recall curves of the function prediction methods RMV and WRMV diverge based on the new evaluation metric, but are roughly indistinguishable based on the conventional metric (Figure 9).

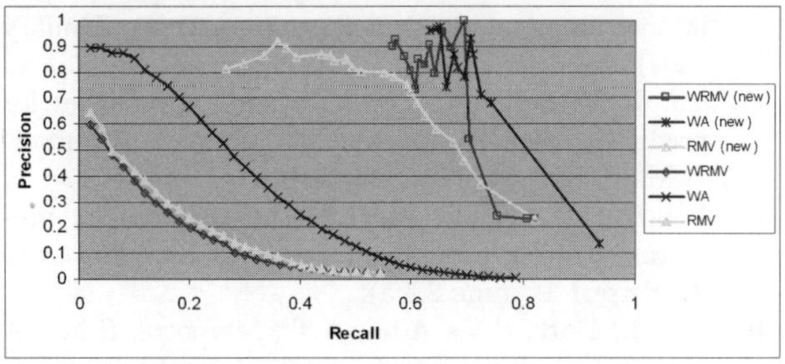

Fig. 9. Comparison of precision-recall analysis of three Majority-Voting function prediction methods using new evaluation metric as described in Eq.(14) and Eq.(15) for 1) Weighted Reliable Majority-Voting method (WRMV new), 2) FS-Weighted Averaging method, (WA new), and 3) Reliable Majority-Voting method (RMV new), and conventional metric as described in Eq.(13) for 4) Weighted Reliable Majority-Voting method (WRMV), 5) FS-Weighted Averaging method, (WA), and 6) Reliable Majority-Voting method (RMV).

4. DISCUSSION AND CONCLUSIONS

Function predictions based on PPI data were performed using two sources of data: GO annotation data and BioGRID PPI data. Previous research on this topic focused on the interaction network inferred from PPI data, while ignoring the topology of the hierarchy representing the annotations. In some cases, only a fraction of the terms were used. Thus the resulting predictions were not comparable. For PPI data, quantitatively assessment of confidence for pairs of proteins becomes a pressing need.

The research described in this paper addresses the above shortcomings. Our significant contributions are:

(1) A mixture model was introduced to model PPI data. The parameters of the model were estimated from the similarity of genes in the PPI data set. This mixture model was used to devise a metric of support for protein-protein interaction data. It is based on the assumption that proteins having similar functional annotations are more likely to interact.
(2) New function prediction methods were proposed to predict the function of proteins in a consistent way based on the entire hierarchical annotation system. Results show that the performance of the predictions was improved significantly by integrating the mixture model described above into the function prediction methods.
(3) A newly proposed evaluation method provides the means by which systematic, consistent, and comprehensive comparison of different function prediction methods is possible.

In this paper, we have mainly focused on introducing a metric of support for the PPI data using GO information, and the application of such a metric in function prediction for uncharacterized proteins. Although the fact that proteins having similar function annotations are more likely to have interactions has been confirmed in the literature, we provide a quantitative measure to estimate the similarity, and to uncover the relationship between the metric and the support of PPI data. GO annotations are generated by integrating information from multiple data sources, many of which have been manually curated by human experts. Thus assessing PPI data using the GO hierarchy is a way in which multiple data sources are integrated. The comprehensive comparison of the method to assess PPI data using GO information and other counterparts as described in section 1 is necessary and will be addressed elsewhere.

Acknowledgments

This research is supported by the program Molecular Assemblies, Genes, and Genomics Integrated Efficiently (MAGGIE) funded by the Office of Science, Office of Biological and Environmental Research, U.S. Department of Energy, under contract number DE-AC02-05CH11231. GN was supported by NIH Grant P01 DA15027-01 and NIH/NIGMS S06 GM008205, and EZ was supported by FIU Dissertation Year Fellowship.

References

1. von Mering, C., Krause, R., Snel, B., Cornell, M., Oliver, S.G., Fields, S., Bork, P.: Comparative assessment of large-scale data sets of protein-protein interactions. Nature **417**(6887) (May 2002) 399–403
2. Sharan, R., Ulitsky, I., Shamir, R.: Network-based prediction of protein function. Molecular Systems Biology **3**(88) (2007) 1–13

3. Tatusov, R.L., Galperin, M.Y., Natale, D.A., Koonin, E.V.: The COG database: a tool for genome-scale analysis of protein functions and evolution. Nucleic Acids Res **28**(1) (January 2000) 33–36
4. Ruepp, A., Zollner, A., Maier, D., Albermann, K., Hani, J., Mokrejs, M., Tetko, I., Güldener, U., Mannhaupt, G., Münsterkötter, M., Mewes, H.W.: The FunCat, a functional annotation scheme for systematic classification of proteins from whole genomes. Nucleic Acids Res **32**(18) (2004) 5539–5545
5. Ashburner, M., Ball, C.A., Blake, J.A., Botstein, D., Butler, H., Cherry, J.M., Davis, A.P., Dolinski, K., Dwight, S.S., Eppig, J.T., Harris, M.A., Hill, D.P., Issel-Tarver, L., Kasarskis, A., Lewis, S., Matese, J.C., Richardson, J.E., Ringwald, M., Rubin, G.M., Sherlock, G.: Gene ontology: tool for the unification of biology. the gene ontology consortium. Nat Genet **25**(1) (May 2000) 25–29
6. Schwikowski, B., Uetz, P., Fields, S.: A network of protein-protein interactions in yeast. Nat Biotechnol **18**(12) (December 2000) 1257–1261
7. Hishigaki, H., Nakai, K., Ono, T., Tanigami, A., Takagi, T.: Assessment of prediction accuracy of protein function from protein–protein interaction data. Yeast **18**(6) (April 2001) 523–531
8. Chua, H.N., Sung, W.K., Wong, L.: Exploiting indirect neighbours and topological weight to predict protein function from proteinprotein interactions. Bioinformatics **22**(13) (July 2006) 1623–1630
9. Vzquez, A., Flammini, A., Maritan, A., Vespignani, A.: Global protein function prediction from protein-protein interaction networks. Nat Biotechnol **21**(6) (June 2003) 697–700
10. Karaoz, U., Murali, T.M., Letovsky, S., Zheng, Y., Ding, C., R.Cantor, C., Kasif, S.: Whole-genome annotation by using evidence integration in functional-linkage networks. Proc Natl Acad Sci U S A **101**(9) (March 2004) 2888–2893
11. Letovsky, S., Kasif, S.: Predicting protein function from protein/protein interaction data: a probabilistic approach. Bioinformatics **19 Suppl 1** (2003) i197–i204
12. Deng, M., Tu, Z., Sun, F., Chen, T.: Mapping gene ontology to proteins based on protein–protein interaction data. Bioinformatics **20**(6) (2004) 895–902
13. Wu, Y., Lonardi, S.: A linear-time algorithm for predicting functional annotations from protein-protein interaction networks. In: Proceedings of the Workshop on Data Mining in Bioinformatics (BIOKDD'07). (2007) 35–41
14. Nabieva, E., Jim, K., Agarwal, A., Chazelle, B., Singh, M.: Whole-proteome prediction of protein function via graph-theoretic analysis of interaction maps. Bioinformatics **21 Suppl 1** (June 2005) i302–i310
15. Jansen, R., Yu, H., Greenbaum, D., Kluger, Y., Krogan, N.J., Chung, S., Emili, A., Snyder, M., Greenblatt, J.F., Gerstein, M.: A Bayesian networks approach for predicting protein-protein interactions from genomic data. Science **302**(5644) (October 2003) 449–453
16. Bader, J.S., Chaudhuri, A., Rothberg, J.M., Chant, J.: Gaining confidence in high-throughput protein interaction networks. Nat Biotechnol **22**(1) (January 2004) 78–85
17. Yu, J., Fotouhi, F.: Computational approaches for predicting protein—protein interactions: A survey. J. Med. Syst. **30**(1) (2006) 39–44
18. Ben-Hur, A., Noble, W.S.: Kernel methods for predicting protein-protein interactions. Bioinformatics **21 Suppl 1** (June 2005)
19. Lee, I., Date, S.V., Adai, A.T., Marcotte, E.M.: A probabilistic functional network of yeast genes. Science **306**(5701) (November 2004) 1555–1558
20. Qi, Y., Bar-Joseph, Z., Klein-Seetharaman, J.: Evaluation of different biological data and computational classification methods for use in protein interaction prediction. PROTEINS: Structure, Function, and Bioinformatics **3**(63) (May 2006) 490–500
21. Resnik, P.: Using information content to evaluate semantic similarity. In: Proceedings of the 14th International Joint Conference on Artificial Intelligence. (1995) 448–453
22. Jiang, J.J., Conrath, D.W.: Semantic similarity based on corpus statistics and lexical taxonomy. In: Proceedings of International Conference on Research in Computational Linguistics. (1997)
23. Lin, D.: An information-theoretic definition of similarity. In: Proceedings of the 15th International Conference on Machine Learning. (1998)
24. Schlicker, A., Domingues, F.S., Rahnenfuhrer, J., Lengauer, T.: A new measure for functional similarity of gene products based on gene ontology. BMC Bioinformatics **7** (June 2006) 302–317
25. Lord, P.W., Stevens, R.D., Brass, A., Goble, C.A.: Semantic similarity measures as tools for exploring the gene ontology. Pac Symp Biocomput (2003) 601–612
26. Murali, T., Wu, C., Kasif, S.: The art of gene function prediction. Nat Biotechnol **24**(12) (2006) 1474–1475
27. Stark, C., Breitkreutz, B.J., Reguly, T., Boucher, L., Breitkreutz, A., Tyers, M.: BioGRID: a general repository for interaction datasets. Nucleic Acids Res **34**(Database issue) (January 2006)
28. Mewes, H., Gruber, F., Geier, C., Haase, B., Kaps, D., Lemcke, A., Mannhaupt, K., Pfeiffer, G., Schuller, F.: MIPS: a database for genomes and protein sequences. Nucleic Acids Res **30**(1) (2002) 31–34

GABORLOCAL: PEAK DETECTION IN MASS SPECTRUM BY GABOR FILTERS AND GAUSSIAN LOCAL MAXIMA

Nha Nguyen
Department of Electrical Engineering, University of Texas at Arlington, TX, USA
Email: nhn3175@exchange.uta.edu

Heng Huang*
Department of Computer Science and Engineering, University of Texas at Arlington, TX, USA
**Email: heng@uta.edu*

Soontorn Oraintara
Department of Electrical Engineering, University of Texas at Arlington, TX, USA
Email: oraintar@uta.edu

An Vo
Department of Electrical Engineering, University of Texas at Arlington, TX, USA
Email: vpnan@msp.uta.edu

Mass Spectrometry (MS) is increasingly being used to discover disease related proteomic patterns. The peak detection step is one of most important steps in the typical analysis of MS data. Recently, many new algorithms have been proposed to increase true position rate with low false position rate in peak detection. Most of them follow two approaches: one is denoising approach and the other one is decomposing approach. In the previous studies, the decomposition of MS data method shows more potential than the first one. In this paper, we propose a new method named GaborLocal which can detect more true peaks with a very low false position rate. The Gaussian local maxima is employed for peak detection, because it is robust to noise in signals. Moreover, the maximum rank of peaks is defined at the first time to identify peaks instead of using the signal-to-noise ratio and the Gabor filter is used to decompose the raw MS signal. We perform the proposed method on the real SELDI-TOF spectrum with known polypeptide positions. The experimental results demonstrate our method outperforms other common used methods in the receiver operating characteristic (ROC) curve.

1. INTRODUCTION

Mass Spectrometry (MS) is an analytical technique has been widely used to discover disease related proteomic patterns. From these proteomic patterns, researchers can identify bio-markers, make a early diagnosis, observe disease progression, response to treatment and so on. Peak detection is one of most important steps in the analysis of mass spectrum because its performance directly effects the other processing steps and final results such as profile alignment [1], bio-marker identification [2] and protein identification [3].

There are two types of peak detection approaches: denoising [4, 5] and non-denoising (or decomposing) [6, 7] approaches. There are several similar steps between these two approaches such as baseline correction, alignment of spectrograms and normalization. They also use local maxima to detect peak positions and use some rules to quantify peaks. Specially, both approaches use the signal to noise ratio (SNR) to remove some small energy peaks whose their SNR values are less than a threshold. However, in the denoising approach, before detecting peaks, a denoising step is added to reduce the noise of mass spectrum data. In the non-denoising approach, a decomposition step is used to analyze mass spectrum into different scales before the peak detection by local maxima. When the smoothing step is applied

*Corresponding author.

into the denoising approach, it possibly removes both noise and signal. If the real peaks are removed by smoothing step, they can never be recovered in the other processing steps. As a result, we lose some important information and introduce error into MS data analysis. Thus, the way we decompose a signal into many scales without denoising is a really better approach with great potentials.

The SNR is used to identify peaks in both denoising and non-denoising methods. In paper [6], P. Du *et al* estimated the SNR in the wavelet space and got much better results than the previous work. But they still failed to detect the peak at 147300 and some peaks with small SNR. This problem came from the SNR value estimation and all previous methods estimated the SNR value by using the relationship between the peak amplitude and the surrounding noise level. Since some sources of noise can also have high amplitudes, the high amplitude peak does not always guarantee to be real peak. On the other hand, some low amplitude peaks can also be real peaks. It is clearly that the way using SNR to quantify peaks is not efficient and accurate. More details of this problem will be discussed in section 3.4. In this paper, we propose a new robust decomposing based MS peak detection approach. We use the Gabor filters to create many scales from one signal without smoothing. The Gaussian local maxima is exploited to detect peaks instead of the local maxima because the Gaussian local maxima method is more robust to the noise of mass spectrum. Finally, we use the maximum rank (MR) of peaks to remove some false peaks instead of the SNR. The real SELDI-TOF spectrum with known polypeptide composition and position is used to evaluate our method. The experimental results show that our new approach can detect both high amplitude and small amplitude peaks with a low false position rate and is much better than the previous methods.

2. METHODS

In this section, we first introduce the basic knowledge of Gabor filters. After that, our proposed method which is a combination of the Gabor filters and the Gaussian local maxima will be detailed. At last, we will use one example to show how our method works.

2.1. Gabor Filters

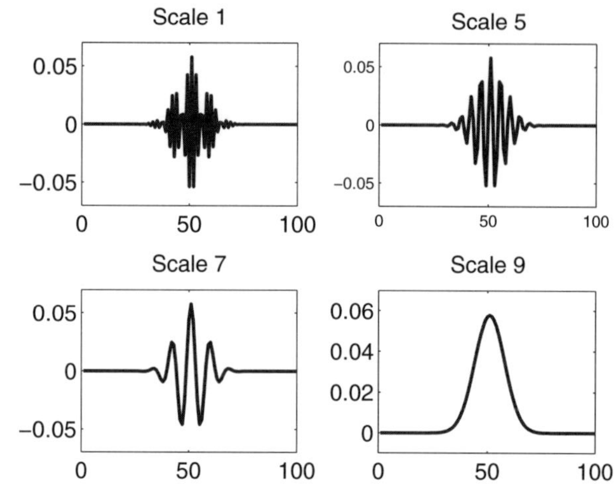

Fig. 1. The real parts of the uniform Gabor filters.

The Gabor filters [8] were developed to create Gaussian transfer functions in the frequency domain. Thus, taking the inverse Fourier transform of this transfer function, we get a filter closely resembling to the Gabor filters. The Gabor filters have been shown to have optimal combined localization in both the spatial and the spatial-frequency domain [9, 10]. In certain applications, this filtering technique has been demonstrated to be robust and fast [11] and the recursive implementation of 1D Gabor filtering has been shown in paper [12]. This recursive algorithm for the Gabor filter achieves the fastest possible implementation. For a signal consisting of N samples, this implementation requires $O(N)$ multiply-and-add (MADD) operations. A generic one dimensional Gabor function and its Fourier transform are given by:

$$h(t) = \frac{1}{\sqrt{2\pi}\sigma} \exp(-\frac{t^2}{2\sigma^2}) \exp(j2\pi F_i t), \tag{1}$$

$$H(f) = \exp(-\frac{(f - F_i)^2}{2\sigma_f^2}), \tag{2}$$

where $\sigma_f = 1/(2\pi\sigma)$ represents the bandwidth of the filter and F_i is the central frequency.

The Gabor filter can be viewed as a Gaussian modulated by a complex sinusoid (with centre frequencies F_i). This filter responds to some frequency, but only in a localized part of the signal. The coefficients of Gabor filters are complex. Therefore, the Gabor filters have one-side frequency support as shown in Fig. 2 and Fig. 4. We also illustrate the real parts of the Gabor filters in Fig. 1 and Fig. 3.

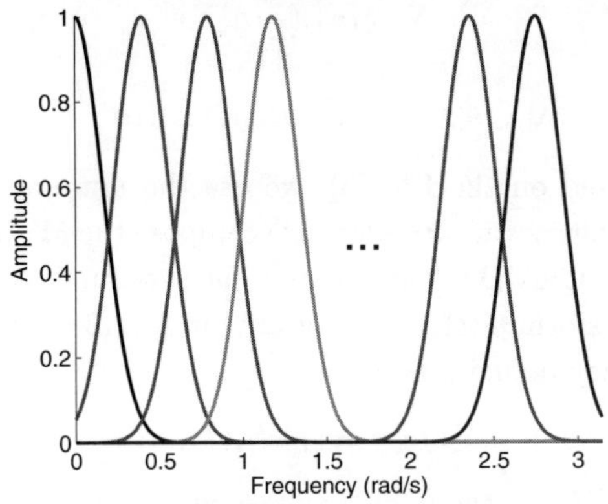

Fig. 2. Frequency supports of the uniform Gabor filters.

Given a certain number of subbands, in order to obtain a Gabor filter bank, the central frequencies F_i and bandwidths σ_f of these filters are chosen to ensure that the half-peak magnitude supports of the frequency responses touch each other as shown in Fig. 2 and Fig. 4. The Gabor filter bank can be designed to be uniform (in Fig. 2) or non-uniform (in Fig. 4). In our experiments, we use the Gabor filter bank with nine non-uniform subbands.

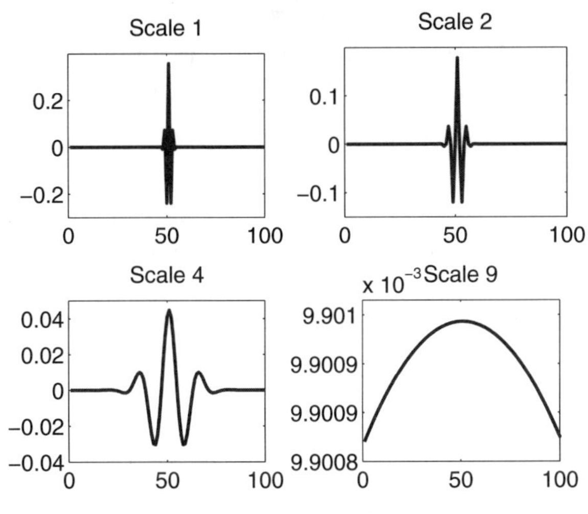

Fig. 3. The real parts of the non-uniform Gabor filters.

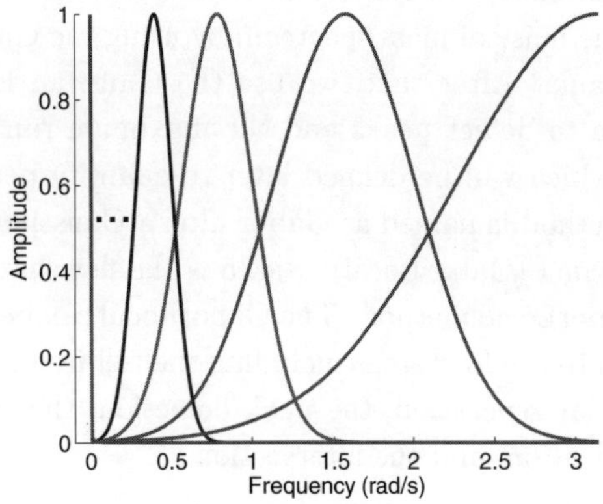

Fig. 4. Frequency supports of the non-uniform Gabor filters.

After decomposing a MS signal, nine subbands are created as follows:

$$y_i(t) = h_i(t) * x(t), \tag{3}$$

where $x(t)$ is the input signal, i = 1,2,...,9, and $*$ is the 1D convolution. This is an over-complete representation with the redundant ratio of 9.

2.2. GaborLocal Method

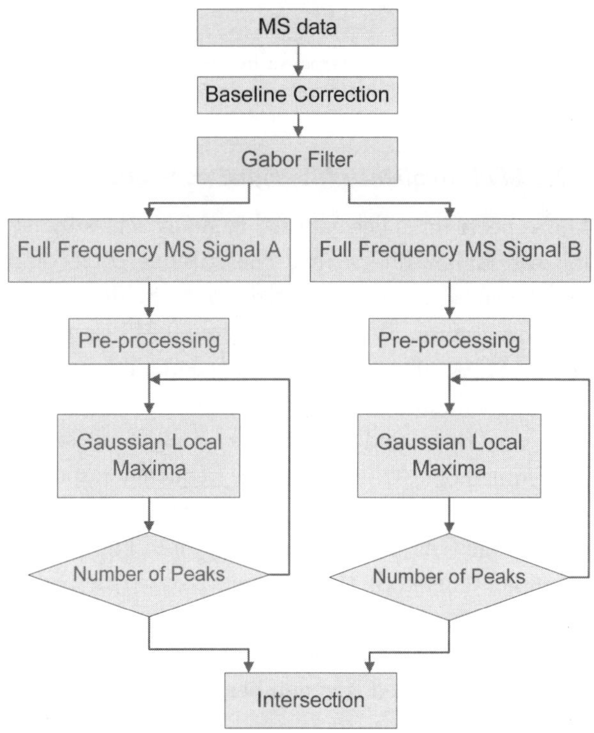

Fig. 5. Flowchart of Gabor-Gaussian local maxima method for peak detection in the MS data.

Our main idea is to amplify the true signal and compress the noise of mass spectrum by using the Gabor filter bank. After that, we use the Gaussian local maxima to detect peaks and the maximum rank of peaks which will be defined later to quantify peaks. This method is named as Gabor filter - Gaussian local maxima (GaborLocal). Fig. 5 is the flowchart of our GaborLocal method. The GaborLocal can be detailed into the four steps including the full frequency MS signal generation, the peak detection, the peak quantification, and the intersection.

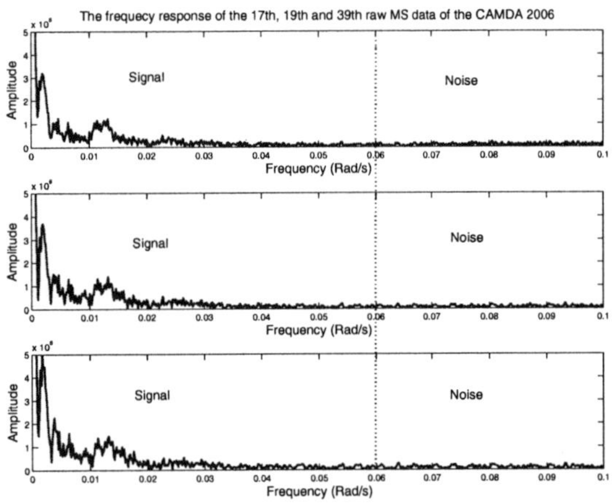

Fig. 6. The frequency response of three raw MS signals - the 17th, 19th and 39th MS data of the CAMDA 2006.

2.2.1. *Full frequency MS signal generation*

Mass spectrum is decomposed to many scales by using the Gabor filters after the baseline correction. Our purpose is to emphasize some hidden peaks buried by noise. When we analyze 60 MS signals of the CAMDA 2006 in the frequency domain, we notice that the valuable information of these signals locate from zero to around 0.06 (rad/s), and the noises locate from 0.06 to π (rad/s). The frequency responses of three raw MS data (the 17th, 19th and 39th MS data of the CAMDA 2006) are shown in Fig. 6 as an example. Therefore, the bandwidth σ_f of the Gabor filters which enhances peaks must be less than 0.06. In our experiments, we use $\sigma_f = 0.01$. If the uniform Gabor filter is used, the number of scales must be

$$N = \frac{\pi}{0.01} \approx 314 \ scales. \tag{4}$$

With 314 scales in (4), we know that the uniform Gabor filter is not efficient. If the non-uniform Gabor filter is used, the number of scales should be calculated as follows:

$$\sigma_f = \frac{\pi}{2^N},$$

$$N = log_2(\frac{\pi}{\sigma_f}),$$

$$N \approx 8.3 \ scales \ with \ \sigma_f = 0.01. \tag{5}$$

Based on the Eq. (5), we use the non-uniform Gabor filters with 9 scales to decompose the MS data (we use CAMDA 2006 data [13] for experiments). If we transform $y_i(t)$, $h_i(t)$ and $x(t)$ in Eq. (3) into the frequency domain, we get

$$Y_i(f) = X(f).H_i(f), \tag{6}$$

where $X(f)$ is the frequency response of the raw MS signal, $H_i(f)$ is the frequency response of the ith Gabor filter, and $Y_i(f)$ is the frequency response of the ith scale. After getting 9 signals according to 9 frequency sub-bands in complex values, the full frequency signal A will be created by summing above signals in complex values first and taking their absolute values at the final. To create the full frequency signal B, we take the absolute values for each sub-band and then sum all these sub-bands. After this step, we have two full frequency signals A and B. Let's denote $y(t)$ and $Y(f)$ as the full frequency signal in time domain and frequency domain, respectively.

$$Y(f) = \sum_{i=N_i} Y_i(f), \tag{7}$$

where N_i are the scales which are used to create the full frequency signal. From Eq. (6) and (7), we get

$$Y(f) = \sum_{i=N_i} X(f)H_i(f)$$

$$= X(f) \sum_{i=N_i} H_i(f) = X(f)H_s(f), \tag{8}$$

where $H_s(f) = \sum_{i=N_i} H_i(f)$ is called the summary filter. From Eq. (2), the summary filter can formulated as follows

$$H_s(f) = \sum_{i=N_i} \exp(-\frac{(f-F_i)^2}{2\sigma_f^2}). \tag{9}$$

Our purpose in this step is to amplify the true signal and compress the noise. The black line in Fig. 7 is

$H_s(w)$ which can amplify the true signal from 0 to 0.06 $\frac{rad}{s}$ and compress noise from 0.06 to π. In this case, if we use $N_i = [1\ 2\ ...\ 9]$ we can get the summarized filter represented by the blue line in Fig. 7. Fig. 9 shows the frequency response of the 19th raw MS signal (blue line) and that of full frequency signal (red line). We can see that the signal from 0 to 0.06 is amplified and the noise from 0.06 to π is compressed. Therefore, in both full frequency MS signal A and B, all peaks have been emphasized to help the next peak detection step. In this step, baseline correction is also used and is detailed as follows

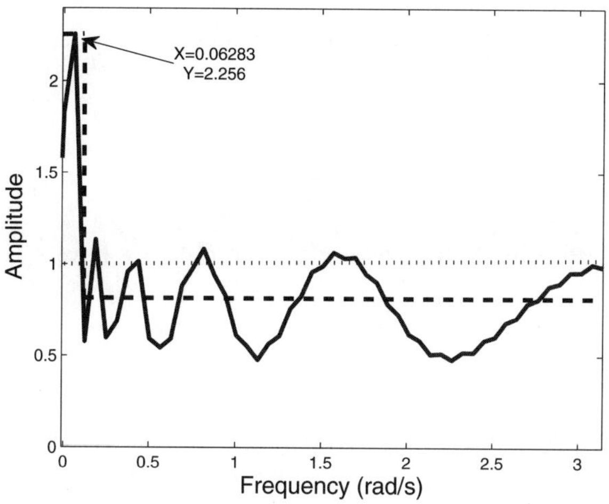

Fig. 7. The frequency response of the summary filter.

Baseline correction The chemical noise or the ion overloading is the main reason causing a varying baseline in mass spectrometry data. Baseline correction is an important step before using Gabor filter to get the full frequency MS signals. The raw MS signal x_{raw} includes some real peaks x_p, the baseline x_b, and the noise x_n.

$$x_{raw} = x_p + x_b + x_n. \tag{10}$$

The baseline correction is used to remove the artifact x_b. In this paper, we use 'msbackadj' function of MATLAB to remove baseline. The msbackadj function estimates a low-frequency baseline first which is hidden among the high-frequency noise and the signal peaks and then subtracts the baseline from the spectrogram. This function follows the algorithms in Andrade *et al.*'s paper [14].

Illustration In order to understand this step easier, one example of the way to create full frequency MS signal is shown in Fig. 8.

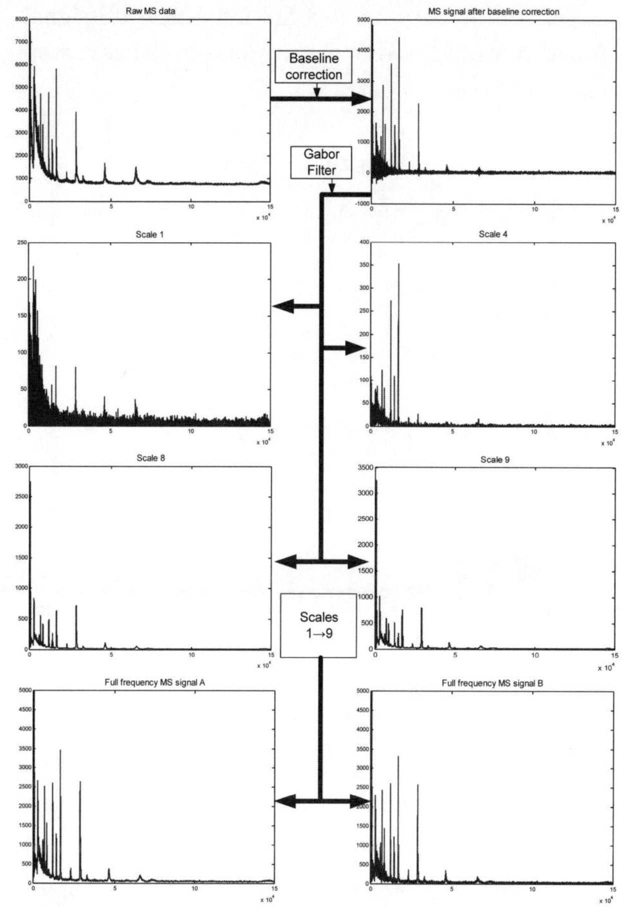

Fig. 8. One example of the step named full frequency MS signal generation. Raw MS data is the 19th MS signal of CAMDA 2006.

In this example, the 19th MS signal of CAMDA 2006 is chosen as raw MS data. After the baseline correction, MS signal is used as the input of the Gabor filters. A Gabor filter bank with 9 non-uniform sub-bands is employed to create 9 MS signals with 9 different frequency sub-bands. In Fig. 8, the signals of scale 1, 4, 8 and 9 are visualized. Some noises in high frequency are separated from the MS signal of the scale 1, 2, ..., 5. In the MS signal under the scales 6, ..., 9, all high intensity peaks are still kept. After combining the MS signals of all scales in two ways, the full frequency MS signal A and B are created. The comparison between the raw MS and full frequency signal in frequency domain is shown in Fig. 9. This figure shows our purpose which amplifies the important signal and compresses the noise has been achieved. We should remember that this is just a compression of noise instead of removing noise. As the outputs, two full frequency MS signal A and B will be used to detect peaks in the next step instead of raw MS data.

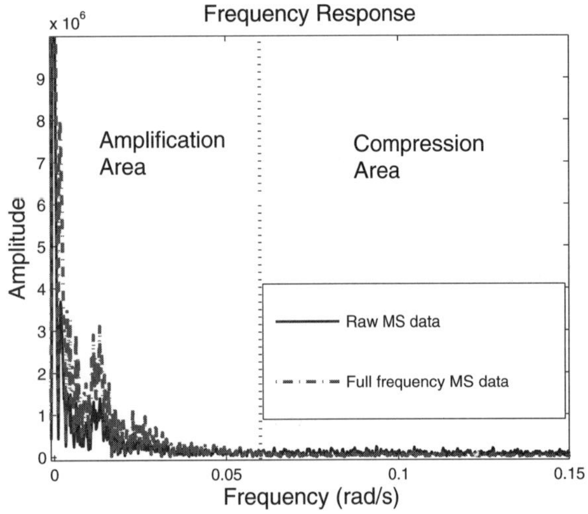

Fig. 9. The frequency response of the 19th MS signal of CAMDA 2006 before and after using the summary filter.

2.2.2. *Peak detection by Gaussian local maxima*

All peaks are detected as many as possible by using Gaussian local maxima with the full frequency MS signal A as well as the full frequency MS signal B. The Gaussian local maxima is used instead of local maxima because Gaussian local maxima is robust with noise in peak detection. Before detecting peaks, pre-processing step is also applied such as elimination peaks in the low-mass region. Now, the Gaussian local maxima will be introduced as follows:

Gaussian local maxima We assume that we want to find local maxima of $y(t)$. We should follow two steps: computing derivative of $y(t)$ and finding zero crossing. The derivative of $y(t)$ is approximated by the finite difference as follows:

$$\frac{d(y(t))}{dt} = \lim_{h->0} \frac{y(t+h) - y(t)}{h} \approx y(t+1) - y(t). \tag{11}$$

At $t = t_0$, if the derivative of $y(t)$ equals to zero and has a change from positive to negative or from negative to positive, we have zero-crossing. If the derivative of $y(t)$ changes from positive to negative at t_0, we have local maxima at t_0. With discrete signal, (11) can be rewritten as follows

$$\frac{d(y(n))}{dn} = y(n+1) - y(n) = y(n) * [1 \quad -1]. \tag{12}$$

Unfortunately, MS data always have noise. Thus, we assume that Gaussian filter $g(t, \sigma)$ is used to handle the denoise in MS data (this is not a denoising step). Finally, derivative of $y(t) * g(t, \sigma)$ will replace the derivative of $y(t)$ as follows

$$\frac{d(y(t) * g(t,\sigma))}{dt} = \frac{d(\int (y(\tau).g(t-\tau,\sigma)d\tau))}{dt}$$

$$= \int (y(\tau).\frac{d(g(t-\tau,\sigma))}{dt} d\tau) = y(t) * \frac{d(g(t,\sigma))}{dt}, \tag{13}$$

where

$$g(t,\sigma) = \exp(-\frac{t^2}{2\sigma^2}). \tag{14}$$

Taking the derivative of $g(t, \sigma)$ in (14), we have

$$\frac{d(g(t,\sigma))}{dt} = \frac{-t}{\sigma^2} \exp(-\frac{t^2}{2\sigma^2}). \tag{15}$$

From (13) and (15), we have

$$\frac{d(y(t) * g(t,\sigma))}{dt} = y(t) * (\frac{-t}{\sigma^2} \exp(-\frac{t^2}{2.\sigma^2})). \tag{16}$$

Instead of finding zero crossing of $\frac{d(y(t))}{dt}$, we find zero-crossing of $\frac{d(y(t)*g(t,\sigma))}{dt}$ by (16). With discrete signal, (16) can be rewritten as follows

$$\frac{d(y(n) * g(n,\sigma))}{dn} = y(n) * v(n), \tag{17}$$

where $v(n)$ is listed in Table 1. Using Gaussian filters makes the Gaussian local maxima method more robust with noise.

2.2.3. *Peak quantification by maximum rank*

After detecting many peaks in full frequency MS signals, a new signal is obtained from these peaks. This new signal will be the input of the next peak detection loop where the Gaussian local maxima method is also applied. Then, many loops are repeated until the number of peaks obtained is less than a threshold. Now, we define the maximum rank of peaks as follows:

Table 1. The value of vector v(n) with different lengths.

length	n = 1	2	3	4	5	6	7	8	9	10
5	0.0007	0.2824	0	-0.2824	-0.0007					
6	0.0007	0.1259	0.7478	-0.7478	-0.1259	-0.0007				
7	0.0007	0.0654	0.6572	0	-0.6572	-0.0654	-0.0007			
8	0.0007	0.0388	0.4398	0.6372	-0.6372	-0.4398	-0.0388	-0.0007		
9	0.0007	0.0254	0.2824	0.7634	0	-0.7634	-0.2824	-0.0254	-0.0007	
10	0.0007	0.0180	0.1851	0.6572	0.5329	-0.5329	-0.6572	-0.1851	-0.0180	-0.0007

Table 2. Definition of maximum rank of peaks. Y means that the peak can be detected at that loop. N means that the peak can not be detected at that loop. The peak with the maximum rank equaling to 1 is able to be detected at all of the loops. The peak with the maximum rank equaling to n only appeared at the first loop.

Maximum Rank	*Loop* 1	*Loop* 2	*Loop* 3	*Loop* 4	... *Loop* $(n-1)$	*Loop* n
1	*Y*	*Y*	*Y*	*Y*	... *Y*	*Y*
2	*Y*	*Y*	*Y*	*Y*	...*Y*	*N*
...	...	...	...	...	...	...
n	*Y*	*N*	*N*	*N*	...*N*	*N*

Maximum rank We assume n loops are used and get m_1 peaks at the loop 1, m_2 peaks at loop 2,...and m_n peaks at the loop n. We have $m_1 > m_2 > ... > m_n$. Maximum peak (MR) is defined as in Table 2.

We have m_n peaks with $MR = 1$, $m_{n-1} - m_n$ peaks with $MR = 2$,...and $m_1 - m_2$ peaks with $MR = n$. In our algorithm, the probability of the true peaks with $MR = i$ is higher than with $MR > i$.

Demonstration Fig. 10 shows an example of the step named the peak quantification by using the maximum rank. First, the full frequency MS signal A is used to detect peaks by using Gaussian local maxima. At the loop 1, we can detect 1789 peaks. From these 1789 peaks, we create a new signal with 1789 positions. At the next loops 2, 3, 4, we can detect 509, 143, 39 peaks, respectively. At the loop 5, 15 peaks can be detected. Because we choose a threshold of 16 and $number\ of\ peaks = 15 < 16$, we stop at the loop 5. Actually, we can select the threshold from 38 to 16 and also get 15 peaks at the final loop. Now, we get 15 peaks with $MR = 1$, $39 - 15 = 24$ peaks with $MR = 2$, $143 - 39 = 104$ peaks with $MR = 3$, $509 - 143 = 366$ peaks with $MR = 4$ and $1789 - 509 = 1280$ peaks with $MR = 5$. In this case, we only keep 15 peaks with $MR = 1$. We also do the same on the full frequency MS signal B and can get 12 peaks with $MR = 1$ at the last loop.

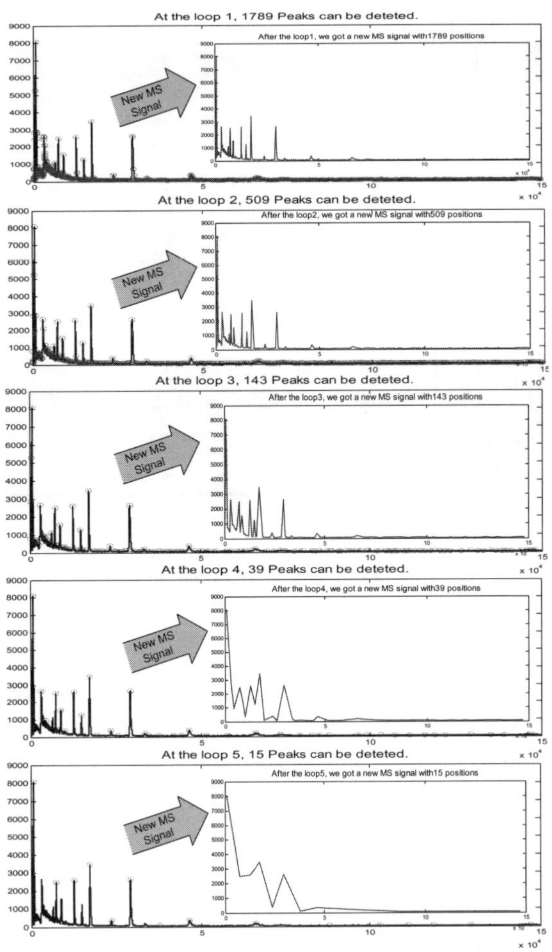

Fig. 10. One example of the step named peak detection and quantification.

2.2.4. *Intersection*

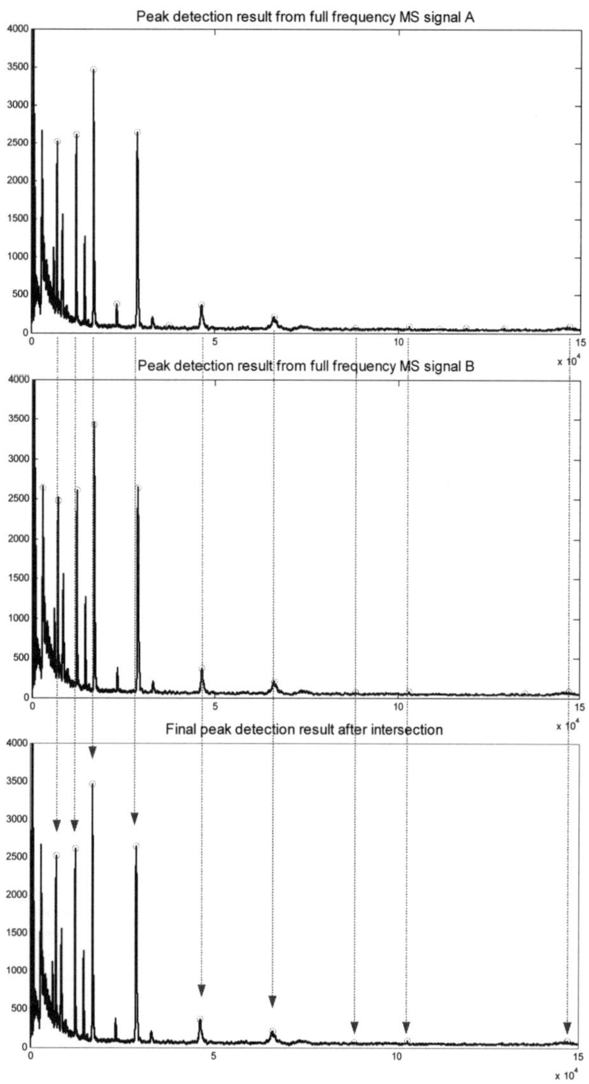

Fig. 11. One example of the step named intersection.

Now, we have two results of peak detection from two full frequency MS signals. The intersection of two above results will be the final result. For example, Fig. 11 shows how to do the intersection of two results. We have 15 peaks in the signal A and 12 peaks in the signal B but we just get 9 peaks as the final result. With this result, we get 7 true peaks and 2 false peaks. This result shows that the true position rate equal to $\frac{7}{7} = 1$ and the false position rate equal to $\frac{2}{9} \approx 0.22$.

3. EXPERIMENTAL RESULTS AND DISCUSSIONS

In this section, our GaborLocal method will be compared to two other most common used methods: the Cromwell [4, 5] and the CWT [6]. We will evaluate the performance of three methods by using the ROC curve that is the standard criterion in this area.

3.1. Cromwell Method

Cromwell method is implemented as a set of MATLAB scripts which can be downloaded from [15]. The algorithms and the performance of the Cromwell were described in [5, 4]. The main idea of the Cromwell method can be summarized as follows

(a) Denoise the individual spectrum using the undecimated discrete wavelet transform. The hard thresholding method was used to reset small wavelet coefficients to zero. In these papers, the authors used the median absolute deviation (MAD) to estimate the thresholding.
(b) Estimate and remove the baseline artifact by using a monotone local minimum curve on the smoothed signal.
(c) Normalize the spectrum by dividing the total ion current, defined to be the mean intensity of the denoised and baseline corrected spectrum.
(d) Identify peaks by using local maxima and signal to noise ratio (SNR).
(e) Match peaks across spectrum and quantify peaks using either the intensity of the local maximum or computing the area under the curve for the region defined to be the peaks.

3.2. CWT Method

The algorithm of CWT method has been implemented in R (called as 'MassSpecWavelet') and the Version 1.4 can be downloaded from [16]. This method was proposed by Pan Du *et al.* [6] in 2006 and can be summarized as follows:

(a) Identify the ridges by linking the local maxima. Continuous wavelet transform (CWT) is used to create many scales from one mass spectrum. The local maxima at each scale is detected. The next step is to link these local maxima as lines.

(b) Identify the peaks based on the ridge lines. There were three rules to identify the major peaks. They are the scale with the maximum amplitude on the ridge line, the SNR being larger than a threshold and the length of ridge being larger than a threshold. We should notice that the SNR is estimated in the wavelet space. This is a nice motivation of this method.
(c) Refine the peak parameter estimation.

3.3. Evaluation Using ROC Curve

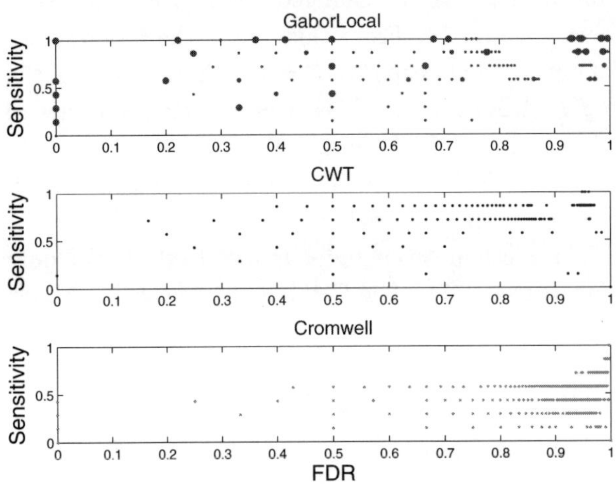

Fig. 12. Detailed receiver operating characteristic (ROC) curves obtained from 60 MS signals using Cromwell, CWT, and our GaborLocal method. The sensitivity is the true position rate.

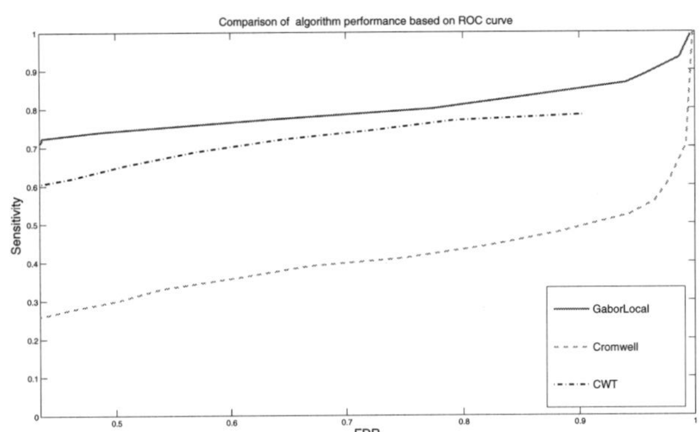

Fig. 13. Average receiver operating characteristic (ROC) curves obtained from 60 MS signals using Cromwell, CWT, and our GaborLocal method. The sensitivity is the true position rate.

The CAMDA 2006 dataset [13] of all-in-1 Protein Standard II (Ciphergen Cat. $\# C100-007$) is used to evaluate three algorithms: the Cromwell, the CWT, and our method. Because we know polypeptide composition and position, we can estimate the true position rate (TPR) and the false position rate (FPR). Another advantage of this dataset is that they are real data and better than the simulated data in evaluation.

The TPR is defined as the number of identified true peaks divided by the total number of true peaks. The FPR is defined as the number of falsely identified peaks divided by the total number of identified peaks. We call an identified peak as true peak if it is located within the error range of 1% of the known m/z value of true peaks. There are seven polypeptides which create seven true peaks at 7034, 12230, 16951, 29023, 46671, 66433 and 147300 of the m/z values. Fig. 12 shows the TPR and the FPR of three above methods with an assumption that there is only one charge. To calculate the ROC curve of Cromwell and CWT methods, the SNR thresholding values are changed. The SNR thresholding values are chosen from 0 to 20 for Cromwell method, from 0 to 65 for CWT method. In our GaborLocal method, the threshold of number of peaks is changed from 2000 to 10 to create the ROC curve. In Fig. 12, the performance of Cromwell method is much worse than CWT and our GaborLocal methods. Most of ROC points of Cromwell method locate at the bottom of right corner and most of ROC points of CWT and GaborLocal methods are well placed on the top regions. In our method, some ROC points appear at the top line with $TPR = 1$ and some ROC points go with $TPR = 1$ and $FPR = 0$. However, it does not happen to the CWT. Therefore, GaborLocal is the best one.

If we take the average of those detailed ROC results of Fig. 12, we get the average ROC curve as in Fig. 13. We should notice that we take average of all ROC points with the same SNR threshold (for Cromwell and CWT) and with the same peak threshold (for our method). From Fig. 13, the results of our method and CWT are much better than the Cromwell's one. Therefore, the decomposing approach without smoothing (both SWT and

GaborLocal) is more efficient than the denoising approach (like Cromwell). At the same FPR, the TPR of our method is consistently higher than the TPR of CWT. Because the maximum rank was used to identify peaks in the GaborLocal method instead of the SNR. It is clear that the utilizing maximum rank to identify peak gives out valuable results. This method has a significant contribution to detect both high energy and small energy peaks. The other advantage of this method is that the threshold of number of peaks can be created easier than the SNR. Therefore, the GaborLocal method is an more efficient and accurate method for real MS data peak detection.

3.4. Examples

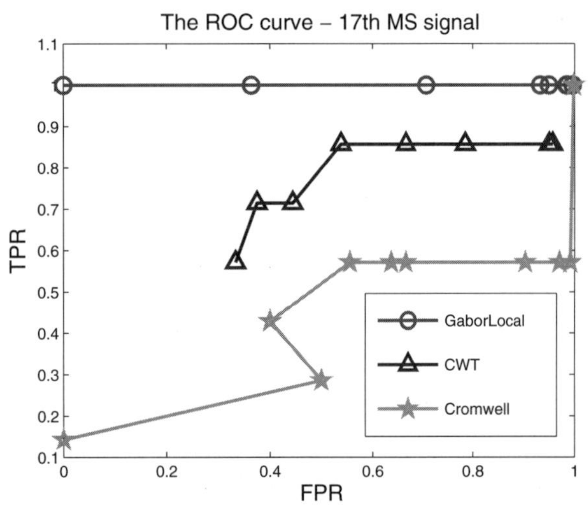

Fig. 14. The ROC curve of three methods such as Cromwell, CWT, and our method with the 17th mass spectrum signal

Now, we study one example shown in Fig. 15 in which the 17th spectrum signal of CAMDA 2006 dataset is picked and tested with three above methods. Fig. 15 (a) includes four sub-figures. The first sub-figure describes the real peak positions and raw data. The second and third sub-figure show the full frequency MS signal A&B with identified peaks. The last sub-figure is the final result after doing intersection.

We get 12 peak candidates from the full frequency MS signal A and 10 peak candidates from the full frequency MS signal B. Finally, we get 7 peak candidates after intersection of 12 and 10 peaks. In the result, our method can detect exactly 7 peaks over 7 true peaks. Fig. 15 (b) shows 9 detectable peaks from CWT method. Among 9 above peaks, there are only 5 true peaks. The CWT loses two peaks at 7034 and 147300 of the m/z values. Fig. 15 (c) shows the result of Cromwell's method. There are three true peaks being detectable by this method. Some peaks with low SNRs can not be detected. Of course, if we decrease the SNR threshold, more peaks can be detected. However, we also get more false peaks and the FPR will be increased dramatically. In general, if the thresholding values of three above methods are changed, we can get the ROC curve in Fig. 14. From this figure, the performance of our method keeps $TPR = 1$ with any value of the FPR (from 1 to 0). However, the TPR's values of Cromwell and CWT methods decrease very quickly when the FPR's value decreases. At the $FPR = 0$, the TPR of Cromwell method equals 0.1429. In CWT method, even the $FPR \approx 1$, the TPR only equals to 0.8571. The CWT and Cromwell methods are limited in peak detection performance because of the way using the SNR to identify peaks. Fig. 14 and Fig. 15, we can prove that

(1) Decomposition of MS data makes peak detection easier.
(2) Using SNR to identify peaks can not detect low SNR peaks.
(3) Using the MR can detect more true peaks than using the SNR.

4. CONCLUSION

In this paper, we proposed a new approach to solve peak detection problem in MS data with promising results. Our GaborLocal method combines the Gabor filter with Gaussian local maxima approach. The maximum rank method is presented and used at the first time to replace the previous SNR method to identify true peaks. With real MS dataset, our method gave out a much better performance in the ROC curve compared to two other most common used peak detection methods. In our future work, we will develop new protein identification method based on our GaborLocal approach.

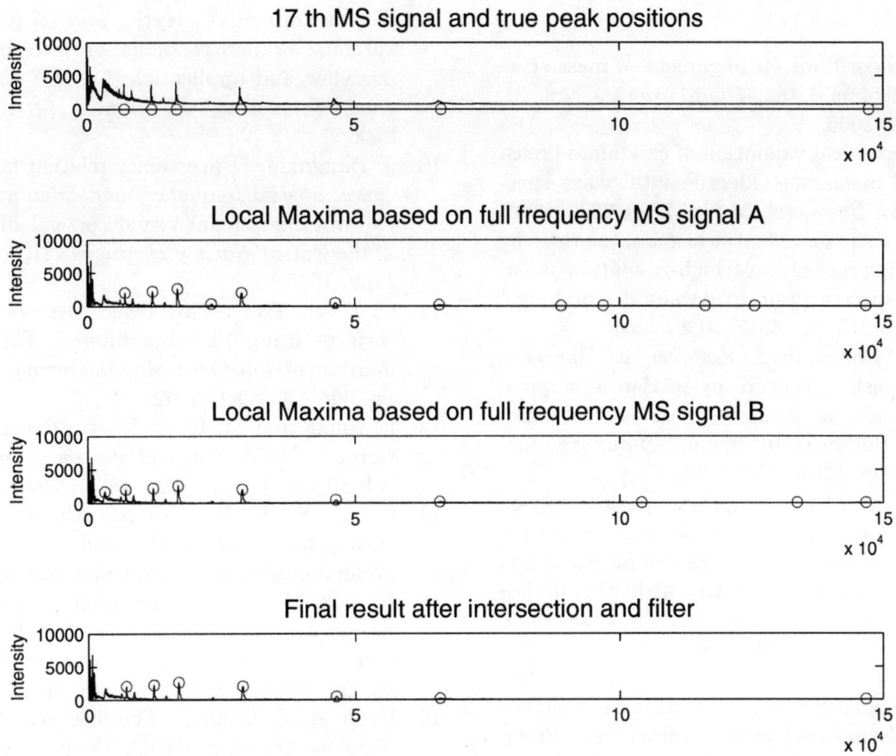

(a)

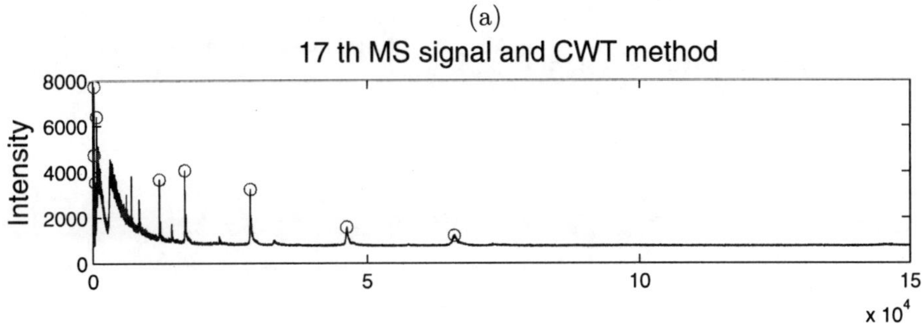

(b)

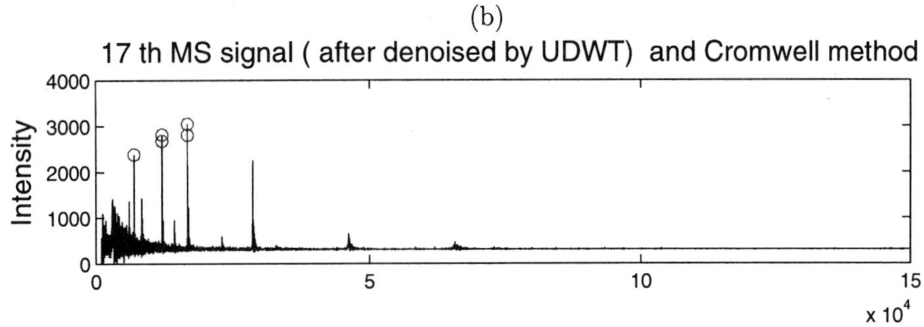

(c)

Fig. 15. Example of peak detection of the 17th mass spectrum signal using (a) our GaborLocal method, (b) CWT method, and (c) Cromwell method.

References

1. N. Jeffries, "Algorithms for alignment of mass spectrometry proteomic data," *Bioinformatics*, vol. 21, pp. 3066–3073, 2005.
2. J. e. Li, "Independent validation of candidate breast cancer serum biomarkers identified by mass spectrometry," *Clin Chem*, vol. 51, pp. 2229–2235, 2005.
3. T. e. Rejtar, "Increased identification of peptides by enhanced data processing of high-resolution maldi tof/tof mass spectra prior to database searching," *Anal Chem*, vol. 76, pp. 6017–6028, 2004.
4. J. Morris, K. Coombes, J. Koomen, K. Baggerly, and R. Kobayashi, "Feature extraction and quantification for mass spectrometry in biomedical applications using the mean spectrum," *Bioinformatics*, vol. 21, no. 9, pp. 1764–1775, 2005.
5. K. Coombes and *et al.*, "Improved peak detection and quantification of mass spectrometry data acquired from surface-enhanced laser desorption and ionization by denoising spectra with the undecimated discrete wavelet transform," *Proteomics*, vol. 5, no. 16, pp. 4107–4117, 2005.
6. P. Du, W. Kibble, and S. Lin, "Improved peak detection in mass spectrum by incorporating continuous wavelet transform-based pattern matching," *Bioinformatics*, vol. 22, no. 17, pp. 2059–2065, 2006.
7. E. Lange and *et al.*, "High-accuracy peak picking of proteomics data using wavelet techniques," in *Processdings of Pacific Symposium on Biocomputing*, 2006, pp. 243–254.
8. D. Gabor, "Theory of communication," *J. Inst. Elec. Engr*, vol. 93, no. 26, pp. 429–457, Nov 1946.
9. J. Kamarainen, V. Kyrki, and H. Kalviainen, "Invariance properties of gabor filter-based features-overview and applications," *IEEE Transactions on Image Processing,*, vol. 15, no. 5, pp. 1088–1099, May 2006.
10. J. Daugman, "Uncertainty relation for resolution in space, spatial frequency, and orientation optimized by two-dimensional visual cortical filters," *Journal of the Optical Society of America A*, vol. 2, pp. 1160–1169, 1985.
11. C. L. D. Tsai, "Fast defect detection in textured surfaces using 1d gabor filters," *The International Journal of Advanced Manufacturing*, vol. 20, no. 9, pp. 664–675, Oct. 2002.
12. I. Young and M. G. L. Vliet, "Recursive gabor filtering," *IEEE Transactions on Signal Processing*, vol. 50, no. 11, pp. 2798–2805, Nov 2002.
13. C. C. F. S. R. Group, "Camda 2006 conference contest datasets." [Online]. Available: http://camda.duke.edu/camda06/datasets/index.html
14. L. Andrade and L. Manolakos, "Signal background estimation and baseline correction algorithms for accurate dna sequencing," *Journal of VLSI, special issue on Bioinformatics*, vol. 35, pp. 229–243, 2003.
15. U. M. A. C. Center, "The new model processor for mass spectrometry data." [Online]. Available: http://bioinformatics.mdanderson.org/cromwell.html
16. P. Du, "Mass spectrum processing by wavelet-based algorithms." [Online]. Available: http://bioconductor.org/packages/2.1/bioc/html/MassSpecWavelet.html

Structural Bioinformatics

OPTIMIZING BAYES ERROR FOR PROTEIN STRUCTURE MODEL SELECTION BY STABILITY MUTAGENESIS

Xiaoduan Ye[1], Alan M. Friedman[2*], and Chris Bailey-Kellogg[1*]

[1] *Department of Computer Science, Dartmouth College*

[2] *Department of Biological Sciences, Markey Center for Structural Biology, Purdue Cancer Center, and Bindley Bioscience Center, Purdue University*

Site-directed mutagenesis affects protein stability in a manner dependent on the local structural environment of the mutated residue; e.g., a hydrophobic to polar substitution would behave differently in the core vs. on the surface of the protein. Thus site-directed mutagenesis followed by stability measurement enables evaluation of and selection among predicted structure models, based on consistency between predicted and experimental stability changes ($\Delta\Delta G°$ values). This paper develops a method for planning a set of individual site-directed mutations for protein structure model selection, so as to minimize the Bayes error, i.e., the probability of choosing the wrong model. While in general it is hard to calculate exactly the multi-dimensional Bayes error defined by a set of mutations, we leverage the structure of "$\Delta\Delta G°$ space" to develop tight upper and lower bounds. We further develop a lower bound on the Bayes error of any plan that uses a fixed number of mutations from a set of candidates. We use this bound in a branch-and-bound planning algorithm to find optimal and near-optimal plans. We demonstrate the significance and effectiveness of this approach in planning mutations for elucidating the structure of the pTfa chaperone protein from bacteriophage lambda.

1. INTRODUCTION

With the extensive development of genome projects, more and more protein sequences are available. Unfortunately, while structural genomics efforts have greatly expanded the set of experimentally determined protein structures, the Protein Data Bank (PDB) still has entries for only about 1% of the protein sequences in UniProt. A significant part of the gap between the sequence and structure determination lies with difficulties in crystallization; among the 75104 targets (45391 cloned) in phase one of the Protein Structure Initiative, only 3311 crystallized and only 1307 of these crystals provided sufficient diffraction[1]. At the same time, it has been suggested that only a small number (perhaps a few thousand[2]) of distinct structural organizations, or "folds," exist among naturally-occurring proteins, and many of them can already be found in the current PDB[3]. Therefore, structure *elucidation* (as opposed to experimental structure *determination*) may soon devolve to selecting the correct model among those generated from existing templates.

Since many more proteins are available for structural studies than can be handled by crystallography, we have been developing integrated computational-experimental methods that use relatively rapid, targeted biochemical/biophysical experiments to select among predicted structure models, based on consistency between predicted and observed experimental measurements[4]. Purely computational protein structure prediction methods[5–8] can often produce models close to the correct structure. However, as the series of Critical Assessment of Structure Prediction (CASP) contests illustrates[9], it remains difficult for any method to always select the correct model, particularly in cases where low sequence identity to templates precludes homology modeling. The best model is often among a pool of highly ranked models, but not necessarily the highest-ranked one. Furthermore, different methods often employ different scoring functions and yield different rankings for the same models. Thus using rapid, planned experiments to select the correct one(s) from a given set of predicted models combines the strengths of both computation and experimentation.

This paper focuses on an approach we call "stability mutagenesis," which exploits the relationship between protein structure and thermodynamic stability to perform model selection. A number of methods[10–15] are available for predicting changes

*Contact authors. CBK: 6211 Sudikoff Laboratory, Hanover, NH 03755, USA; cbk@cs.dartmouth.edu. AMF: Lilly Hall, Purdue University, West Lafayette, IN 47907, USA; afried@purdue.edu.

in unfolding free energy (ΔΔG°) upon site-directed mutagenesis (i.e., substitution of one amino acid for another at a specific position). These prediction methods provide good accuracy in the aggregate or for defined subsets of mutations, e.g., the FOLD-X method achieved a global correlation of 0.83 between the predicted and experimental ΔΔG° values for 95% of more than 1000 point mutations, with a standard deviation of 0.81 kcal/mol[13]. Since different structure models place some of their equivalent residues in different environments, they yield different predicted ΔΔG° values for those residues. The consistency between predicted and experimentally determined ΔΔG° values thus allows selecting the correct model(s) from a given set.

This paper develops a method for planning the most informative stability mutagenesis experiments for selecting among a given set of protein structure models. In particular, we seek to minimize the expected probability of choosing a wrong model, i.e., the Bayes error. It is difficult to compute exactly the Bayes error in multiple dimensions (here, for sets of mutations), and the general problem of estimating and bounding it has received considerable attention[16–19]. We take advantage of the particular structure of our mutagenesis planning problem in order to derive tight upper and lower bounds on the Bayes error for "ΔΔG° space." In order to efficiently find an optimal set of mutations, we develop a lower bound on the Bayes error of any plan that uses a fixed number of mutations from a set of candidates, along with a branch-and-bound algorithm to identify optimal and near-optimal plans.

2. METHODS

2.1. Bayes Error Bounds

Let $S = \{s_1, s_2, ...s_n\}$ be a given set of predicted protein structure models, and X be a vector of random variables representing the experimental ΔΔG° values with Normal errors (as is standardly assumed). Then each model can be represented as a conditional distribution in the "ΔΔG° space" (Fig. 1), in which each dimension has the ΔΔG° value for one mutation. That is,

$$p(X|s_i) = \mathcal{N}(\boldsymbol{\mu}_i, \sigma^2 I) \tag{1}$$

where $\boldsymbol{\mu}_i$ is the vector of expected ΔΔG° values for model i, and the variance $\sigma^2 I$ (where I is the identity matrix) is mutation independent and model independent.

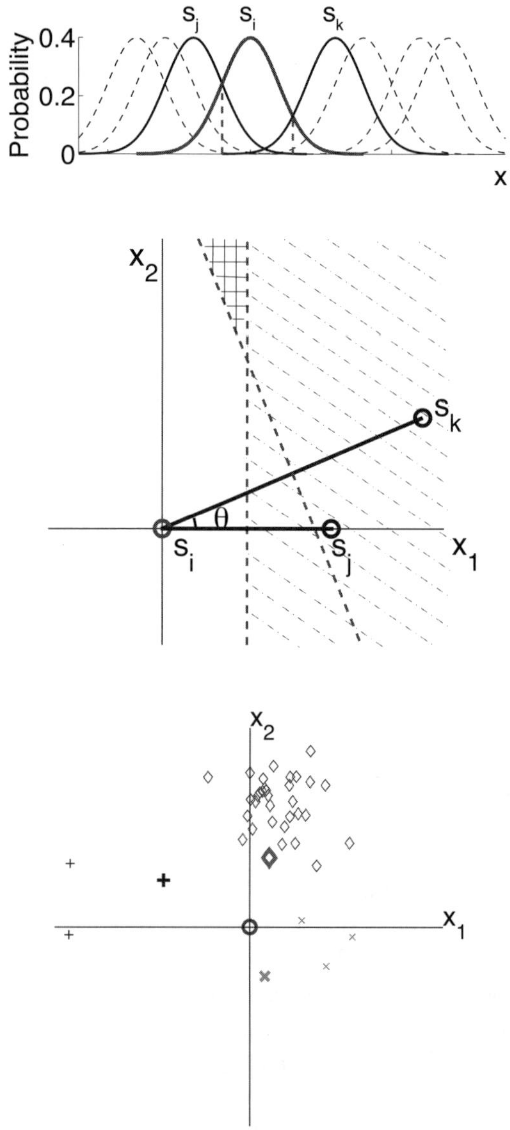

Fig. 1. Intuition for upper bound on ε_i, the Bayes error conditioned on model s_i. (top) In the 1D case, ε_i is determined by s_j and s_k, the closest neighbors on each side of s_i, with no effect from other models (dashed curves). (middle) In higher-dimensional cases, multiple models are unlikely to be collinear. However, if the angle between $\overrightarrow{s_i s_j}$ and $\overrightarrow{s_i s_k}$ is small and s_k is not closer to s_i than s_j is, adding s_k will only increase ε_i by a small amount, the integral of $p(X|s_i)$ over the "#" shaded area. (bottom) To find representative models that are "closest" to s_i, other models are represented as vectors from s_i and hierarchically clustered w.r.t. their angles. Here there are three clusters (different markers), each represented by the model closest to s_i (bold markers) for the purposes of error bounds.

Once the experimental ΔΔG° values have been measured, we will choose the model with the maximum posterior probability. In considering a possible set of mutations during experiment planning, we don't know what the resulting experimental ΔΔG° values will be. Thus we integrate over the possible values, computing the Bayes error ε, formulated as:

$$\varepsilon = \sum_{i=1}^{n} P(s_i)\varepsilon_i \tag{2}$$

where $P(s_i)$ is the prior probability of model s_i and ε_i is the conditional error given that model s_i is correct. By "correct" we mean that the distribution of the measurements X w.r.t. this model is very similar to that for the "true" protein structure. For simplicity, we assume a uniform prior for models, but all discussion applies to the case of non-uniform priors. We define ε_i as:

$$\varepsilon_i = P_i\{p(X|s_i) < \max_{j\neq i} p(X|s_j)\} \tag{3}$$

Here and in the following we use notation $P_i\{e\}$ for the probability of predicate e w.r.t. model s_i:

$$P_i\{e\} = \int p(X|s_i) \cdot I\{e\} \mathrm{d}X \tag{4}$$

where the integral is taken over all possible ΔΔG° values and the indicator function $I\{e\}$ returns 1 if predicate e is true, or 0 if false. The predicate in Eq. 3 evaluates whether a wrong model is selected because the experimental data X is more consistent with it than with the correct model. Weighted integration over all possible datasets then calculates the total probability of error.

Straightforward initial bounds on ε_i can be derived from the Bonferroni inequalities[20]:

$$\varepsilon_i \le \sum_{j\neq i} P_i\{p(X|s_i) < p(X|s_j)\} \tag{5}$$

$$\varepsilon_i \ge \sum_{j\neq i} P_i\{p(X|s_i) < p(X|s_j)\} - \sum_{j<k\neq i} P_i\{p(X|s_i) < \min(p(X|s_j), p(X|s_k))\} \tag{6}$$

The union bound (Eq. 5) evaluates the probability that at least one of the wrong models beats the correct one, which is at most the sum of the probabilities of each individual wrong model beating the correct one. Eq. 6 subtracts out the potential "double-counting" in the union bound, when multiple wrong models beat the correct one but some are better than others. Both bounds are easy to calculate, but are too loose for our purposes here.

Since we assume a common variance for all mutations in all models, the error probability is completely determined by the relative distances among the distribution means. The top and middle panels of Fig. 1 illustrate that in cases where the means are nearly collinear, ε_i is much less than the sum of the individual error probabilities (i.e., the union bound). Conditioning on model s_i, we shift the coordinate system so that $\boldsymbol{\mu}_i$ is at the origin and the rest of the models are represented as vectors from the origin. We cluster these vectors (Fig. 1, bottom) into disjoint sets C_t for $t = 1, 2, \ldots$. We discuss our clustering method below, but for any set of clusters, the following inequality holds:

$$\varepsilon_i \le \sum_{t} P_i\{p(X|s_i) < \max_{j\in C_t} p(X|s_j)\} \tag{7}$$

The difference between Eq. 7 and Eq. 5 is that in Eq. 7 the Bonferroni inequality is applied on clusters instead of individual models. Choosing a representative model s_{j_t} from cluster C_t, we have

$$\begin{aligned} &P_i\{p(X|s_i) < \max_{j\in C_t} p(X|s_j)\} \\ &= P_i\{p(X|s_i) < p(X|s_{j_t})\} \\ &\quad + P_i\{p(X|s_{j_t}) < p(X|s_i) < \max_{j\in C_t, j\neq j_t} p(X|s_j)\} \quad (8) \\ &\le P_i\{p(X|s_i) < p(X|s_{j_t})\} \\ &\quad + \sum_{j\in C_t, j\neq j_t} P_i\{p(X|s_{j_t}) < p(X|s_i) < p(X|s_j)\} \quad (9) \end{aligned}$$

Eq. 8 is just a rewriting of the probability; either model s_{j_t} beats s_i or some other models in cluster C_t beat it. Eq. 9 is obtained by applying the union bound on the second term of Eq. 8, where the first and second terms correspond to the integral of $p(X|s_i)$ over the stripe-shaded area and the "#" shaded area in the middle panel of Fig. 1, respectively.

Turning to the lower bound, we note that Eq. 6 could be very loose (even negative) if models are highly dependent, because the number of pairwise joint probabilities subtracted out could be much larger than the number of individual probabilities added in. For example, consider a variation of the top panel in Fig. 1 where s_j and the models to the left of it have nearly identical distributions and similarly for s_k and the models to the right of it, such that $P_i\{p(X|s_i) < p(X|s_j)\} \approx \epsilon$ for all wrong models. Then Eq. 6 gives a lower bound of -2ϵ (one added and two pairs subtracted on each side). However, we can obtain a tighter lower bound by using a subset of the models that are highly independent;

in the example, one from the left and one from the right. More generally, still conditioning on s_i, let $S' \subset S - \{s_i\}$ be a subset of the remaining models. Then

$$\varepsilon_i \geq P_i\{p(X|s_i) < \max_{j \in S'} p(X|s_j)\} \quad (10)$$

$$\geq \sum_{j \in S'} P_i\{p(X|s_i) < p(X|s_j)\}$$

$$- \sum_{j<k \in S'} P_i\{p(X|s_i) < \min(p(X|s_j), p(X|s_k))\} \quad (11)$$

Eq. 10 holds because the probability that a model in a superset beats s_i is always at least the probability that a model in a subset does. Eq. 11 is just the Bonferroni inequality applied to S'.

We now have lower and upper bounds that are tighter than simply applying the Bonferroni inequalities. The tightness depends on the choices of clustering method (upper bound) and model subset (lower bound). In fact, we can readily trade off tightness and computation, using more, finer clusters and more trial subsets in order to obtain tighter bounds. For the results presented below, we employ an agglomerative approach to cluster models, with distance between two clusters defined as the maximum angle between any two vectors in them. A cutoff θ determines the number of clusters, and then the model with the smallest distance to s_i in each cluster is selected as the representative model for the cluster (bottom panel of Fig. 1). We also use the representative models as the model subset for the lower bound, because these models are likely to be relatively independent and thus the pairwise joint probabilities are smaller and the lower bound tighter. Since the quality of the bounds depends on the choice of θ and the best choice could be model specific and different for the upper bound and the lower bound, we simply try three different values: $\pi/4$, $\pi/3$, and $\pi/2$. The running time is only three times that of using a fixed cutoff, and we found that the result is significantly improved in practice.

2.2. Planning Algorithm

If there are only a few candidate mutations, or a few are to be selected for a plan, we can enumerate all possible plans, calculate their upper and lower bounds, and choose a good one. In terms of Bayes error, plan A is definitely better than plan B if the upper bound for A is less than the lower bound for B. In practice, the computational complexity of such a brute force method becomes prohibitive for even a modest number of mutations.

In cases where the exhaustive method is infeasible, we can use a greedy approach to minimize the upper bound on the Bayes error—select mutations one by one, minimizing the upper bound on Bayes error at each step. A tight upper bound will allow us to identify a set of selected mutations guaranteed to be of high quality. However, we still do not know how close a plan is to the optimal one, and the greedy plan may be far from optimal.

To evaluate the optimality of a given plan M, we compute a lower bound on how its Bayes error compares to that of the best possible (though unknown) plan that uses the same number of mutations from a set of candidates C:

$$\text{Optimality}(M, C) \geq \frac{lb(C, |M|)}{ub(M)} \quad (12)$$

where $ub(M)$ is the upper bound we previously discussed (Eq. 7, Eq. 9) and we develop below $lb(C, |M|)$, a lower bound on the Bayes error of the optimal plan. An Optimality score close to 1 indicates that the plan is guaranteed to be near optimal. A plan with a lower score may still be good, but we just cannot prove it with our bounds. The Optimality score also supports the branch-and-bound algorithm we develop below: we can ignore all plans chosen from mutations in C if the score in Eq. 12 is greater than 1 for some plan M.

To derive $lb(C, |M|)$, the lower bound on the optimal plan, we start from a lower bound on the Bayes error based on pairwise risk functions developed for multi-hypothesis testing[19]:

$$\varepsilon \geq \left(\frac{2}{n}\right)^2 \cdot \sum_{i=1}^{n-1} \sum_{j=i+1}^{n} \varepsilon_{ij} \quad (13)$$

We can also prove the following lemma.

Lemma 1. *Let $d^2 = \sum_{i=1}^{n} d_i^2$ be the sum of squares of n positive real numbers d_i, $i = 1, 2, ..., n$, and let ε_i be the cumulative density of Normal distribution $\mathcal{N}(0, \sigma)$ at point $-d_i/2$. Then for a fixed value of d^2, $\sum_{i=1}^{n} \varepsilon_i$ is minimized when $d_i = d_j$ for $1 \leq i, j \leq n$.*

Proof. Suppose we can find $d_i = b$ and $d_j = a$ that are not equal, say $0 < b < a$, and let $c = \sqrt{(a^2 + b^2)/2}$ be new equal values for d_i and d_j, so that the sum of squared values d^2 is not changed. ε_i

decreases more than ε_j increases:

$$\frac{1}{\sqrt{2\pi}}\int_{b/2}^{c/2} e^{-\frac{x^2}{2}}\mathrm{d}x > \frac{1}{\sqrt{2\pi}}\int_{c/2}^{a/2} e^{-\frac{x^2}{2}}\mathrm{d}x \qquad (14)$$

This follows from the fact that $c - b > a - c$, so that the first integral region is larger than the second; along with the fact that the density is higher in the first region because it is closer to the mean. Thus equalizing a pair reduces the total error, and if we could equalize all pairs, we would minimize the sum. □

Combining Lemma 1 and Eq. 13, we have

$$\varepsilon \geq \frac{2(n-1)}{n}\int_{-\infty}^{-r} \frac{1}{\sqrt{2\pi}\sigma} e^{-x^2/2\sigma^2}\mathrm{d}x \qquad (15)$$

where $r = \frac{1}{2}\sqrt{d^2/\binom{n}{2}}$ and d^2 is the sum of squared distances among all model distribution means:

$$d^2 = \sum_{i<j}\sum_{k}(\mu_{ki} - \mu_{kj})^2 = \sum_{k}\sum_{i<j}(\mu_{ki} - \mu_{kj})^2 \qquad (16)$$

where μ_{ki} is the predicted $\Delta\Delta G°$ value of the kth mutation according to model s_i. Since the inner sum on the right-hand side of Eq. 16 is for only one mutation, we can easily maximize d^2 by independently choosing mutations according to their sums of squared distances over all models. With d^2 thus maximized, the lower bound in Eq. 15 is minimized and becomes a lower bound for any plan of the same size, including the optimal plan.

With the lower bound on the optimal plan in place, let us now turn to the branch-and-bound search for an experiment plan (set of mutations). First let us define the structure of the search tree we will use. In this tree, a node corresponds to an index into the list of possible mutations. All mutations on the path to the root have been eliminated, all those with smaller indices but not on the path to the root have been selected, and all those with larger indices are still candidates. In the example in Fig. 2, at the starred node (index 4), mutations A2G, F3A and R4A (indices 1, 2 and 4) have been eliminated, mutation F3L (index 3) has been selected, and mutations R4G and M5A (indices 5 and 6) are still candidates.

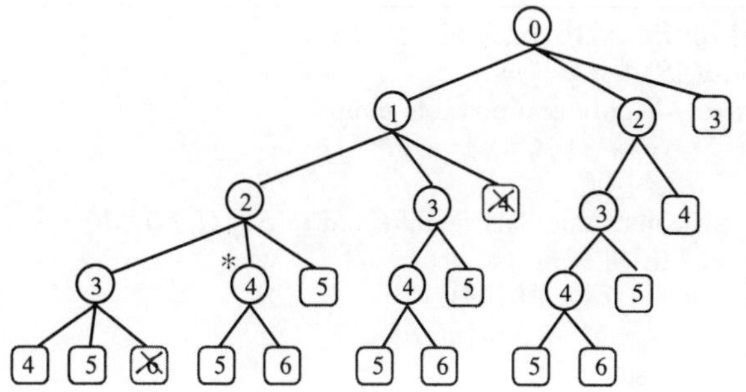

Fig. 2. Example branch and bound search tree structure for choosing two from six mutations at four positions, {A2G, F3A, F3L, R4A, R4G, M5A}. Circles are internal nodes and squares leaf nodes (i.e., plans of the desired size). 'X's indicate violation of a constraint allowing at most one mutation per position.

We now develop our branch-and-bound algorithm, MutPlanBB (Fig. 3), to explore this search tree, using the lower bound on the Bayes error of the optimal plan, along with our earlier upper bound on the Bayes error of a specific plan. The algorithm prunes a node (and its subtree) if the Optimality with respect to the best plan found so far (and thus also to the best one) is more than a user-specified threshold λ. If $\lambda > 1.0$, sub-optimal plans will also be listed. If $\lambda < 1.0$, some good plans might be missed in order to speed up the algorithm, but plans that are "really" good are guaranteed to be kept. For example, if $\lambda = 0.5$, any plan with Bayes error at most half of ub^*, the best upper bound found in the search, cannot be missed. Since ub^* bounds the Bayes error of the optimal plan, this means that we would find all plans with Bayes error no more than half the upper bound on the optimal error. The default value of λ is 1.0, so that all good plans will be listed.

An additional function in the search, *constraintsSatisfied*, checks user-specified constraints, e.g., indicating that at most one mutation per position can be selected. In our branch-and-bound algorithm, we assume that a constraint is monotonic—it is violated by a superset of mutations if it is violated by any subset—as is the case with the one-mutation-per-position constraint. In fact, we can avoid visiting right siblings if any monotonic constraint is violated (a simple modification to the pseudocode in Fig. 3). We can also modify the algorithm in Fig. 3 to handle non-monotonic constraints: only check the satisfaction of constraints on complete plans.

```
MutPlanBB(m, λ, ub*, Ψ, S, C)
  if |S| + |C| = m
    # only one possible plan
    S ← S ⋃ C
    C ← ∅
  if constraintsSatisfied(S) and lb(S ⋃ C, m)/ub* ≤ λ
    if |S| = m
      Ψ ← Ψ ⋃ {S}
      ub* ← min{ub(S), ub*}
    else
      for i from 1 to m − |S| + 1
        # discard C[i] for the i^th child
        S' ← S ⋃ C[1..i − 1]
        C' ← C[i + 1..|C|]
        [ub*, Ψ] ← MutPlanBB(m, λ, ub*, Ψ, S', C')
  return [ub*, Ψ]
```

Fig. 3. Branch and bound algorithm for mutagenesis planning. The inputs include the desired size of plan (m), pruning cutoff (λ), the best upper bound (ub^*) and good plans (Ψ) so far (initially from the greedy approach), and sets of selected and candidate mutations (S and C) at the current node.

There is clearly an exponential number of nodes in the search tree; practical efficiency is attained by effective pruning high up in the tree, so that many nodes need be explicitly visited. As discussed, if desired, the bounds are "tunable"—at more computational cost, we can obtain tighter bounds and thus better pruning. In addition, in order to increase the pruning rate, we initially sort all mutations in ascending order of upper bound on Bayes error, which is easy to calculate in the 1D (single mutation) case. This heuristic[21] structures the search so as to try to exclude good mutations first, so that the error of the remaining mutations is larger, as is the chance of pruning left subtrees, which are larger (see again the tree in Fig. 2). In practice, we found that this reordering improves the pruning rate significantly. Although we can reorder mutations at each level of the search tree, the cost of the sorting may not be worth the benefit, which is not likely to be as significant as the initial sorting.

The cost of visiting an internal node is a table lookup of the Normal cumulative density function (Eq. 15), to compute the lower bound on the optimal plan ($lb(S \bigcup C, m)$). Visiting a leaf node is more expensive, as it requires computing the upper bound ($ub(S)$), by numerical integration in 2D space (Eq. 9).

2.3. Accounting for Bias

While $\Delta\Delta G°$ predictors are based on general determinants of protein stability, some proteins are naturally easier or harder to destabilize than others are. This could lead to bias in the experimental data, which, without care, could result in selection of the incorrect model. For example, if the plan included mutations in which one model was predominantly predicted to be more destabilized than the others, that model would tend to be favored if the protein were relatively easy to destabilize independent of mutation choice. If we knew the bias for a protein, as a single number or a distribution, we could incorporate it into the prediction distribution $p(X|s_i)$. We assume, however, that we only know the range of a constant bias (i.e., a constant offset to $\Delta\Delta G°$, from anywhere in a specified range), because that is a fairly realistic situation in practice.

Conditioning on model s_i (so that its predictions should be biased, as it reflects the native state), its distribution is moved from $\boldsymbol{\mu}_i$ to $\boldsymbol{\mu}'_i = \boldsymbol{\mu}_i + \delta \cdot \mathbf{1}$, for δ in the specified range. Significantly, the error bound expressions (Eq. 9, Eq. 11) are all in terms of only two or three models. Thus, the vector $\overrightarrow{\boldsymbol{\mu}_i\boldsymbol{\mu}'_i}$ can be decomposed into two perpendicular vectors, one parallel and the other orthogonal to line $\boldsymbol{\mu}_i\boldsymbol{\mu}_j$ or plane $\boldsymbol{\mu}_i\boldsymbol{\mu}_j\boldsymbol{\mu}_k$. Since the orthogonal vector does not provide any information for model discrimination (distributions $p(X|s_i)$, $p(X|s_j)$ and $p(X|s_k)$ have identical projections in that direction), such projections lose no information for discrimination. In our implementation, we try bias values within the range $[-2, 2]$ kcal/mol at a resolution of 0.1 and use the worst case (maximum upper bound of Bayes error) as the robustness measurement of a plan. A robust plan will have a biased error bound close to the unbiased one.

3. RESULTS

We employ one representative $\Delta\Delta G°$ prediction method, FOLD-X[13], which predicts stability by developing an empirical effective energy potential including van der Waals, solvation, hydrogen bonding, and electrostatics, and training its parameters and weights using stability data from wild-type and site-directed mutants. We use WHAT-IF[22] to model the mutant structures and version 2.5 of FOLD-X[13] to calculate the stability of mutant and wild-type proteins, and thereby $\Delta\Delta G^\circ$.

In a planning-based framework, we have the luxury of considering only those experiments which we believe to be most reliable. Thus we exclude substitutions involving Cys and Pro, at the first residue position, and in poorly modeled regions. We also adopt FOLD-X's restriction allowing only "non-augmenting" mutations, those whose mutant structures are easy to predict because they involve either a substitution to a smaller sidechain (e.g., Ile → Val) or direct replacement of atoms (e.g., Asp → Asn).

We evaluated the tightness of our bounds using models deposited for a number of different CASP targets. Fig. 4 shows the bounds for four representative test cases from CASP 5, each consisting of the top 10 models by GDT_TS z-score. (Other test cases displayed similar behavior.) Our upper bound is much less than the union bound, and quite close to our lower bound. We have tightly bracketed the Bayes error.

We put our planning mechanism into practice on the pTfa protein of bacteriophage lambda. Lambda pTfa is a small 194 amino acid protein, and, except for our cross-linking work[4, 23], no structural information is available for it or any homolog. The pTfa fragment from residues 1 to 108 forms a stable well-expressed protein that unfolds cooperatively in urea by a two-step mechanism[24]. We previously constructed three high-quality threading models of pTfa 1–108[4], with templates from chaperone DnaK substrate binding domain (PDB id 1dkz), heme chaperone Ccme (PDB id 1liz), and mRNA capping enzyme (PDB id 1ckm). There are altogether 2052 possible substitutions, 19 at each of 108 positions. After applying the restrictions described above, we were left with 192 possible mutations at 77 positions.

Our algorithm first finds a plan by greedily selecting mutations. As Fig. 5 illustrates, the Bayes error has converged fairly well by about 6 mutations (intuitively, 3 mutations distinguishing each pair of models). The Optimality score (Eq. 12) of the six-mutation greedy plan is about 0.6, which means that the Bayes error is within a factor of two of the optimal value. Therefore, we expect a high pruning rate in the branch-and-bound algorithm using this greedy plan as the initial solution. The greedy plan is good in the unbiased case, with a Bayes error of 1.4%. However, we found that with a bias range of $[-2, 2]$ kcal/mol, the Bayes error goes up to 17%.

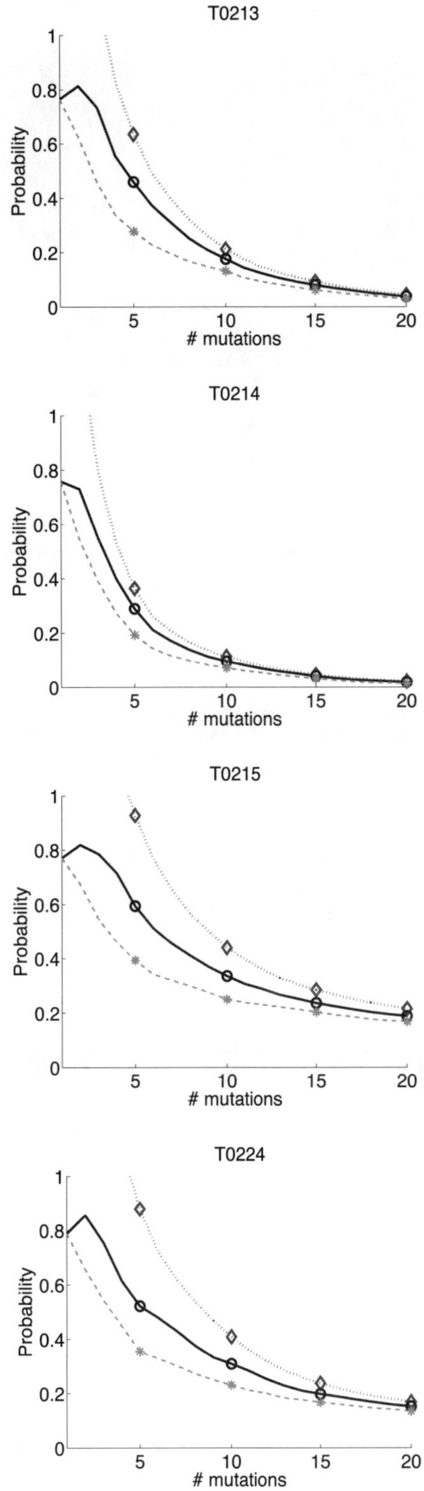

Fig. 4. Error probabilities for greedy plans for four CASP targets: union bound (dotted), tight upper bound (solid), and tight lower bound (dashed).

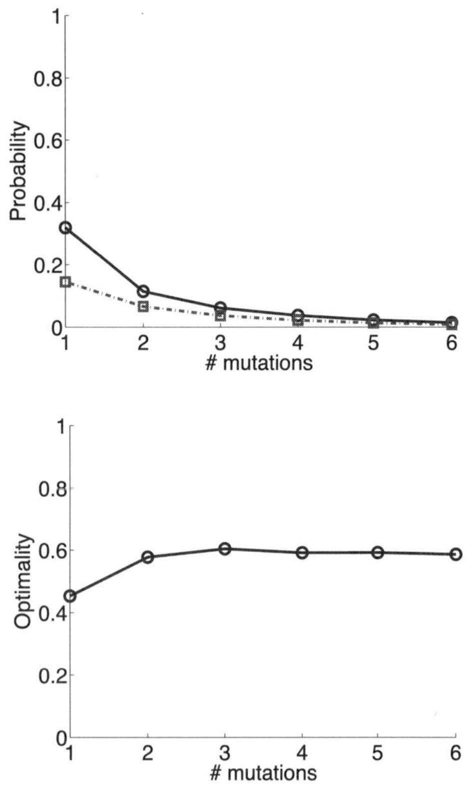

Fig. 5. Greedy plans for three pTfa models. (top) Bayes error of greedy plans (solid line, circles) and lower bound of the optimal plan of the same size (dash-dotted line, squares). (bottom) Optimality, as defined in Eq. 12, of greedy plans.

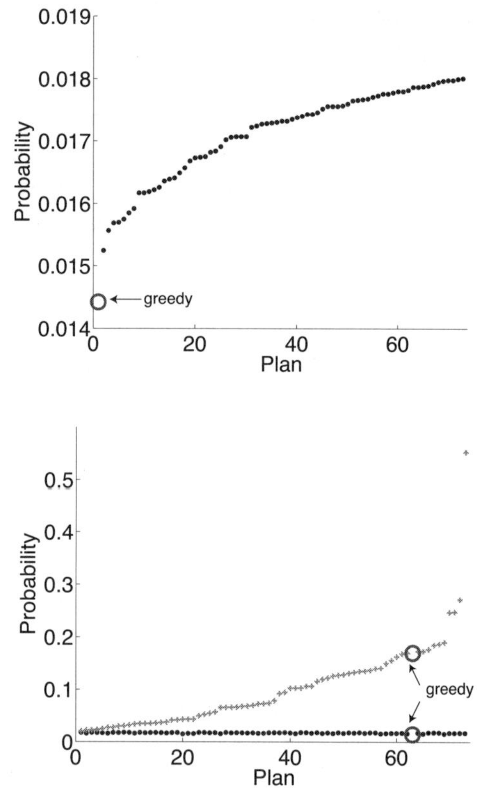

Fig. 6. Bayes error of the six-mutation plans selected by MutPlanBB for three pTfa models. (top) unbiased; (bottom) biased by −2 to 2 kcal/mol. The big circles indicate the Bayes error of the greedy plan in each case.

In order to be robust to a constant bias in how easily pTfa is unfolded, we use our branch-and-bound search to generate a number of good plans, and apply our robustness analysis. With a total of 192 candidate mutations at 77 positions, there are about 5.7×10^{10} possible combinations of six mutations. Our search was much more efficient, visiting a total of 15942 nodes in about 2 hours and identifying 73 good plans at $\lambda = 1$ (the value that ensures finding the optimal plan). Fig. 6 summarizes the Bayes errors for the identified plans, assuming either no bias or bias between −2 and 2 kcal/mol. While the greedy plan happens to be the best if there were no systematic experimental offset, it is much worse in the presence of such possible bias.

Table 1 details three particular plans: the selected plan, the initial greedy plan, and the worst among all plans identified by our branch-and-bound search ("worst-of-bb"). These three plans are comparable without bias, with Bayes errors of 0.018 (selected), 0.014 (greedy), and 0.017 (worst-of-bb). However, the selected plan stands out in the presence of bias, with a significantly smaller Bayes error of 0.020, compared to 0.170 for the greedy and 0.553 for the worst-of-bb. The difference in the presence of bias comes down to a certain "balance" among the selected mutations, in terms of how they discriminate among models. Both D83G (selected) and R10G (greedy) have quite different predictions for the first model and the third model, with a difference of 2.08 kcal/mol for D83G and −2.85 kcal/mol for R10G. Mutation Y77G, common to both plans, also differs for these two models, with a difference of −2.92 kcal/mol. Significantly, this difference has the same sign as that of R10G, but is opposite from

Table 1. Three six-mutation pTfa plans, with the predicted $\Delta\Delta G^\circ$ values for the three models

mut	$\Delta\Delta G^\circ_p$		
N22D	0.68	-3.26	0.02
Y77G	-3.26	0.23	-0.34
T75G	-2.98	-0.75	0.27
F3A	0.15	-0.09	-2.71
D83G	1.73	-1.29	-0.35
T11G	-0.56	0.19	-2.34

selected

mut	$\Delta\Delta G^\circ_p$		
N22D	0.68	-3.26	0.02
Y77G	-3.26	0.23	-0.34
T75G	-2.98	-0.75	0.27
F3A	0.15	-0.09	-2.71
R10G	-1.35	-0.84	1.50
N16D	-0.71	-2.77	-0.17

greedy

mut	$\Delta\Delta G^\circ_p$		
N22D	0.68	-3.26	0.02
Y77G	-3.26	0.23	-0.34
T75G	-2.98	-0.75	0.27
R10G	-1.35	-0.84	1.50
N16D	-0.71	-2.77	-0.17
V79S	-2.45	-2.93	-0.50

worst-of-bb

D83G. Thus a systematic bias would be balanced out in the selected plan but not in the greedy plan.

Two mutations differ between the greedy and the selected plans. In fact, the plans selected by the branch-and-bound search do tend to overlap, as shown in Fig. 7. There are a relatively small number of informative mutations, and the search eliminates the rest while identifying ways to combine the good ones so as to optimize the overall Bayes error.

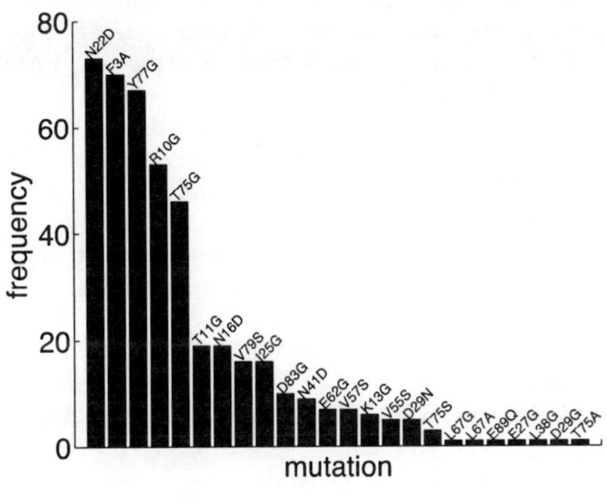

Fig. 7. Frequencies of the 24 unique mutations involved in the six-mutation plans identified by MutPlanBB for three pTfa models.

4. CONCLUSION

Bayes error provides a powerful criterion for evaluating the quality of an experiment plan, assessing how likely we are to make the wrong decision once we have collected the data. Since it is hard to compute Bayes error exactly, we develop here tight error bounds to estimate it for the case of selecting among predicted protein structure models by mutagenesis followed by stability evaluation. We use these error bounds in a branch-and-bound algorithm to optimize experiment plans for model selection. To allow for systematic bias in the experimental data (since proteins vary in how easy or hard they are to destabilize, overall), we consider the predicted performance of possible plans under a range of possible offsets in stability measurement. We demonstrated the tightness of our bounds on several test sets of models, and the effectiveness of our planning mechanism on a system of particular interest to us. Our experimental results for stability mutagenesis will be published separately[24], but we believe the present computational contribution stands on its own as a new solution to the important challenge of planning experiments optimizing Bayes error.

Our approach readily supports several extensions. (1) A mutation may have reliable $\Delta\Delta G^\circ$ predictions in some models but not in all of them. In the calculation of error bounds, what matters is the differences between predictions in different models; thus we set to zero the differences involving missing values, so that they convey no information. (2) In selecting plans, the constraint check can incorporate additional criteria such as the dispersion of selected mutations in the sequence or 3D structure. (3) In a sequential experiment plan, we can seek in each round of experiments to select a "top group" of models rather than a single best; then a subsequent round can focus on selecting among the top models. We can modify our error bounds (Eq. 9 and Eq. 11) so that the correct model will be included in the top group with high probability. To choose a top group of size t, we should ignore the closest $t-1$ neighbors in calculating the error bounds; we will then bound the probability that more than $t-1$ models beat the correct one.

Acknowledgments

This work was supported in part by a grant from NSF SEIII (IIS-0502801) to CBK, AMF, and Bruce Craig.

References

1. Natl. Inst. Gen. Med. Sci. The Protein Structure Initiative. `http://www.structuralgenomics.org`.
2. S. Govindarajan, R. Recabarren, and R. A. Goldstein. Estimating the total number of protein folds. *Proteins*, 35:408–414, 1999.
3. Y. Zhang and J. Skolnick. The protein structure prediction problem could be solved using the current PDB library. *PNAS*, 102:1029–1034, 2005.
4. X. Ye, P. K. O'Neil, A. N. Foster, M. J. Gajda, J. Kosinski, M. A. Kurowski, J. M. Bujnicki, A. M. Friedman, and C. Bailey-Kellogg. Probabilistic cross-link analysis and experiment planning for high-throughput elucidation of protein structure. *Protein Sci.*, 13:3298–3313, 2004.
5. K. T. Simons, C. Kooperberg, E. Huang, and D. Baker. Assembly of protein tertiary structures from fragments with similar local sequences using simulated annealing and Bayesian scoring functions. *J. Mol. Biol.*, 268:209–225, 1997.
6. D. Kihara, H. Lu, A. Kolinski, and J. Skolnick. TOUCHSTONE: an *ab initio* protein structure prediction method that uses threading-based tertiary restraints. *PNAS*, 98:10125–10130, 2001.
7. A. Godzik. Fold recognition methods. *Methods Biochem. Anal.*, 44:525–546, 2003.
8. M. A. Kurowski and J. M. Bujnicki. Genesilico protein structure prediction meta-server. *Nucleic Acids Res.*, 31(13):3305–3307, 2003. `http://genesilico.pl/meta`.
9. A. Kryshtafovych, C. Venclovas, K. Fidelis, and J. Moult. Progress over the first decade of CASP experiments. *Proteins*, 61:225–236, 2005.
10. C. M. Topham, N. Srinivasan, and T. L. Blundell. Prediction of the stability of protein mutants based on structural environment-dependent amino acid substitution and propensity tables. *Protein Eng.*, 10(1):7–21, 1997.
11. D. Gilis and M. Rooman. PoPMuSiC, an algorithm for predicting protein mutant stability changes. application to prion proteins. *Protein Eng.*, 12:849–856, 2000.
12. C. W. Carter Jr., B. C. LeFebvre, S. A. Cammer, A. Tropsha, and M. H. Edgell. Four-body potentials reveal protein-specific correlations to stability changes caused by hydrophobic core mutations. *J. Mol. Biol.*, 311:621–638, 2001.
13. R. Guerois, J. E. Nielsen, and L. Serrano. Predicting changes in the stability of proteins and protein complexes: a study of more than 1000 mutations. *J. Mol. Biol.*, 320:369–387, 2002.
14. V. Parthiban, M. M. Gromiha, and D. Schomburg. CUPSAT: prediction of protein stability upon point mutations. *Nucleic Acids Res.*, 34:W239–W242, 2006.
15. H. Kamisetty, E.P. Xing, and C.J. Langmead. Free energy estimates of all-atom protein structures using generalized belief propagation. In *Proc. RECOMB*, pages 366–380, 2007.
16. D. G. Lainiotis. A class of upper bounds on the probability of error for multihypothesis pattern recognition. *IEEE Trans. Info. Theory*, 15:730–731, 1969.
17. G. T. Toussaint. Bibliograph on estimation of misclassification. *IEEE Trans. Info. Theory*, 20:472–479, 1974.
18. K. Fukunaga and T. E. Flick. Classification error for a very large number of classes. *IEEE Trans. Pattern Analysis and Machine Intelligence*, PAMI-6:779–788, 1984.
19. F. D. Garber and A. Djouadi. Bounds on the bayes classification error based on pairwise risk functions. *IEEE Trans. Pattern Analysis and Machine Intelligence*, 10:281–288, 1988.
20. L. Comtet. *Advanced Combinatorics: The Art of Finite and Infinite Expansions.* Springer, 2001.
21. K. Fukunaga. *Introduction to Statistical Pattern Recognition.* Morgan Kaufmann, second edition, 1990.
22. G. Vriend. WHAT IF: a molecular modeling and drug design program. *J. Mol. Graph.*, 8:52–56, 1990.
23. P. O'Neil et al., in preparation.
24. A.N. Foster et al., in preparation.

FEEDBACK ALGORITHM AND WEB-SERVER FOR PROTEIN STRUCTURE ALIGNMENT

Zhiyu Zhao*
Department of Computer Science, University of New Orleans,
New Orleans, LA 70148, USA
**Email: zzha2@cs.uno.edu*

Bin Fu
Department of Computer Science, University of Texas–Pan American,
Edinburg, TX 78539, USA
Email: binfu@cs.panam.edu

Francisco J. Alanis
Department of Computer Science, University of Texas–Pan American,
Edinburg, TX 78539, USA
Email: fjalanis@gmail.com

Christopher M. Summa
Department of Computer Science, University of New Orleans,
New Orleans, LA 70148, USA
Email: csumma@cs.uno.edu

We have developed a feedback algorithm for protein structure alignment between two protein backbones. A web portal implementing this method has been constructed and is freely available for use at http://fpsa.cs.uno.edu/ with a mirror site at http://fpsa.cs.panam.edu/FPSA/. We compare our algorithm with three other, commonly used methods: CE, DaliLite and SSM. The results show that in most cases our algorithm outputs a larger number of aligned positions when the (C_α) $RMSD$ is comparable. Also, in many cases where the number of aligned positions is larger or comparable, our learning method is able to achieve a smaller (C_α) $RMSD$ than the other methods tested. This trend of larger number of aligned positions and smaller (C_α) $RMSD$ is observed more frequently in cases where the similarity between protein structures is weak.

1. INTRODUCTION

Protein structure alignment attempts to compare the structural similarity between protein backbone chains. A protein molecule can have one or more protein chains, and each chain consists of a series of amino acid residues connected by peptide bonds. Protein structural similarity can be used to infer evolutionary relationships, or in classifying protein structures into more generalized groups. Typically, in protein structure comparison process, each protein chain is represented by an ordered set of 3-D points where each point corresponds to an alpha-carbon (C_α) atom in an amino acid residue. To compare the structural similarity between these "backbone" representations, a protein structure alignment algorithm seeks an optimal transformation by which chains are matched as closely as possible. An alignment is characterized by (1) how many positions are matched, (2) where these positions are, and (3) how well they are matched. (1) and (2) are available once an alignment is determined. For (3), a transformation based alignment algorithm usually calculates (C_α) $RMSD$, the root mean square distance between aligned positions.

The alignment problem is non-trivial – in fact, the problem of finding the optimal global alignment between protein structures has been shown to be NP-hard [12, 6]. Therefore, there have been a number of protein structure alignment algorithms presented in the past years (e.g. Refs. 1, 3, 4, 7, 8, 9, 11, 13, 14, 15, 16, 17, 18, 19, 21, 22, 23, 24, 25), among

*Corresponding author.

them: DALI [7] (distance matrix based method), SSM [11] (secondary structure matching), CE [16] (the combinatorial extension method), and FATCAT [21] (protein structure alignment based on flexible transformation) are commonly used. We have developed a feedback algorithm for pairwise protein structure alignment and our web alignment tool is available for public access. Our algorithm is named SLIPSA, which stands for Self Learning and Improving Protein Structure Alignment. SLIPSA is self learning in that it has a feedback loop which sends the current alignment result back to its input in order to learn a better result in the next stage. In addition, SLIPSA accepts any reasonable upper-bound (C_α) $RMSD$ value as one of the inputs, and outputs an alignment result with an (C_α) $RMSD$ never greater than that value. Like CE, DALI and SSM, the SLIPSA alignment method is based on rigid body transformation, as opposed to flexible transformation-based algorithms such as the one described in Ref. 21.

Our paper is organized as follows: section 2 presents the SLIPSA algorithm; section 3 describes the framework and procedures used in SLIPSA; section 4 reports the experimental results of SLIPSA and compares it with some well known algorithms such as CE, DaliLite (the pairwise version of DALI), and SSM, each of them having a public website; section 4.3 discusses the results and concludes the paper.

2. SLIPSA: AN ALGORITHM WITH FEEDBACK

SLIPSA can be traced to a preliminary algorithm that we reported previously in Ref. 25, but the former has proceeded far beyond the latter in terms of maturity, stability, efficiency and availability. The SLIPSA algorithm first searches all the locally similar sub-chain pairs from two protein backbone chains. Such sub-chain pairs are called local alignments. Next, consistent local alignments are grouped into global alignment candidates called "double-center stars" and a currently optimal global alignment is chosen from all the candidates. Then this output is sent back to its own input in order to learn from itself. We call this a feedback. Such feedback is repeated to obtain improved results, until finally an optimal alignment (i.e. a result with as many as possible aligned C_α pairs and an acceptable (C_α) $RMSD$). SLIPSA can also learn from other algorithms when they are available.

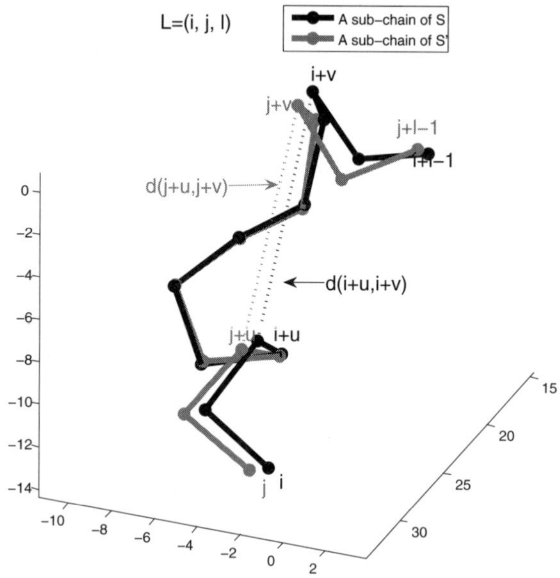

Fig. 1. Local alignment L=(i, j, l)

2.1. General Algorithmic Concepts

As shown in Figure 1, local alignments are discovered by checking the distance difference between corresponding C_α pairs. A local alignment $L = (i, j, l)$ is defined as the longest consecutive stretch of C_α pairs starting from position i in protein backbone chain $S = p_1 \cdots p_n$ and position j in backbone chain $S' = q_1 \cdots q_m$ and having length l, such that $|d(p_{i+u}, p_{i+v}) - d(q_{j+u}, q_{j+v})| \leq 2\epsilon$ for any $0 \leq u, v \leq l-1$ and $u \neq v$, where $d(p, q)$ is the Euclidean distance between two 3-D points p and q, and ϵ is a small constant. A local alignment has to be long enough to make sense.

After local alignments are discovered, they are organized into groups. Ideally, only consistent local alignments should be added to the same group. Suppose there are two local alignments $L_1 = (i_1, j_1, l_1)$ and $L_2 = (i_2, j_2, l_2)$, the point set $P = \{p_{i_1}, \cdots, p_{i_1+l_1-1}, p_{i_2}, \cdots, p_{i_2+l_2-1}\}$ is all the aligned points in the first chain, including those in L_1 and L_2, and $Q = \{q_{j_1}, \cdots, q_{j_1+l_1-1}, q_{j_2}, \cdots, q_{j_2+l_2-1}\}$ is all the aligned points in the second chain, also including those in L_1 and L_2. We say that local alignments L_1 and L_2 are consistent if, after applying a rigid body transformation to Q, the (C_α) $RMSD$ between P and transformed Q is small enough. In other words, if we have a set of local alignments, we conclude that all these local alignments are consistent if all the local alignments share a common rigid body transformation which makes them consistent with each other. Therefore a global alignment can be defined as such a set of consistent local alignments with a common transformation and an acceptable (C_α) $RMSD$.

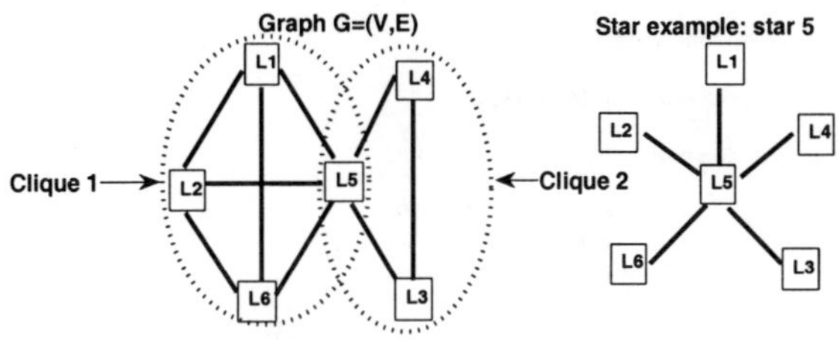

Fig. 2. An example star

The consistency relationship between local alignments can be represented as a graph. Given $A_L = \{L_1, L_2, \cdots, L_w\}$ where each $L_u = (i_u, j_u, l_u)$ $(1 \leq u \leq w)$ is a local alignment. A graph $G = (V, E)$ is defined accordingly, where each local alignment is a vertex of the graph, $V = A_L$ is the vertex set and E is the edge set. Edge $e_{uv}, e_{vu} \in E$ if and only if L_u and L_v are consistent. With this representation, grouping mutually consistent local alignments is equivalent to finding cliques in a graph, which is an NP-complete problem. A possible simplification to this problem is to look for "stars" rather than cliques in a graph. A star is a set of vertices including a center and all the other vertices that are connected to the center vertex. Since any clique must be included in some star, for our particular problem this simplification will not miss useful vertices. Figure 2 shows a graph, two cliques and an example star. There are 6 stars in the graph since $|V|$=6. They are $Star_1 = Star_2 = Star_6 = \{L_1, L_2, L_5, L_6\}$, $Star_3 = Star_4 = \{L_3, L_4, L_5\}$ and $Star_5 = \{L_1, L_2, L_3, L_4, L_5, L_6\}$. A set of all the unique stars is $Stars = \{Star_1, Star_3, Star_5\}$. Note that each star is finally a set of local alignments and each local alignment is a set of C_α pairs.

For each unique star, a corresponding global alignment candidate is calculated by deleting badly aligned C_α pairs involved in that star. Then all the candidates are compared and the optimal one is chosen. An example global alignment between protein chains 1ATP:E and 1PHK is shown in Figure 3, where N_{mat} is the number of aligned C_α pairs, (C_α) $RMSD$ is the root mean square distance between the aligned pairs, and the rigid body transformation used to align the two chains is T (the translation vector) and R (the rotation matrix).

The "star" approach has been used in a preliminary version of this algorithm [25], which has one center for each of its stars and shows some instability for aligning large proteins. We introduce the double-center method to group the local alignments and it is described in section 2.2. This greatly improves the reliability of the algorithm. Another crucial new technical development of this paper is the learning strategy based on feedback, which is described in section 2.3. The combination of two new methods greatly improves speed, reliability, and accuracy of the algorithm.

2.2. Introduction of Double-Center "Stars"

The single-center star method is not flawless. It works well when the two protein chains match well or the chain diameters are small. However, we have found that it is less stable when the chains do not match very well or the chain diameters are large. This is caused by deleting badly matched C_α pairs from each star, a method applied to obtain a global alignment candidate. When local alignments are grouped into an initial star, there may exist point pairs which do not match well. An initial transformation is calculated and the worst matched pair based on that transformation is first deleted, then the transformation is recalculated to select the second worst pair. This process is repeated. In this way the well matched pairs survive and the (C_α) $RMSD$ becomes smaller and smaller, until an acceptable (C_α) $RMSD$ is achieved. The effect of deleting bad point pairs relies on a good initial transformation, which in turn depends on the star center selection. With a single star center, the initial transformation has

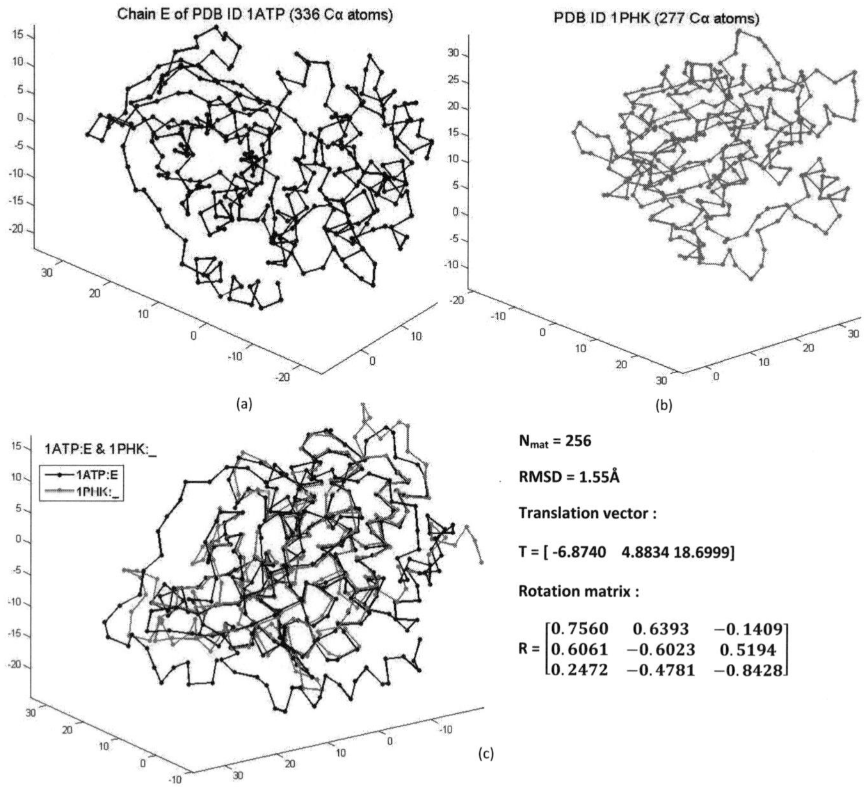

Fig. 3. A global alignment

great freedom to move and rotate in the point pair deletion process, thus the deletion may go along a more unpredictable way. This is more obvious when the local alignments are relatively short, which usually happens when the chains do not match very well or the chain diameters are large. Based on this observation, we consider grouping local alignments into double-center "stars".

A "double-center star" is, as suggested by its name, a "star" with two centers. Each single-center star can be extended to a corresponding double-center star, while the latter is much more stable. In a single-center star, each local alignment consistent with the center is added to the star, while in a double-center star, a local alignment can be added only when it is consistent with both centers. The first center of a double-center star is exactly the one in a single-center star, and the second center is selected from that star. The selection of the second center satisfies the following conditions: (1) it is consistent with the first center, (2) it is long enough to make sense, and (3) it is as far as possible from the first center. Figure 4 illustrates a double-center star corresponding to star 5 in Figure 2, suppose L_2 is the second center.

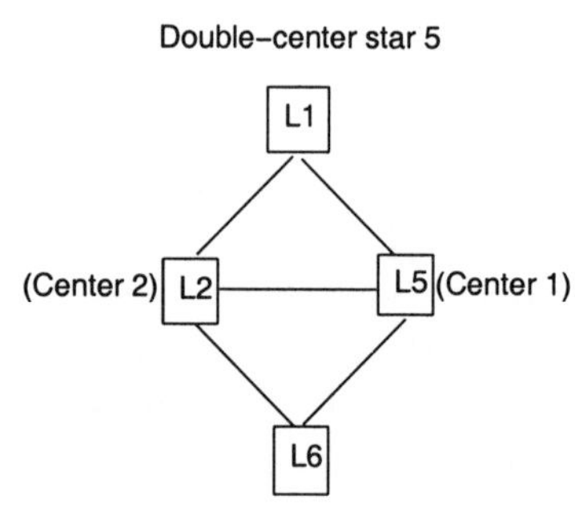

Fig. 4. A double-center star

Each local alignment in a single-center star is consistent with the center, however, this does not automatically guarantee that all the local alignments in the star are consistent. The consistency relationship is not necessarily transitive. To reduce the probability of adding inconsistent local alignments to the star, a double-center star accepts local alignments in a more prudent way. It rejects the local alignments originally surviving in the single-center star on a weak basis, therefore local alignments in the double-center star are more likely to be those very good ones. To some extent, the presence of the second center has the effect of "extending" the local alignment in the first center. With such a long "local alignment" as the center, the star will be more stable because points in it have much less freedom to move or rotate. From another aspect, with this improvement the extent to which the initial transformation will change along with the deletion of bad point pairs is reduced significantly - the initial transformation will be closer to the final one, and thus the deletion will cause less unpredictability. Furthermore, the filtering of unpromising local alignments reduces their negative contribution to the overall transformation (as well as the number of point pairs involved in the initial star), speeding up the deletion process and resulting in a faster and better global alignment.

2.3. Development of Learning Ability

2.3.1. *Self-learning*

As we have mentioned, good star centers produce promising stars and have a greater probability of generating good global alignments. However, thus far the selection of star centers has been naïve: any local alignment with sufficient length can be the center of a star. The double-center method helps remove some unpromising local alignments from a star when the first center is determined, but it contributes nothing to the selection of the first center. If the first center of a star can be selected intelligently rather than by arbitrarily picking up a local alignment, then the star may yield a better global alignment. This intelligence may be difficult to achieve without any a priori knowledge on the global structural similarity between the two chains. However, when such knowledge is available, it is possible to improve the alignment by way of a self-learning strategy.

Once a currently optimal global alignment is output, we are able to know approximately where the aligned positions are. We organize the consecutively aligned point pairs into groups, and each group of consecutive point pairs is called a global alignment segment. A global alignment segment looks exactly like a local alignment, while as a part of a good global alignment, it should be a good "local alignment". Here local alignment is quoted because global alignment segments are not output of the local alignment phase, although there is no substantial difference between both definitions. To take advantage of these good global alignment segments, we apply a feedback mechanism to teach our alignment algorithm how to improve itself. The self-learning is implemented *via* the iterative utilization of its own output. When a global alignment is ready, consecutive alignment segments are extracted, then each segment is used as a new star center and local alignments consistent with the center are added to its group. This global alignment phase is repeated with a few new stars obtained from the currently optimal global alignment, until the alignment output converges (i.e. no changes are found between two iterations).

2.3.2. *Learning from others*

When a global alignment from another algorithm is available, the global alignment segments in that result can work as initial star centers. These centers are likely to be better than our own local alignments because they are from an optimal alignment result obtained from another algorithm. With these centers, our global alignment searching starts from a very good jumping-off point, therefore it is possible to output a result better than without learning. Learning from other algorithms may be more effective in the cases our algorithm performs worse than others. When it performs better than other algorithms even without learning, this learning may be less necessary, however it is never harmful, because if it results in a worse global alignment, its results can simply be disregarded. Therefore the combination of self-learning and learning-from-others will never output an alignment worse than the one of another algorithm. In the worst case it outputs nothing different. For this reason, our algorithm can also be used to improve the result of any other algorithm. We call this a refinement to that algorithm.

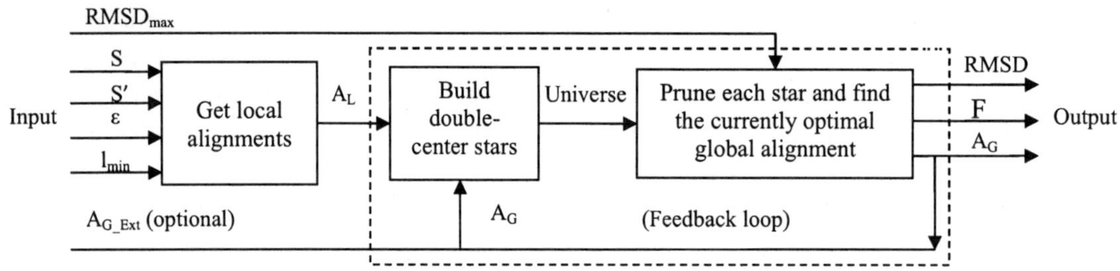

Fig. 5. The SLIPSA framework

3. THE FORMAL DESCRIPTION OF SLIPSA

We give the formal description of SLIPSA. Combining the double-center star, the self-learning and the learning-from-others methods which use feedback, we greatly improve our earlier work [25] and have found interesting results when comparing SLIPSA with some other algorithms. The SLIPSA framework is shown in Figure 5. This system takes six parameters: protein chains S and S', $RMSD_{max}$ (a user specified maximum (C_α) $RMSD$), distance constant ϵ, minimum local alignment length l_{min}, and an optional external global alignment A_{G_Ext}. Parameters S and S' are determined by the user, $RMSD_{max}$ is either determined by the user or obtained from another algorithm, ϵ and l_{min} are selected empirically according to the user input, and A_{G_Ext} is either empty or also obtained from the external algorithm. The system outputs an optimal global alignment result consisting of A_G (a set of global alignment segments), (C_α) $RMSD$ (a value not greater than $RMSD_{max}$) and F (a rigid body transformation corresponding to the final global alignment). The following sub-sections describe the details of the SLIPSA algorithm.

3.1. Getting Local Alignments

The calculation of local alignments has been reviewed in section 2.1. The procedure used to get local alignments can be from either Ref. 25 or other related papers (e.g. Ref. 21). The procedure body is omitted.

Get-Local-Alignments($S, S', \epsilon, l_{\min}$)
Input: protein backbone chains $S = p_1 \cdots p_n$, $S' = q_1 \cdots q_m$, distance constant ϵ and minimum local alignment length $l_{\min}$, where each p_i or q_i is a 3-D point corresponding to a C_α atom in a protein backbone.
Output: $A_L = \{L_1, L_2, \cdots, L_w\}$, a set containing all the local alignments of length $\geq l_{\min}$ between S and S'.

3.2. Building up Stars from Local Alignments

The improved procedure outputs double-center stars. The input is star centers from a set of global alignment segments, or from a local alignment set when the former is empty. The non-center nodes in a star are still chosen from the local alignment set.

Build-Double-Center-Stars(A_L, A_G)
Input: $A_L = \{L_1, L_2, \cdots, L_w\}$ and $A_G = \{L_{1'}, L_{2'}, \cdots, L_{w'}\}$, where A_L is a set of local alignments and A_G is a set of global alignment segments.
Output: $Universe = \{Star_1, Star_2, \cdots, Star_k\}$, a set of all the unique double-center stars.

```
begin
    Universe ← {} (the empty set);
    if (A_G = {}) then A ← A_L;
     otherwise A ← A_G;
    for (each local alignment L_u in A)
        find L_u', the second center based on L_u,
         in A;
        Star_u ← {L_u, L_u'};
        for (each local alignment L_v in A_L)
            if (L_v is consistent with both L_u and
              L_u') then Star_u ← Star_u ∪ {L_v};
        end for
        if (Star_u ∉ Universe)
```

then $Universe \leftarrow Universe \cup \{Star_u\}$;
end for
return $Universe$;
end

3.3. Finding a Global Alignment from the Stars

In each iteration of our algorithm, a global alignment is output and used as an input of the next iteration. We describe how to prune the set of aligned pairs in a star and obtain the global alignment which has an (C_α) $RMSD$ not greater than that specified by the user. We refine a similar idea that was used in our original algorithm [25], which does not use feedback.

Prune-One-Star($Star, RMSD_{max}$)
Input: a $Star$ and $RMSD_{max}$ (a user specified maximum $RMSD$).
Output: $(A_S, RMSD_S, F_S, l_S)$, where $A_S = \{L_{1''}, L_{2''}, \cdots, L_{w''}\}$ is a set of global alignment segments which share a common transformation F_S with $RMSD_S \leq RMSD_{max}$, and l_S is the number of aligned point pairs in A_S.

```
begin
    A_S ← Star;
    l_S ← the number of point pairs involved in A_S;
    calculate transformation F_S and RMSD_S for all
      the point pairs involved in A_S;
    while (RMSD_S > RMSD_max)
        delete point pair (p, q) with the largest
          d(p, F_S(q)) in A_S;
        l_S ← l_S − 1;
        recalculate transformation F_S and RMSD_S
          for all the point pairs involved in A_S;
    end while
    return (A_S, RMSD_S, F_S, l_S);
end
```

In the following function Find-Global-Alignment(), we apply the Prune-One-Star() procedure to each of the stars in the universe which is built from Build-Double-Center-Stars(). The alignment that contains the largest number of aligned pairs will be returned.

Find-Global-Alignment($Universe, RMSD_{max}$)
Input: $Universe = \{Star_1, Star_2, \cdots, Star_k\}$ and $RMSD_{max}$ (a user specified maximum $RMSD$).
Output: $(A_G, RMSD, F)$, where $A_G = \{L_{1'}, L_{2'}, \cdots, L_{w'}\}$ is a set of global alignment segments which share a common transformation F with $RMSD \leq RMSD_{max}$.

```
begin
    sort Universe by a descending order of the
      number of 3-D point pairs involved in each star;
    l_max ← 0;
    for (each Star_u in Universe)
        (A_S, RMSD_S, F_S, l_S) ← Prune-One-Star
          (star_u, RMSD_max);
        if (l_S > l_max) then A_G ← A_S; RMSD ←
          RMSD_S; F ← F_S; l_max ← l_S;
    end for
    return (A_G, RMSD, F);
end
```

3.4. The Feedback Procedure

This is the main procedure of SLIPSA. It calls Get-Local-Alignments in the first step, then Build-Double-Center-Stars and Find-Global-Alignment are called repeatedly. A global alignment output by the current iteration serves as the input of the next iteration. The procedure terminates when the global alignment ceases to change (i.e. converges).

SLIPSA($S, S', \epsilon, l_{min}, RMSD_{max}, A_{G\_Ext}$)
Input: S, S', ϵ, l_{min}, $RMSD_{max}$ and A_{G_Ext}, where A_{G_Ext} can be either empty or a set of global alignment segments obtained from an external algorithm.
Output: $(A_G, RMSD, F)$.

```
begin
    A_L ← Get-Local-Alignments(S, S', ε, l_min);
    A_G ← A_G_Ext;
    do
        A'_G ← A_G;
        Universe ← Build-Double-Center-Stars
          (A_L, A'_G);
        (A_G, RMSD, F) ← Find-Global-Alignment
          (Universe, RMSD_max);
    while (A_G ≠ A'_G);
    return (A_G, RMSD, F);
end
```

When no external alignment is available, procedure SLIPSA is called by way of SLIPSA(S, S', ϵ, l_{min}, $RMSD_{max}$, $\{\}$). When it is available, SLIPSA can be called as SLIPSA($S, S', \epsilon, l_{min}, RMSD_{max}, A_{G\_Ext}$). We call this a refinement to external alignment A_{G_Ext}. To independently test the performance of our algorithm,

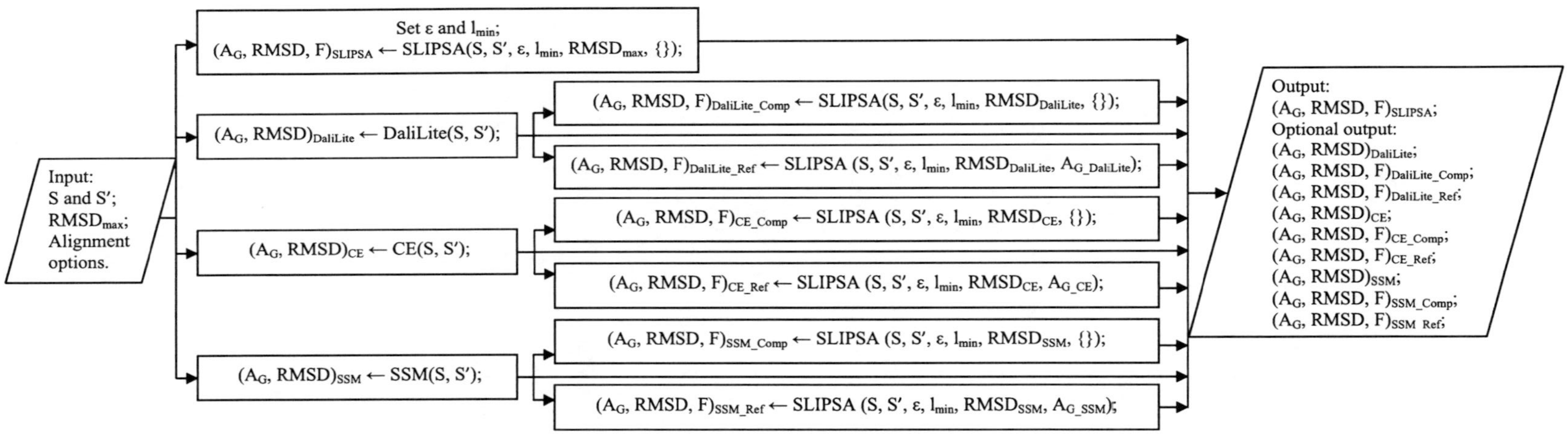

Fig. 6. The web alignment work flow

none of the experiments reported in section 4 uses any external alignment as our input.

4. EXPERIMENTAL ENVIRONMENT AND RESULTS

4.1. Our Web Alignment Tool

We have developed a web alignment tool based on the SLIPSA algorithm. The website is available for public access at http://fpsa.cs.uno.edu/ with a mirror site at http://fpsa.cs.panam.edu/FPSA/. It is not only a SLIPSA alignment tool but also an alignment comparison tool between SLIPSA and DaliLite, CE and SSM, some commonly used protein structure alignment algorithms with public websites.

The data used for protein alignment are the PDB files downloaded from the RCSB Protein Data Bank. The files have been moved to the Worldwide Protein Data Bank (wwPDB) by the time we wrote this paper. As of March 2008, there were over 49,000 protein structures with over 100,000 chains discovered.

Our website is built on an Intel dual-Xeon 3G Hz PC server with 3GB memory. The web development tools we have used include Apache HTTP server with PHP support, ActivePerl and MySQL database server. The SLIPSA algorithm is written in MATLAB. See Refs. 20 and 2 for the rigid body transformation method that we have used in SLIPSA.

The work flow of our website is shown in Figure 6. Besides a maximum value for (C_α) $RMSD$, it accepts either PDB IDs or user uploaded PDB files as input. It is optional to compare SLIPSA with DaliLite, CE or SSM. When a comparing option is chosen, our tool automatically submits alignment request to and retrieves result from DaliLite, CE or SSM website, and performs SLIPSA alignment according to the retrieved (C_α) $RMSD$ value. The website outputs the following alignment results. Beyond the first result listed, all others are optional depending on the user choices. Note that SLIPSA outputs A_G (a set of global alignment segments), (C_α) $RMSD$ and F (a rigid body transformation).

(1) $(A_G, RMSD, F)_{\text{SLIPSA}}$: the SLIPSA result with a user specified $RMSD_{max}$.
(2) $(A_G, RMSD)_{DaliLite}$: the DaliLite result retrieved automatically from its website.
(3) $(A_G, RMSD, F)_{DaliLite\_Comp}$: the SLIPSA result with an $RMSD$ retrieved from DaliLite website as input. This result is used to compare SLIPSA with DaliLite.
(4) $(A_G, RMSD)_{CE}$: the CE result retrieved automatically from its website.
(5) $(A_G, RMSD, F)_{CE\_Comp}$: the result used to compare SLIPSA with CE.
(6) $(A_G, RMSD)_{SSM}$: the SSM result retrieved automatically from its website.
(7) $(A_G, RMSD, F)_{SSM\_Comp}$: the result used to compare SLIPSA with SSM.

4.2. Experimental Results

We have collected 224 alignment cases to test the performance of our algorithm. The test cases were originally proposed by various papers for various testing purposes. They include No. 1 - No. 20 (see Table III in Ref. 16), No. 21 - No. 88 (see Table I in Ref. 5), No. 89 (see Tables I and II in Ref. 16), No. 90 - No. 92 (supplement to Table III in Ref. 16), No. 93 (see Figure 5 in Ref. 16), No. 94 - No. 101 (see Table IV in Ref. 16), No. 102 - No. 111 (see Table V in Ref. 16), No. 112 - No. 120 (supplement to Table V in Ref. 16), No. 121 - No. 124 (see Table VII in Ref. 16), No. 125 - No. 143 (see Table 1 in Ref. 15), No. 144 - No. 183 (see Table 1 in Ref. 22) and No. 184 - No. 224 (see Table 2 in Ref. 22). Due to the space limit, the PDB IDs of those proteins are not listed in this paper and they can be provided upon request.

Based on this test set, we compare SLIPSA with DaliLite, CE and SSM in terms of N_{mat} (the number of aligned positions) and (C_α) $RMSD$. Common protein alignment scoring methods such as Z-score, Q-score, P-score and geometric measures proposed in Ref. 10 all take N_{mat} and (C_α) $RMSD$ into account. Because of the $RMSD$ flexibility of SLIPSA, it is easy to compare SLIPSA with DaliLite, CE and SSM on the basis of N_{mat} because in most cases SLIPSA outputs an equal (C_α) $RMSD$. In each test case SLIPSA outputs an (C_α) $RMSD$ not greater than that of DaliLite, CE, or SSM. If N_{mat} of SLIPSA is larger than N_{mat} of DaLiLite, CE, or SSM, we call it an N_{mat} increment. Similarly, if the (C_α) $RMSD$ of SLIPSA is smaller than the (C_α) $RMSD$ of DaLiLite, CE or SSM, we call it a

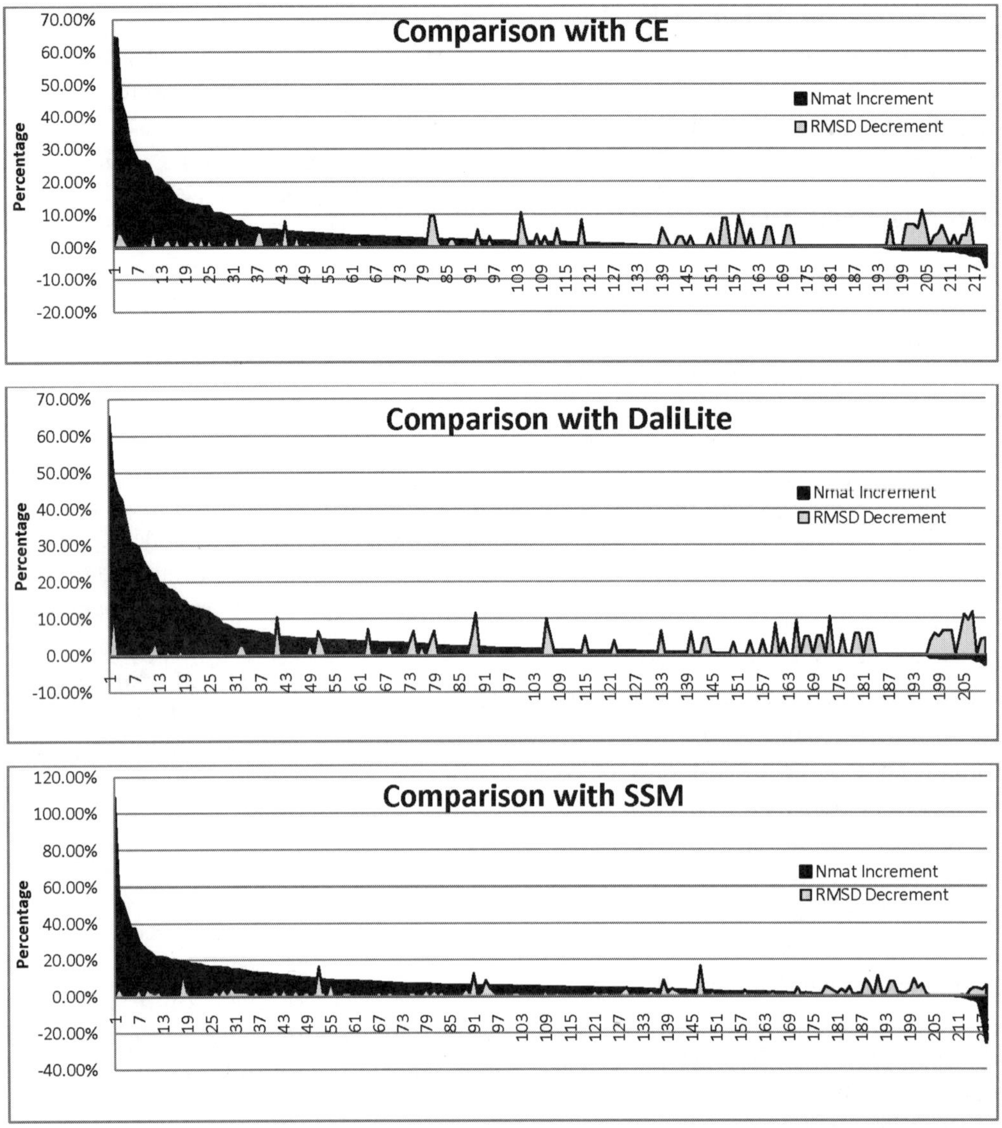

Fig. 7. Comparing SLIPSA with DaliLite, CE and SSM

(C_α) $RMSD$ decrement. The N_{mat} increment rate is calculated by $(N_{mat\_\mathrm{SLIPSA}} - N_{mat\_X})$ / N_{mat_X} and the (C_α) $RMSD$ decrement rate is calculated by $(RMSD_X - RMSD_{\mathrm{SLIPSA}})$ / $RMSD_X$, where X is DaliLite, CE or SSM. Figure 7 illustrates such increments and decrements in percentage. For the convenience of illustration, the results are sorted in a descending order of the N_{mat} increment rate. Due to the space limit, the detailed result data are not listed in this paper. They can be provided upon request. It should be mentioned that, (1) no SSM comparison was performed in our earlier paper [25]; (2) in a few cases that we could not find results on the DaliLite, CE or SSM websites, we marked the cases as "n/a" and did not compare SLIPSA with them; (3) from the time we completed this paper, it is possible to see result changes on any of the alignment websites and we have observed minor changes on some of them; (4) the SLIPSA experiments did not use any external alignment as input, although our algorithm is able to refine the alignment results retrieved from other web servers.

4.3. Discussion on the Results

Table 1 shows some statistical data based on the results in Figure 7. For each case in which an

Table 1. Statistics on the experimental results

	DaliLite	CE	SSM
Number of valid cases	210	220	218
Cases with larger N_{mat} by SLIPSA	149(66.67%)	136(61.82%)	189(86.70%)
Cases with smaller N_{mat} by SLIPSA	14(6.67%)	26(11.82%)	8(3.67%)
Maximum N_{mat} increment by SLIPSA	49	56	51
Maximum N_{mat} decrement by SLIPSA	2	9	12
Maximum N_{mat} increment rate by SLIPSA	65.33%	64.58%	109.09%
Maximum N_{mat} decrement rate by SLIPSA	2.74%	6.45%	25.53%
Average N_{mat} increment by SLIPSA	4.15	3.63	7.24
Average N_{mat} increment rate by SLIPSA	4.56%	4.13%	7.37%
Cases with smaller $RMSD$ by SLIPSA	56(26.67%)	64(29.09%)	177(81.19%)
Maximum $RMSD$ decrement by SLIPSA	0.7	0.4	0.52
Maximum $RMSD$ decrement rate by SLIPSA	13.21%	11.11%	16.56%
Average $RMSD$ decrement by SLIPSA	0.04	0.04	0.05
Average $RMSD$ decrement rate by SLIPSA	1.55%	1.42%	2.07%

Table 2. Comparison based on weak similarity

	DaliLite		CE		SSM	
	Valid Cases	Avg. N_{mat} Inc.	Valid Cases	Avg. N_{mat} Inc.	Valid Cases	Avg. N_{mat} Inc.
$RMSD \geq 5.0$	12	26.48%	14	21.62%	0	/
$RMSD \geq 4.0$	20	23.48%	41	14.75%	9	17.50%
$RMSD \geq 3.0$	77	10.09%	102	7.64%	51	12.15%

alignment result was missing from either DaliLite, CE or SSM, we did not compare it with SLIPSA. Also, since DaliLite, CE and SSM may have different (C_α) $RMSD$ values for a given test case, they were not compared mutually. In our experiments, when compared with DaliLite, CE and SSM respectively, SLIPSA outputs a larger N_{mat} in 66.67%, 61.82% and 86.70% of the cases; The maximum N_{mat} increment rate of SLIPSA is 65.33%, 64.58% and 109.09%; Averagely, SLIPSA increases 4.56%, 4.13%, and 7.37% of the N_{mat}; In 26.67%, 29.09% and 81.19% of the cases SLIPSA outputs a smaller (C_α) $RMSD$ with the maximum (C_α) $RMSD$ decrement rate being 13.21%, 11.11% and 16.56%. To sum up, in most cases we see SLIPSA results with a larger or same N_{mat} and a same or smaller (C_α) $RMSD$. In some cases that SLIPSA outputs a smaller N_{mat}, we also see a smaller (C_α) $RMSD$.

We also attempt to compare SLIPSA with DaliLite, CE and SSM in the cases of weak similarities. To simplify the comparison process, we tentatively define a weak similarity as a large (C_α) $RMSD$ between aligned chains. This definition may be incomplete, however, we have already observed some interesting results. For example, when compared with DaliLite and CE, the average N_{mat} increment rates of SLIPSA are 4.56% and 4.13% respectively, while in the cases with (C_α) $RMSD \geq 5.0$, the numbers are 26.48% and 21.62%, much higher than the overall average values. See Table 2 for details. In brief, SLIPSA obtains high average N_{mat} increment rate in weak similarity cases, and the larger the (C_α) $RMSD$, the higher the average N_{mat} increment rate.

The running time of each algorithm was recorded. The average running time of DaliLite, CE and SSM is 16.86s, 6.14s and 9.15s, respectively. When compared with them (i.e. using the $RMSD$ from the best fit from the comparison algorithms as the $RMSD$ upper-bound in SLIPSA), the average running time of SLIPSA is 105.97s, 69.89s and 81.43s, respectively. In about 50% of the cases the SLIPSA average time is below the DaliLite average, and the corresponding numbers for CE and SSM are about 25% and 28%, respectively. Possible ways to reduce the running time are discussed below.

(1) The web server was built on a slow machine. We have also tested the algorithm on an IBM ThinkPad laptop computer with Core2 Duo

1.8GHz CPUs. This machine was much slower than the mainstream web server machines, while the same results took only $\frac{1}{2}$ to $\frac{2}{3}$ of the time used on our current web server. It is possible to improve the speed to great extent by using a machine with high computational performance. (2) We used Matlab to implement the algorithm. Matlab facilitates the proof-of-concept development of complicated scientific programs, however, according to our experience it is possible to speed up algorithms at least several times if they are implemented in other languages such as C, C++ and Java. In addition, parallel and distributed programming on high performance computational resources can also help reduce the execution time. (3) The algorithm is slower when the proteins are long and/or the (C_α) $RMSD$ is large. In such cases the number of local alignments are large and the graph complexity is high. However, the algorithm can be optimized to reduce the complexity. Possible methods include reducing the dimension of data, removing unpromising local alignments as early as possible, limiting the number of times of feedback, and so on.

References

1. Chew LP, Huttenlocher D, Kedem K, Kleinberg J. Fast detection of common geometric substructure in proteins. *Journal of Computational Biology* 1999; **6(3-4)**: 313–325.
2. Lorusso A, Eggert DW, Fisher RB. A comparison of four algorithms for estimating 3-D rigid transformations. *British Machine Vision Conference* 1995; 237–246.
3. Falicov A, Cohen FE. A surface of minimum area metric for the structureal comparison of proteins. *Journal of Molecular Biology* 1996; **258**: 871–892.
4. Fischer D, Nussinov R, Wolfson H. 3D substructure matching in protein molecules. *Proc. 3rd Intl Symp. Combinatorial Pattern Matching, LNCS* 1992; **644**: 136–150.
5. Fischer D, Elofsson A, Rice D, Eisenberg D. Assessing the performance of fold recognition methods by means of a comprehensive benchmark. *Proc. 1st Pacific Symposium on Biocomputing* 1996; 300–318.
6. Godzik A. The structural alignment between two proteins: Is there a unique answer? *Protein Science* 1996; **5**: 1325–1338.
7. Holm L, Sander C. Protein structure comparison by alignment of distance matrices. *Journal of Molecular Biology* 1993; **233**: 123–138.
8. Ilyin VA, Abyzov A, Leslin CM. Structural alignment of proteins by a novel TOPOFIT method, as a superimposition of common volumes at a topomax point. *Protein Science* 2004; **13**: 1865–1874.
9. Kolodny R, Linial N, Levitt M. Approximate protein structural alignment in polynomial time. *Proc. Natl. Acad. Sci. USA* 2004; **101(33)**: 12201-12206.
10. Kolodny R, Koehl P, Levitt M. Comprehensive evaluation of protein structure alignment methods: scoring by geometric measures. *Journal of Molecular Biology* 2005; **346(4)**: 1173-1188.
11. Krissinel E, Henrick K. Secondary-structure matching (SSM), a new tool for fast protein structure alignment in three dimensions. *Acta Cryst.* 2004; **D60**: 2256–2268.
12. Lathrop RH. The protein threading problem with sequence amino acid interaction preferences is NP-complete. *Protein Engineering* 1994; **7**: 1059–1068.
13. Lessel U, Schomburg D. Similarities between protein 3-D structures. *Protein Engineering* 1994; **7(10)**: 1175–87.
14. Madej T, Gibrat JF, Bryant SH. Threading a database of protein cores. *Proteins* 1995; **23**: 356–369.
15. Ortiz AR, Strauss CEM, Olmea O. MAMMOTH (matching molecular models obtained from theory): an automated method for model comparison. *Protein Science* 2002; **11**: 2606–2021.
16. Shindyalov IN, Bourne PE. Protein structure alignment by incremental combinatorial extension (CE) of the optimal path. *Protein Engineering* 1998; **11**: 739–747.
17. Singh AP, Brutlag DL. Hierarchical protein superposition using both secondary structure and atomic representation. *Proc. Intelligent Systems for Molecular Biology* 1997; 284–293.
18. Taylor WR, Orengo CA. Protein structure alignment. *Journal of Molecular Biology* 1989; **208(1)**: 1–22.
19. Taylor WR. Protein structure comparison using iterated double dynamic programming. *Protein Science* 1999; **9**:654–665.
20. Umeyama S. Least-squares estimation of transformation parameters between two point patterns. *IEEE Tran. on Pattern Analysis and Machine Intelligence* 1991; **13(4)**: 376–380.
21. Ye Y, Godzik A. Database searching by flexible protein structure alignment. *Protein Science* 2004; **13(7)**: 1841–1850.
22. Ye J, Janardan R, Liu S. Pairwise protein structure alignment based on an orientation-independent backbone representation. *Journal of Bioinformatincs and Computational Biology* 2005; **4(2)**: 699–717.
23. Yona G, Kedem K. The URMS-RMS hybrid algorithm for fast and sensitive local protein structure alignment. *Journal of Computational Biology* 2005; **12**:12–32.
24. Zhang Y, Skolnick J. TM-align: a protein structure alignment algorithm based on the TM-score. *Nucleic Acids Research* 2005; **33**: 2302–2309.
25. Zhao ZY, Fu B. A Flexible algorithm for pairwise protein structure alignment. *the 2007 International Conference on Bioinformatics and Computational Biology* 2007; 16–22.

PREDICTING FLEXIBLE LENGTH LINEAR B-CELL EPITOPES

Yasser EL-Manzalawy*[1,2,5], Drena Dobbs[3,4,5], and Vasant Honavar[1,2,4,5]

[1] *Artificial Intelligence Laboratory*
[2] *Department of Computer Science*
[3] *Department of Genetics, Development and Cell Biology*
[4] *Bioinformatics and Computational Biology Graduate Program*
[5] *Center for Computational Intelligence, Learning, and Discovery*
Iowa State University
Ames, IA 50010, USA.
email: {yasser, ddobbs, honavar}@iastate.edu

Identifying B-cell epitopes play an important role in vaccine design, immunodiagnostic tests, and antibody production. Therefore, computational tools for reliably predicting B-cell epitopes are highly desirable. We explore two machine learning approaches for predicting flexible length linear B-cell epitopes. The first approach utilizes four sequence kernels for determining a similarity score between any arbitrary pair of variable length sequences. The second approach utilizes four different methods of mapping a variable length sequence into a fixed length feature vector. Based on our empirical comparisons, we propose FBCPred, a novel method for predicting flexible length linear B-cell epitopes using the subsequence kernel. Our results demonstrate that FBCPred significantly outperforms all other classifiers evaluated in this study. An implementation of FBCPred and the datasets used in this study are publicly available through our linear B-cell epitope prediction server, BCPREDS, at: http://ailab.cs.iastate.edu/bcpreds/.

1. INTRODUCTION

B-cell epitopes are antigenic determinants that are recognized and bound by receptors (membrane-bound antibodies) on the surface of B lymphocytes [1]. The identification and characterization of B-cell epitopes play an important role in vaccine design, immunodiagnostic tests, and antibody production. As identifying B-cell epitopes experimentally is time-consuming and expensive, computational methods for reliably and efficiently predicting B-cell epitopes are highly desirable [2].

There are two types of B-cell epitopes: (i) linear (continuous) epitopes which are short peptides corresponding to a contiguous amino acid sequence fragment of a protein [3, 4]; (ii) conformational (discontinuous) epitopes which are composed of amino acids that are not contiguous in primary sequence but are brought into close proximity within the folded protein structure. Although it is believed that a large majority of B-cell epitopes are discontinuous [5], experimental epitope identification has focused primarily on linear B-cell epitopes [6]. Even in the case of linear B-cell epitopes, however, antibody-antigen interactions are often conformation-dependent. The conformation-dependent aspect of antibody binding complicates the problem of B-cell epitope prediction, making it less tractable than T-cell epitope prediction. Hence, the development of reliable computational methods for predicting linear B-cell epitopes is an important challenge in bioinformatics and computational biology [2].

Previous studies have reported correlations between certain physicochemical properties of amino acids and the locations of linear B-cell epitopes within protein sequences [7–11]. Based on that observation, several amino acid propensity scale based methods have been proposed. For example, methods in [8–11] utilized hydrophilicity, flexibility, turns, and solvent accessibility propensity scales, respectively. PREDITOP [12], PEOPLE [13], BEPITOPE [14], and BcePred [15] utilized groups of physicochemical properties instead of a single property to improve the accuracy of the predicted linear B-cell epitopes. Unfortunately, Blythe and Flower [16] showed that propensity based methods can not be used reliably for predicting B-cell epitopes. Using a dataset of 50 proteins and an exhaustive assessment of 484 amino acid propensity scales, Blythe and Flower [16] showed

*Corresponding author.

that the best combinations of amino acid propensities performed only marginally better than random. They concluded that the reported performance of such methods in the literature is likely to have been overly optimistic, in part due to the small size of the data sets on which the methods had been evaluated.

Recently, the increasing availability of experimentally identified linear B-cell epitopes in addition to Blythe and Flower results [16] motivated several researchers to explore the application of machine learning approaches for developing linear B-cell epitope prediction methods. BepiPred [17] combines two amino acid propensity scales and a Hidden Markov Model (HMM) trained on linear epitopes to yield a slight improvement in prediction accuracy relative to techniques that rely on analysis of amino acid physicochemical properties. ABCPred [18] uses artificial neural networks for predicting linear B-cell epitopes. Both feed-forward and recurrent neural networks were evaluated on a *non-redundant* data set of 700 B-cell epitopes and 700 non-epitope peptides, using 5-fold cross validation tests. Input sequence windows ranging from 10 to 20 amino acids, were tested and the best performance, 66% accuracy, was obtained using a recurrent neural network trained on peptides 16 amino acids in length. In the method of Söllner and Mayer [19], each epitope is represented using a set of 1487 features extracted from a variety of propensity scales, neighborhood matrices, and respective probability and likelihood values. Of two machine learning methods tested, decision trees and a nearest-neighbor method combined with feature selection, the latter was reported to attain an accuracy of 72% on a data set of 1211 B-cell epitopes and 1211 non-epitopes, using a 5-fold cross validation test [19]. Chen et al. [20] observed that certain amino acid pairs (AAPs) tend to occur more frequently in B-cell epitopes than in non-epitope peptides. Using an AAP propensity scale based on this observation, in combination with a support vector machine (SVM) classifier, they reported prediction accuracy of 71% on a data set of 872 B-cell epitopes and 872 non-B-cell epitopes, estimated using 5-fold cross validation. In addition, [20] demonstrated an improvement in the prediction accuracy, 72.5%, when the APP propensity scale is combined with turns accessibility, antigenicity, hydrophilicity, and flexibility propensity scales.

Existing linear B-cell epitope prediction tools fall into two broad categories. Tools in the first category, residue-based predictors, take as input a protein sequence and assign binary labels to each individual residue in the input sequence. Each group of neighboring residues with predicted positive labels define a variable length predicted linear B-cell epitope. Residue-based prediction methods scan the input sequence using a sliding window and assign a score to the amino acid at the center of the window based on the mean score of a certain propensity scale (e.g., flexibility or hydrophilicity). The target residue is predicted positive if its score is greater than a predetermined threshold. Unfortunately, it has been shown that the performance of these methods is marginally better than random [16]. PepiPred [17] used the information extracted using the sliding window to train a HMM and combined it with two propensity scale based methods. BcePred [15] combined several propensity scales and showed that the performance of the combined scales is better than the performance of any single scale.

The second category of linear B-cell prediction tools consist of the epitope-based predictors. An example of such predictors is the ABCPred server [18]. For this server, the input is a protein sequence and an epitope length (should be in $\{20, 18, .., 10\}$). The server then applies a sliding window of the user specified length and passes the extracted peptides to a neural network classifier trained using epitope dataset in which all the epitope sequences have been set to the specified epitope length via trimming and extending longer and shorter epitopes, respectively. A limitation of this approach is that the user is forced to select one of the available six possible epitope lengths and can not specify a different epitope length.

Because linear B-cell epitopes can vary in length over a broad range (see Figure 1), it is natural to train classifiers using the experimentally reported epitope sequences without trimming or extending them. Such an approach will allow us to provide a linear B-cell epitope prediction tool that allows the user to experiment with virtually any arbitrary epitope length. In this work, we explore two machine learning approaches for predicting flexible length linear B-cell epitopes. The first approach utilizes several sequence kernels for determining a similarity score between any arbitrary pair of variable length sequences. The second approach utilizes many different methods of mapping a variable length sequence into a fixed length feature vector. Based on our empirical comparisons, we propose FBCPred, a novel method for predicting flexible length linear B-cell epitopes using the subsequence kernel. Our results demonstrate that FBCPred significantly outperforms all other classifiers evaluated in this study. An implementation of FBCPred and the datasets used in this study are publicly available through our linear B-cell epitope prediction server, BCPREDS, at: http://ailab.cs.iastate.edu/bcpreds/.

2. MATERIALS AND METHODS

2.1. Data

We retrieved 1223 unique linear B-cell epitopes of lengths more than 3 amino acids from Bcipep database [21]. To avoid over-optimistic performance of classifiers evaluated on the set of unique epitopes, we applied a homology reduction procedure proposed by Raghava [22] for reducing sequence similarity among flexible length major histocompatibility complex class II (MHC-II) peptides. Briefly, given two peptides p_1 and p_2 of lengths l_1 and l_2 such that $l_1 \leq l_2$, we compare p_1 with each l_1-length subpeptide in p_2. If the percent identity (PID) between p_1 and any subpeptide in p_2 is greater than 80%, then the two peptides are deemed to be similar. For example, to compute the PID between (ACDEFGHIKLMNPQRST) and (DEFGGIKLMN), we compare (DEFGGIKLMN) with (ACDEFGHIKL), (CDEFGHIKLM), ..., (IKLMNPQRST). The PID between (DEFGGIKLMN) and (DEFGHIKLMN) is 90% since nine out of 10 residues are identical.

Applying the above homology reduction procedure to the set of 1223 unique variable length linear B-cell epitopes yields a *homology-reduced* set of 934 epitopes. Two datasets of flexible length linear B-cell epitopes have been constructed. An *original* dataset constructed from the set of 1223 unique epitopes as the positive examples and 1223 non-epitopes randomly extracted from SwissProt [23] and a *homology-reduced* dataset constructed from *homology-reduced* set of 934 epitopes as positive examples and an equal number of negative examples extracted randomly form SwissProt sequences. In both datasets two selection criteria have been applied to the randomly extracted non-epitopes: (i) the length distribution in the negative data is identical to the length distribution in the positive data; (ii) none of the non-epitopes appears in the set of epitopes.

2.2. Support vector machines and kernel methods

Support vector machines (SVMs) [24] are a class of supervised machine learning methods used for classification and regression. Given a set of labeled training data (x_i, y_i), where $x_i \in R^d$ and $y_i \in \{+1, -1\}$, training an SVM classifier involves finding a hyperplane that maximizes the geometric margin between positive and negative training data samples. The hyperplane is described as $f(x) = \langle w, x \rangle + b$, where w is a normal vector and b is a bias term. A test instance, x, is assigned a positive label if $f(x) > 0$, and a negative label otherwise. When the training data are not linearly separable, a kernel function is used to map nonlinearly separable data from the input space into a feature space. Given any two data samples x_i and x_j in an input space $X \in R^d$, the kernel function K returns $K(x_i, x_j) = \langle \Phi(x_i), \Phi(x_j) \rangle$ where Φ is a nonlinear map from the input space X to the corresponding feature space. The kernel function K has the property that $K(x_i, x_j)$ can be computed without explicitly mapping x_i and x_j into the feature space, but instead, using their dot product $\langle x_i, x_j \rangle$ in the input space. Therefore, the kernel trick allows us to train a linear classifier, e.g., SVM, in a high-dimensional feature space where the data are assumed to be linearly separable without explicitly mapping each training example from the input space into the feature space. This approach relies implicitly on the selection of a feature space in which the training data are likely to be linearly separable (or nearly so) and explicitly on the selection of the kernel function to achieve such separability. Unfortunately, there is no single kernel that is guaranteed to perform well on every data set. Consequently, the SVM approach requires some care in selecting a suitable kernel and tuning the kernel parameters (if any).

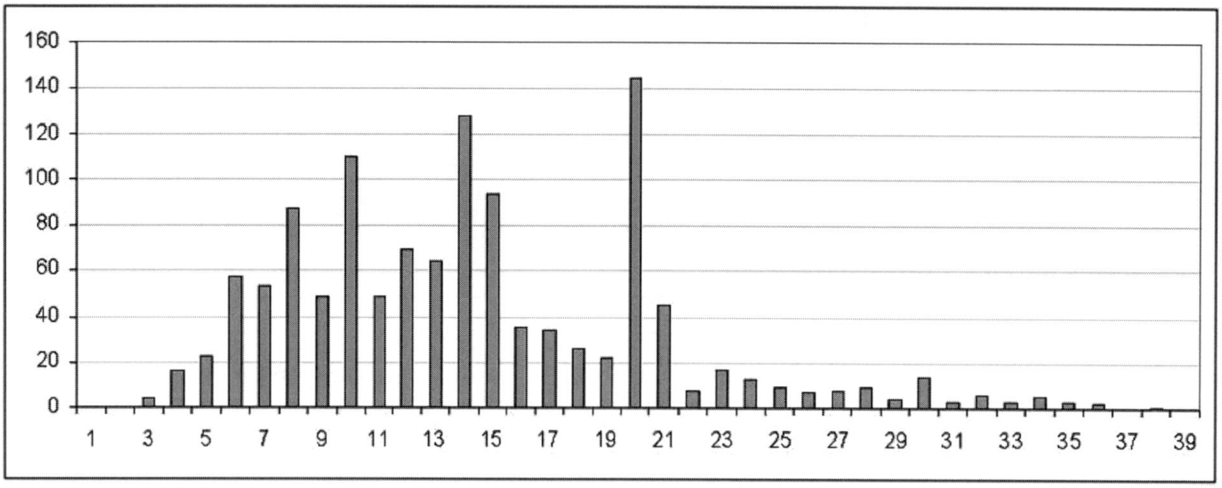

Fig. 1. Length distribution of unique linear B-cell epitopes in Bcipep database.

2.3. Sequence kernel based methods

String kernels [25–29] are a class of kernel methods that have been successfully used in many sequence classification tasks [25, 26, 28, 30–32]. In these applications, a protein sequence is viewed as a string defined on a finite alphabet of 20 amino acids. In this work, we explore four string kernels: spectrum [25], mismatch [26], local alignment [28], and subsequence [27], in predicting linear B-cell epitopes. A brief description of the four kernels follows.

2.3.1. *Spectrum kernel*

Let A denote a finite alphabet, e.g., the standard 20 amino acids. x and y denote two strings defined on the alphabet A. For $k \geq 1$, the k-spectrum is defined as [25]:

$$\Phi_k = (\phi_\alpha(x))_{\alpha \in A^k} \tag{1}$$

where ϕ_α is the number of occurrences of the k-length substring α in the sequence x. The k-spectrum kernel of the two sequences x and y is obtained by taking the dot product of the corresponding k spectra:

$$K_k^{spct}(x, y) = \langle \Phi_k(x), \Phi_k(y) \rangle \tag{2}$$

The k-spectrum kernel captures a simple notion of string similarity: two strings are deemed similar (i.e., have a high k-spectrum kernel value) if they share many of the same k-length substrings.

2.3.2. *Mismatch kernel*

The mismatch kernel [26] is a variant of the spectrum kernel in which inexact matching is allowed. Specifically, the (k, m)-mismatch kernel allows up to $m \leq k$ mismatches to occur when comparing two k-length substrings. Let α be a k-length substring, the (k, m)-mismatch feature map is defined on α as:

$$\Phi_{(k,m)}(\alpha) = (\phi_\beta(\alpha))_{\beta \in A^k} \tag{3}$$

where $\phi_\beta(\alpha) = 1$ if $\beta \in N_{(k,m)(\alpha)}$, where β is the set of k-mer substrings that differs from α by at most m mismatches. Then, the feature map of an input sequence x is the sum of the feature vectors for k-mer substrings in x:

$$\Phi_{(k,m)}(x) = \sum_{k-mers\ \alpha\ in\ x} \Phi_{(k,m)}(\alpha) \tag{4}$$

The (k, m)-mismatch kernel is defined as the dot product of the corresponding feature maps in the feature space:

$$K_{(k,m)}^{msmtch}(x, y) = \langle \Phi_{(k,m)}(x), \Phi_{(k,m)}(y) \rangle \tag{5}$$

It should be noted that the $(k, 0)$-mismatch kernel results in a feature space that is identical to that of the k-spectrum kernel. An efficient data structure for computing the spectrum and mismatch kernels in $O(|x|+|y|)$ and $O(k^{m+1}|A|^m(|x|+|y|))$, respectively, is provided in [26].

2.3.3. *Local alignment kernel*

Local alignment (LA) kernel [28] is a string kernel adapted for biological sequences. The LA kernel measures the similarity between two sequences by summing up scores obtained from gapped local alignments of the sequences. This kernel has several parameters: the gap opening and extension penalty parameters, d and e, the amino acid mutation matrix s, and the factor β, which controls the influence of suboptimal alignments on the kernel value. Detailed formulation of the LA kernel and a dynamic programming implementation of the kernel with running time complexity in $O(|x||y|)$ are provided in [28].

2.3.4. *Subsequence kernel*

The subsequence kernel (SSK) [27] generalizes the k-spectrum kernel by considering a feature space generated by the set of all (contiguous and non-contiguous) k-mer subsequences. For example, if we consider the two strings *"act"* and *"acctct"*, the value returned by the spectrum kernel with $k = 3$ is 0. On the other hand, the $(3, 1)$-mismatch kernel will return 3 because the 3-mer substrings *"acc"*, *"cct"*, and *"tct"* have at most one mismatch when compared with *"act"*. The subsequence kernel considers the set (*"ac − t"*, *"a − ct"*, *"ac − − − t"*, *"a − c − −t"*, *"a − − − ct"*) of non-contiguous substrings and returns a similarity score that is weighted by the length of each non-contiguous substring. Specifically, it uses a decay factor, $\lambda \leq 1$, to penalize non-contiguous substring matches. Therefore, the subsequence kernel with $k = 3$ will return $2\lambda^4 + 3\lambda^6$ when applied to *"act"* and *"acctct"* strings. More precisely, the feature map $\Phi_{(k,\lambda)}$ of a string x is given by:

$$\Phi_{(k,\lambda)}(x) = \Big(\sum_{i:u=x[i]} \lambda^{l(i)}\Big)_{u \in A^k} \tag{6}$$

where $u = x[i]$ denotes a substring in x where $1 \leq i_1 < \cdots < i_{|u|} \leq |x|$ such that $u_j = s_{i_j}$, for $j = 1, \ldots, |u|$ and $l(i) = i_{|u|} - i_1 + 1$ is the length of the subsequence in x. The subsequence kernel for two strings x and y is determined as the dot product of the corresponding feature maps:

$$\begin{aligned} K(x,y)^{sub}_{(k,\lambda)} &= \langle \Phi_{(k,\lambda)}(x), \Phi_{(k,\lambda)}(y) \rangle \\ &= \sum_{u \in A^k} \sum_{i:u=x[i]} \lambda^{l(i)} \sum_{j:u=y[j]} \lambda^{l(j)} \\ &= \sum_{u \in A^k} \sum_{i:u=x[i]} \sum_{j:u=y[j]} \lambda^{l(j)+l(j)} \end{aligned} \tag{7}$$

This kernel can be computed using a recursive algorithm based on dynamic programming in $O(k|x||y|)$ time and space. The running time and memory requirements can be further reduced using techniques described in [33].

2.4. Sequence-to-features based methods

This approach has been previously used for protein function and structure classification tasks [34–37] and the classification of flexible length MHC-II peptides. The main idea is to map each variable length amino acid sequence into a feature vector of fixed length. Once the variable length sequences are mapped to fixed length feature vectors, we can apply any of the standard machine learning algorithms to this problem. Here, we considered SVM classifiers trained on the mapped data using the widely used RBF kernel.

We explored four different methods for mapping a variable length amino acid sequence into a fixed length feature vector: (i) amino acid composition; (ii) dipeptide composition; (iii) amino acid pairs propensity scale; (iv) composition-transition-distribution. A brief summary of each method is given below.

2.4.1. *Amino acid and dipeptide composition*

Amino acid composition (AAC) represents a variable length amino acid sequence using a feature vector of 20 dimensions. Let x be a sequence of $|x|$ amino acids. Let A denote the set of the standard 20 amino acids. The amino acid composition feature mapping is defined as:

$$\Phi_{AAC}(x) = (\phi_\beta(x))_{\beta \in A} \tag{8}$$

where $\phi_\beta(x) = \frac{number\ of\ occurrences\ of\ amino\ acid\ \beta\ in\ x}{|x|}$.

A limitation of the amino acid composition feature representation of amino acid sequences is that we lose the sequence order information. Dipeptide composition (DC) encapsulates information about the fraction of amino acids as well as their local order. In dipeptide composition each variable length amino acid sequence is represented by a feature vector of 400 dimensions defined as:

$$\Phi_{DC}(x) = (\phi_\alpha(x))_{\alpha \in A^2} \quad (9)$$

where $\phi_\alpha(x) = \frac{number\ of\ occurrences\ of\ dipeptide\ \alpha\ in\ x}{total\ number\ of\ all\ possible\ dipeptides\ in\ x}$.

2.4.2. *Amino acid pairs propensity scale*

Amino acid pairs (AAPs) are obtained by decomposing a protein/peptide sequence into its 2-mer subsequences. [20] observed that some specific AAPs tend to occur more frequently in B-cell epitopes than in non-epitope peptides. Based on this observation, they developed an AAP propensity scale defined by:

$$\theta(\alpha) = log(\frac{f_\alpha^+}{f_\alpha^-}) \quad (10)$$

where f_α^+ and f_α^- are the occurrence frequencies of AAP α in the epitope and non-epitope peptide sequences, respectively. These frequencies have been derived from Bcipep [21] and Swissprot [23] databases, respectively. To avoid the dominance of an individual AAP propensity value, the scale in Eq. (10) has been normalized to a $[-1, +1]$ interval through the following conversion:

$$\theta(\alpha) = 2(\frac{\theta(\alpha) - min}{max - min}) - 1 \quad (11)$$

where max and min are the maximum and minimum values of the propensity scale before the normalization.

The AAP feature mapping, Φ_{AAP}, maps each amino acid sequence, x, into a 400-dimentional feature space defined as:

$$\Phi_{AAP}(x) = (\phi_\alpha(x) \cdot \theta(\alpha))_{\alpha \in A^2} \quad (12)$$

where $\phi_\alpha(x)$ is the number of occurrences of the 2-mer α in the peptide x.

2.4.3. *Composition-Transition-Distribution*

The basic idea behind the Composition-Transition-Distribution (CTD) method [38, 39] is to map each variable length peptide into a fixed length feature vector such that standard machine learning algorithms are applicable. From each peptide sequence, 21 features are extracted as follows:

- First, each peptide sequence p is mapped into a string s_p defined over an alphabet of three symbols, $\{1, 2, 3\}$. The mapping is performed by grouping amino acids into three groups using a physicochemical property of amino acids (see Table 3). For example the peptide (AIRHIPRRIR) is mapped into (2312321131) using the hydrophobicity division of amino acids into three groups (see Table 3).
- Second, for each peptide string s_p, three descriptors are derived as follows:
  - Composition (C): three features representing the percent frequency of the symbols, $\{1, 2, 3\}$, in the mapped peptide sequence.
  - Transition (T): three features representing the percent frequency of i followed by j or j followed by i, for $i, j \in \{1, 2, 3\}$.
  - Distribution (D): five features per symbol representing the fractions of the entire sequence where the first, 25, 50, 75, and 100% of the candidate symbol are contained in s_p. This yields an additional 15 features for each peptide.

Table 1 shows division of the 20 amino acids, proposed by Chinnasamy et al. [40] , into three groups based on hydrophobicity, polarizability, polarity, and Van der Waal's volume properties. Using these four properties, we derived 84 CTD features from each peptide sequence. In our experiments, we trained SVM classifiers using RBF kernel and peptide sequences represented using their amino acid sequence composition (20 features) and CTD descriptors (84 features).

Table 1. Categorization of amino acids into three groups for a number of physicochemical properties.

Proporty	Group 1	Group 2	Group 3
Hydrophobicity	RKEDQN	GASTPHY	CVLIMFW
Polarizability	GASCTPD	NVEQIL	MHKFRYW
Polarity	LIFWCMVY	PATGS	HQRKNED
Van der Waala volume	GASDT	CPNVEQIL	KMHFRYW

2.5. Performance evaluation

We report the performance of each classifier using the average of 10 runs of 5-fold cross validation tests. Each classifier performance is assessed by both threshold-dependent and threshold-independent metrics. For threshold-dependent metrics, we used accuracy (ACC), sensitivity (S_n), specificity (S_p), and correlation coefficient (CC). The CC measure has a value in the range from -1 to +1 and the closer the value to +1, the better the predictor. The S_n and S_p summarize the accuracies of the positive and negative predictions respectively. ACC, S_n, S_p, and CC are defined in Eq. (13-15) where TP, FP, TN, FN are the numbers of true positives, false positives, true negatives, and false negatives respectively.

For threshold-independent metrics, we report the Receiver Operating Characteristic (ROC) curve. The ROC curve is obtained by plotting the true positive rate as a function of the false positive rate or, equivalently, sensitivity versus (1-specificity) as the discrimination threshold of the binary classifier is varied. Each point on the ROC curve describes the classifier at a certain threshold value and hence a particular choice of tradeoff between true positive rate and false negative rate. We also report the area under ROC curve (AUC) as a useful summary statistic for comparing two ROC curves. AUC is defined as the probability that a randomly chosen positive example will be ranked higher than a randomly chosen negative example. An ideal classifier will have an AUC = 1, while a classifier performs no better than random will have an AUC = 0.5, any classifier performing better than random will have an AUC value that lies between these two extremes.

2.6. Implementation and SVM parameter optimization

We used Weka machine learning workbench [41] for implementing the spectrum, mismatch, and LA kernels (RBF and SSK kernels are already implemented in Weka). We evaluated the k-spectrum kernel, K_k^{spct}, for $k = 1, 2$, and 3. The (k, m)-mismatch kernel was evaluated at (k,m) equals $(3, 1) and (4, 1)$. The subsequence kernel, $K_{(k,\lambda)}^{sub}$, was evaluated at $k = 2, 3$, and 4 and the default value for λ, 0.5. The LA kernel was evaluated using the BLOSUM62 substitution matrix, gap opening and extension parameters equal to 10 and 1, respectively, and $\beta = 0.5$. For the SVM classifier, we used the Weka implementation of the SMO [42] algorithm. For the string kernels, the default value of the C parameter, $C = 1$, was used for the SMO classifier. For methods that uses the RBF kernel, we found that tuning the SMO cost parameter C and the RBF kernel parameter γ is necessary to obtain satisfactory performance. We tuned these parameters using a 2-dimensional grid search over the range $C = 2^{-5}, 2^{-3}, \ldots, 2^3, \gamma = 2^{-15}, 2^{-13}, \ldots, 2^3$.

3. RESULTS AND DISCUSSION

Table 2 compares the performance of different SVM based classifiers on the *original* dataset of unique flexible length linear B-cell epitopes. The SVM classifier trained using SSK with $k = 4$ and $\lambda = 0.5$, $k_{(4,0.5)}^{sub}$, significantly (using statistical paired t-test [43] with p-value = 0.05) outperforms all other classifiers in terms of the AUC. The two classifiers based on the mismatch kernel have the worst AUC. The classifier trained using k_3^{spct} is competitive to those trained using the LA kernel and $k_{(2,0.5)}^{sub}$. The last four classifiers belong to the sequence-to-feature approach. Each of these classifiers has been trained using an SVM classifier and the RBF kernel but on different data representation. The results suggest that representation of

$$ACC = \frac{TP + TN}{TP + FP + TN + FN} \quad (13)$$

$$S_n = \frac{TP}{TP + FN} \text{ and } S_p = \frac{TN}{TN + FP} \quad (14)$$

$$CC = \frac{(TP \times TN) - (FP \times FN)}{\sqrt{(TN + FN)(TN + FP)(TP + FN)(TP + FP)}} \quad (15)$$

the peptides using their dipeptide composition performs better than other feature representations on the *original* dataset. Figure 2 shows the ROC curves for different methods on *original* dataset of unique flexible length linear B-cell epitopes. The ROC curve of $K^{sub}_{(4,0.5)}$ based classifier almost dominates all other ROC curves (i.e., for any choice of specificity value, the $K^{sub}_{(4,0.5)}$ based classifier almost has the best sensitivity) .

Table 3 reports the performance of the different SVM based classifiers on the *homology-reduced* dataset of flexible length linear B-cell epitopes. We note that the performance of each classifier is considerably worse than its performance on the *original* dataset of unique epitopes. This discrepancy can be explained by the existence of epitopes with significant pairwise sequence similarity in the *original* dataset. Interestingly, the SVM classifier based on the $k^{sub}_{(4,0.5)}$ kernel still significantly outperforms all other classifiers at 0.05 level of significance. Figure 3 shows the ROC curves for different methods on *homology-reduced* dataset of flexible length linear B-cell epitopes. Again, the ROC curve of $K^{sub}_{(4,0.5)}$ based classifier almost dominates all other ROC curves.

Comparing results on Table 2 and Table 3 reveals two important issues that to the best of our knowledge have not been addressed before in the literature on B-cell epitope prediction. First, our results demonstrate that performance estimates reported on the basis of the *original* dataset of unique linear B-cell epitopes is overly optimistic compared to the performance estimates obtained using the *homology-reduced* dataset. Hence, we suspect that the actual performance of linear B-cell epitope prediction methods on *homology-reduced* datasets is somewhat lower than the reported performance on the original dataset of unique peptides. Second, our results suggest that conclusions regarding how different prediction methods compare to each other drawn on the basis of datasets of unique epitopes may be misleading. For example, from the reported results in Table 2, one may conclude that k^{spct}_3 outperforms k^{spct}_1 and k^{spct}_2 while results on the *homology-reduced* dataset (see Table 3) demonstrate that the three classifiers are competitive with each other. Another example of misleading conclusions drawn from results in Table 2 is that dipeptide composition features is a better representation than amino acid composition representation of the data. This conclusion is contradicted by results in Table 3 which show that the classifier constructed using the amino acid composition representation of the data slightly outperforms the classifier constructed using the dipeptide composition of the same data.

The results in Table 2 and Table 3 show that the classifier that used the amino acid composition features outperforms the classifier that used CTD features. This is interesting because the set of amino acid composition features is a subset of the CTD features. Recall that CTD is composed of 20 amino acid composition features plus 84 physicochemical features, we conclude that the added physicochemical features did not yield additional information that was relevant for the classification task. In addition, we observed that the classifier that used the dipeptide composition outperforms the classifier that used the AAP features. This is interesting because AAP features as defined in Eq. (12) can be viewed as dipeptide composition features weighted by the amino acid propensity of each dipeptide.

3.1. Web server

An implementation of FBCPred is available as a part of our B-cell epitope prediction server (BCPREDS) [44] which is freely accessible at http://ailab.cs.iastate.edu/bcpreds/. Because it is often valuable to compare predictions of multiple methods, and consensus predictions are more re-

Table 2. Performance of different SVM based classifiers on original dataset of unique flexible length linear B-cell epitopes. Results are the average of 10 runs of 5-fold cross validation.

Method	ACC	Sn	Sp	CC	AUC
K_1^{spct}	62.86	61.76	63.95	0.257	0.680
K_2^{spct}	63.29	63.84	62.74	0.266	0.683
K_3^{spct}	65.36	79.28	51.44	0.320	0.720
$K_{(3,1)}^{msmtch}$	47.88	48.42	47.33	-0.042	0.480
$K_{(4,1)}^{msmtch}$	58.93	57.79	60.07	0.179	0.618
LA	65.41	63.36	67.46	0.308	0.716
$K_{(2,0.5)}^{sub}$	65.58	65.08	66.09	0.312	0.710
$K_{(3,0.5)}^{sub}$	70.56	71.05	70.07	0.411	0.778
$K_{(4,0.5)}^{sub}$	**73.37**	**74.08**	72.67	**0.468**	**0.812**
AAC	65.61	68.41	62.81	0.313	0.722
DC	70.55	68.28	**72.83**	0.411	0.750
AAP	65.65	66.20	65.11	0.313	0.717
CTD	63.21	63.15	63.28	0.264	0.686

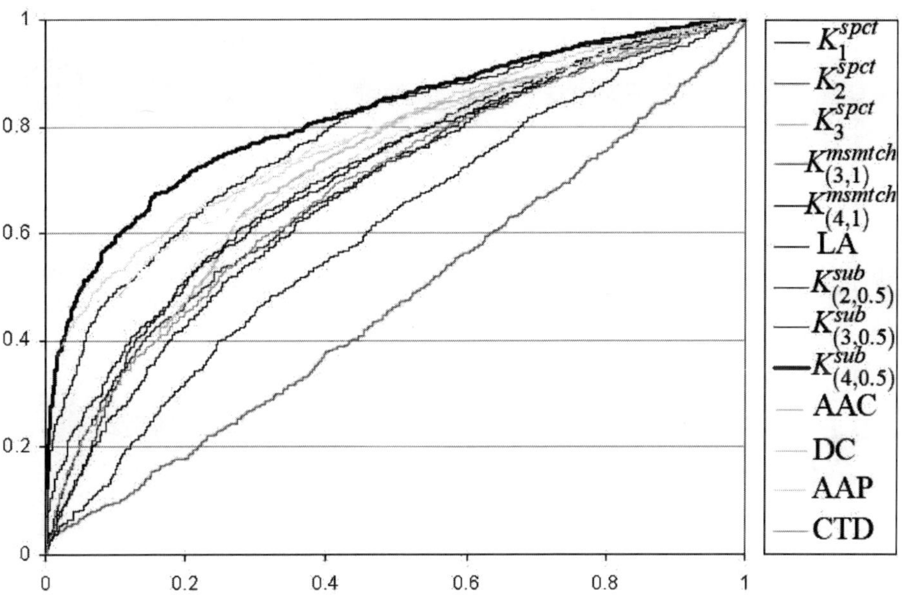

Fig. 2. ROC curves for different methods on *original* dataset of unique flexible length linear B-cell epitopes. The ROC curve of $K_{(4,0.5)}^{sub}$ based classifier almost dominates all other ROC curves.

liable than individual predictions, the BCPREDS server aims at providing predictions using several B-cell epitope prediction methods. The current implementation of BCPREDS allows the user to select among three prediction methods: (i) Our implementation of AAP method [20]; (ii) BCPred [44], a method for predicting linear B-cell epitope using the subsequence kernel; (iii) FBCPred, the method introduced in this study for predicting flexible length B-cell epitopes. The major difference between FBCPred and the other two methods is that FBCPred can predict linear B-cell epitopes of virtually any arbitrary length while for the other two methods the length has to be one of possible six values, $\{12, 14, \ldots, 22\}$.

Another goal of BCPREDS server is to serve as a repository of benchmark B-cell epitope datasets. The datasets used for training and evaluating BCPred and the two datasets used in this study can

Table 3. Performance of different SVM based classifiers on homology-reduced dataset of flexible length linear B–cell epitopes. Results are the average of 10 runs of 5-fold cross validation.

Method	ACC	Sn	Sp	CC	AUC
K_1^{spct}	58.22	56.70	59.74	0.165	0.621
K_2^{spct}	60.26	61.04	59.49	0.205	0.642
K_3^{spct}	60.86	62.45	59.27	0.217	0.635
$K_{(3,1)}^{msmtch}$	46.42	46.34	46.50	-0.072	0.451
$K_{(4,1)}^{msmtch}$	54.35	54.75	53.95	0.087	0.561
LA	61.38	60.41	62.35	0.228	0.658
$K_{(2,0.5)}^{sub}$	60.09	60.52	59.66	0.202	0.647
$K_{(3,0.5)}^{sub}$	63.85	65.05	62.65	0.277	0.701
$K_{(4,0.5)}^{sub}$	**65.49**	68.36	62.61	**0.310**	**0.738**
AAC	63.31	**70.90**	55.73	0.269	0.683
DC	63.78	63.05	**64.51**	0.276	0.667
AAP	61.42	62.85	60.00	0.229	0.658
CTD	60.32	59.66	60.98	0.206	0.639

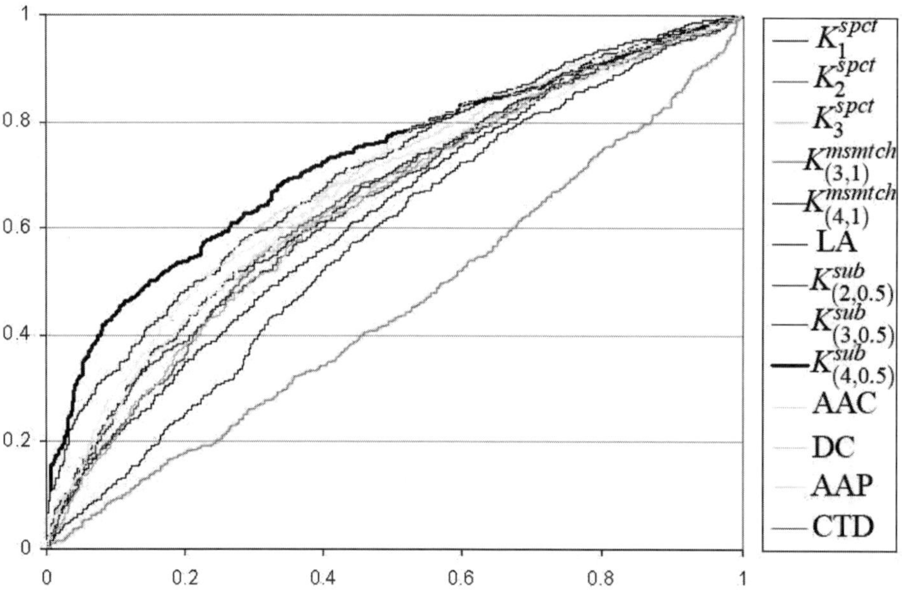

Fig. 3. ROC curves for different methods on *homology-reduced* dataset of flexible length linear B-cell epitopes. The ROC curve of $K_{(4,0.5)}^{sub}$ based classifier almost dominates all other ROC curves.

be freely downloaded from the web server.

4. SUMMARY AND DISCUSSION

We explored two machine learning approaches for predicting flexible length linear B-cell epitopes. The first approach utilizes sequence kernels for determining a similarity score between any arbitrary pair of variable length sequences. The second approach utilizes several methods of mapping a variable length sequence into a fixed length feature vector. Our results demonstrated a superior performance of the subsequence kernel based SVM classifier compared to other SVM classifiers examined in our study. Therefore, we proposed FBCPred, a novel method for predicting flexible length linear B-cell epitopes using the subsequence kernel. An implementation of FBCPred and the datasets used in this study are publicly available through our linear B-cell prediction server, BCPREDS, at: http://ailab.cs.iastate.edu/bcpreds/.

Previous methods for predicting linear B-cell epitopes (e.g., [15, 17, 19, 18, 20]) have been evaluated on datasets of unique epitopes without applying any homology reduction procedure as a pre-processing step on the data. We showed that performance estimates reported on the basis of such datasets is considerably over-optimistic compared to performance estimates obtained using the *homology-reduced* datasets. Moreover, we showed that using such *non homology-reduced* datasets for comparing different prediction methods may lead to false conclusions regarding how these methods compare to each other.

4.1. Related work

Residue-based prediction methods [7–11, 15, 17] assign labels to each residue in the query sequence and therefore are capable of predicting linear B-cell epitopes of variable length. However, most of these methods have been shown to be of low to moderate performance [16].

AAP method [20] maps each peptide sequence into a set of fixed length numeric features and therefore it can be trained using datasets of flexible length sequences. However, the performance of this method had been reported using a dataset of 20-mer peptides.

Söllner and Mayer [19] introduced a method for mapping flexible length epitope sequences into feature vectors of 1478 attributes. This method has been evaluated on a dataset of flexible length linear B-cell epitopes. However, no homology reduction procedure was applied to remove highly similar sequences from the data. In addition, the implementation of this method is not publicly available.

Recently, two methods [45, 39] have been successfully applied to the problem of predicting flexible length MHC-II binding peptides. The first method [45] utilized the LA kernel [28] for developing efficient SVM based classifiers. The second method [39] mapped each flexible length peptide into the set of CTD features employed in our study in addition to some extra features extracted using two secondary structure and solvent accessibility prediction classifiers. In our study we could not use these extra features due to the unavailability of these two programs.

Acknowledgments

This work was supported in part by a doctoral fellowship from the Egyptian Government to Yasser EL-Manzalawy and a grant from the National Institutes of Health (GM066387) to Vasant Honavar and Drena Dobbs.

References

1. GB Pier, JB Lyczak, and LM Wetzler. *Immunology, infection, and immunity,* 1st ed. ASM Press, PL Washington. 2004.
2. JA Greenbaum, PH Andersen, M Blythe, HH Bui, RE Cachau, J Crowe, M Davies, AS Kolaskar, O Lund, S Morrison, et al. Towards a consensus on datasets and evaluation metrics for developing B-cell epitope prediction tools. *J. Mol. Recognit.* 2007; **20**:75–82.
3. DJ Barlow, MS Edwards, JM Thornton, et al. Continuous and discontinuous protein antigenic determinants. *Nature* 1986; **322**:747–748.
4. JP Langeveld, J martinez Torrecuadrada, RS boshuizen, RH Meloen, and CJ Ignacio. Characterisation of a protective linear B cell epitope against feline parvoviruses. *Vaccine* 2001; **19**:2352–2360.
5. G Walter. Production and use of antibodies against synthetic peptides. *J. Immunol. Methods* 1986; **88**:149–61.
6. DR Flower. *Immunoinformatics: Predicting immunogenicity in silico*, 1st ed. Humana, Totowa NJ. 2007.
7. JL Pellequer, E Westhof, and MH Van Regenmortel. Predicting location of continuous epitopes in proteins from their primary structures. *Meth. Enzymol.* 1991; **203**:176–201.
8. JMR Parker and Hodges RS Guo, D and. New hydrophilicity scale derived from high-performance liquid chromatography peptide retention data: correlation of predicted surface residues with antigenicity and x-ray-derived accessible sites. *Biochemistry* 1986; **25**:5425–5432.
9. PA Karplus and GE Schulz. Prediction of chain flexibility in proteins: a tool for the selection of peptide antigen. *Naturwiss.* 1985; **72**:21–213.
10. EA Emini, JV Hughes, DS Perlow, and J Boger. Induction of hepatitis A virus-neutralizing antibody by a virus-specific synthetic peptide. *J. Virol.* 1985; **55**:836–839.
11. JL Pellequer, E Westhof, and MH Van Regenmortel. Correlation between the location of antigenic sites and the prediction of turns in proteins. *Immunol. Lett.* 1993; **36**:83–99.
12. JL Pellequer and E Westhof. PREDITOP: a program for antigenicity prediction. *J. Mol. Graph.* 1993; **11**:204–210.

13. AJ Alix. Predictive estimation of protein linear epitopes by using the program PEOPLE. *Vaccine* 1999; **18**:311–314.
14. M Odorico and JL Pellequer. BEPITOPE: predicting the location of continuous epitopes and patterns in proteins. *J. Mol. Recognit.* 2003; **16**:20–22.
15. S Saha and GP Raghava. BcePred: Prediction of continuous B-cell epitopes in antigenic sequences using physico-chemical properties. *Artificial Immune Systems, Third International Conference (ICARIS 2004), LNCS*, 2004; **3239**:197–204.
16. MJ Blythe and DR Flower. Benchmarking B cell epitope prediction: Underperformance of existing methods. *Protein Sci.* 2005; **14**:246–248.
17. JE Larsen, O Lund, and M Nielsen. Improved method for predicting linear B-cell epitopes. *Immunome Res.* 2006; **2**:2.
18. S Saha and GP Raghava. Prediction of continuous B-cell epitopes in an antigen using recurrent neural network. *Proteins* 2006; **65**:40–48.
19. J Söllner and B Mayer. Machine learning approaches for prediction of linear B-cell epitopes on proteins. *J. Mol. Recognit.* 2006; **19**:200–208.
20. J Chen, H Liu, J Yang, and KC Chou. Prediction of linear B-cell epitopes using amino acid pair antigenicity scale. *Amino Acids* 2007; **33**:423–428.
21. S Saha, M Bhasin, and GP Raghava. Bcipep: a database of B-cell epitopes. *BMC Genomics* 2005; **6**:79.
22. GPS Raghava. *MHCBench: Evaluation of MHC Binding Peptide Prediction Algorithms.* datasets available at `http://www.imtech.res.in/raghava/mhcbench/`.
23. A Bairoch and R Apweiler. The SWISS-PROT protein sequence database and its supplement TrEMBL in 2000. *Nucleic Acids Res.* 2000; **28**:45–48.
24. VN Vapnik. *The nature of statistical learning theory* 2nd Ed. Springer-Verlag New York Inc. New York, USA. 2000.
25. C Leslie, E Eskin, and WS Noble. The spectrum kernel: A string kernel for SVM protein classification. *Proceedings of the Pacific Symposium on Biocomputing* 2002; **7**:566–575.
26. CS Leslie, E Eskin, A Cohen, J Weston, and WS Noble. Mismatch string kernels for discriminative protein classification. *Bioinformatics* 2004; **20**:467–476.
27. H Lodhi, C Saunders, J Shawe-Taylor, N Cristianini, and C Watkins. Text classification using string kernels. *J. Mach. Learn. Res.* 2002; **2**:419–444.
28. H Saigo, JP Vert, N Ueda, and T Akutsu. Protein homology detection using string alignment kernels. *Bioinformatics* 2004; **20**:1682–1689.
29. D Haussler. Convolution kernels on discrete structures. *UC Santa Cruz Technical Report UCS-CRL-99-10*, 1999.
30. NM Zaki, S Deris, and R Illias. Application of string kernels in protein sequence classification. *Appl. Bioinformatics* 2005; **4**:45–52.
31. H Rangwala, K DeRonne, G Karypis, and Minnesota Univ. Minneapolis Dept. of Computer Science. *Protein structure prediction using string kernels.* Defense Technical Information Center. 2006.
32. F Wu, B Olson, D Dobbs, and V Honavar. Comparing kernels for predicting protein binding sites from amino acid sequence. *International Joint Conference on Neural Networks (IJCNN06)* 2006; 1612–1616.
33. AK Seewald and F Kleedorfer. Lambda pruning: An approximation of the string subsequence kernel. Technical report, Technical Report, Osterreichisches Forschungsinstitut fur Artificial Intelligence, Wien, TR-2005-13, 2005.
34. S. Hua and Z. Sun. Support vector machine approach for protein subcellular localization prediction. *Bioinformatics* 2001; **17**:721–728.
35. P.D. Dobson and A.J. Doig. Distinguishing Enzyme Structures from Non-enzymes Without Alignments. *J. Mol. Biol.* 2003; **330**:771–783.
36. F. Eisenhaber, C. Frommel, and P. Argos. Prediction of secondary structural content of proteins from their amino acid composition alone. II. The paradox with secondary structural class. *Proteins*, 1996; **25**:169–79.
37. R. Luo, Z. Feng, and J. Liu. Prediction of protein structural class by amino acid and polypeptide composition. *FEBS J.* 2002; **269**:4219–4225.
38. CZ Cai, LY Han, ZL Ji, X Chen, and YZ Chen. SVM-Prot: web-based support vector machine software for functional classification of a protein from its primary sequence. *Nucleic Acids Res.* 2003; **31**:3692–3697.
39. J Cui, LY Han, HH Lin, ZQ Tan, L Jiang, ZW Cao, and YZ Chen. MHC-BPS: MHC-binder prediction server for identifying peptides of flexible lengths from sequence-derived physicochemical properties. *Immunogenetics* 2006; **58**:607–613.
40. A. Chinnasamy, WK Sung, and A. Mittal. Protein structure and fold prediction using tree-augmented naive Bayesian classifier. *Pac. Symp. Biocomput.* 2004; **387**:98.
41. IH Witten and E Frank. *Data mining: Practical machine learning tools and techniques*, 2nd ed. Morgan Kaufmann, San Francisco, USA. 2005.
42. J Platt. *Fast training of support vector machines using sequential minimal optimization.* MIT Press, Cambridge, MA, USA. 1998.
43. C. Nadeau and Y. Bengio. Inference for the generalization error. *J. Mach. Learn. Res.* 2003; **52**:239–281.
44. Y. EL-Manzalawy, D. Dobbs, and V. Honavar. Predicting linear B-cell epitopes using string kernels. *J. Mol. Recognit.* to appear.
45. J Salomon and DR Flower. Predicting class II MHC-peptide binding: a kernel based approach using similarity scores. *BMC Bioinformatics* 2006; **7**:501.

FAST AND ACCURATE MULTI-CLASS PROTEIN FOLD RECOGNITION WITH SPATIAL SAMPLE KERNELS

Pavel Kuksa, Pai-Hsi Huang, and Vladimir Pavlovic*

Department of Computer Science, Rutgers University, Piscataway, NJ 08854, USA

Email: {pkuksa;paihuang;vladimir}@cs.rutgers.edu

Establishing structural or functional relationship between sequences, for instance to infer the structural class of an unannotated protein, is a key task in biological sequence analysis. Recent computational methods such as profile and neighborhood mismatch kernels have shown very promising results for protein sequence classification, at the cost of high computational complexity. In this study we address the multi-class sequence classification problems using a class of string-based kernels, the sparse spatial sample kernels (SSSK), that are both biologically motivated and efficient to compute. The proposed methods can work with very large databases of protein sequences and show substantial improvements in computing time over the existing methods. Application of the SSSK to the multi-class protein prediction problems (fold recognition and remote homology detection) yields significantly better performance than existing state-of-the-art algorithms.

1. INTRODUCTION

Protein homology detection and structure annotation are fundamental problems in computational biology. With the advent of large-scale sequencing techniques, experimental elucidation of an unknown protein sequence function and structure becomes an expensive and tedious task. Currently, there are more than 61 million DNA sequences in GenBank [1], and approximately 349, 480 annotated and 5.3 million unannotated sequences in UNIPROT [2]. The rapid growth of sequence databases makes development of computational aids for functional and structural annotation a critical and timely task.

The goals of remote homology detection and remote fold detection are to infer functional and structural information based only on the primary sequence of an unknown protein. In this study, we address these two problems in the context of Structural Classification of Proteins (SCOP) [3]. In SCOP, a manually curated protein data set derived from PDB [4], sequences are grouped into a tree hierarchy containing classes, folds, superfamilies, and families, from root to leaf. The difficulty of the remote homology and structural similarity detection tasks arises from low sequence identities among proteins on the superfamily and fold levels.

Early approaches to computationally-aided homology detection, such as BLAST [5] and FASTA [6], rely on pairwise alignment. Later methods, such as profiles [7] and profile hidden Markov models (profile HMM) [8], collect aggregate statistics from a group of sequences known to belong to the same family or superfamily. Studies have shown that these *generative* methods are accurate in detecting close homology (family detection) with moderate sequence identities. However, when sequence identities are low, which is typical for superfamilies (remote homology) and folds (structural similarity), generative method becomes insufficient and therefore *discriminative* methods are necessary. For protein remote homology detection, several types of discriminative kernel methods were proposed, for example, SVMFisher [9] by Jaakkola *et al.* and the class of string kernels [10, 11] by Leslie *et al.* Both classes of kernels demonstrated improved discriminative power over generative methods. Most of the studies formulated *binary-class* problems. In a different task, fold recognition [12–15], studies formulated *multi-class* learning problems. Ding *et al.* , in [12], proposed to extract features based on amino acid compositions and physico-chemical properties and in [13], Ie *et al.* extended the profile kernel [11] framework with adaptive codes for fold recognition. Both fold recognition methods showed promising results on detecting structural similarities based on primary sequences only.

Protein classification problems are typically characterized by few *positive training* sequences accompanied by a large number of negative training examples, which may result in weak classifiers. Enlarging the training sample size by experimentally labeling the sequences is

*Corresponding author.

costly, leading to the need to leverage *unlabeled data* to refine the decision boundary. The profile [16] and the mismatch neighborhood [17] kernels use large unlabeled datasets and show significant improvements over the sequence classifiers trained under the supervised setting. However, the promising results are offset by a significant increase in computational complexity, hindering wide application of such powerful tools.

In this study, we consider a new family of string-based kernel methods, the sparse spatial sample kernels (SSSK), for the multi-class sequence classification tasks. The proposed kernels are induced by the features that sample the sequences at different resolutions while taking mutations, insertions and deletions into account. These features are low-dimensional and their evaluation incurs low computational cost. Such characteristics open the possibility for analyzing very large unlabeled datasets under the semi-supervised setting with modest computational resources. The proposed methods perform significantly better and run substantially faster than existing state-of-the-art algorithms, including the profile [16, 11] and neighborhood mismatch [17] kernels, for both remote homology and fold detection problems on three well-known benchmark datasets. Moreover, in a multi-class setting, use of SSSK does not incur the need for formulating a complex optimization problem, as suggested in [13, 14]; we obtain our performance in a straightforward manner with no parameter adjusting.

2. BACKGROUND

In this section, we briefly review existing state-of-the-art methods, under supervised and semi-supervised learning paradigms. We also briefly discuss the multi-class learning problem.

Supervised Methods: The spectrum-like kernels, the state-of-the-art string kernels in the supervised setting, *implicitly* map a sequence X to a $|\Sigma|^d$-dimensional vector, where Σ is the alphabet set. The $mismatch(k, m)$ kernel [10] relaxes exact string matching by allowing up to m mismatches, or substitutions, to accommodate the mutation process. The main drawback of the mismatch kernel is the exponential size of the induced feature space and the presence of mismatches, both of which, when combined, incur high computational cost.

Semi-supervised Methods: The performance of the supervised methods depends greatly on the availability and quality of the labeled training data. In the presence of limited number of labeled training sequences, the performance of the classifiers estimated under such setting, though promising [9, 10], is still sub-optimal. Enlarging the size of the training set will improve the accuracy of the classifiers; however, the cost of doing so by experimentally obtaining functional or other group labels for large numbers of protein sequences may be prohibitive, but the *unlabeled data* can still be leveraged to refine and potentially improve the decision boundary. Recent advances in computational methods for remote homology prediction have relied heavily on the use of such data sources [11, 17, 13]. The profile kernel [11] uses unlabeled data directly by constructing a profile and using local information in the profile to estimate the *mutation neighborhood* of all k-mers. Construction of profiles for each sequence may incur high computational cost since highly diverged regions in profiles may result in a mutational neighborhood size exponential in the number of k-mers.

Multi-class classification: One way to solve the multi-class learning problem is to directly formulate a multi-class optimization problem, as done in [18, 19]. An alternative is to combine *binary* predictors using one-vs-one or one-vs-rest schemes. For instance, in a one-vs-rest scheme, $|Y|$ classifiers are estimated, where Y is the output space, and the predicted class, $\hat{y}$, corresponds to the highest scoring classifier output (Equation 1) where f_y denotes the binary classifier for class y. In contrast to the simple decision rule, Ie *et al.* in [14] proposed to use the adaptive codes to tackle the multi-class fold recognition problem with decision rule in Equation 2 where $\times$ denotes component-wise multiplication, $\mathbf{f}(x)$ a 1-by-$(n_f + n_s)$ output vector from the binary classifiers and C_y a binary code matrix and W the parameters to estimate. Under such framework, one needs to train at least $n_f + n_s + n_{fa}$ independent binary classifiers, where n_f, n_s, and n_{fa} denote the number of folds, superfamilies and families, respectively.

$$\hat{y} = \operatorname*{argmax}_{y \in Y} f_y(x), \tag{1}$$

$$\hat{y} = \operatorname*{argmax}_{y \in Y} (W \times \mathbf{f}(x))C_y, \tag{2}$$

The practice of using one-vs-rest classifiers (Equation 1) has received both favorable and unfavorable comments. In [20], showed that formulating complex optimization problems, such as error-correcting codes [21], does not offer any advantage over the simple decision rules, for example, Equation 1. In contrast, Ie *et al.* in [14] argued that the simple decision rule can only cope with problems with small number of classes. At present, no clear evidence indicates one decision rule dominates the other. In this study, we use one-vs-rest scheme (Equation 1) and only estimate n_f or n_s binary classifiers for fold and superfamily detection, respectively.

3. THE SPARSE SPATIAL SAMPLE KERNELS

Sparse spatial sample kernels (SSSK) present a new class of string kernels that effectively model complex biological transformations (such as highly diverse mutation, insertion and deletion processes) and can be efficiently computed. The SSSK family of kernels, parametrized by three positive integers, assumes the following form:

$$K^{(t,k,d)}(X,Y) = \sum_{\substack{(a_1,d_1,\ldots,d_{t-1},a_t)\\ a_i\in\Sigma^k, 0\le d_i<d}} \begin{matrix} C(a_1,d_1,\cdots,a_{t-1},d_{t-1},a_t|X)\cdot \\ C(a_1,d_1,\cdots,a_{t-1},d_{t-1},a_t|Y) \end{matrix}, \quad (3)$$

where $C(a_1,d_1,\cdots,a_{t-1},d_{t-1},a_t|X)$ denotes the number of times we observe substring $a_1 \overset{d_1}{\leftrightarrow} a_2, \overset{d_2}{\leftrightarrow}, \cdots, \overset{d_{t-1}}{\longleftrightarrow} a_t$ (a_1 separated by d_1 characters from a_2, separated by d_2 characters from a_3, etc.) in X. This is illustrated in Figure 1.

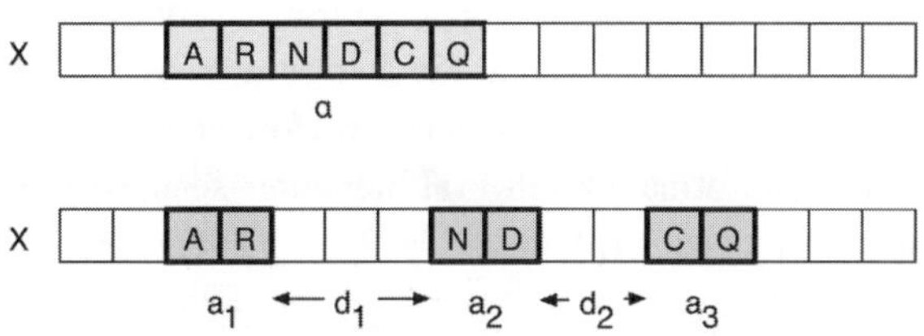

Fig. 1. Contiguous k-mer feature α of a traditional spectrum/mismatch kernel (top) contrasted with the sparse spatial samples of the proposed kernel (bottom).

The new kernel implements the idea of sampling the sequences at different resolutions and comparing the resulting spectra; similar sequences will have similar spectrum at one or more resolutions. This takes into account possible mutations, as well as insertions/deletions. Each sample consists of t spatially-constrained probes of size k, each of which lie no more than d positions away from its neighboring probes. In the proposed kernels, the parameter k controls the individual probe size, d controls the locality of the sample, and t controls the cardinality of the sampling neighborhood. In this work, we use short samples of size 1 (i.e., $k = 1$), and set t to 2 (i.e. features are pairs of monomers) or 3 (i.e. features are triples). These sample string kernels, unlike the family of spectrum kernels [10, 11], not only take into account the feature counts, but also include spatial configuration information, *i.e.* how the features are positioned in the sequence. The spatial information can be critical in establishing similarity of sequences under complex transformations such as the evolutionary processes in protein sequences. The addition of the spatial information experimentally demonstrates very good performance, even with very short sequence features (*i.e.* k=*1*), as we will show in Section 4; the use of short features also leads to significantly lower computational complexity of the kernel evaluations as shown in Section 5.1.

The use of short features can also lead to significantly lower computational complexity of the kernel evaluations. The dimensionality of the features induced by the proposed kernel is $|\Sigma|^t d^{t-1}$ for our choice of $k = 1$. As a result, for triple-(1,3) ($t = 3$, $k = 1$, $d = 3$) and double-(1,5) ($t = 2$, $k = 1$, $d = 5$) feature sets, the dimensionalities are $72,000$ and $2,000$, respectively, compared to $3,200,000$ for the spectrum-(k), mismatch(k,m) [10], and profile(k,σ) [11] kernels with the common choice of $k = 5$. The proposed kernels can be efficiently computed using sorting and counting. To compute the kernel values, we first extract the features from the sequences and sort the extracted features in linear time with counting sort. Then we count the number of distinct features and update the kernel matrix. For N sequences of the longest length n and u distinct features, computing the kernel matrix takes linear $O(dnN + min(u, dn)N^2)$ time. We provide a comprehensive comparison of the computational complexity and running times in Section 5.

3.1. Spatial sample neighborhood kernels

The proposed kernel can be extended to accommodate the semi-supervised learning setting for sequence classification, for example, as in [17]. Denote $\Phi^{orig}(X)$ as the original representation of the sequence X and $N(X)$ the *neighborhood* of X[a]; then, the smoothed

[a] We discuss how to define $N(X)$ in Section 4.2.

re-representation $\Phi^{new}(X)$ using the unlabeled data is defined as Equation 4 with the corresponding kernel, $K^{nbhd}(X,Y)$, in Equation 5:

$$\Phi^{new}(X) = \frac{1}{|N(X)|} \sum_{X' \in N(X)} \Phi^{orig}(X'), \quad (4)$$

$$K^{nbhd}(X,Y) = \sum_{X' \in N(X), Y' \in N(Y)} \frac{K(X',Y')}{|N(X)||N(Y)|}. \quad (5)$$

Note that in this setting, unlike in the traditional semi-supervised learning setting, both *training* and *test* sequences assume a new representation. Weston *et al.* in [17] showed that the discriminative power of the neighborhood mismatch kernel improves significantly; however such neighborhood kernel evaluation requires substantially longer computational time, as indicated by the Weston *et al.* in [16, 17].

4. EXPERIMENTAL RESULTS

In this section, we evaluate the proposed methods on multi-class remote fold recognition and multi-class remote homology detection in Sections 4.3 and 4.4, and multi-class fold as well as superfamily prediction in Section 4.5. We use three standard benchmark datasets to allow comparison with previously published results. The datasets used in our experiments and supplementary data are made available at `http://seqam.rutgers.edu/projects/bioinfo/spatial-kernels/csb08.html`.

4.1. Datasets

Ding and Dubchak dataset[12]: Ding *et al.* designed a challenging fold recognition data set [b], which has been used as a benchmark in many studies, for example [13]. The data set contains sequences from 27 folds divided into two *independent* sets, such that the training and test sequences share less than 35% sequence identities and within the training set, no sequences share more than 40% sequence identities.

Remote fold detection data set [14]: Melvin *et al.* derived this data set from SCOP 1.65 [3] for the tasks of multi-class remote fold detection. The data set contains 26 folds, 303 superfamilies and 652 families for training with 46 superfamilies held out for testing to model remote fold recognition setting.

Remote homology detection data set [14]: Ie *et al.* prepared a different data set for remote homology detection in a similar fashion. The derived data set contains 74 superfamilies and 544 families for training with from 110 families held out for testing.

The three datasets lead to a 27-, a 26- and a 74-way multi-class classification problems.

4.2. Settings, parameters used and performance measures

For all kernel methods, we perform kernel normalization to remove the dependencies between the kernel values and sequence lengths:

$$K'(X,Y) = \frac{K(X,Y)}{\sqrt{K(X,X)K(Y,Y)}}.$$

In all experiments, we use an existing implementation of SVM from a standard machine learning package SPIDER[c] with the linear kernel and default SVM parameters. For kernel smoothing (Equation 5) on each sequence X, we query the unlabeled database with 2 PSI-BLAST iterations and recruit the sequences with low e-values (≤ 0.05) to form the neighborhood $N(X)$. We use Swiss-Prot [22] (moderate size) and the *non-redundant* (large size) sets as unlabeled databases.

To evaluate our classifiers, we use zero-one and balanced error rates, as well as top-5 error rates, as in [12–14]. In addition, we also use standard performance measures from the information retrieval literature: sensitivity (recall) (r), precision (p) and $F1 = 2pr/(p+r)$ scores.

4.3. Comparison on Ding and Dubchak benchmark

We compare the performance of our methods under supervised and semi-supervised settings with previously published methods on Ding and Dubchak benchmark data set in Table 1. As can be seen from the table, our spatial kernels achieve higher performance compared to the state-of-the-art classifiers (we highlight the best performance in each category). Under the supervised setting, the triple spatial features consistently demonstrate better overall performance. The observed higher precision of the mismatch and profile kernels is achieved at

[b] `http://ranger.uta.edu/~chqding/bioinfo.html`
[c] `http://www.kyb.tuebingen.mpg.de/bs/people/spider`

the cost of lower recall rates; the F1 score, a function of both recall and precision measures, suggests that the triple spatial features achieve better performance. Under semi-supervised learning, we again observe better overall performance of spatial kernels. We also note that the spatial kernels strongly outperform the profile kernel evaluated in the same setting with 2 PSI-BLAST iterations as used by all our methods.

To demonstrate the benefit of leveraging unlabeled data, in Figure 2 we contrast the confusion matrices under the supervised (Figure 2(a)) and semi-supervised (Figure 2(b)) settings using the triple(1,3) feature set. Similar to Damoulas *et al.* [23], we observe that in the supervised setting, testing examples in classes with fewer training examples tend to be incorrectly assigned to two overly represented folds (7 and 16). However, such problem is alleviated in the semi-supervised setting when we enlarge the training sets with neighboring sequences which, in turn, reinforces class assignments of the target proteins. Under the supervised setting, we observe 3 folds with accuracy higher than 90%, one of which achieves 100% accuracy, while under the semi-supervised setting, we observe 9 folds with accuracy higher than 90%, 7 of which achieves 100% accuracy.

4.4. Remote fold and remote homology detection

We report the performance on remote fold and homology detection problems on the SCOP 1.65 benchmark data set in this section.

Table 2 corresponds to the remote fold prediction problem and we highlight the best performance in each category. The spatial kernels consistently outperform the one-vs-rest mismatch and profile kernels, as well as the best profile NR with adaptive codes, under both supervised and semi-supervised settings.

We also note that the performance of profile NR with adaptive codes was obtained by solving a complex optimization problem of minimizing the balanced error, whereas all other methods use the simple one-vs-rest decision rule (Equation 1). The use of the adaptive and error-correcting codes offers no clear evidence of advantage over the simple decision rule.

Table 3 summarizes results of the remote homology (superfamily) detection problem. Under the supervised learning, the spatial kernels consistently outperform the state-of-the-art mismatch(5,1) kernel. We make the same observation under the semi-supervised setting with the Swiss-Prot data set. With the non-redundant database, the spatial kernels consistently outperform the profile(5,7.5) kernel and, in particular, the triple (1,3) kernel shows the best top-5 error rate of 8.6%. We similarly outperform the highly optimized one-vs-rest profile NR by all measures. Further, compared to the profile NR, estimated using adaptive codes with complex optimization, our triple kernel with the simple one-vs-rest decision rule achieves better top-5 error and comparable error rates.

Substantial improvement in performance comes from the use of unlabeled data. We observe that use of a large unlabeled data set (non-redundant) results in significantly better performance over a Swiss-Prot data set, which is only moderate in size. The non-redundant data set provides a richer neighborhood for the data sequences, improving sensitivity of the methods. We take a further look at this improvement in Section 5.3.

4.5. Multi-class protein fold and superfamily recognition

For the remote homology and remote structural similarity detection, the goal is to detect a possibly new subclass within the class of interest. Another, simpler goal may be to match an unknown sequence to one of the known classes. We consider the direct multi-class protein fold and superfamily prediction on the SCOP 1.65 datasets and evaluate our classifiers using a 10-fold cross validation. We show the classification performance on the multi-class fold recognition and superfamily prediction tasks in Tables 4 and 5, respectively. The high error rates of 45% and 37% of the methods under the supervised setting highlight the difficulty of the problem. Tables 4 and 5, in both supervised and semi-supervised settings, indicate that the spatial kernels consistently outperform the mismatch and profile kernels on the multi-class fold recognition task.

5. DISCUSSION

The low computational complexity and running times are distinct characteristics of the spatial kernels, which we discuss in the following section. We also contrast the induced features of our kernel and those of traditional string kernels. Finally, we illustrate potential benefits of SSSK in view of the kernel-induced data manifolds on different string kernels.

Table 1. Comparison on Ding and Dubchak benchmark data set

Method	Error	Top 5 Error	Balanced Error	Top 5 Balanced Error	Recall	Top 5 Recall	Precision	Top 5 Precision	F1	Top5 F1
Supervised										
SVM(D&D)†	-	-	56.5	-	-	-	-	-	-	-
Mismatch(5,1)	51.17	22.72	53.22	28.86	46.78	71.14	**90.52**	**95.25**	61.68	81.45
Double(1,5)	44.13	23.50	46.19	23.92	53.81	76.18	61.90	79.85	57.57	77.97
Triple (1,3)	**41.51**	**18.54**	**44.99**	**21.09**	**55.01**	**78.91**	80.42	89.19	**65.33**	**83.74**
Semi-supervised (Swiss-Prot)										
Profile(5,7.5)	36.03	16.19	37.78	18.46	62.22	81.54	**88.39**	**94.53**	73.03	87.56
Double(1,5)	27.42	16.45	**21.81**	**13.26**	**76.57**	86.74	77.73	86.07	77.15	86.4
Triple(1,3)	**25.33**	**13.05**	22.72	13.27	76.27	**86.74**	84.48	92.05	**80.17**	**89.31**
Semi-supervised (Non-redundant data set)										
Profile(5,7.5)	31.85	15.14	32.17	16.73	67.83	83.27	**89.49**	**94.9**	77.16	88.71
Double(1,5)	28.72	14.99	24.74	**11.6**	75.26	**88.4**	76.02	86.86	75.63	87.62
Triple(1,3)	**24.28**	**12.79**	**22.38**	11.79	**77.62**	88.21	84.02	91.45	**80.69**	**89.8**
Profile NR(Perceptron)‡	-	-	26.5	-	-	-	-	-	-	-

All measures are presented as percentages.

†: quoted from [12]

‡: quoted from [14]

Table 2. Multi-class **remote** fold prediction

Method	Error	Top 5 Error	Balanced Error	Top 5 Balanced Error	Recall	Top 5 Recall	Precision	Top 5 Precision	F1	Top5 F1
Supervised										
Mismatch(5,1)	53.75	29.15	82.75	52.4	17.25	47.6	16.61	70	16.92	56.67
Double (1,5)	50.98	25.73	**70.77**	**37.6**	29.23	**62.49**	33.27	67.04	**31.12**	**63.26**
Triple (1,3)	**48.7**	**25.08**	73.04	44.05	**26.96**	55.95	**35.28**	**70.46**	30.57	62.37
Semi-supervised (Swiss-Prot)										
Profile(5,7.5)	49.35	20.36	76.67	35.28	23.33	64.72	29.47	71.82	26.05	68.09
Double (1,5)	**43**	19.22	**60.94**	27.76	**39.06**	72.24	**45.8**	70.48	**42.16**	71.35
Triple(1,3)	43.97	**15.64**	62.29	**26.77**	37.71	**73.23**	40.74	**77.04**	39.17	**75.08**
Semi-supervised (Non-redundant data set)										
Profile(5,7.5)	45.11	15.8	66.88	31.55	28.73	68.45	36.98	**84.63**	32.34	75.68
Double (1,5)	**36.65**	13.36	**47.87**	23.61	**49.59**	76.39	55.86	78.66	**52.54**	77.51
Triple (1,3)	37.13	**10.91**	49.34	20.07	47.19	**79.93**	**55.91**	82.78	51.18	**81.33**
Profile NR (one-vs-rest)‡	46.3	14.5	62.8	23.5	-	-	-	-	-	-
Profile NR (Adaptive codes)‡	37.0	11.4	49.9	**15.5**	-	-	-	-	-	-

‡: quoted from [14]

Table 3. Multi-class **remote** superfamily prediction

Method	Error	Top 5 Error	Balanced Error	Top 5 Balanced Error	Recall	Top 5 Recall	Precision	Top 5 Precision	F1	Top5 F1
Supervised										
Mismatch(5,1)	65.34	35.91	85.32	63.31	14.68	36.69	17.47	46.93	15.96	41.18
Double (1,5)	64.71	36.91	**79.16**	**49.34**	**20.84**	**50.66**	**19.98**	52.83	**20.4**	51.72
Triple (1,3)	**58.85**	**29.55**	80.34	52.05	19.66	47.95	18.93	**61.14**	19.29	**53.75**
Semi-supervised (Swiss-Prot)										
Profile(5,7.5)	41.4	19.58	66.07	39.51	33.93	60.49	34.63	64.34	34.28	62.36
Double (1,5)	32.04	15.09	46.99	**25.75**	53.01	**74.25**	53.37	76.21	53.37	**76.21**
Triple(1,3)	**28.8**	**14.34**	**46.98**	27.14	**53.07**	72.86	**55.84**	**76.28**	**54.39**	74.53
Semi-supervised (Non-redundant data set)										
Profile(5,7.5)	37.16	14.22	58.96	28.86	41.05	71.15	46.95	72.19	43.80	72.19
Double(1,5)	22.44	9.85	39.21	18.69	**60.79**	81.31	56.85	80.58	58.76	80.94
Triple(1,3)	22.94	**8.6**	39.86	17.29	60.14	**82.72**	**61.62**	**83.76**	**60.87**	**83.24**
Profile NR (one-vs-rest)‡	27.1	10.5	44.5	19.7	-	-	-	-	-	-
Profile NR (Adaptive codes)‡	**21.7**	10.3	**32.0**	**15.3**	-	-	-	-	-	-

‡: quoted from [14]

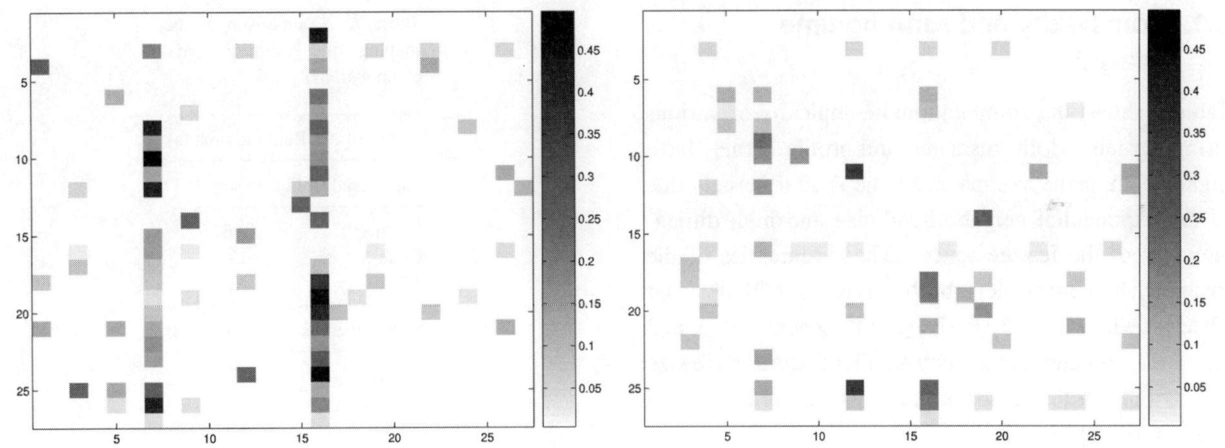

(a) (b)

Fig. 2. (a) Confusion matrix using the triple(1,3) feature set under the supervised setting. (b) Confusion matrix using the triple(1,3) feature set under the semi-supervised setting using the non-redundant (NR) unlabeled sequence database. In both figures, we remove the main diagonal terms to emphasize the differences in off-diagonal (error) terms.

Table 4. Multi-class protein fold recognition (10-fold cross-validation)

Method	Error	Top 5 Error	Balanced Error	Top 5 Balanced Error	Recall	Top 5 Recall	Precision	Top 5 Precision	F1	Top5 F1
Supervised										
Mismatch(5,1)	23.81 (1.25)	6.86 (1.32)	44.54 (1.94)	16.29 (2.47)	55.46	83.71	84.87	97.43	67.09	90.05
Double (1,5)	23.52 (1.38)	9.09 (0.63)	38.76 (3.18)	14.46 (1.95)	61.25	85.54	69.03	89.01	64.91	87.24
Triple (1,3)	**19.79** (1.04)	**5.78** (0.61)	**37.85** (2.94)	**12.41** (2.75)	**62.15**	**87.59**	**86.67**	**97.46**	**72.39**	**92.26**
Semi-supervised (Swiss-Prot)										
Profile(5,7.5)	12.20 (1.49)	4.09 (1.01)	22.05 (3.92)	9.38 (2.48)	77.95	90.62	**95.12**	**98.42**	85.68	94.36
Double (1,5)	9.64 (1.61)	4.14 (0.86)	15.36 (2.72)	6.53 (1.59)	84.64	**93.48**	89.22	95.10	86.87	94.28
Triple(1,3)	**8.60** (1.39)	**3.57** (0.56)	**15.21** (3.09)	**8.48** (1.69)	**84.79**	91.52	94.52	97.71	**89.39**	**94.51**
Semi-supervised (Non-redundant data set)										
Profile (5,7.5)	9.74 (1.32)	2.72 (0.65)	16.72 (2.11)	6.52 (1.98)	82.36	93.49	95.79	**98.74**	88.57	96.04
Double(1,7)	6.61 (0.86)	2.87 (0.80)	10.20 (1.66)	4.07 (1.46)	89.43	95.93	92.81	96.69	91.09	96.31
Triple (1,3)	**5.78** (0.98)	**1.76** (0.45)	**10.06** (1.84)	**3.35** (1.19)	**89.61**	**96.65**	**96.29**	98.73	**92.83**	**97.68**

*standard deviation for cross-validation error rates is indicated in parenthesis

Table 5. Multi-class superfamily prediction (10-fold cross-validation)

Method	Error	Top 5 Error	Balanced Error	Top 5 Balanced Error	Recall	Top 5 Recall	Precision	Top 5 Precision	F1	Top5 F1
Supervised										
Mismatch(5,1)	21.86 (1.28)	9.42 (1.49)	37.12 (1.70)	18.77 (2.61)	62.88	81.23	80.24	91.62	70.51	86.11
Double (1,5)	23.04 (1.67)	9.98 (1.81)	35.16 (2.09)	15.07 (2.55)	64.84	84.93	70.43	88.13	67.52	86.5
Triple (1,3)	**18.09** (1.48)	**7.17** (1.12)	**31.31** (2.45)	**14.01** (2.15)	**68.69**	**85.99**	**82.21**	**93.37**	**74.84**	**89.53**
Semi-supervised (Swiss-Prot)										
Profile(5,7.5)	8.69 (1.86)	4.12 (1.02)	14.67 (3.68)	8.33 (2.51)	85.33	91.69	93.15	96.07	89.07	93.82
Double (1,5)	6.03 (1.13)	3.05 (0.60)	**8.02** (2.18)	4.04 (1.39)	**91.98**	95.95	93.59	96.48	92.77	96.22
Triple (1,3)	**5.43** (0.69)	**2.25** (0.39)	8.31(1.39)	**4.03** (1.09)	91.69	**95.97**	**95.76**	**97.97**	**93.68**	**96.96**
Semi-supervised (Non-redundant dataset)										
Profile(5,7.5)	6.20 (1.36)	2.59 (0.51)	10.91 (2.60)	5.27 (1.52)	89.09	94.73	94.79	97.63	91.85	96.16
Double (1,5)	3.9 (0.95)	1.85 (0.37)	**5.14** (1.45)	2.38 (0.82)	**94.86**	97.62	95.42	97.93	95.14	97.77
Triple(1,3)	**3.39** (0.83)	**1.39** (0.59)	5.24 (1.55)	**2.10** (1.19)	94.76	**97.89**	**97.00**	**98.89**	**95.86**	**98.39**

*standard deviation for cross-validation error rates is indicated in parenthesis

5.1. Complexity and running time analysis

Table 6 shows the computational complexity of various string kernels. Both mismatch and profile kernels have higher complexity compared to the spatial kernels due to the exponential neighborhood size and high dimensionality of the feature space. The cardinalities of the mismatch and profile neighborhoods are $O(k^m|\Sigma|^m)$ and $O(M_\sigma)$, with $k^m|\Sigma|^m \leq M_\sigma \leq |\Sigma|^k$, where $k \geq 5$, and $|\Sigma| = 20$, compared to a much smaller feature space size of $d^{t-1}|\Sigma|^t$ for the sample kernels, where t is 2 or 3, and d is 3 or 5, respectively.

Table 6. Complexity of computations

Method	Time complexity
Supervised methods	
Triple kernel	$O(d^2nN + d^2|\Sigma|^3N^2)$
Double kernel	$O(dnN + d|\Sigma|^2N^2)$
Mismatch	$O(k^{m+1}|\Sigma|^m nN + |\Sigma|^k N^2)$
Semi-supervised methods	
Triple kernel	$O(d^2HnN + d^2|\Sigma|^3N^2)$
Double kernel	$O(dHnN + d|\Sigma|^2N^2)$
Mismatch	$O(k^{m+1}|\Sigma|^m HnN + |\Sigma^k|N^2)$
Profile kernel	$O(kM_\sigma nN + |\Sigma|^k N^2)$

Notations used in the table:
N-number of sequences, n-sequence length,
H is the sequence neighborhood size, $|\Sigma|$ is the alphabet size
k, m are the mismatch kernel parameters
($k = 5, 6$ and $m = 1, 2$ in most cases)
M_σ is the profile neighborhood size,
$k^m|\Sigma|^m \leq M_\sigma \leq |\Sigma^k|$
d is the distance parameter for the spatial kernel.

This complexity difference leads to order-of-magnitude improvements in the running times, as shown in Table 7, of the spatial kernels over the standard string kernels (mismatch and profile).

The running time measurements are obtained on a single 2.8GHz CPU machine. The code used for evaluation of the competing methods has been highly optimized to perform on par or better than the published spectrum/mismatch kernel code. We also used an existing implementation of the profile kernel provided by Kuang *et al.* in [16].

The neighborhood mismatch kernel becomes substantially more expensive to compute for large datasets, as indicated in [16, 17]. Table 8 summarizes the size of the unlabeled datasets and the mean, median, and maximum number of neighbors used for kernel smoothing.

Table 7. Comparison of the running time (kernel matrix computations)

Method	Running time (s)
Supervised (fold data set)	
Mismatch	396
Double	22
Triple	52
Semi-supervised (fold data set)	
Profile	1633
Double	165
Triple	701
Mismatch	-

Table 8. Data set characteristics

Data set	# Seq.	# Neighbors (mean/median/max)	
		Superfamily	Fold
Swiss-Prot	101602	42/30/244	30/17/174
NR	534936	79/58/356	52/28/360

5.2. Comparison of the features induced by different string kernels

Compared to mismatch/profile kernels, the feature sets induced by our kernels cover segments of variable length (e.g., $2-6$ residues in the case of the double-$(1,5)$ kernel), whereas the mismatch and profile kernels cover segments of the fixed length (e.g., 5 or 6 residues long) as illustrated in Figure 1. Sampling at different resolutions also allows to capture similarity in the presence of more complex substitution, insertion, and deletion processes, whereas sampling at a fixed resolution, the approach used in mismatch and spectrum kernels, limits the sensitivity in the case of multiple insertions/deletions or substitutions. Increasing the parameter m (number of mismatches allowed) to accommodate the multiple substitutions, in the case of mismatch/spectrum kernels, leads to an exponential growth in the neighborhood size, and results in high computational complexity.

To further illustrate the differences and the trade-off between different features, we consider an example of modeling a slightly diverged region using the mismatch and spatial kernel similarity measures. We first compare the spectrum-induced features with our proposed spatial features, extracted from a string S = 'HKYNQLIM', in Figure 3(a). The symbol 'x' in the mismatch-(k,m) fea-

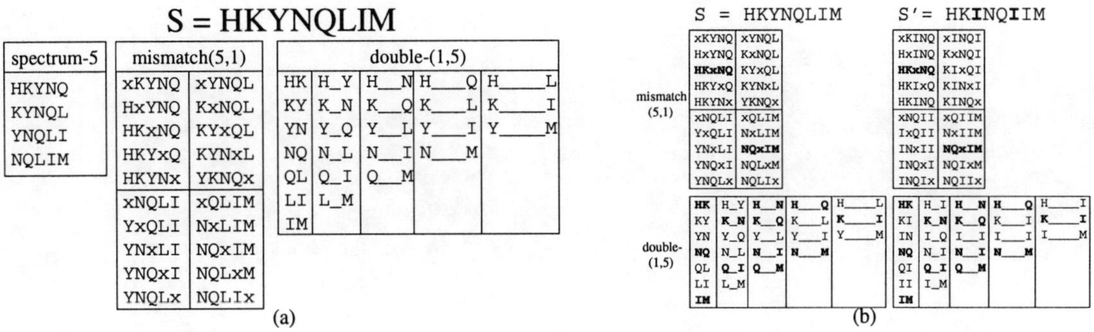

Fig. 3. (a) Comparison of features extracted by the spectrum-like and spatial kernels. In the mismatch features, each symbol 'x' represent an arbitrary symbol in the alphabet set. As a result, each feature basis corresponds to $|\Sigma|$ features. (b) Differences in handling substitutions by the mismatch and spatial features. We represent all common features between the original and the mutated strings, S and S', with bold fonts.

tures corresponds to an arbitrary symbol in Σ. As a result, each mismatch feature basis in the figure corresponds to $|\Sigma|$ features. With such representation, the number of neighboring k-mers induced by an observation grows exponentially with the choice of m, the number of mismatches allowed. In contrast, the spatial features scans the strings at different resolutions and the number of features induced is small. Furthermore, as shown in Figure 3(b), for two slightly diverged strings, S and S', very few common features (bold font) are observed in the case of the mismatch kernel, leading to low similarity scores. On the other hand, the larger subset of common features indicates that the spatial kernels are still able to capture the similarity between the two sequences. For the mismatch features to handle such a slightly diverged region, one needs to increase the number of allowed mismatches m, at the expense of an increase in computational effort.

5.3. Kernel-induced data manifolds

To shed more light on the causes of improved performance of SSSK, we compare the data manifolds induced by different kernels in both supervised and semi-supervised settings[d]. We show the kernel-induced manifolds for the *double-stranded beta-helix* (b.82) fold in Figures 4(a) and 4(b) for the supervised setting and in Figures 4(c) and 4(d) for the semi-supervised setting. The fold contains proteins carrying out a diverse range of functions and participating in many biological processes. Each node in the graph represents a sequence, with darker nodes corresponding to the training sequences and lighter nodes corresponding to the test sequences (superfamily b.82.3). Each cluster (box) represents a superfamily in the fold. We normalize the kernel as discussed in Section 4.2 to remove the dependencies between kernel values and sequence length. We draw an edge between two sequences, X and Y, if $K(X,Y) > \delta$, (δ is chosen so that the total number of nodes outside the fold having similarity values above the threshold with nodes inside the fold is small).

In the supervised setting (Figures 4(a) and 4(b)), we observe a slightly more connected graph induced by the triple kernel compared to the mismatch kernel. Similarly, in the semi-supervised setting (Figures 4(c) and 4(d)) with the non-redundant set, compared to the profile kernel the triple kernel induces a data manifold with stronger connectivity, suggesting better sensitivity of the spatial kernels (on this fold, the triple and profile kernels achieve 91.67% and 83.33% recall rate, both with 100% precision). This, in turn, leads to lower error rates of classifiers with the SSSK.

6. CONCLUSIONS

We present a new family of sparse spatial sample kernels that demonstrate state-of-the-art performance for multi-class protein fold and remote homology prediction problems. The key component of the method is the spatially-constrained sample kernel for efficient sequence comparison which, combined with kernel smoothing using unlabeled data, leads to efficient and accurate semi-supervised protein remote homology detection and remote fold recognition. We show that our methods can work with large, unlabeled databases of protein sequences, taking full advantage of all available data and

[d]We use the *fdp* package in *Graphviz* `http://graphviz.org` for visualization.

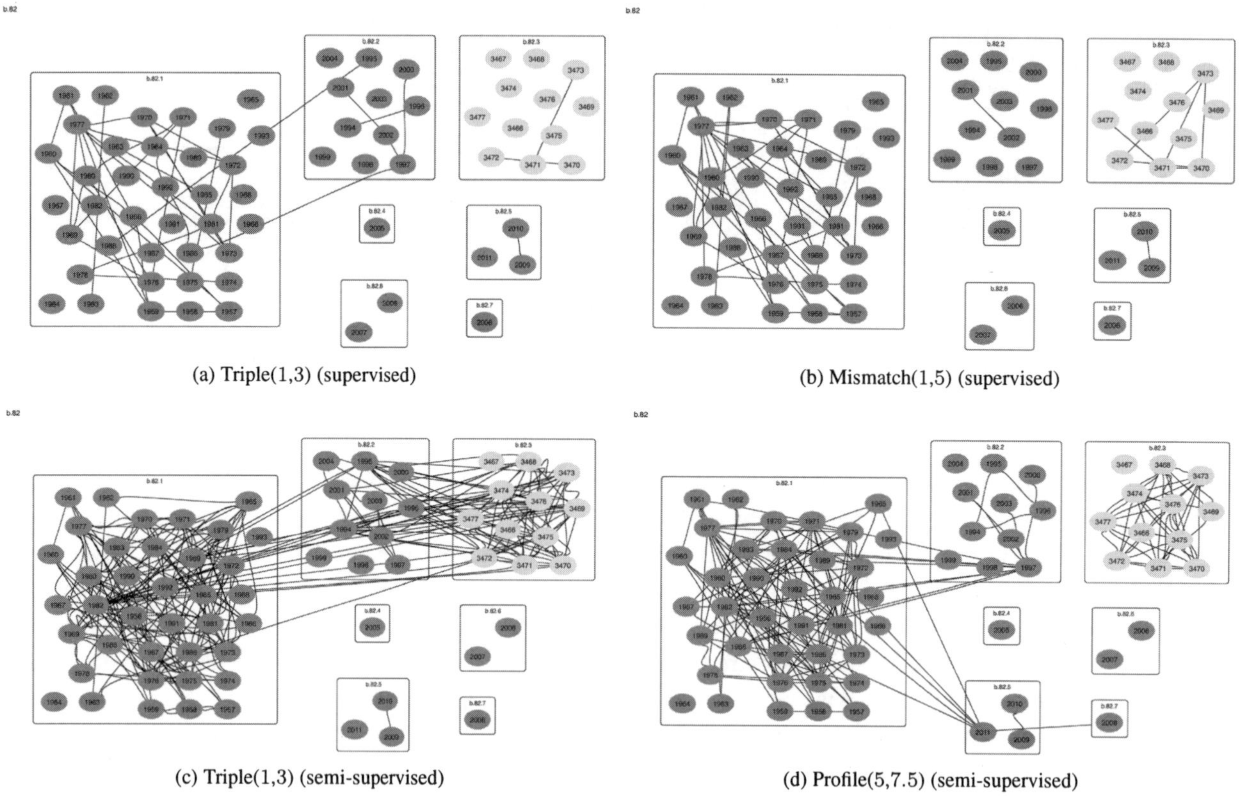

(a) Triple(1,3) (supervised)

(b) Mismatch(1,5) (supervised)

(c) Triple(1,3) (semi-supervised)

(d) Profile(5,7.5) (semi-supervised)

Fig. 4. Kernel-induced data manifold for fold $b.82$, with 7 superfamilies, under the supervised and semi-supervised settings. The darker and lighter nodes are the training and testing sequences, respectively. The numbers in the nodes index the sequences in the database.

substantially improving the classification accuracy. This opens the possibility for the proposed methodology to be readily applied to other challenging problems in biological sequence analysis.

References

1. Dennis A. Benson, Ilene Karsch-Mizrachi, David J. Lipman, James Ostell, and David L. Wheeler. Genbank. *Nucl. Acids Res.*, 33(suppl-1):D34–38, 2005.
2. Amos Bairoch, Rolf Apweiler, Cathy H. Wu, Winona C. Barker, Brigitte Boeckmann, Serenella Ferro, Elisabeth Gasteiger, Hongzhan Huang, Rodrigo Lopez, Michele Magrane, Maria J. Martin, Darren A. Natale, Claire O'Donovan, Nicole Redaschi, and Lai-Su L. Yeh. The Universal Protein Resource (UniProt). *Nucl. Acids Res.*, 33(suppl-1):D154–159, 2005.
3. L. Lo Conte, B. Ailey, T.J. Hubbard, S.E. Brenner, A.G. Murzin, and C. Chothia. SCOP: a structural classification of proteins database. *Nucleic Acids Res.*, 28:257–259, 2000.
4. H. M. Berman, J. Westbrook, Z. Feng, G. Gilliland, T.N. Bhat, H. Weissig, I. N. Shindyalov, and P. E. Bourne. The Protein Data Bank. *Nucleic Acids Research*, 28:235–242, 2000.
5. S.F. Altschul, W. Gish, W. Miller, E.W. Myers, and D.J. Lipman. Basic Local Alignment Search Tool. *Journal of Molecular Biology*, pages 403–410, 1990.
6. W.R. Pearson and D.J. Lipman. Improved tools for biological sequence comparison. *Proceedings of the National Academy of Sciences of the United States of America*, 85:2444–2448, 1988.
7. M. Gribskov, A.D. McLachlan, and D. Eisenberg. Profile analysis: detection of distantly related proteins. *Proceedings of the National Academy of Sciences*, 84:4355–4358, 1987.
8. SR Eddy. Profile hidden Markov models. *Bioinformatics*, 14(9):755–763, 1998.
9. Tommi Jaakkola, Mark Diekhans, and David Haussler. A discriminative framework for detecting remote protein homologies. In *Journal of Computational Biology*, volume 7, pages 95–114, 2000.
10. Christina S. Leslie, Eleazar Eskin, Jason Weston, and William Stafford Noble. Mismatch string kernels for svm protein classification. In *NIPS*, pages 1417–1424, 2002.
11. Rui Kuang, Eugene Ie, Ke Wang, Kai Wang, Mahira Siddiqi, Yoav Freund, and Christina Leslie. Profile-based string kernels for remote homology detection and motif extraction. In *CSB '04: Proceedings of the 2004 IEEE Computational Systems Bioinformatics*

Conference (CSB'04), pages 152–160, August 2004. http://www.cs.columbia.edu/compbio/profile-kernel.

12. Chris H.Q. Ding and Inna Dubchak. Multi-class protein fold recognition using support vector machines an d neural networks . *Bioinformatics*, 17(4):349–358, 2001.
13. Eugene Ie, Jason Weston, William Stafford Noble, and Christina Leslie. Multi-class protein fold recognition using adaptive codes. In *ICML '05: Proceedings of the 22nd international conference on Machine learning*, pages 329–336, New York, NY, USA, 2005. ACM.
14. Iain Melvin, Eugene Ie, Jason Weston, William Stafford Noble, and Christina Leslie. Multi-class protein classification using adaptive codes. *J. Mach. Learn. Res.*, 8:1557–1581, 2007.
15. Jianlin Cheng and Pierre Baldi. A machine learning information retrieval approach to protein fold recognition. *Bioinformatics*, 22(12):1456–1463, June 2006.
16. Rui Kuang, Eugene Ie, Ke Wang, Kai Wang, Mahira Siddiqi, Yoav Freund, and Christina Leslie. Profile-based string kernels for remote homology detection and motif extraction. *J Bioinform Comput Biol*, 3(3):527–550, June 2005.
17. Jason Weston, Christina Leslie, Eugene Ie, Dengyong Zhou, Andre Elisseeff, and William Stafford Noble. Semi-supervised protein classification using cluster kernels. *Bioinformatics*, 21(15):3241–3247, 2005.
18. J. Weston and C. Watkins. Support vector machines for multiclass pattern recognition. In *Proceedings of the Seventh European Symposium On Artificial Neural Networks*, 4 1999.
19. Vladimir N. Vapnik. *Statistical Learning Theory*. Wiley-Interscience, September 1998.
20. Ryan Rifkin and Aldebaro Klautau. In defense of one-vs-all classification. *J. Mach. Learn. Res.*, 5:101–141, 2004.
21. Thomas G. Dietterich and Ghulum Bakiri. Solving multiclass learning problems via error-correcting output codes. *Journal of Artificial Intelligence Research*, 2:263–286, 1995.
22. B. Boeckmann, A. Bairoch, R. Apweiler, M.C. Blatter, A. Estreicher, E. Gasteiger, M.J. Martin, K. Michoud, C. O'Donovan, I. Phan, S. Pilbout, and M. Schneider. The SWISS-PROT protein knowledgebase and its supplement TrEMBL in 2003. *Nucleic Acids Res*, 31:365–370, 2003.
23. Theodoros Damoulas and Mark A. Girolami. Probabilistic multi-class multi-kernel learning: on protein fold recognition and remote homology detection. *Bioinformatics*, 24(10):1264–1270, 2008.

DESIGNING SECONDARY STRUCTURE PROFILES FOR FAST NCRNA IDENTIFICATION

Yanni Sun* and Jeremy Buhler

Department of Computer Science and Engineering, Washington University, St. Louis, MO 63130, USA

Email: (yanni,jbuhler)@cse.wustl.edu

Detecting non-coding RNAs (ncRNAs) in genomic DNA is an important part of annotation. However, the most widely used tool for modeling ncRNA families, the covariance model (CM), incurs a high computational cost when used for search. This cost can be reduced by using a filter to exclude sequence that is unlikely to contain the ncRNA of interest, applying the CM only where it is likely to match strongly. Despite recent advances, designing an efficient filter that can detect nearly all ncRNA instances while excluding most irrelevant sequences remains challenging.

This work proposes a systematic procedure to convert a CM for an ncRNA family to a *secondary structure profile (SSP)*, which augments a conservation profile with secondary structure information but can still be efficiently scanned against long sequences. We use dynamic programming to estimate an SSP's sensitivity and FP rate, yielding an efficient, fully automated filter design algorithm. Our experiments demonstrate that designed SSP filters can achieve significant speedup over unfiltered CM search while maintaining high sensitivity for various ncRNA families, including those with and without strong sequence conservation. For highly structured ncRNA families, including secondary structure conservation yields better performance than using primary sequence conservation alone.

1. INTRODUCTION

Non-coding RNAs (ncRNAs) are transcribed but are not translated into protein. Annotating common ncRNAs, such as tRNAs and microRNAs, as well as non-coding structures like riboswitches in mRNAs, is important because of their functions in many biological processes[1]. The function of an ncRNA is determined not only by its sequence but also by its secondary structure. Exploiting this structural signal can improve ncRNA homology detection[2].

The state-of-the-art method to recognize an ncRNA of known family is to align it to a covariance model (CM). A CM is a stochastic context-free grammar (profile SCFG)[3, 4] that describes an ncRNA family's sequence and secondary structure conservation. Aligning an RNA to a CM uses a probabilistic variant of the well-known CYK parsing algorithm[5]. CM alignment has been implemented in the INFERNAL software suite[2]. In conjunction with a database of CMs, such as Rfam[6], INFERNAL can be used to annotate ncRNAs in genomic DNA.

A major challenge of CM alignment is its high computational cost. Probabilistic CYK is a cubic-time algorithm with a significant constant factor. For example, Weinberg et al. estimated that it would take about 1 CPU year to locate all tRNAs in an 8-Gbase DNA database on a 2.8 GHz Intel Pentium 4 PC[16]. Although CPUs have gotten faster, and INFERNAL implements optimizations designed to lower the cost of CYK, speeding up CM alignment remains a major problem for annotating large sequences or for classifying many sequences into one of the many known ncRNA families.

One approach to accelerate CM alignment is to use a filter to exclude "unpromising" sequences. Sequences that pass the filter have a higher probability of containing the target ncRNA and so are aligned with the full CM. Careful filter design can effectively accelerate pairwise DNA sequence comparison[8–11] as well as alignment of a sequence to a profile hidden Markov model (pHMM)[12, 13].

Several filtering strategies have been proposed to speed up CM search[7, 14–18]. Construction of the Rfam database uses primary sequence comparison with BLAST to exclude sequences unlikely to be part of an ncRNA family[6]. Weinberg and Ruzzo developed a pHMM-based, lossless filtering strategy for arbitrary ncRNA families[7] as well as a faster but lossy strategy that designs a pHMM from a CM for a family[15]. A disadvantage of all these filters is that they forgo the opportunity to exploit RNA structural conservation. Moreover, while pHMMs can be

*Corresponding author.

scanned against a database much more efficiently than CMs, their computational cost remains an issue for large database searches.

Other types of filter exploit RNA structural conservation. Weinberg and Ruzzo used a sub-CM structure [16] to improve the filtering sensitivity for ncRNA families with low sequence conservation. However, the filter design process is expensive (1 to 50 CPU hours per family), and using the resulting filters leads to a slow filtration process. Zhang et al.[17, 20] used the (k, w)-stack as the basis for their filter design. It is not clear whether their method can be used to design filters for a large database of ncRNA families because the authors need to choose optimal filters empirically by trying different parameters[20]. In addition, using (k, w)-stacks may not be optimal for many families with strong sequence conservation.

Recently, Zhang et al.[18] designed a *chain filter*, based on a collection of short matches to conserved words in a CM, that can sensitively and efficiently identify riboswitch elements. The chain filter does not consider structural conservation either. Moreover, the design of such filters requires specifying score thresholds for matches to the various words in the filter. The procedure for selecting thresholds was not described in Ref. 18, so the design of chain filters appears less than fully automated.

In this work, we describe a robust, efficient approach to fully automated filter design for ncRNA search. We show how to design filters starting from a CM using *secondary structure profiles (SSPs)*, which recognize both primary sequence and secondary structure conservation. The main properties of our filters and filter design program are:

- SSP matching is a simple extension of the standard profile matching algorithm and has linear time complexity;
- Designing SSPs from CMs is efficient;
- SSP-based filters generalize to ncRNA families of all types;
- The match score threshold for an SSP can be automatically computed from its CM, using a practically accurate model-based estimate of its specificity in background DNA sequence.

SSPs were first used in the ERPIN program[19] to characterize RNA signatures. They generalize profiles (a.k.a. position specific score matrices) by incorporating probability distributions for conserved base pairs. The main difference between our data structure and ERPIN's is that our SSP can accommodate gaps inside stacks, such as bulges. Also, we use different methods for SSP design and scanning.

Our method constructs a list of candidate SSPs from a given CM, then uses dynamic programming, first to assign a threshold to each SSP to control its false positive (FP) rate and then to estimate each SSP's sensitivity. The candidate SSP that maximizes sensitivity subject to an upper bound on its FP rate is chosen as the final filter. The sensitivity and FP rate computed via dynamic programming are typically good predictors of a filter's performance on real sequences, so their computation allows us to fully automate selection of SSPs and their associated score thresholds. We extend our filtering strategy to use multiple SSPs to improve the trade-off between sensitivity and FP rate.

Our results demonstrate that automatically designed SSP filters have an average speedup of about 200x over INFERNAL 0.7 without filtration yet detect almost all ($\geq$ 99% of) occurrences of most ncRNA families we tested. For highly structured ncRNA families with limited sequence conservation, such as tRNAs and 5S rRNA, we show that including secondary structure conservation in an SSP yields a better sensitivity/FP rate tradeoff than relying on primary sequence conservation alone.

The remainder of this paper is organized as follows. Section 2 briefly reviews CMs and formally defines SSPs. Section 3 describes how to construct SSPs from a CM and how to evaluate an SSP's performance. In Section 4, we first demonstrate the advantages of using SSPs versus primary conservation profiles on ncRNA families drawn from BRAliBase III[21]. We then measure the sensitivity, FP rate, and speedup obtained for 233 ncRNA families from the Rfam database. We also compare SSPs with other types of filter. Finally, Section 5 concludes and suggests directions for future work.

2. CMS AND SSPS

This section briefly reviews CMs and formally defines our secondary structure profiles (SSPs). To distin-

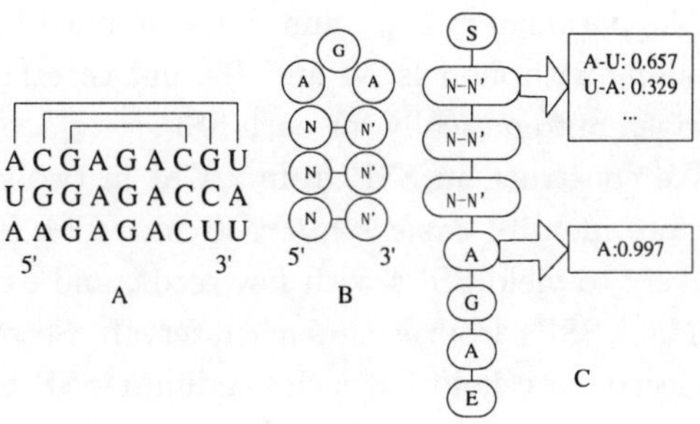

Fig. 1. (A) ungapped alignment of RNAs with three conserved base pairings; (B) corresponding secondary structure; (C) CM describing the structure.

guish an SSP, which includes structural information, from a profile that describes only the primary sequence conservation at a series of sequence positions, we call the latter a *regular profile* hereafter.

2.1. Covariance models

A CM consists of a series of groups of states, which are associated with base pairs and unpaired bases in an RNA structure.

Figure 1 shows the structure of a simple CM built from an RNA multiple sequence alignment annotated with three base-pair interactions. This CM contains start state S, end state E, three states emitting base pairs, and three states emitting unpaired bases. Each state is associated with an emission distribution; for example, the top paired state emits A-U and U-A pairs only. States are connected by transitions with associated probabilities. All transitions have probability 1 in the example, but insertions and deletions in a structure can be modeled by states with lower-probability in-transitions. The key observations for our work are that (1) the emitting states of a CM encode both primary sequence conservation and the locations of paired bases, and (2) the transition probabilities between these states encode how often a given state is present in a randomly chosen path through the CM.

More detailed descriptions of CMs and probabilistic CYK can be found in Ref. 4.

2.2. Secondary structure profiles

SSPs augment regular profiles by characterizing base pair frequencies in an RNA structure. Hence, unlike a regular profile, we must tell an SSP which pairs of

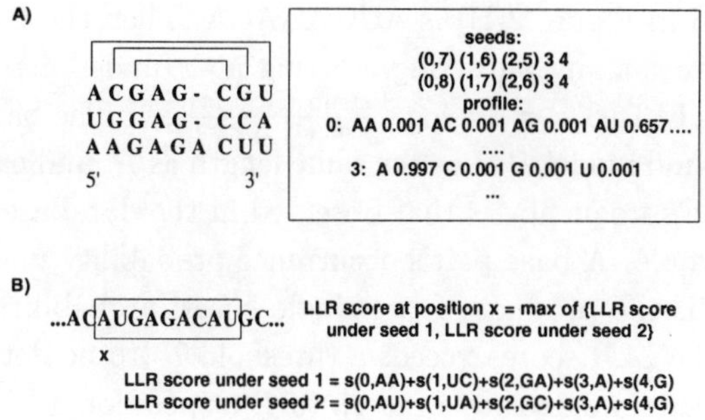

Fig. 2. (A) gapped alignment of RNAs with three base pairing interactions, with a corresponding SSP. Two seeds handle the possibility of an insertion after position 4. Column 0 may pair with column 7 or 8, resulting in seed pairs $(0, 7)$ and $(0, 8)$; (B) computation of LLR score at offset x in an input sequence.

bases it inspects are expected to be complementary. Figure 2(A) shows an example of an SSP.

An SSP $\mathcal{P}$ consists of two components. The first component contains one or more *seeds* that designate paired and unpaired base positions. A seed π of length ℓ is an ordered list of ℓ single or paired values. A single value π_i denotes that the ith base relative to the start of the SSP is unpaired, while a pair of values (π_i^1, π_i^2), with $\pi_i^1 < \pi_i^2$, indicates that positions π_i^1 and π_i^2 are paired. To describe common variations in the locations of paired bases caused by insertion and deletion, an SSP may include multiple seeds. Note that the set of positions described by a seed need not be contiguous in the sequence.

The second component of an SSP describes emission distributions for bases or base pairs in the alignment. For example, the probability of A-U at offsets specified by the first element of both seeds is $\mathcal{P}_{0,AU} = 0.657$. Note that all seeds have the same length as the number of rows in the profile, since each SSP has only one profile.

During search, an SSP $\mathcal{P}$ is aligned with a sequence S at each possible offset $0 \leq x < |S|$. The hypothesis that the bases of S matched by some seed π at offset x are generated from the emission distributions of $\mathcal{P}$ is compared to the null hypothesis that the positions come from a background model $\mathcal{P}^0$, using a log likelihood ratio (LLR). Starting at any offset x in S, we extract bases of S using positions specified in a seed π. Define the concatenation of those bases as $\text{substr}(S, x, \pi)$. For example, the substring starting at x under the first seed in Figure 2(B) is AUGAGACA. Then the LLR score for any substring starting at x under a seed π is $\text{LLR}(S, x, \pi) = \log \frac{\Pr(\text{substr}(S,x,\pi)|\mathcal{P})}{\Pr(\text{substr}(S,x,\pi)|\mathcal{P}^0)}$. The background model $\mathcal{P}^0$ has the same length as $\mathcal{P}$, and each base's frequency is that observed in the database as a whole. A base pair's occurrence probability under $\mathcal{P}^0$ is the product of two single bases' probabilities. If the LLR score exceeds a threshold T (to be determined), we declare a match to $\mathcal{P}$ at position x.

$\text{LLR}(S, x, \pi)$ can be computed as the sum of LLR scores of individual bases or pairs in $\text{substr}(S, x, \pi)$. For bases a and a', let $s(i, aa')$ and $s(i, a)$ be the LLR scores of base pair aa' or base a at the ith column of the SSP. For example, $s(0, \text{A-U}) = \log \frac{\mathcal{P}_{0,AU}}{\mathcal{P}^0_{0,AU}}$. Figure 2(B) shows the computation of LLR scores starting at x in S under two input seeds. Considering all the seeds $\pi \in \mathcal{P}$ and all the offsets $0 \leq x < |S|$, we define the final LLR score between S and $\mathcal{P}$ as

$$\text{LLR}(S, \mathcal{P}) = \max_{x,\pi} \text{LLR}(S, x, \pi).$$

3. DESIGNING SSPS FROM A CM

This section describes our algorithm to derive SSPs from a CM. We begin by formally defining the problem. For an SSP filter $\mathcal{P}$ and associated threshold T derived from a CM $\mathcal{M}$, $\mathcal{P}$'s *sensitivity* to $\mathcal{M}$ is defined as the probability that $\mathcal{P}$ matches a sequence generated by $\mathcal{M}$: $Pr_{S\sim\mathcal{M}}(\text{LLR}(S, \mathcal{P}) \geq T)$. The *false positive (FP) rate* for $\mathcal{P}$ at threshold T is defined as the probability of seeing an SSP match with score $\geq T$ at a random position in a background sequence. Thus, the FP rate is $Pr_{S\sim\mathcal{P}^0}(\text{LLR}(S, \mathcal{P}) \geq T)$, where background model $\mathcal{P}^0$ is the same as in Section 2.2 and $|\mathcal{P}^0| = |\mathcal{P}|$.

Many SSPs can be constructed from a CM. Our objective is to choose an SSP with high sensitivity and specificity to its CM. We also wish to keep the length of the designed SSP short to maximize the efficiency of scanning it against a large sequence database. The design problem is therefore as follows:

Given CM $\mathcal{M}$ and null model $\mathcal{P}^0$, construct an SSP $\mathcal{P}$ of length at most $L_{\max}$ and an associated threshold T so as to maximize $\mathcal{P}$'s sensitivity to $\mathcal{M}$ while keeping its FP rate relative to $\mathcal{P}^0$ no larger than a specified value τ.

The parameters $L_{\max}$ and τ are supplied by the user along with models $\mathcal{M}$ and $\mathcal{P}^0$, but threshold T is derived automatically for each SSP.

We construct an SSP from a CM in two steps. First, we identify *gapless intervals* in a CM, which are likely to yield SSPs with few seeds, and extract candidate SSPs from each such interval. Secondly, we select a threshold for each candidate SSP to ensure its sensitivity with a bounded false positive rate, then select the best SSP(s) to act as filter for the CM.

3.1. Selecting candidate SSPs

Although our SSPs can handle gaps caused by insertions or deletions, variable-length gaps cause seeds to proliferate and so slow down the search process and increase the FP rate. We therefore design SSPs only in *gapless intervals* of the CM, which are regions without either deletions or more than two consecutive insertions. For a given CM, we calculate the length distributions of insertions and deletions starting from each state via dynamic programming. If an insertion state can generate more than two contiguous random bases with high probability, we call it a *split point*. Similarly, if a deletion state can be hit with high probability, it forms a split point. The positions between a pair of split points constitute a gapless interval.

We extract SSPs from a gapless interval as follows. Let i be a position inside the interval. When there is no base pairing, an SSP of length $L \leq L_{\max}$ is constructed starting at i using the emission probabilities of the match states associated with single stranded bases i to $j = i + L - 1$. The corresponding seed is $0 \ldots L-1$. If positions x and y are paired, with $i \leq x < y \leq j$, then (x, y) forms a base pair in the SSP. That is, we keep only base pairs inside the same gapless interval. Bases that pair with a base outside the interval are treated as single-stranded.

When a base pair to be included in an SSP spans a gap whose length is not fixed, the resulting SSP contains multiple seeds, reflecting the different possible distances between the pair's endpoints. While the number of seeds can be exponential in the length of interval spanned by the SSP, we generate many fewer seeds in practice and could, if needed, arbitrarily limit the number of seeds generated.

3.2. Choosing the best SSP

The gapless intervals in a CM may generate a large number of candidate SSPs. For each candidate $\mathcal{P}_i$, we compute a threshold T_i to achieve an FP rate of at most τ, then compute the candidate's sensitivity given this threshold. The candidate SSP with the highest sensitivity is chosen as the final filter. More precisely, we select a threshold T_i for each candidate $\mathcal{P}_i$ (of length L) that satisfies the constraint

$$\Pr_{S\sim\mathcal{P}^0 \text{ and } |S|=L} (\text{LLR}\,(S,\mathcal{P}_i) \geq T_i) \leq \tau, \qquad (1)$$

then choose the candidate SSP $\mathcal{P}_i$ and associated T_i that maximize

$$\Pr_{S\sim\mathcal{P}_i} (\text{LLR}\,(S,\mathcal{P}_i) \geq T_i)\,. \qquad (2)$$

We note that, although we wish to judge whether a given SSP $\mathcal{P}_i$ will detect sequences drawn from a CM $\mathcal{M}$, we use the base distribution of the SSP itself, rather than that of the full CM, to estimate its sensitivity. This estimate may be inaccurate in two ways. On one hand, a path sampled from $\mathcal{M}$ might omit the CM states corresponding to the SSP $\mathcal{P}_i$, in which case the corresponding sequence lacks the portion that *should* match the SSP with a high score. On the other hand, $\mathcal{P}_i$ might happen to match some *other* portion of the CM with a high score. In theory, neglecting these two events results in an inaccurate estimate of the match probability.

Empirically, however, we find that the match probability is well approximated even if the above two events are ignored. For 117 ncRNA families chosen at random from Rfam, we compared our simplified sensitivity, computed via Eq. (2), to sensitivity as measured on a large set of Monte Carlo samples from the family's CM. The simplified and Monte Carlo estimates were highly correlated ($R^2 = 0.9901$), as desired. A detailed comparison of the two estimates is given in our supplementary data[a].

3.2.1. *Computing sensitivity and FP rate*

In our previous work[13], we developed a dynamic programming algorithm to compute the sensitivity and FP rate for a regular profile constructed from a profile HMM. In this work, we extend that algorithm to apply to an SSP constructed from a CM, which may include secondary structure conservation as well.

Following the definition of sensitivity in Eq. (2), we compute the sensitivity of an SSP $\mathcal{P}$, $\Pr_{S\sim\mathcal{P}}(\text{LLR}\,(S,\mathcal{P}) \geq T)$ as follows:

$$\Pr_{S\sim\mathcal{P}} (\text{LLR}\,(S,\mathcal{P}) \geq T) = \sum_{\theta=T}^{A^*} \Pr_{S\sim\mathcal{P}} (\text{LLR}\,(S,\mathcal{P}) = \theta),$$

where A^* is the highest possible LLR score for a sequence produced by $\mathcal{P}$. Let $\mathcal{P}_{1..j}$ be a sub-SSP consisting of the first j values in a seed and the corresponding emission profile columns (unpaired *or* paired) for $\mathcal{P}$. The sensitivity in Eq. (2) is given by $\sum_{\theta=T}^{A^*} \Pr(|\mathcal{P}|,\theta)$, where $|\mathcal{P}|$ is the SSP's length.

For convenience below, let $\Pr(\ell,\theta)$ denote the probability $\Pr_{S\sim\mathcal{P}_{1..\ell}}(\text{LLR}\,(S,\mathcal{P}_{1..\ell}) = \theta)$. Let $\mathcal{P}_{i,a}$ be the emission probability of unpaired base a at column i. Similarly, let $\mathcal{P}_{i,a_1a_2}$ be the emission probability of base pair a_1a_2 at column i. Two dynamic programming cases are needed, depending on whether column ℓ describes an unpaired base or a base pair.

When column ℓ describes the frequency distribution of an unpaired base,

$$\Pr(\ell,\theta) = \sum_{a\in\Sigma} \mathcal{P}_{\ell,a} \Pr\left(\ell-1, \theta - \log\frac{\mathcal{P}_{\ell,a}}{\mathcal{P}^0_a}\right),$$

where $\mathcal{P}^0_a$ is the probability of the residue a in the background model $\mathcal{P}^0$. When column ℓ describes the frequency distribution of a base pair,

$$\Pr(\ell,\theta) = \sum_{a_1a_2\in\Sigma^2} \mathcal{P}_{\ell,a_1a_2} \Pr\left(\ell-1, \theta - \log\frac{\mathcal{P}_{\ell,a_1a_2}}{\mathcal{P}^0_{a_1}\mathcal{P}^0_{a_2}}\right).$$

Initially, for each base $a \in \Sigma$ or $a_1a_2 \in \Sigma^2$, $\Pr(1, \log\frac{\mathcal{P}_{1,a}}{\mathcal{P}^0_a}) = \mathcal{P}_{1,a}$ if column 1 is created from an unpaired base, or $\Pr(1, \log\frac{\mathcal{P}_{1,a_1a_2}}{\mathcal{P}^0_{a_1}\mathcal{P}^0_{a_2}}) = \mathcal{P}_{1,a_1a_2}$ if it is created from base pair.

If we let S be sampled from $\mathcal{P}^0$ rather than from $\mathcal{P}$, the above algorithm can be modified to compute the FP rate against $\mathcal{P}^0$. The FP rate for $\mathcal{P}$ is

$$\Pr_{S\sim\mathcal{P}^0} (\text{LLR}\,(S,\mathcal{P}) \geq T) = \sum_{\theta=T}^{A^*} \Pr_{S\sim\mathcal{P}^0} (\text{LLR}\,(S,\mathcal{P}) = \theta).$$

For a given FP threshold τ, the score threshold T chosen for $\mathcal{P}$ is computed as

$$\text{T} = \text{argmin}_{T'} \left(\sum_{\theta=T'}^{A^*} \Pr_{S\sim\mathcal{P}^0} (\text{LLR}\,(S,\mathcal{P}) = \theta) \leq \tau\right).$$

[a]http://www.cse.wustl.edu/~yanni/ncRNA

Let s_{max} be the maximum possible LLR score for a single position of the SSP (one base or base pair). Similarly, let s_{min} be the minimum such score. The time complexity of our dynamic programming algorithm is $\Theta(|\Sigma|L^2(s_{max} - s_{min}))$. Because only short intervals (we set $L_{\max} = 25$) are used to produce SSPs, the range of possible scores, and hence the running time, is limited. It typically takes only seconds to compute an SSP's score threshold and sensitivity.

3.3. Using SSPs vs. regular profiles

For many ncRNA families, particularly those with high primary sequence conservation, filtering with a regular profile produces fewer false positives than using an SSP. Regular profiles generally look at shorter intervals of the sequence than equally sensitive SSPs because the latter often need to span long loops to "see" significant stems whose two sides are widely separated in the primary sequence. Long loops tend to have variable length, so the SSP needs more distinct seeds to encode the range of possible loop lengths and hence has a higher chance of matching unrelated sequences purely by chance. On the other hand, for some ncRNA families with low primary conservation, the secondary structure encoded by SSPs may be the only available evidence on which to base a filter.

To best exploit both primary and secondary conservation, our filter design procedure selects between an SSP and a regular profile for each RNA family. When designing a filter for a family, we first design a regular profile without secondary structure information. If this regular profile achieves sensitivity ≤ 0.9 to sequences from the CM according to our dynamic programming estimate, we instead design a full SSP for the family allowing base pairing. This approach applies the extra complexity of secondary structure filtering only where it is clearly needed.

3.4. Using multiple SSPs to improve sensitivity

A sensitive SSP is usually constructed from a well-conserved region within a CM. When multiple such regions exist in one CM, we can improve overall search sensitivity by designing a filter that is a union of SSPs from all well-conserved regions. For a query sequence S and a filter Φ that contains m SSPs $\mathcal{P}_1, ..., \mathcal{P}_m$, Φ matches S iff at least one component $\mathcal{P}_i \in \Phi$ matches S. The total FP rate for Φ is at worst the sum of rates for its component filters $\mathcal{P}_i$.

Our SSP design algorithm can be extended to multiple SSPs. Instead of choosing the single SSP with the highest sensitivity under a specified FP rate threshold, we choose the top m non-overlapping SSPs by estimated sensitivity. When two SSPs overlap, only the one with higher sensitivity is kept.

4. RESULTS

In this section, we first show that SSPs with secondary structure conservation exhibit a better empirical sensitivity/false positive rate tradeoff than regular profiles for detecting structured ncRNA families, such as tRNA and 5S rRNA, in the BRAliBase III benchmark database[21]. We then apply our automated filter design methods to a large number of ncRNA families from the Rfam database and quantify the resulting filters' sensitivity, FP rate, speedup when used in search, and their dependence on secondary structural conservation. We also compare SSPs and other filter types from related work. Finally, we investigate a small set of Rfam families on which our designed filters exhibit low sensitivity.

4.1. SSP utility for structured RNAs

To demonstrate and quantify SSPs' ability to exploit secondary structure, we first tested our heuristics on BRAliBase III[21], a database containing 602 5S rRNAs, 1114 tRNAs, and 235 U5 spliceosomal RNAs. BRAliBase III has been used as a benchmark for comparing ncRNA detection tools, including INFERNAL 0.7. We compared SSPs to regular profiles with no secondary structural information. We also tested a restricted form of SSP that was permitted only a single seed and so fixed the separation of all base pairs. Single-seed SSPs were tested to quantify the importance of handling variable-length gaps as part of SSP filter design.

We used the same methods as Ref. 21 to evaluate the sensitivity and FP rate of SSPs. A total of 40 sequence sets were sampled from each of the three ncRNA types; each tRNA set contained 60 sequences, while each rRNA and U5 set contained 20

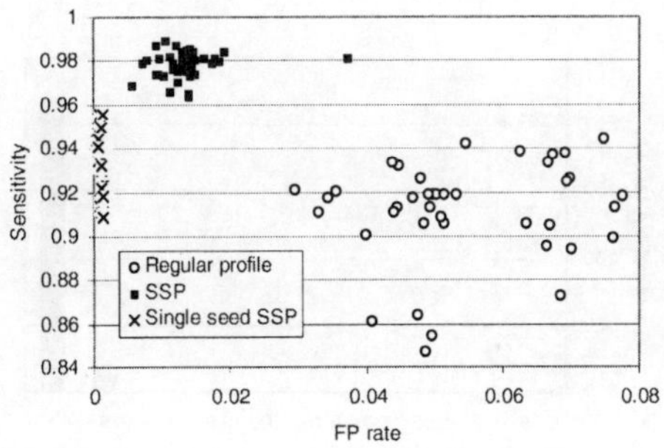

Fig. 3. Performance comparison for three types of filter designed for CMs built from tRNAs from BRAliBase III. Each CM was built from 60 sequences with pairwise identities between 0 and 0.6.

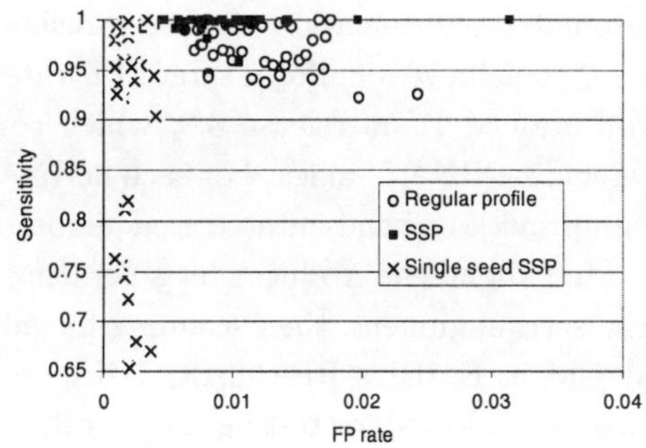

Fig. 4. Performance comparison for three types of filter designed for CMs built from 5S rRNAs from BRAliBase III. Each CM was built from 20 sequences with pairwise identities between 0.4 and 0.6.

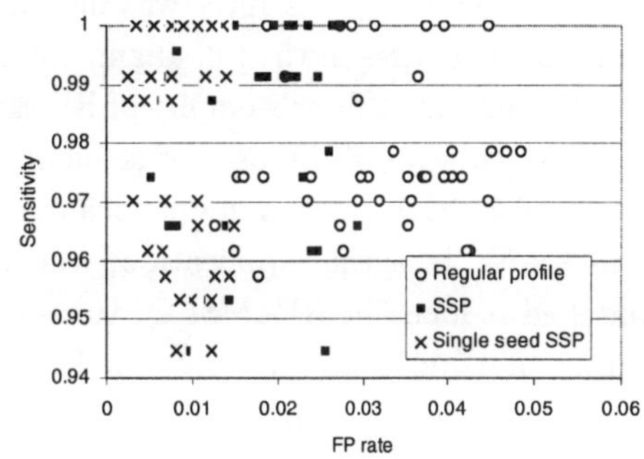

Fig. 5. Performance comparison for three types of filter designed for CMs built from U5 RNAs from BRAliBase III. Each CM was built from 20 sequences with pairwise identities between 0.4 and 0.6.

sequences. Sets were chosen so that no two sequences in a set aligned with greater than 60% identity. Each sampled set was used to train a CM. We designed heuristic filters from each CM, then tested the sensitivity of each filter on all sequences of the corresponding type in the database (e.g. 1114 tRNAs for tRNA-derived filters). For a CM $\mathcal{M}$, let $H_{\mathcal{M}}$ be the test set for $\mathcal{M}$, and let $S^{\mathcal{P}}_{H_{\mathcal{M}}}$ be the subset of sequences in $H_{\mathcal{M}}$ which contain a match to filter $\mathcal{P}$. $\mathcal{P}$'s sensitivity is defined as $|S^{\mathcal{P}}_{H_{\mathcal{M}}}|/|H_{\mathcal{M}}|$. FP rate was measured, as in Ref. 21, on a shuffled version of the test set that was ten-fold larger than the original. We note that we tested only the filters, rather than the underlying CMs, because experiments in Ref. 21 showed that CM search is already highly sensitive and specific for this database; hence, few if any true positives from a filter would be discarded by the CM, and nearly all false positives *would* be discarded.

Figures 3, 4, and 5 plot the sensitivities and FP rates of 40 designed regular profiles, SSPs, and single-seed SSPs for tRNA, 5S rRNA, and U5 spliceosomal RNA. Using SSP filters for tRNAs and 5S rRNAs consistently boosted sensitivity compared to regular profiles while reducing FP rate. Improvements for U5 RNAs were more uneven. Using multiple seeds in the SSP consistently improved sensitivity relative to single-seed SSPs, usually from < 0.95 to 0.98-0.99, at a cost to FP rate. Overall, incorporating secondary structure in our filter significantly improved its performance on these RNA families.

Variations in improvement observed with SSPs vs. regular profiles across these families can be explained by looking more closely at their conservation. The average sequence lengths for tRNAs, 5S rRNAs, and U5 RNAs are respectively 73, 117, and 119 bases, while the average number of annotated base pairs in their training sets are 21, 18, and 4. SSPs performed best on the tRNAs, which exhibit the highest density of base pairing, and worst on the U5 RNAs, with by far lowest such density.

4.2. Evaluation on Rfam database

In order to test our filter design methods on diverse ncRNA families with a wide range of sequence conservation levels, we applied the methods to families from the Rfam ncRNA database[b]. The filters in these tests came from our fully automated design pipeline, including automatic selection between regular profiles and SSPs as described in Section 3.3,

[b] http://www.sanger.ac.uk/Software/Rfam/

and automatic determination of score thresholds for each filter to achieve a uniform target FP rate.

We obtained Rfam release 8.0, which contains 574 non-coding RNA families. For each ncRNA family, Rfam provides a hand-curated seed alignment, as well as a full alignment produced by generating a CM from the seed alignment, then scanning for matches to that CM in EMBL's RFAMSEQ DNA sequence database. We selected for testing those ncRNA families with at least five sequences in the seed alignment (used to train the CM and hence our filters) and ten sequences in the full alignment (used to quantify sensitivity below). These criteria reduced our test set to 233 ncRNA families.

Empirical sensitivity of a filter was measured as the fraction of sequences in the full alignment that it matched. To measure a filter's empirical FP rate, we used the filter to scan 65 Mbases of sequence sampled from RFAMSEQ, using a simple scanning tool written in C++. In actual application, whenever a filter matched a locus in RFAMSEQ, the sequence surrounding that locus would be scanned using the full CM for the family. The filter's FP rate was therefore computed, following Ref. 18, as the ratio of the total length of sequences selected for scanning by the CM to the total length of the database.

More precisely, let $\mathcal{P}$ be the filter designed for CM $\mathcal{M}$. Let the average length of the sequences matched by $\mathcal{M}$ be L, and suppose that $\mathcal{P}$ matches the database D at m distinct positions. Then each match to $\mathcal{P}$ results in applying the CM to a region of length L around the match. $\mathcal{P}$'s FP rate vs. data set D was therefore estimated as $(m \times 2 \times L)/|D|$, where $|D|$ is the total length of D. For an CM $\mathcal{M}$ in INFERNAL, L is the mean length of a match to $\mathcal{M}$.

Our filter designs used a theoretical FP rate upper bound of $\tau = 5 \times 10^{-6}$ and allowed multiple SSPs or profiles per family. As discussed in Section 3.3, we prefer to use regular profiles to SSPs when our theoretical estimate of sensitivity suggests that a regular profile would detect nearly all instances of an ncRNA family. Of the 233 ncRNA families tested, our methods produced regular profiles with theoretical sensitivity at least 0.9 for 220; for the remaining 13 families, we used SSPs to capture secondary structure information as well.

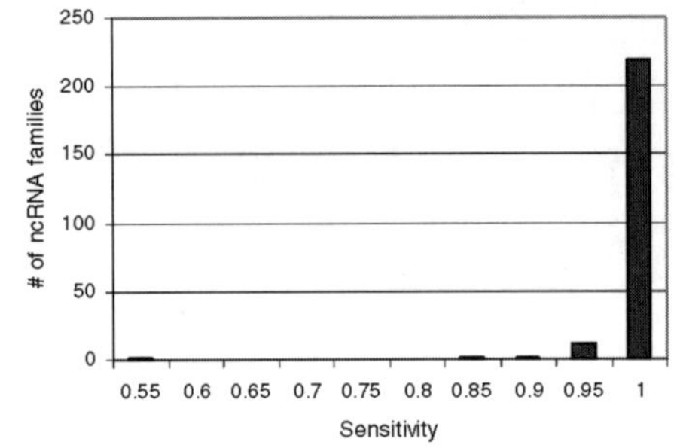

Fig. 6. Empirical sensitivities of filters for 233 ncRNA families from Rfam, measured on sequences in each family's full alignment.

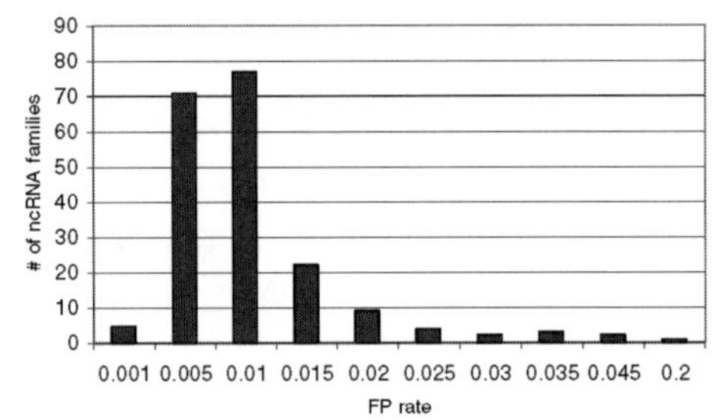

Fig. 7. Empirical FP rates of filters with sensitivity ≥ 0.99 for 196 ncRNA families from Rfam, measured on 65 Mbases of genomic sequence from RFAMSEQ.

4.2.1. *Sensitivity and FP rate*

Figure 6 shows the sensitivities for our designed filters. Of the 233 filters designed, 196 (84%) had empirical sensitivity ≥ 0.99. For these ncRNA families, the average number of sequences in the test set is 321. Only 3 filters (1.3%) had sensitivity less than 0.9. For the 196 filters, Figure 7 shows their empirical false positive rates on our 65-Mbase test database. The average FP rate observed was 0.008.

We note that the observed FP rate is several orders of magnitude greater than our theoretical FP rate τ. This is because τ measures the filter match probability at a random position in a database. Thus, for a sequence database D, the expected number of matches is $|D| \times \tau$, and the empirical FP rate is $(|D| \times \tau \times 2L)/|D| = \tau \times 2L$. L is on the order of hundreds for typical ncRNA-family CMs, so the observed FP rate is expected to be of the order shown in Figure 7 given our choice of τ.

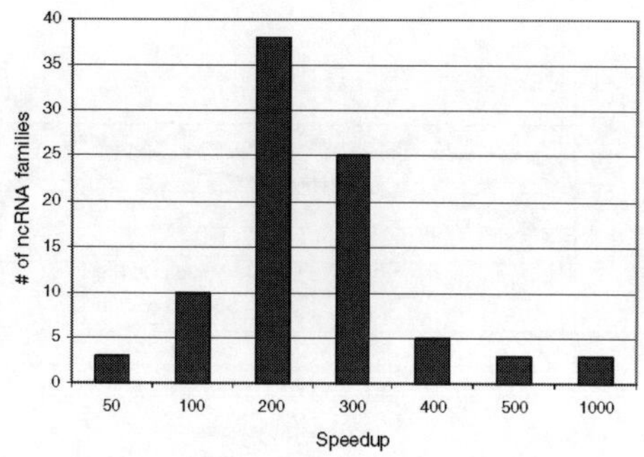

Fig. 8. Speedup distribution for 88 randomly chosen ncRNA families using filters composed of at most four profiles or SSPs.

4.2.2. *Acceleration ability*

The key reason to place a filter in front of CM search is the efficiency of filtration compared to CM scanning. The average time (over 233 ncRNA families) to scan a family's CM against one Mbase of genomic DNA using INFERNAL's *cmsearch* tool was about 8701 seconds; in contrast, the average time to scan the same length with one of our filters was only 0.67 seconds. In this section, we first analyze the relationship between FP rate and observed speedup, then show the empirical speedups obtained by our filters.

Let T_o be the time to run `cmsearch -noalign` against a database D. Let T_s be the time to scan a filter against database D, and let S_D^P be the set of matched substrings output by the filter. We estimate the time T_f to scan a database with filtering enabled as $T_f = T_s + (|S_D^P|/|\mathrm{D}|) \times T_o$, where $|S_D^P|/|\mathrm{D}|$ is the FP rate of the filter. The speedup of filtered over unfiltered search is then T_o/T_f. To save time, we estimated the times T_s and T_o for our 65-Mbase RFAMSEQ sample from times measured on a 1-Mbase synthetic DNA sequence, since the cost of the filtration and CM scanning algorithms is insensitive to the actual sequence content. However, we used accurate empirical estimates of $|S_D^P|$ from our FP rate measurements of the previous section.

Figure 8 shows estimated speedups for 88 ncRNA families sampled at random from our set of 233. The average speedup for these families is 222x.

To validate our speedup estimates, we directly measured speedup on our 65-Mbase database for three ncRNA families, whose filters had FP rates ranging from 0.003 to 0.012. Table 1 gives both the estimated and observed speedups for these three families. These observations suggest that our estimates actually *underestimate* the speedup conferred by filtration. The reason is that $(|S_D^P|/|\mathrm{D}|) \times T_o$ empirically overestimates the cost of running *cmsearch* on the sequences emitted by the filter (data not shown). Consequently, the results shown in Figure 8 are conservative estimates of the actual speedups obtained by filtration.

Table 1. Estimated vs. observed speedups with filtration for 3 ncRNA families.

Rfam ID	T_o **(s)**	T_f **(s)**	**Est. speedup**	**Obs. speedup**
RF00476	149502	1631	79	91
RF00490	131625	354	281	371
RF00167	262475	1217	149	215

4.3. Comparison with other filters

In this section, we compare our filters to two related works on filtered ncRNA search. Our first comparison is to Zhang et al.'s chain filters (CFs)[18], which were tested on a set of twelve riboswitch sub-families. We designed ten sets of regular profiles and two sets of SSPs for these sub-families. The results of our comparison are given in Table 2; the false positive rates shown are measured on the same synthetic data set described in Ref. 18. The average sensitivities observed for CFs and our filters were respectively 0.998 and 0.993, and the corresponding FP rates were 0.0353 and 0.0106. Overall, our automatically designed filters exhibited similar performance to CFs, which in Ref. 18 required manual intervention to choose numerical cut-offs for each filter.

Our second comparison is to the profile HMM-based filters of Ref. 15. The performance of HMM-based filters was tested using *cmsearch* in INFERNAL with option `-hmmfilter`. We compared our methods with HMM-based filters on two datasets: BRAliBase III, and 88 randomly selected ncRNA families from Rfam 8.0. Table 3 presents the median sensitivity and FP rates for the three types of ncRNA families in BRAliBase III. According to these experiments, the sensitivity of the two filter types is comparable, and the FP rate of HMM-based filters is smaller than that of SSP-based filters. However, because searching for HMM matches is much more

Table 2. Comparison of SSPs and chain filters (CFs) on 12 riboswitch sub-families.

Rfam ID	CF sen	CF FP	SSP sen	SSP FP
RF00050	1	0.013	0.993	0.0034
RF00059	1	0.063	0.994	0.02314
RF00080	1	0.15	0.990	0.0043
RF00162	1	0.018	1	0.0038
RF00167	0.99	0.038	0.991	0.0034
RF00168	0.99	0.015	0.986	0.0046
RF00174	1	0.063	0.995	0.0052
RF00234	1	0.013	1	0.0118
RF00379	1	0.012	1	0.0024
RF00380	1	0.012	1	0.0030
RF00442	1	0.0017	1	0.0020
RF00504	1	0.025	0.967	0.0598

Table 3. Comparison of SSPs and HMM filters.

Name	SSP sensitivity median	HMM sensitivity median	SSP FP rate median	HMM FP rate median
tRNA	0.979	0.983	0.013	0.002
rRNA	0.998	1	0.012	0.0
U5	0.991	0.972	0.020	0.0

expensive than searching for profile matches, SSP-based filters yield better speedup. Figure 9 quantifies the advantage of SSP-based filters in the form of box plots describing the distribution of speedups for ncRNA families tRNA, rRNA, and U5 in BRAliBase.

In order to compare SSPs and HMM-based filters in a larger dataset, we then tested the sensitivity, FP rate, and actual search time using these

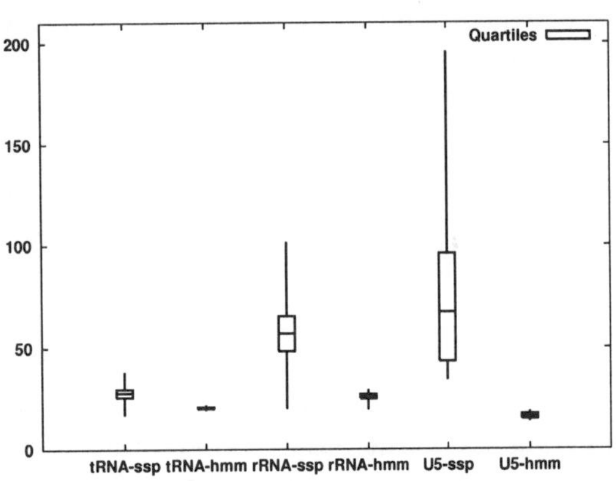

Fig. 9. Speedup comparison between HMM- and SSP-based filters for tRNA, rRNA, and U5 in BRAliBase III database. X-axis shows the names of ncRNA families and the used filters. Y-axis measures speedups.

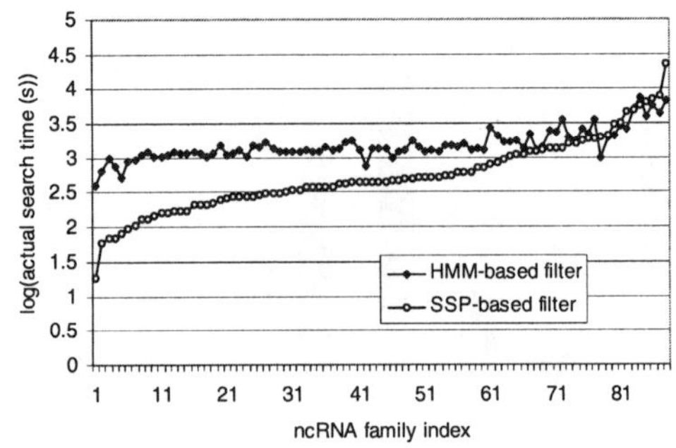

Fig. 10. Speed comparison between HMM- and SSP-based filters for 88 ncRNA families from Rfam. Y-axis measures actual ncRNA search time on logarithmic scale.

two types of filters on 88 randomly chosen ncRNA families from Rfam. The experimental setting was as in Section 4.2. The actual search time comparison is summarized in Figure 10. As we can see, INFERNAL runs significantly faster using SSP-based filters than using HMM-based filters for most of the tested ncRNA families. The HMM-based filter proved faster for only 7 out of 88 families, for which the SSP filter exhibited a high FP rate (around 0.02). According to our experimental results on over 200 ncRNA families in Section 4.2, the average FP rate of SSP-based filters is 0.008, which is small enough to ensure a better acceleration ability for a majority of SSP-based filters.

4.4. Analysis of low-sensitivity SSPs

For 35 of the 233 Rfam ncRNA families tested, our filters' empirical sensitivity was < 0.99. We divide these filters into two groups: those for which our theoretical estimates accurately predicted their low sensitivity (difference from empirical < 0.05), and those for which we predicted sensitivity ≥ 0.99, but the empirical result was < 0.95. Our supplementary data gives examples of RNA families in both groups.

All but nine of the 35 "bad" families fall into the first category; while these cases illustrate limitations of our filtering heuristic, we can detect them during design and opt to use a less aggressive filter or no filter at all, depending on the user's tolerance for missed ncRNA occurrences.

For the remaining nine bad families, the high theoretical but low empirical sensitivity of their filters would result in unexpected loss of matches to the

CM. We therefore investigated these failures more closely. Because the CMs used to design our filters are trained only on seed alignments, filter quality depends heavily on whether a family's seed alignment accurately reflects the range of variation in its full alignment. A close look at one bad family, RF00266, reveals that the full alignment contains much shorter sequences than those in the seed alignment, with long deletions that are not described by the CM. As a result, SSPs constructed from the CM do not attempt to avoid these deletions. For three other families, the full alignment has much lower primary conservation than the seed alignment; hence, high predicted sensitivity on the CM's output is misleading as a predictor of empirical sensitivity. For a further three ncRNA families, low empirical sensitivity was an artifact of the family's small test set. For example, the filter for family RF00002 missed only one of 15 sequences in its test set, but this yielded empirical sensitivity of only 0.93.

In the above seven cases, the apparent "badness" appears to be either an artifact of a small test set or a limitation in how representative the seed alignment is of the full family. There are only two cases (RF00008 and RF00013) where we cannot yet explain the discrepancies between the theoretical and experimental sensitivities.

5. CONCLUSIONS

Covariance models are a state-of-the-art method to identify and annotate non-coding RNAs. However, their high computational cost impedes their use with large sequence databases. Our automatically designed SSP filters encode both primary sequence and (optionally) secondary structure conservation in an ncRNA family, yet they can scan a large sequence database efficiently. 84% of our designed filters have sensitivity at least 0.99, and their average FP rate is 0.008. Our filters obtain an average speedup of 222x over search using CMs alone on Rfam.

There is considerable room to improve the sensitivity and design efficiency of SSP filters. We plan to study more systematic methods to choose a set of SSPs so as to maximize their combined sensitivity. We also plan to design chain filters using SSPs as components. The lengths of the component SSPs can be shorter than the typical lengths of the filters in this work because all or most must match to yield a chain filter match. We expect that collections of short filters would be most effective for ncRNA families whose alignment contains frequent gaps, preventing the appearance of long gapless intervals.

Acknowledgments

This work was supported by NSF CAREER award DBI-0237903.

References

1. S. R. Eddy. Noncoding RNA genes. *Curr. Opin. Genet. Dev.* 1999; **9**:695–9.
2. S. R. Eddy. A memory-efficient dynamic programming algorithm for optimal alignment of a sequence to an RNA secondary structure. *BMC Bioinformatics* 2002; **3**:3–18.
3. S. R. Eddy and R. Durbin. RNA sequence analysis using covariance models. *Nucleic Acids Res.* 1994; **22**:2079–88.
4. R. Durbin, S. R. Eddy, A. Krogh, and G. Mitchison. Biological Sequence Analysis Probabilistic Models of Proteins and Nucleic Acids. UK: Cambridge U. Press, 1998.
5. D. H. Younger. Recognition and parsing of context-free languages in time n^3. *Information and Control* 1967;**10**:189–208.
6. S. Griffiths-Jones, S. Moxon, M. Marshall, A. Khanna, S. R. Eddy, and A. Bateman. Rfam: annotating non-coding RNAs in complete genomes. *Nucleic Acids Res.* 2005; *33*:D121–4.
7. Weinberg Z, Ruzzo WL. Faster genome annotation of non-coding RNA families without loss of accuracy. In *Proc. 8th Ann. Int. Conf. on Res. in Comp. Mol. Bio. (RECOMB '04)*. 2004; 243–51.
8. B. Brejova, D. G. Brown, and T. Vinar. Optimal spaced seeds for hidden Markov models, with application to homologous coding regions. *14th Ann. Symp. Combinatorial Pattern Matching*. 2003; 42–54.
9. J. Buhler, U. Keich, and Y. Sun. Designing seeds for similarity search in genomic DNA. *Proc. 7th Ann. Int. Conf. Comp. Mol. Bio.* 2003; 67–75.
10. M. Li, B. Ma, D. Kisman, and J. Tromp. PatternHunter II: highly sensitive and fast homology search. *J. Bioinf. and Comp. Bio.* 2004; **2**:417–39.
11. L. Noe and G. Kucherov. Improved hit criteria for DNA local alignment. *BMC Bioinformatics* 2004; **5**.
12. Portugaly E., Ninio M. HMMERHEAD – accelerating HMM searches on large databases (poster). *Proc. 7th Ann. Int. Conf. Comp. Mol. Bio.* 2003.
13. Sun Y, Buhler J. Designing patterns and profiles for profile HMM search. *IEEE/ACM Trans. Comp. Bio. and Bioinf.* 2008.

14. Lowe T, Eddy S. tRNAscan-SE: a program for improved detection of transfer RNA genes in genomic sequence. *Nucleic Acids Res.* 1997; **25**:955–64.
15. Weinberg Z, Ruzzo WL. Sequence-based heuristics for faster annotation of non-coding RNA families. *Bioinformatics* 2006; **22**:35–9.
16. Weinberg Z, Ruzzo WL. Exploiting conserved structure for faster annotation of non-coding RNAs without loss of accuracy. *Bioinformatics* 2004; **20 suppl. 1**:i334–40.
17. Zhang S, Haas B, Eskin E, Bafna V. Searching genomes for noncoding RNA using FastR. *IEEE/ACM Transactions on Comp. Bio. and Bioinf.* 2005; **2**:366–79.
18. Zhang S, Borovok I, Aharonowitz Y, Sharan R, Bafna V. A sequence-based filtering method for ncRNA identification and its application to searching for riboswitch elements. *Bioinformatics* 2006; **22**:e557-65.
19. Gautheret D, Lambert A. Direct DNA motif definition and identification from multiple sequence alignments using secondary structure profiles. *J. Mol. Bio.* 2001; **313**:1003–11.
20. Bafna V, Zhang S. FastR: fast database search tool for non-Coding RNA. *Proc. 2004 IEEE Comp. Systems Bioinf. Conf.* 2004; 52–61.
21. Freyhult EK, Bollback JB, Gardner PP. Exploring genomic dark matter: a critical assessment of the performance of homology search methods on noncoding RNA. *Genome Res.* 2006; **17**:117–25.

MATCHING OF STRUCTURAL MOTIFS USING HASHING ON RESIDUE LABELS AND GEOMETRIC FILTERING FOR PROTEIN FUNCTION PREDICTION

Mark Moll [1] and Lydia E. Kavraki [1,2,3]

[1] *Department of Computer Science, Rice University, Houston, TX 77005, USA,*
[2] *Department of Bioengineering, Rice University, Houston, TX 77005, USA,*
[3] *Structural and Comp. Biology and Molec. Biophysics, Baylor College of Medicine, Houston, TX 77005, USA*
Email: {mmoll,kavraki}@cs.rice.edu

There is an increasing number of proteins with known structure but unknown function. Determining their function would have a significant impact on understanding diseases and designing new therapeutics. However, experimental protein function determination is expensive and very time-consuming. Computational methods can facilitate function determination by identifying proteins that have high structural and chemical similarity. Our focus is on methods that determine binding site similarity. Although several such methods exist, it still remains a challenging problem to quickly find all functionally-related matches for structural motifs in large data sets with high specificity. In this context, a structural motif is a set of 3D points annotated with physicochemical information that characterize a molecular function. We propose a new method called LabelHash that creates hash tables of n-tuples of residues for a set of targets. Using these hash tables, we can quickly look up partial matches to a motif and expand those matches to complete matches. We show that by applying only very mild geometric constraints we can find statistically significant matches with extremely high specificity in very large data sets and for very general structural motifs. We demonstrate that our method requires a reasonable amount of storage when employing a simple geometric filter and further improves on the specificity of our previous work while maintaining very high sensitivity. Our algorithm is evaluated on 20 homolog classes and a non-redundant version of the Protein Data Bank as our background data set. We use cluster analysis to analyze why certain classes of homologs are more difficult to classify than others. The LabelHash algorithm is implemented on a web server at http://kavrakilab.org/labelhash/.

1. INTRODUCTION

High-throughput methods for structure determination have greatly increased the number of proteins with known structure in the Protein Data Bank[1]. Determining the function of all these proteins would greatly impact drug design. Unfortunately, functional annotation has not kept up with the pace of structure determination. Sequence-based methods are an established way for automated functional annotation[2–5], but sequence similarity does not always imply functional similarity and vice versa. Structural analysis allows for the discovery of similar function in proteins with very different sequences and even different folds[6]. For an overview of current approaches in sequence- and structure-based methods see Refs. 7 and 8.

Structure-based methods can be divided into several categories, such as methods that compare fold similarity[9, 10], methods that model pockets and clefts[11–13], and search algorithms based on active sites and templates (see section 2). The combination of structural and phylogenetic information can be used to identify residues that are of structural or functional importance[14–16]. Several web servers exist that use a combination of several sequence- and structure-based methods[17, 18] to provide an aggregate of information.

The method in this paper falls in the template search category. We will describe a new method for partial structure comparison. In partial structure comparison, the goal is to find the best geometric and chemical similarity between a set of 3D points called a *motif* and a subset of a set of 3D points called the *target*. Both the motif and targets are represented as sets of labeled 3D points. A motif is ideally composed of the functionally most-relevant residues in a binding site. The labels denote the type of residue. Motif points can have multiple labels to denote that substitutions are allowed. Any subset of the target that has labels that are compatible with the motif's labels is called a *match*. The aim is to find statistically significant matches to a structural motif. Our method preprocesses, in a fashion that borrows ideas from a well-known technique called geometric hashing[19], a background database of targets such as a non-redundant subset of the Protein Data Bank. It does this in such a way that we can look up in constant time partial matches to a motif. Using a variant of the previously described match augmentation algorithm[20], we obtain complete matches to our motif. The nonparametric statistical model developed in Refs. 21 and 22 corrects for any bias introduced by our algorithm. This bias is introduced by excluding matches that do not satisfy certain geometric constraints for efficiency reasons.

The contributions of this paper are as follows. Our new method is based on hashing of residue labels and geometric constraints, an approach that proves to be efficient, highly sensitive, and highly specific. It further improves the already high specificity of our previous work. It removes the requirement of needing an ordering of the importance of the points of the motifs. Using cluster analysis, we provide a more complete picture of match results and we illustrate the difficulty of matching certain functional homologs. Last but not least, our approach can be easily adapted to use different motif types or incorporate different constraints. Although not discussed in detail in this paper, we can optionally include partial matches or multiple matches per target in the match results. Before we will describe our method, we will first give an overview of related methods.

2. RELATED WORK

Over the years several algorithms have been proposed for the motif matching problem. In its generality, this problem has a chemical, a geometric, and a statistical component. First, points in our motif need to be matched with chemically compatible points. This can translate into simply matching the same residue types, but can also be defined in terms of a more general classification of physicochemical properties[23, 24]. Geometrically, we want to solve the partial structure comparison problem: find all correspondences between a motif and groups of points in the targets that are chemically compatible. Solving issues associated with the high complexity of the problem are discussed in Ref. 25. Most existing methods employ heuristics to find only matches that are close under the Least Root Mean Square Deviation (LRMSD) metric, since these matches are most likely functionally related to the motif. This brings us to the statistical component of the problem: there is no universal LRMSD threshold that can be used to separate functional homologs from other matches, and thus statistical analysis is needed to decide whether a match is functionally related to a motif and unlikely to occur due to chance.

In table 1 we have summarized some selected related work that we will discuss in more detail below. A direct comparison of our work with other methods is challenging for several reasons: (1) there are several ways to represent structural motifs, (2) most of the methods included in the table solve a slightly different version of the problem discussed in this paper, and (3) for most systems there is no freely available or web-accessible implementation with which we could perform experiments similar to our own.

Geometric hashing[19, 31, 32] is a technique to preprocess the targets that will be used for matching and create index tables that facilitate fast matching. These tables only need to be computed once for a set of targets. They are based on geometric characteristics. One has to carefully pick the geometric constraints to balance the potentially enormous storage requirements with completeness of the matching phase. The application of geometric hashing to motif matching was first introduced in Ref. 19 and has been refined in subsequent years. TESS[27] is an algorithm that uses geometric hashing to match structural motifs. By focusing on a specific class of motifs (catalytic triads), TESS can create space-efficient hashing tables. More recent work on geometric hashing[31] uses several "pseudo-centers" per residue to represent physicochemical properties to achieve more accurate matching.

In Ref. 23 a graph-based approach is used. Residues are represented by a pair of pseudo-atoms. The pseudo-atoms are the vertices of the graph, and edges are defined by the distances between them. The matching is then solved by solving the subgraph isomorphism problem[33]. In Ref. 34 distance constraints on C_α and C_β atoms are used to guide a depth-first search for matches. Unlike much of the previous work, this paper also introduced a statistical model to determine the significance of a match. Matching results were fitted to an extreme value distribution and allowed for matching of catalytic triads

Table 1. Overview of selected related work.

Name	Physicochemical information	Geometric algorithm	Statistical model	Demonstrated application
ASSAM[23]	pairs of pseudo-atoms per residue	subgraph isomorphism	—	catalytic triads
FEATURE[26]	supervised learning of many physicochemical and geometric features		nonparametric model, Bayesian scoring	ATP-binding, S-S sites, Mg^{2+} binding sites in RNA
TESS[27]	all atoms of selected residues	geometric hashing	—	His-based catalytic triads
Jess[28]	user-defined constraints on atoms	constraint satisfaction + match augmentation	mixture of two Gaussians	HTH motifs
PINTS[29]	reduced # res., 1 pseudocenter per res.	depth-first search w/distance constraints	extreme value dist. on weighted RMSD	catalytic triads, salt bridges, S-S sites
DRESPAT[30]	C_α's, C_β's, and functional atoms	graph based on distance constraints, max. complete subgraph detection	significance estimated from algorithm parameters & output	detection of many motifs (e.g., catalytic triads, EF-hand)
SiteEngine[31]	pseudocenters	geometric hashing	—	finding and comparing functional sites
MASH[20]	evolutionary importance, residue-labeled C_α's	match augmentation	nonparametric model	matching motifs of ~5–15 residues against large data sets
LabelHash [this paper]	residue-labeled C_α's	hash tables of res. labels + match augmentation	nonparametric model	matching motifs of ~5–15 residues against large data sets

and zinc fingers[29]. More recently, in Ref. 30 a graph-based method was described that automatically detects repeating patterns in subgraphs of graph representations of proteins. This is reduced to a graph clique enumeration problem, a well-known, very difficult problem in general, but by taking advantage of the structure of the underlying data, this method can avoid the worst-case complexity.

The FEATURE algorithm[26] takes a radically different approach to matching. It uses supervised learning to characterize the active sites of proteins. Many attributes can be defined and the learning algorithm will automatically learn the salient features. More recently, this algorithm has been applied to ATP-binding and disulfide bond-forming sites[35] and magnesium binding sites in RNA[36]. Although in is original form the FEATURE algorithm worked directly on structural data, later work showed that it is able to construct structural motifs from sequence-based motifs[37]. The FEATURE algorithm is accessible through a public web server[38]. The representation of motifs is very different, making comparison with other methods challenging.

In Ref. 39 a parametrized statistical model is proposed to determine the significance of the LRMSD of a match. The model parameters are obtained by fitting the model to the data. This model is part of the PINTS server[29], which uses a distance constraint-based matching algorithm similar to the one described in Ref. 34. The PINTS server used to allow matching against a non-redundant subset of the PDB, but at the time of writing this option was no longer available, making a comparison with our method difficult. In Ref. 28 a more general matching framework is proposed, where user-defined constraints can be associated with a number of residues. The residues and constraints together form a template. A mixture of two Gaussians is used to model the distribution of the LRMSD's of matches. The same template-based approach was successfully applied to finding DNA-binding proteins that contain the helix-turn-helix (HTH) motif[40]. This last work also showed that for finding HTH matches, 3D templates could be used to detect similarity between many different HTH motifs, while a sequence-based approach based on Hidden Markov Models could not.

Recent work on template matching[41] argues in favor of using a heuristic similarity measure rather than LRMSD to rank matches. This similarity measure is a function of the number of overlapping atoms and a residue mutation score. It is shown to eliminate many false positives in certain cases. This paper introduces so-called reverse templates, which are conceptually similar to geometric hashing's notion of reference sets.

In Ref. 20 the MASH matching algorithm is introduced. It is based on a depth-first search that finds matches to motif points in order of decreasing importance ranking. Our approach is most compatible with this algorithm. In our algorithm we preprocess the targets to speed up matching, remove the need for importance ranking, and improve specificity. Further improvements can be made to the MASH algorithm by explicitly representing cavities[42] and by creating composite motifs in case several instances of a functional site are known[43].

3. METHODS

We are interested in matching a structural motif against a set of targets. The structural motif is defined by the backbone C_α coordinates of a number of residues and (optionally) allowed residue substitutions for each motif residue which are encoded as labels. Previous work[44, 39, 20, 14] has established that this is a feasible approach. There is no fundamental reason why other points cannot be used as well.

The method presented below is called LabelHash. It builds hash tables for n-tuples of residues that occur in a set of targets. In spirit the method is reminiscent of the geometric hashing technique[19], but the particulars of the approach are different. The n-tuples are hashed based on the residues' labels. Each n-tuple has to satisfy certain geometric constraints. Using this table we can look up partial matches of size n in constant time. These partial matches are augmented to full matches with an algorithm similar to MASH[20]. Compared to geometric hashing[19], our method significantly reduces the storage requirements. Relative to MASH, we further improve the specificity. Also, in the LabelHash algorithm it is no longer required to use importance ranking of residues to guide the matching. In our previous work, this ranking was obtained using Evolutionary Trace (ET) information[45]. The LabelHash algorithm was designed to improve the (already high) accuracy of MASH and push the envelope of matching with only very few geometric constraints. For this work we wanted motifs to be as general as possible to allow for future extensions and to facilitate motif design through a variety of methods. The input should be easy to generate from "raw data" such as PDB files, and the output should be easy to post-process and visualize. Although the ideal of functional annotation is full automation, an exploratory process of iterative and near-interactive motif design and refinement will be extremely valuable. Our simple-to-use and extensible LabelHash algorithm can be a critical component of this process. The LabelHash algorithm consists of two stages: a preprocessing stage and a stage where matches are computed from the preprocessed data.

3.1. Preprocessing Stage

The preprocessing stage has to be performed only once for a given set of targets. Every motif can be matched against the same preprocessed data. During the preprocessing stage we aim to find possible candidate partial matches. This is done by finding all n-tuples of residues that satisfy certain geometric constraints. We will call these n-tuples *reference sets.* All valid reference sets for all targets are stored in a hash map, a data structure for key/value pairs that allows for constant time insertions and lookups (on average). In our case, each key is a sorted n-tuple of residue labels, and the value is a list of reference sets that contain residues with these labels in any order. So for any reference set in a motif we can instantly find all occurrences in all targets. Notice that in contrast to geometric hashing[19] we do not store copies of the targets for each reference set, which allows us to store many more reference sets in the same amount of memory.

In our current implementation the geometric constraints apply to the C_α coordinates of each residue, but there is no fundamental reason preventing other control points from being used instead. We have defined the following four constraints:

- Each C_α in a reference set has to be within a distance $d_{\text{maxmindist}}$ from its nearest neighboring C_α.
- The maximum distance between any two C_α's within a reference set is restricted to be less than d_{diameter}.
- Each residue has to be within distance d_{maxdepth}

from the molecular surface. The distance is measured from the atom closest to the surface.

- At least one residue has to be within distance $d_{\text{maxmindepth}}$ from the surface.

The first pair of constraints requires points in reference sets to be within close proximity of each other, and the second pair requires them to be within close proximity of the surface. The distance parameters that define these constraints should be picked large enough to allow for at least one reference set for each motif that one is willing to consider, but small enough to restrict the number of seed matches in the targets. One would perhaps expect that the storage requirements would be prohibitively expensive, but, in fact, in the experiments described in section 4 we used very generous settings, and the storage used was still very reasonable.

3.2. Matching Stage

For a valid reference set in a motif, we look up the matching reference sets in the hash table. Next, we run a variant of the match augmentation algorithm[20] that consists of the following steps. First, it solves the residue label correspondence between a motif reference set and the matching reference sets stored in the hash map. (If more than one correspondence exists, we will consider all of them.) Next, we augment the match one residue at a time, each time updating the optimal alignment that minimizes the LRMSD. If a partial match has an LRMSD greater than some threshold ε, it is rejected. For a given motif point, we find all residues in a target that are within some threshold distance (after alignment). This threshold is for simplicity usually set to ε. The ε is set to be sufficiently large (7Å in our experiments) so that no interesting matches are missed. The value ε also affects the computation of the statistical significance of a match. It can be shown that for a motif of n residues our statistical model computes the *exact* p-value of matches with LRMSD less than $\varepsilon/\sqrt{n}$, i.e., their p-value would not change if no ε threshold was used [22, 21]. For example, for a 6-residue motif and $\varepsilon = 7$Å, the p-values of all matches within 2.3Å of the motif are exact.

The algorithm recursively augments each partial match with the addition of each candidate target residue. The residues added to a match during match augmentation are *not* subject to the geometric constraints of reference sets. In other words, residues that are not part of a reference set are allowed to be further from each other and more deeply buried in the core. For small-size reference sets, the requirement that a motif contains at least one reference set is therefore only a very mild constraint. Nevertheless, as we will see in the next section, our approach is still highly sensitive and specific.

For a given motif, we generate all the valid reference sets for that motif. Any of these reference sets can be used as a starting point for matching. However, those reference sets that have the smallest number of matching reference sets in the hash map may be more indicative of a unique function. Reference sets with a large number of matches are more likely to be common structural elements or due to chance. We could exhaustively try all possible reference sets, but for efficiency reasons we only process a fixed number of least common reference sets. Note that the selection of reference sets as seed matches is based only on frequency. In contrast, in our previous work, only one seed match was selected based on importance ordering frequently based on evolutionary importance[20]. Because of the preprocessing stage it now becomes feasible to expand matches from many different reference sets. The hash map files have embedded within them a "table of contents," so that during matching only the relevant parts of the hash map need to be read from disk.

The matching algorithm is flexible enough to give users full control over the kind of matches that are returned. It is possible to keep partial matches that match at least a certain minimum number of residues. This can be an interesting option for larger motifs where the functional significance of each motif point is not certain. In such a case, a 0.5Å LRSMD partial match of, say, 9 residues, might be preferable over a 2Å complete match of 10 residues. With partial matches, the matches can be ranked by a scoring function that balances the importance of LRMSD, and the number of residues matched. One can also choose between keeping only the LRMSD match per target or all matches for a target, which may be desirable if the top-ranked matches for targets have very similar LRMSD's. Keeping partial matches or multiple matches per target complicates the determination of the statistical significance of each match. This is an issue we plan to investigate in future work. Finally, the number of motif reference sets that the algorithm uses for match augmentation can also be varied. Usually most matches are found with the first couple reference sets, but occasionally a large number of reference sets need to be tried before the LRMSD match for each target is found.

4. RESULTS

4.1. Data Sets

LabelHash was tested on a diverse set of previously identified motifs. The motifs we used in our experiments are listed in table 2. Some were determined experimentally, others were determined using the Evolutionary Trace (ET) method[45]. More information on the function of these motifs and how they were obtained can be found in Refs. 20 and 42. Although the performance of the matching algorithm depends critically on the motif, the focus in this paper is on the motif matching *method* and not on motif *design*. Any motif of a similar type can be used by our method. For each motif we have listed the residue sequence numbers, followed by the 1-letter residue name and possible substitutions. The substitutions were in some cases determined using ET, but any reasonable set is accepted (sometimes experiments or intuition give the substitutions). It is important to note that our algorithm is "neutral" with respect to how a motif is obtained; importance ranking or other very specific information on the motif is not required.

Table 2. Motifs used in experiments.

PDB ID	Residue ID's with alternate labels
1acb	$42^{GSN}, 57, 58^{SKV}, 102, 194^{QE}, 195, 214^{AT}$
1ady	$81^{D}, 83, 112^{S}, 130^{D}, 264^{L}, 311^{NKQ}$
1ani	$51^{A}, 101^{E}, 102, 166^{CS}, 331^{G}, 412^{NQ}$
1ayl	$249, 250, 251, 253, 254, 255$
1b7y	$149^{GA}, 178^{Q}, 180^{T}, 206^{ER}, 218, 258^{NY}, 260^{Y}$
1czf	$178, 180, 201, 256^{H}, 258, 291$
1did	$25, 53, 56, 93, 136, 182$
1dww	$194, 346, 363, 366, 367^{F}, 371, 376^{D}$
1ep0	$53^{TA}, 61^{A}, 64, 73, 90, 172$
1ggm	$188^{T}, 239^{T}, 341, 311^{L}, 359^{S}, 361^{A}$
1jg1	$97^{DNQ}, 99, 101^{AL}, 160^{NS}, 179^{VI}, 183^{NE}$
1juk	$53, 89, 91, 233, 182, 110$
1kp3	$106, 139, 202^{S}, 286, 288, 331$
1kpg	$17, 72, 74, 75, 76, 200$
1lbf	$51, 56, 57, 89, 91, 112, 159, 180, 211, 233$
1nsk	$12^{RL}, 13, 52^{HL}, 105^{H}, 115, 118^{P}$
1ucn	$12, 13, 92, 105, 115, 118$
2ahj	$53, 120, 127, 190, 193, 196$
7mht	$80, 81, 85^{T}, 119^{L}, 163, 165$
8tln	$120^{WL}, 143^{A}, 144^{VI}, 157^{SL}, 231^{L}$

The targets consisted of a non-redundant snapshot of the Protein Database (PDB), taken on February 21, 2008. We used the automatically generated 95% sequence identity filtered version of the PDB. Each chain was inserted separately in the hash map. This resulted in roughly 18,000 targets. Molecular surfaces were computed with the MSMS software[46]. We chose to use reference sets of size 3. The following parameter values were used for the reference sets:

$$d_{\text{maxmindist}} = 16\text{Å}, \quad d_{\text{diameter}} = 25\text{Å},$$
$$d_{\text{maxmindepth}} = 1.6\text{Å}, \quad d_{\text{maxdepth}} = 3.1\text{Å}.$$

These values were chosen such that the motifs in table 2 contained at least one reference set of size 3. They are very generous in the sense that most motifs contain many reference sets. If reference sets of more than 3 residues are used, the values of the distance parameters need to be increased to guarantee that each motif contains at least one reference set. The advantage of larger reference sets is that we instantly match a larger part of a motif. However, increasing these values also quickly increases the number of reference sets in the targets. So the number of reference sets to perform match augmentation on will also quickly increase. Finally, the storage required for the hash tables grows rapidly with reference set size. After the preprocessing phase the total hash map size given the settings described above was 32GB (split into several files).

4.2. Matching Results

The results of matching the motifs from table 2 against the targets is shown in table 3. We evaluated the performance using the PDB entries with the corresponding Enzyme Classification (EC) code or corresponding Gene Ontology (GO) molecular function term as the set of positive matches. Typically, there is more than one GO molecular function term associated with one PDB entry. We picked the most

Table 3. Matching results with a p-value of 0.001.

PDB ID	Enzyme classification TP	FP	Gene Ontology TP	FP	time (s)
1acb	87.50% (28)	0.08% (13)	—	—	27541
1ady	100.00% (22)	0.07% (13)	68.00% (17)	0.08% (14)	10268
1ani	75.61% (62)	0.06% (11)	—	—	12673
1ayl	100.00% (19)	0.07% (12)	—	—	3006
1b7y	40.00% (8)	0.07% (12)	40.00% (4)	0.07% (12)	15744
1czf	100.00% (21)	0.04% (7)	100.00% (13)	0.06% (9)	1078
1did	100.00% (152)	0.02% (2)	100.00% (108)	0.02% (2)	181
1dww	88.94% (209)	0.04% (5)	95.31% (183)	0.04% (5)	1635
1ep0	100.00% (39)	0.05% (8)	100.00% (21)	0.05% (8)	2308
1ggm	81.82% (9)	0.07% (12)	33.33% (5)	0.07% (13)	12620
1jgl	100.00% (17)	0.06% (11)	100.00% (13)	0.07% (13)	44982
1juk	100.00% (12)	0.06% (10)	—	—	1211
1kp3	100.00% (36)	0.06% (10)	100.00% (35)	0.07% (11)	637
1kpg	84.62% (11)	0.06% (8)	84.62% (11)	0.06% (8)	126
1lbf	100.00% (12)	0.05% (8)	77.78% (7)	0.06% (9)	2650
1nsk	72.91% (148)	0.00% (0)	—	—	7128
1ucn	81.77% (166)	0.01% (2)	—	—	851
2ahj	35.90% (14)	0.06% (10)	33.33% (11)	0.07% (12)	420
7mht	90.91% (10)	0.08% (10)	—	—	2130
8tln	95.08% (58)	0.08% (14)	—	—	1989

specific term (i.e., the one with the fewest PDB entries). For some motifs no GO annotation for molecular function is available, which is indicated by a '—'. The true and false positives are given as percentages followed by the absolute number of matches between parentheses. In most cases our method finds close to 100% of all known true positives with a p-value of 0.001, and only very few false positives. Even in absolute terms the number of false positives is very small. For the 1acb motif, which represents the catalytic triad, we only counted α-chymotrypsin as a true positive. This excludes several other members of the corresponding EC class. An additional complication for this motif is that it sometimes spans several (but not all) chains in a complex. In this case we manually separated chymotrypsin from its inhibitor.

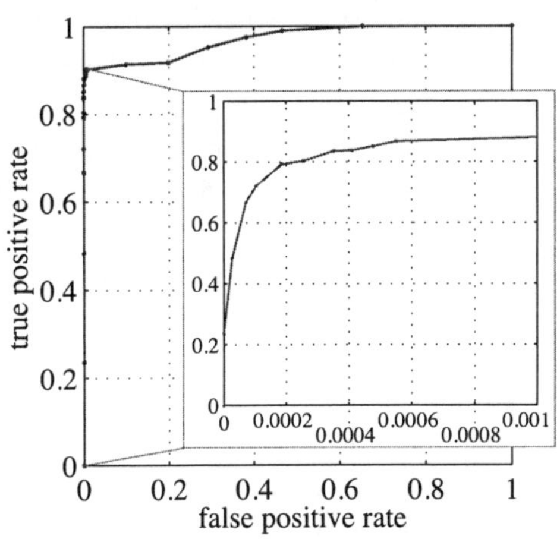

Fig. 1. ROC curve. The true positive rate and false positive rate are averaged over all motifs at a given p-value. The inset plot shows the performance for very small false positive rates.

Figure 1 shows the false positive rate and true positive rate as we vary the p-value. The true positive rate and false positive rate are averaged over all motifs at a given p-value. With MASH, our previous algorithm, we could achieve on average a 83.7% true positive rate at a 0.98% false positive rate. Now, at the same false positive rate, we achieve 90% sensitivity. Or, alternatively, at the same true positive rate, we now achieve a 0.04% false positive rate. The improvement in false positive rate is especially significant. Since in our case the number of targets is so much larger than the number of homologs, a small false positive rate can still mean that the number of false positives is many times larger than the number of true positives. For example, for the 8tln motif the false positive rate went from 9.1% with MASH to 0.08% with LabelHash. In absolute terms, the *number* of false positive matches went from 168 with MASH to 14 with LabelHash. *In almost all cases the number of false positives is now less than the number of true positive matches.*

The p-values of matches are computed using a so-called point-weight correction[22, 21]. This is a sta-

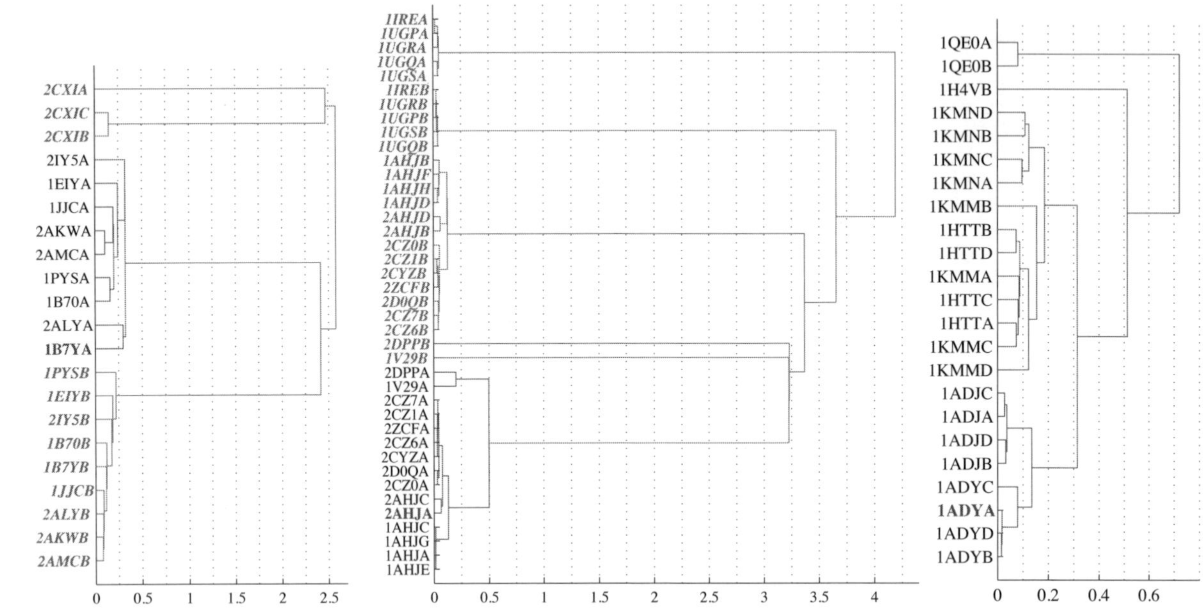

(a) 1b7y, EC 6.1.1.20 (b) 2ahj, EC 4.2.1.84 (c) 1ady, EC 6.1.1.21

Fig. 2. Clustering of matches in EC classes for three motifs. Matches in bold italics are likely to be missed because they are in a cluster that is very different from the cluster that contains the motif (shown in bold).

tistical correction for the bias introduced by only considering matches in a small neighborhood of motif points. While using the neighborhood heuristic typically preserves biologically relevant matches, eliminating biologically irrelevant matches can affect the accuracy of thresholds provided by the statistical models of sub-structural similarity. Statistical models depend on an unbiased representation of matches to yield the most accurate thresholds. During the match augmentation phase of the algorithm we only considered matching points in targets that were up to $\varepsilon = 7$Å away, but other matching points may exist. These other matches tend to be in right-hand side of the RMSD distribution of matches. The existence of these matches can be determined by simply looking at residue frequencies for each target. The point-weight represents these matches in the p-value determination. This can significantly improve the accuracy, especially for small ε. For a relatively large value of $\varepsilon = 7$Å, the effect is relatively small: *with* the point-weight the average sensitivity for the motifs in table 2 is 86.0%, but *without* the point-weight this drops to 82.7%. The specificity is relatively unaffected: it changes from 99.94% (with point-weight) to 99.96% (without). However, if a small $\varepsilon = 3$Å threshold is used, the sensitivity with point-weight is 85.7%, and without point-weight it is 32.9%. Again, specificity is relatively unaffected: 99.94% with point-weight and 99.996% without. The reason one may want to use a small value for ε is that it significantly reduces the runtime. The total time for matching all of the motifs in table 2 can be reduced by almost 60% by changing ε from 7Å to 3Å. The accuracy improvements over MASH observed at $\varepsilon = 7$Å are also observed at smaller ε levels.

To better understand what happens when a homolog is classified as false negative, let us now consider the homolog matches for three motifs. Suppose we take all the homolog matches for a given motif, compute all pairwise LRMSD distance between the matches, and cluster the results based on these distances. We expect that matches that end up in a different cluster than the motif's cluster, are more likely to be misclassified. This is indeed what appears to be the case for our motifs. Figure 2 shows dendrograms computed for three motifs. For the 1b7y motif and corresponding homologs in the EC 6.1.1.20 class of homologs there are two very distinct large clusters consisting of the 'A' and 'B' chains, respectively, and one small cluster for the outlier protein 2cxi. The 'B' chains of enzymes in EC 6.1.1.20 are very different from the 'A' chains. The assigned function for this class is really a property of the complex, rather than a single chain. It is therefore not surprising that the 'A' chains do not match the 'B' chains very well. For the 2ahj motif the situation is more complex (see figure 2(b)). Again, there are very distinct classes, but this time it is not obvious why this is the case. The last example, for 1ady and homologs, shows a dendrogram for a case where our matching algorithm found all homologs. Now all homologs are very close to each other and the clusters are not well-separated. This suggests that cluster analysis on match results can provide additional insight into whether matches are likely to be functionally related to a motif.

The runtime of matching each motif against 18,000 targets in the non-redundant PDB is shown in the last column of table 2. The time is highly variable: it ranges from a couple of minutes to several hours. The variability is due to the size of the motif, the type of residues, and—most importantly—the number of alternate labels. For instance, for the 1jgl motif the number of alternate labelings for the entire motif is $4 \times 1 \times 3 \times 3 \times 3 \times 3 = 324$. Although we do not match each alternate labeling separately, the increased branching factor during match augmentation still exponentially increases the runtime. Compared to MASH, our previous algorithm, the runtime has increased by a factor 5. This is due mostly because LabelHash algorithm performs match augmentation on many reference sets (up to 40 per motif in our experiments), whereas MASH only used one reference set, because its definition of the reference set was based on the availability of importance ranks for the residues. We expect that further parameter optimization and code profiling will allow LabelHash to run at comparable speed, but with superior accuracy. Comparison with other approaches was attempted, but it was impossible to complete due to reasons given in section 2. In particular, the problems solved are not always the same, or it is not possible to translate our motifs, or compare performance results. In an effort to help in solution of this problem in the future, a web server that will enable the community to use our work has been implemented and is accessible at http://kavrakilab.org/labelhash. More demanding users can also download a command line version that offers more options. We have also developed a prototype match visualization plugin for Chimera[47]. It superimposes the selected match with the motif and shows some additional information such as the corresponding EC and GO terms. On demand, additional information from PDBsum[48] is displayed. This will give the user an incredible wealth of information about a match. The ViewMatch plugin is also available at the LabelHash web site.

The runtime is measured by running the matching on a single CPU core of a 2.2GHz Dual Core AMD Opteron processor. Multi-core processors and distributed computing clusters are increasing commonplace, and naturally we would like to take advantage of that. Both the preprocessing stage and the matching stage are trivially parallelized, and a near-linear speed-up with the number of CPU cores can be expected. In the preprocessing phase we divide the targets into a number of groups and build a hash map for each. Each core can be responsible for building a number of hash maps. This requires no communication. Matching can also easily be parallelized. Each core can match a given motif against a set of targets independently. Once matching is finished, the match results can be aggregated into one output file by one of the cores.

5. CONCLUSION AND FUTURE WORK

We have presented LabelHash, a new algorithm for partial structural alignment. It quickly matches a motif consisting of residue positions, and possible residue types to large sets of targets. We have shown that LabelHash achieves very high sensitivity and specificity with 20 motifs matched against a background data set consisting of the non-redundant PDB filtered at 95% sequence identity. Accuracy is further improved due to a nonparametric statistical model that corrects for systematic bias in our algorithm. Typically, the number of false positive matches is much smaller than the number of true positive matches, despite the large number of targets in our background database. This greatly speeds up the analysis of match results. Our algorithm uses only a small number of parameters whose meaning is easy to understand. We have shown that clustering of matches can provide useful clues about the functional similarity between a motif and a match.

Extensibility was an important factor in the design of the LabelHash implementation. Our program is easily extended to incorporate additional constraints or use even conceptually different types of motifs. For instance, matching based on physicochemical pseudo-centers[23, 24] could easily be incorporated, and we plan to offer this functionality in the future. Input and output are all in XML format, which enables easy integration with other tools or web services. It is also not hard to imagine LabelHash as part of a multi-stage matching pipeline. The matches produced by LabelHash can be given to the next program, which can apply additional constraints to eliminate more false positives. As long as the set of matches produced by LabelHash include all functional homologs, this seems to be a viable strategy. Of course, the output of LabelHash can also easily be passed on to any clustering algorithm (as was done for figure 2) or a visualization front-end.

As mentioned at the end of section 3, our matching algorithm has the capability to keep partial matches and multiple matches per target. This makes the statistical analysis significantly more complicated. Currently, we just disable the p-value computation when either option is selected, but we plan to investigate the modeling of the statistical distribution of these matches.

Acknowledgements

The project upon which this publication is based was performed pursuant to Baylor College of Medicine Grant No. DBI-054795 from the National Science Foundation. LK has also been supported by a Sloan Fellowship. The computers used to carry out experiments of this project were funded by NSF CNS 0454333 and NSF CNS-0421109 in partnership with Rice University, AMD and Cray.

The authors are indebted to V. Fofanov for many useful discussions on the use of statistical analysis and for his comments on LabelHash. They are also deeply grateful for the help of B. Chen and D. Bryant with MASH. This work has benefited from earlier contributions by O. Lichtarge, M. Kimmel, D. Kristensen and M. Lisewski within the context of an earlier NSF funded project. The authors would also like to thank the members of the Kavraki Lab for proofreading this paper.

References

1. Berman, H. M., Westbrook, J., Feng, Z., Gilliland, G., Bhat, T. N., Weissig, H., Shindyalov, I. N., and Bourne, P. E. (2000) The protein data bank. *Nucleic Acids Research,* **28**, 235–242.
2. Altschul, S., Gish, W., Miller, W., Myers, E., and Lipman, D. (1990) Basic local alignment search tool. *J. Mol. Biol,* **215**(3), 403–410.
3. Thompson, J. D., Higgins, D. G., and Gibson, T. J. (1994) CLUSTAL W: improving the sensitivity of progressive multiple sequence alignment through sequence weighting, position-specific gap penalties and weight matrix choice. *Nucleic Acids Research,* **22**(22), 4673–4680.
4. Eddy, S. R. (1996) Hidden markov models. *Curr Opin Struct Biol,* **6**(3), 361–365.
5. Finn, R. D., Tate, J., Mistry, J., Coggill, P. C., Sammut, S. J., Hotz, H.-R., Ceric, G., Forslund, K., Eddy, S. R., Sonnhammer, E. L. L., and Bateman, A. (2008) The Pfam protein families database. *Nucleic Acids Res,* **36**(Database issue), D281–8.
6. Hermann, J. C., Marti-Arbona, R., Fedorov, A. A., Fedorov, E., Almo, S. C., Shoichet, B. K., and Raushel, F. M. (2007) Structure-based activity prediction for an enzyme of unknown function. *Nature,* **448**(7155), 775–779.
7. Watson, J., Laskowski, R., and Thornton, J. (2005) Predicting protein function from sequence and structural data. *Current Opinion in Structural Biology,* **15**(3), 275–284.
8. Zhang, C. and Kim, S. H. (2003) Overview of structural genomics: from structure to function. *Current Opinion in Chemical Biology,* **7**(1), 28–32.
9. Holm, L. and Sander, C. (1993) Protein structure comparison by alignment of distance matrices.. *J Mol Biol,* **233**(1), 123–138.
10. Zhu, J. and Weng, Z. (2005) FAST: a novel protein structure alignment algorithm.. *Proteins,* **58**(3), 618–627.
11. Binkowski, T. A., Freeman, P., and Liang, J. (2004) pvSOAR: detecting similar surface patterns of pocket and void surfaces of amino acid residues on proteins. *Nucleic Acids Res,* **32**, W555–W558.
12. Dundas, J., Ouyang, Z., Tseng, J., Binkowski, A., Turpaz, Y., and Liang, J. (2006) CASTp: computed atlas of surface topography of proteins with structural and topographical mapping of functionally annotated residues. *Nucleic Acids Research,* **34**(Web Server issue), W116–W118.
13. Laskowski, R. A. (1995) SURFNET: a program for visualizing molecular surfaces, cavities, and intermolecular interactions. *J Mol Graph,* **13**(5), 323–330.
14. Kristensen, D. M., Ward, R. M., Lisewski, A. M., Chen, B. Y., Fofanov, V. Y., Kimmel, M., Kavraki, L. E., and Lichtarge, O. (2008) Prediction of enzyme function based on 3D templates of evolutionary im-

portant amino acids. *BMC Bioinfomatics,* **9**(17).
15. Glaser, F., Rosenberg, Y., Kessel, A., Pupko, T., and Ben-Tal, N. (2005) The ConSurf-HSSP database: the mapping of evolutionary conservation among homologs onto PDB structures. *Proteins,* **58**(3), 610–617.
16. Chakrabarti, S. and Lanczycki, C. (2007) Analysis and prediction of functionally important sites in proteins. *Protein Science,* **16**(1), 4.
17. Laskowski, R. A., Watson, J. D., and Thornton, J. M. (2005) ProFunc: a server for predicting protein function from 3D structure. *Nucleic Acids Research,* **33**, W89–W93.
18. Pal, D. and Eisenberg, D. (2005) Inference of protein function from protein structure. *Structure,* **13**(1), 121–130.
19. Nussinov, R. and Wolfson, H. J. (1991) Efficient detection of three-dimensional structural motifs in biological macromolecules by computer vision techniques.. *Proc Natl Acad Sci U S A,* **88**(23), 10495–10499.
20. Chen, B. Y., Fofanov, V. Y., Bryant, D. H., Dodson, B. D., Kristensen, D. M., Lisewski, A. M., Kimmel, M., Lichtarge, O., and Kavraki, L. E. (2007) The MASH pipeline for protein function prediction and an algorithm for the geometric refinement of 3D motifs. *J. Comp. Bio.,* **14**(6), 791–816.
21. Fofanov, V. Y. Statistical Models in Protein Structural Alignments PhD thesis Department of Statistics, Rice University Houston, TX (2008).
22. Fofanov, V. Y., Chen, B. Y., Bryant, D., Moll, M., Lichtarge, O., Kavraki, L., and Kimmel, M. (2008) Correcting systematic bias caused by algorithmic thresholds in statistical models of protein sub-structural similarity. *BMC Biology Direct,* Submitted.
23. Artymiuk, P. J., Poirrette, A. R., Grindley, H. M., Rice, D. W., and Willett, P. (1994) A graph-theoretic approach to the identification of three-dimensional patterns of amino acid side-chains in protein structures. *Journal of Molecular Biology,* **243**(2), 327–344.
24. Schmitt, S., Kuhn, D., and Klebe, G. (2002) A new method to detect related function among proteins independent of sequence and fold homology. *J Mol Biol,* **323**(2), 387–406.
25. Shatsky, M. The Common Point Set Problem with Applications to Protein Structure Analysis PhD thesis School of Computer Science, Tel Aviv University (June, 2006).
26. Bagley, S. C. and Altman, R. B. (1995) Characterizing the microenvironment surrounding protein sites. *Protein Sci,* **4**(4), 622–635.
27. Wallace, A. C., Borkakoti, N., and Thornton, J. M. (1997) TESS: A geometric hashing algorithm for deriving 3D coordinate templates for searching structural databases. application to enzyme active sites. *Protein Science,* **6**(11), 2308.
28. Barker, J. A. and Thornton, J. M. (2003) An algorithm for constraint-based structural template matching: application to 3D templates with statistical analysis. *Bioinformatics,* **19**(13), 1644–1649.
29. Stark, A. and Russell, R. B. (2003) Annotation in three dimensions. PINTS: Patterns in non-homologous tertiary structures. *Nucleic Acids Research,* **31**(13), 3341–3344.
30. Wangikar, P. P., Tendulkar, A. V., Ramya, S., Mali, D. N., and Sarawagi, S. (2003) Functional sites in protein families uncovered via an objective and automated graph theoretic approach. *J Mol Biol,* **326**(3), 955–978.
31. Shulman-Peleg, A., Nussinov, R., and Wolfson, H. J. (June, 2004) Recognition of functional sites in protein structures. *J Mol Biol,* **339**(3), 607–633.
32. Wolfson, H. J. and Rigoutsos, I. (1997) Geometric hashing: an overview. *IEEE Computational Science and Engineering,* **4**(4), 10–21.
33. Ullmann, J. R. (1976) An algorithm for subgraph isomorphism. *J. of the ACM,* **23**(1), 31–42.
34. Russell, R. B. (1998) Detection of protein three-dimensional side-chain patterns: new examples of convergent evolution. *Journal of Molecular Biology,* **279**(5), 1211–1227.
35. Wei, L. and Altman, R. B. (2003) Recognizing complex, asymmetric functional sites in protein structures using a Bayesian scoring function. *J Bioinform Comput Biol,* **1**(1), 119–138.
36. Banatao, D. R., Altman, R. B., and Klein, T. E. (2003) Microenvironment analysis and identification of magnesium binding sites in RNA. *Nucleic Acids Research,* **31**(15), 4450–4460.
37. Liang, M. P., Brutlag, D. L., and Altman, R. B. (2003) Automated construction of structural motifs for predicting functional sites on protein structures.. In *Pacific Symposium on Biocomputing.* pp. 204–215.
38. Liang, M. P., Banatao, D. R., Klein, T. E., Brutlag, D. L., and Altman, R. B. (2003) WebFEATURE: an interactive web tool for identifying and visualizing functional sites on macromolecular structures. *Nucleic Acids Research,* **31**(13), 3324–3327.
39. Stark, A., Sunyaev, S., and Russell, R. B. (2003) A model for statistical significance of local similarities in structure. *Journal of Molecular Biology,* **326**(5), 1307–1316.
40. Jones, S., Barker, J. A., Nobeli, I., and Thornton, J. M. (2003) Using structural motif templates to identify proteins with DNA binding function. *Nucleic Acids Research,* **31**(11), 2811–2823.
41. Laskowski, R. A., Watson, J. D., and Thornton, J. M. (2005) Protein function prediction using local 3D templates. *Journal of Molecular Biology,* **351**(3), 614–626.
42. Chen, B. Y., Bryant, D. H., Fofanov, V. Y., Kristensen, D. M., Cruess, A. E., Kimmel, M., Lichtarge, O., and Kavraki, L. E. (April, 2007) Cavity scaling: Automated refinement of cavity-aware motifs in pro-

tein function prediction. *Journal of Bioinformatics and Computational Biology,* **5**(2), 353–382.
43. Chen, B. Y., Bryant, D. H., Cruess, A. E., Bylund, J. H., Fofanov, V. Y., Kimmel, M., Lichtarge, O., and Kavraki, L. E. (2007) Composite motifs integrating multiple protein structures increase sensitivity for function prediction. In *Comput Syst Bioinformatics Conf.*
44. Yao, H., Kristensen, D. M., Mihalek, I., Sowa, M. E., Shaw, C., Kimmel, M., Kavraki, L., and Lichtarge, O. (Feb, 2003) An accurate, sensitive, and scalable method to identify functional sites in protein structures. *J Mol Biol,* **326**(1), 255–261.
45. Lichtarge, O., Bourne, H. R., and Cohen, F. E. (Mar, 1996) An evolutionary trace method defines binding surfaces common to protein families. *J Mol Biol,* **257**(2), 342–358.
46. Sanner, M. F., Olson, A. J., and Spehner, J. C. (1996) Reduced surface: an efficient way to compute molecular surfaces.. *Biopolymers,* **38**(3), 305–320.
47. Pettersen, E. F., Goddard, T. D., Huang, C. C., Couch, G. S., Greenblatt, D. M., Meng, E. C., and Ferrin, T. E. (2004) UCSF Chimera—a visualization system for exploratory research and analysis. *Journal of Computational Chemistry,* **25**(13), 1605–1612.
48. Laskowski, R. A. (2001) PDBsum: summaries and analyses of PDB structures. *Nucleic Acids Research,* **29**(1), 221–222.

A HAUSDORFF-BASED NOE ASSIGNMENT ALGORITHM USING PROTEIN BACKBONE DETERMINED FROM RESIDUAL DIPOLAR COUPLINGS AND ROTAMER PATTERNS

Jianyang (Michael) Zeng and Chittaranjan Tripathy
Department of Computer Science, Duke University, Durham, NC 27708, USA

Pei Zhou
Department of Biochemistry, Duke University Medical Center, Durham, NC 27708, USA

Bruce R. Donald*†
Department of Computer Science, Duke University, Department of Biochemistry, Duke University Medical Center, Durham, NC 27708, USA
**Email: www.cs.duke.edu/brd*

High-throughput structure determination based on solution Nuclear Magnetic Resonance (NMR) spectroscopy plays an important role in structural genomics. One of the main bottlenecks in NMR structure determination is the interpretation of NMR data to obtain a sufficient number of accurate distance restraints by assigning nuclear Overhauser effect (NOE) spectral peaks to pairs of protons. The difficulty in automated NOE assignment mainly lies in the ambiguities arising both from the resonance degeneracy of chemical shifts and from the uncertainty due to experimental errors in NOE peak positions. In this paper we present a novel NOE assignment algorithm, called *HAusdorff-based NOE Assignment* (HANA), that starts with a high-resolution protein backbone computed using only two residual dipolar couplings (RDCs) per residue[37, 39], employs a Hausdorff-based pattern matching technique to deduce similarity between experimental and back-computed NOE spectra for each rotamer from a statistically diverse library, and drives the selection of optimal position-specific rotamers for filtering ambiguous NOE assignments. Our algorithm runs in time $O(tn^3 + tn\log t)$, where t is the maximum number of rotamers per residue and n is the size of the protein. Application of our algorithm on biological NMR data for three proteins, namely, human ubiquitin, the zinc finger domain of the human DNA Y-polymerase Eta (pol η) and the human Set2-Rpb1 interacting domain (hSRI) demonstrates that our algorithm overcomes spectral noise to achieve more than 90% assignment accuracy. Additionally, the final structures calculated using our automated NOE assignments have backbone RMSD < 1.7 Å and all-heavy-atom RMSD < 2.5 Å from reference structures that were determined either by X-ray crystallography or traditional NMR approaches. These results show that our NOE assignment algorithm can be successfully applied to protein NMR spectra to obtain high-quality structures.

1. INTRODUCTION

High-throughput structure determination based on X-ray crystallography and Nuclear Magnetic Resonance (NMR[a]) spectroscopy are key steps towards the era of structural genomics. Unfortunately, structure determination by either approach is generally time-consuming. In X-ray crystallography, growing a good quality crystal is in general a difficult task, while in NMR structure determination, the bottleneck lies in the processing and analysis of NMR data, and in interpreting a sufficient number of accurate distance restraints from experimental Nuclear Overhauser Enhancement Spectroscopy (NOESY) spectra, which exploit the dipolar interaction of nuclear spins, called *nuclear Overhauser effect* (NOE), for through-space correlation of protons. The intensity (or volume) of an NOE peak in a NOESY spectrum is converted into a distance restraint by calibrating the

*Corresponding author.
†This work is supported by the following grant to B.R.D.: National Institutes of Health (R01 GM-65982).
[a]Abbreviations used: NMR, Nuclear Magnetic Resonance; ppm, parts per million; RMSD, root mean square deviation; NOESY, Nuclear Overhauser Enhancement SpectroscopY; HSQC, Heteronuclear Single Quantum Coherence spectroscopy; NOE, Nuclear Overhauser Effect; RDC, Residual Dipolar Coupling; PDB, Protein Data Bank; pol η, zinc finger domain of the human DNA Y-polymerase Eta; hSRI, human Set2-Rpb1 interacting domain; POF, Principal Order Frame; CCD, Cyclic Coordinate Descent; SA, Simulated Annealing; MD, Molecular Dynamics; □, Q.E.D.; SM, Supplementary Material.

intensity (or volume) vs. distance curve or classifying all NOESY peaks into different bins.[12, 16, 38] Traditional NMR structure determination approaches use NOE distance restraints as the main source of information to compute the structure of a protein, a problem known to be strongly NP-hard,[30] essentially due to the local nature of the restraints. Rigorous approaches to solve this problem using NOE data, such as the distance geometry method,[10] require exponential time in the worst-case (see discussion in Ref. 39). While substantial progress has been made to design practical algorithms for structure determination,[3, 12–14, 24, 28, 31] most algorithms still rely on heuristic techniques such as molecular dynamics (MD) and simulated annealing (SA), which use NOE data plus other NMR data to compute a protein structure. The NOE distances used by these distance-based structure determination protocols must be obtained by *assigning* NOE data, i.e., for every NOE, we must determine the associated pair of interacting protons in the primary sequence. This is called the *NOE assignment problem.*

While much progress has been made in automated NOE assignment,[12, 14, 16, 21, 24, 27, 28] most NOE assignment algorithms have a SA/MD-based or a distance geometry-based structure determination protocol sitting in a tight inner loop, which is invoked many times to filter ambiguous assignments. Since distance geometry methods have exponential worst-case time complexity, and SA/MD-based structure determination protocols lack combinatorial precision and have no guarantees on solution quality or running time, these NOE assignment algorithms suffer from the same drawbacks, in addition to the inherent difficulties in the interpretation of NOESY spectra. Therefore, it is natural to ask if there exists a *provably polynomial-time* algorithm for the NOE assignment problem, which can guarantee solution quality—this will pave new ways for better understanding and interpretation of experimental data, and for developing robust protocols with *both* theoretical guarantees and good practical performance.

In Ref. 39, a new linear time algorithm was developed, based on Refs. 37 and 36, to determine protein backbone structure accurately using a minimum amount of residual dipolar coupling (RDC) data. RDCs provide *global* orientational restraints on internuclear vectors, for example, backbone NH and CH bond vectors with respect to a global frame of reference. The algorithm in Refs. 37, 36, and 39 computes the backbone conformation by solving, in closed form, systems of low-degree polynomial equations formulated using the RDC restraints. The algorithm is combinatorially-precise and employs a systematic search strategy to compute the backbone structure in polynomial time. The accurately-computed backbone conformations enable us to propose a new strategy for NOE assignment. In Ref. 38, for example, an NOE assignment algorithm was proposed to filter ambiguous NOE assignments based on an ensemble of distance intervals computed using intra-residue vectors mined from a rotamer database, and inter-residue vectors from the backbone structure determined from Refs. 37, 36, and 39. The algorithm in Ref. 38 uses a *triangle-like inequality* between the intra-residue and inter-residue vectors to prune incorrect assignment for side-chain NOEs. However, the algorithm in Ref. 38 has the following deficiencies: (a) it does not exploit the diversity of the rotamers in the library, (b) uncertainty in NOE peak position, and other inherent difficulties in interpreting NOESY spectra suggest a probabilistic model with provable properties which Ref. 38 does not capture, and (c) it does not exploit rotamer pattern structure in NOESY spectra.

To address the shortcomings in Ref. 38 and other previous work, our algorithm, *HAusdorff-based NOE Assignment* (HANA), uses a novel pattern-directed framework for NOE assignment, that combines a combinatorially-precise, algebraic geometry-based approach for computing *high-resolution* protein backbones from residual dipolar coupling data, with a framework that uses a statistically diverse library of rotamers and the Hausdorff distance to measure similarity between experimental and back-computed NOE spectra, and drives the selection of optimal position-specific rotamers to prune ambiguous NOE assignments. Our Hausdorff-based framework views the NOE assignment problem as a *pattern-recognition* problem, where the objective is to establish a match by choosing the correct rotamers between the experimental NOESY spectrum and the back-computed NOE pattern. By explicitly modeling the uncertainty in NOE peak positions and the probability of mismatches between NOE patterns, we provide a rigorous means of analyzing and evaluating the algorithmic benefits and the quality of assignments.

We first compute a high-resolution protein backbone from RDC data using the algorithms in Refs. 37, 36, and 39. Using this backbone structure, an assigned resonance list, and a library of rotamers[25], the NOE pattern for each rotamer can be back-computed (Figure 1B). By measuring the match of the back-computed NOE patterns with experimental NOESY spectrum, we choose an ensemble of top rotamers according to the match scores for each residue. Then, we construct an initial low-resolution protein structure by combining the high-resolution backbone and the chosen approximate rotamers together. The low-resolution structure is then used to *filter* ambiguous NOE assignments. Finally, our NOE assignments are fed to a structure calculation program, e.g., XPLOR/CNS[3] which outputs the final ensemble of structures. The experimental results, based on our NMR data for three proteins, viz., human ubiquitin, the zinc finger domain of the human DNA Y-polymerase Eta (pol η) and the human Set2-Rpb1 interacting domain (hSRI) show that HANA achieves an assignment accuracy of more than 90%. In summary, our main contributions in this paper are:

(1) Development of a novel framework that combines a combinatorially-precise, algebraic geometry-based linear time algorithm for high-resolution backbone structure determination with the Hausdorff distance measure, and exploits the statistical diversity of a rotamer library to infer accurate NOE assignments for both backbone and side-chain NOEs from 2D and 3D NOESY spectra.
(2) Introduction of Hausdorff distance-based pattern matching technique to measure the similarity between experimental NOE spectra and back-computed NOE spectra, and modeling uncertainties arising both from false random matches and from experimental deviations in NOE peak positions.
(3) A fully-automated $O(tn^3 + tn \log t)$ time NOE assignment algorithm, where t is the maximum number of rotamers in a residue and n is the number of residues in the protein.
(4) Derivation of provable properties, viz. soundness in rotamer selection.
(5) Application of our algorithm on three real biological NMR data sets to demonstrate high assignment accuracy ($> 90\%$), and fast running times (< 2 minutes).

2. PRELIMINARIES AND PROBLEM DEFINITION

In NMR spectra, each proton or atom is identified by its *chemical shift* (or *resonance*), which is obtained by mapping atom names in the known primary sequence of the protein to the corresponding frequencies from triple-resonance or other NMR spectra; this process is referred to as *resonance assignment.* Substantial progress has been made in designing efficient algorithms[1, 20, 22, 26] for automatic resonance assignment. Given the chemical shift of each proton, the *NOE assignment problem* in two dimensions[b] is to assign each NOESY peak to each pair of protons that are correlated through a dipole-dipole NOE interaction.

Formally, let $\{a_1, \ldots, a_q\}$ denote the set of proton names (e.g., H^α of Arg56), where $q = \Theta(n)$ is the total number of protons and n is the number of residues in a protein. Let $\omega(a_i)$ denote the chemical shift for proton a_i determined from resonance assignment, $1 \le i \le q$. An NOE peak (a.k.a. *cross-peak*) with respective frequencies x and y for a pair of protons, is denoted by the point (x, y) on the plane of NOESY spectrum. Given a set of known chemical shifts $L = \{\omega(a_i), \ldots, \omega(a_q)\}$ for all protons $\{a_1, \ldots, a_q\}$ and a list of NOESY peaks (i.e., a set of points on the plane of NOESY spectrum), the *NOE assignment problem* is to map each NOE cross-peak (x, y) to an interacting proton pair (a_i, a_j) such that $\|\omega(a_i) - x\| \le \delta_x$ and $\|\omega(a_j) - y\| \le \delta_y$, where δ_x and δ_y encode the uncertainty in the peak position due to experimental errors.

[b]The problem for 3D and 4D cases can be defined in an analogous manner. Here the 2D case is explained for clarity. Our NOE assignment algorithm has been tested on both 2D and 3D spectra, and extends easily to handle 4D NOESY spectra.

In a hypothetical ideal case without any experimental error and noise, this would be an easy problem. However, for most proteins, two pairs of interacting protons can produce overlapping NOE peaks in a NOESY spectrum. The chemical shift differences of different protons are often too small to resolve experimentally, a phenomenon often referred to as *chemical shift degeneracy*. Also, due to experimental noise, artifact NOE peaks might occur from either manual or automated peak picking. These factors lead to more than one possible NOE assignment for a 2D NOESY spectrum which are called *ambiguous NOE assignments*.[12, 21] Hence, one or more additional dimensions are generally introduced to relieve the congestion of NOE peaks. In a 3D NMR experiment, for example, each NOE peak is labeled with chemical shifts of a triple of atoms, viz., dipole-dipole interacting protons plus the heavy atom nucleus such as ^{15}N or ^{13}C bonded to the second proton. Even for 3D spectra, the interpretation and assignment of NOESY cross-peaks still remains hard, and poses a difficult computational challenge to obtain a unique NOE assignment. Manual assignment of NOESY peaks take months of time on average, requires significant expertise, and is prone to human errors. In structure determination, even a few incorrect NOE assignments can result in incorrect structures.[5] Hence, it is critical to develop highly efficient and fully automated NOE assignment algorithms to aid high-throughput NMR structure determination.

3. PREVIOUS WORK

Protein structure determination using NOE distance restraints is strongly NP-hard,[30] essentially due to sparsity of the experimental data and local nature of the constraints. While rigorous approaches to solve this problem using distance intervals from NOE data, such as the distance geometry method,[10] require exponential time in the worst-case; heuristic approaches such as SA/MD, while providing practical ways of solving this problem, lack combinatorial precision, and have no guarantees on running time or solution quality. Previous approaches for NOE assignment[12, 14, 16, 21, 24, 27, 28] follow an iterative strategy, in which an initial set of relatively unambiguous NOEs is used to generate an ensemble of structures, which are then used to filter ambiguous and inconsistent NOE assignments. This iterative assignment process is repeated until no further improvements in NOE assignments or structures can be obtained. What makes such approaches loose guarantees on the running time and assignment accuracy is their tight coupling with a heuristic structure determination protocol, which sits in a tight inner-loop of the assignment algorithm.

NOAH,[27, 12] for example, uses the structure determination package DYANA,[14] and follows the previously mentioned iterative strategy starting with an initial set of NOE assignments with supposedly one or two possible assignments. ARIA[28, 24] and CANDID[14] improved on NOAH by incorporating better modeling of ambiguous distance constraints. In AUTO-STRUCTURE[16] more experimental data such as dihedral angle restraints from TALOS[8] and slow H-D exchange data are used to improve assignment accuracy. In PASD[21] several strategies were proposed to reduce the chance of invoking the structure calculation into a biased path due to the incorrect initial global fold. Since all these iterative NOE assignment programs invoke SA/MD-based structure determination protocols such as XPLOR/CNS[3], they may converge to a local, but not a global minimum to obtain a best-fit of the data; therefore, the NOE assignments might not be correct.

An alternative approach for automated NOE assignment proposed by Wang and Donald in Ref. 38, based on Refs. 37, 36, and 39, uses a rotamer ensemble and residual dipolar couplings, and is the first polynomial-time algorithm for automated NOE assignment. However, Ref. 38 does not exploit the pattern structure of NOESY spectrum to model the uncertainty in peak positions probabilistically using a library of rotamers; therefore, assignment accuracy is reduced while processing NOESY spectra with many noisy peaks.

Our algorithm HANA retains the paradigm of Ref. 38, and develops a novel framework using the algebraic geometry-based linear time algorithm developed in Ref. 39 to compute high-resolution protein backbones from residual dipolar couplings, and then uses this backbone and a library of rotamers to do NOE assignments. Viewing the NOE assignment problem as a pattern-recognition problem, our algorithm uses an extended Hausdorff distance-

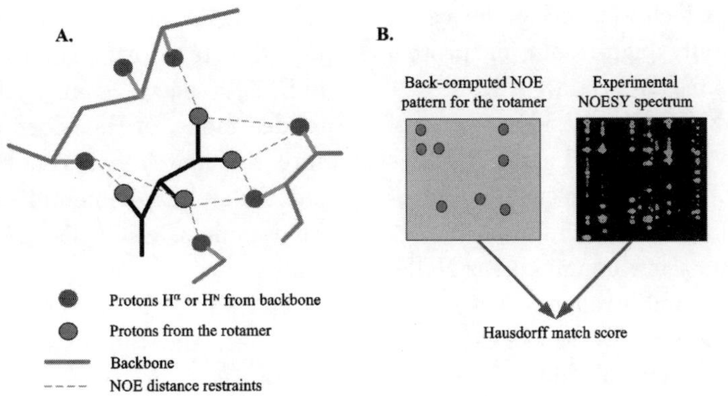

Fig. 1. Schematic illustration of the NOE assignment approach.

based probabilistic framework to model the uncertainties in NOE peak positions and the probability of mismatches between NOE patterns. In contrast to previous heuristic algorithms[12, 14, 16, 21, 24, 27, 28] for NOE assignment, HANA has the advantages of being combinatorially precise with a running time of $O(tn^3 + tn \log t)$, where t is the maximum number of rotamers per residue and n is the size of the protein, and runs extremely fast in practice to compute high quality NOE assignments ($>$ 90% assignment accuracy).

4. NOE ASSIGNMENT BASED ON ROTAMER PATTERNS

4.1. Overview of our approach

Our goal is to assign pairs of proton names[c] to crosspeaks in NOESY data. Figure 1 illustrates the basic idea of our algorithm. The NOE assignment process can be divided into three phases, viz. initial NOE assignment (phase 1), rotamer selection (phase 2), and filtration of ambiguous NOE assignments (phase 3). The initial NOE assignment (phase 1) is done by considering all pairs of ambiguous NOEs assigned to a NOESY cross peak if the resonances of corresponding atoms fall within a tolerance window around the NOE peak. In the rotamer selection phase, we first compute the backbone structure from RDCs (see Section 4.2), and then place all the rotamers at each residue into backbone and compute all expected NOEs within the upper-bound limit of NOE distance (Figure 1A). Based on the set of all expected NOEs and the resonance assignment list, we back-compute the expected NOE peak pattern for each rotamer (Figure 1B). By matching the back-computed NOE pattern with the experimental NOESY spectrum using an extended model of the Hausdorff distance,[17, 19] we measure how well a rotamer fits the real side-chain conformation when interpreted in terms of the NOESY data. We then select the top k rotamers with highest fitness scores at each residue, and obtain a "low-resolution" structure,[d] by combining the high-resolution backbone structure and the approximate ensemble of side-chain conformations at each residue. The low-resolution structure is then used (in phase 3) to filter ambiguous NOE assignments. The details of filtering ambiguous NOE assignments using the low-resolution structure are provided in Supplementary Material (**SM**) **Section 4** available online in Ref. 40.

4.2. Protein backbone structure determination from residual dipolar couplings

Residual dipolar coupling[33, 34] data provide global orientational restraints on the internuclear bond vectors, such as, backbone NH and CH bond vectors with respect to a global coordinate frame. In solution NMR, RDCs can be recorded with high precision, and assigned much faster than NOEs. In Refs. 39 and 37, the authors proposed the first polynomial-

[c]We will use terms *proton name* and *proton* interchangeably in this paper.

[d]The "low resolution" structure generally has approximately 2.0–3.0 Å (all heavy atom) RMSD from the reference structures solved by X-ray or traditional NMR approaches.

time *de novo* algorithm, which we henceforth refer to as RDC-EXACT, to compute high-resolution protein backbone structures from RDC data. RDC-EXACT takes as input (a) two RDCs per residue (e.g., assigned NH RDCs in two media or NH and CH RDCs in a single medium), (b) delimited α-helices and β-sheets with known hydrogen bond information between paired strands, and a few unambiguous NOEs (used to pack the helices and strands). Note that, these sparse set of NOEs used by RDC-EXACT can usually be assigned using chemical shift information alone[37, 39] without requiring any sophisticated NOE assignment algorithm. Our algorithm HANA uses the high-resolution backbones computed by RDC-EXACT. Loops with missing RDCs are computed using an enhanced version of robotics-based cyclic coordinate descent (CCD) algorithm.[4, 32] The details of RDC-EXACT and modeling of loops (in case of missing RDCs) are provided in **SM[40] Section 1**.

4.3. NOE pattern matching based on the Hausdorff distance measure

Given two finite sets of points $B = \{b_1, \ldots, b_m\}$ and $Y = \{y_1, \ldots, y_n\}$ in Euclidean space, the Hausdorff distance between B and Y is defined as $H(B,Y) = \max\{h(B,Y), h(Y,B)\}$, where $h(B,Y) = \max_{b \in B} \min_{y \in Y} \|b - y\|$, and $\|b - y\|$ measures the normed distance (e.g., L_2-norm) between points b and y. Intuitively, the Hausdorff distance $H(B,Y)$ finds the point in one set that is farthest from any point in the other set, and thus measures the degree of mismatch between the two point sets B and Y. The Hausdorff distance has been widely used in the image processing and computer vision problems, such as visual correspondence,[17] pattern recognition,[19] and shape matching,[18] etc. Unlike many other pattern-recognition algorithms, Hausdorff-based algorithms are combinatorially precise, and provide a robust method for measuring the similarity between two point sets or image patterns[18, 19] in the presence of noise and positional uncertainties.

In the NOE assignment problem, let B denote a back-computed NOE pattern, i.e., the set of back-computed NOE peaks, and let Y denote the set of experimental NOESY peaks. Generally, the size of a back-computed NOE pattern is much smaller than the total number of experimental NOESY peaks. Therefore, we only consider the *directed* Hausdorff distance from B to Y, namely, $h(B,Y) = \max_{b \in B} \min_{y \in Y} \|b - y\|$. We apply an extended model of Hausdorff distance[18, 19, 17] to measure the match between the back-computed NOE pattern and experimental NOESY spectrum. Below, we assume 3D NOESY spectra without loss of generality.

Given the back-computed NOE pattern B with m peaks, and the set of NOESY peaks Y with w peaks, the τ*-th Hausdorff distance* from B to Y is defined as

$$h_\tau(B,Y) = \tau\text{th} \min_{b \in B} \min_{y \in Y} \|b - y\|,$$

where τth is the τ-th largest of m values. We call $f = \tau/m$ the *similarity score* between the back-computed NOE pattern B and the experimental peak set Y, after fixing the Hausdorff distance $h_\tau(B,Y) = \delta$, which is the error tolerance in the NOESY spectra. The similarity score for a rotamer given δ can be computed using a scheme similar to Ref. 17:

$$s = \frac{|B \cap Y_\delta|}{|B|}, \tag{1}$$

where Y_δ denotes the union of all balls obtained by replacing each point in Y with a ball of radius δ, $B \cap Y_\delta$ denotes the intersection of sets B and Y_δ, and $|\cdot|$ denotes the size of a set.

We incorporate two types of uncertainty in the calculation of the similarity score in Equation (1) for the match between the back-computed NOE pattern and experimental NOESY spectrum: (a) possibility of a false random match[17] in the NOESY spectra; (b) uncertainty of NOE peak positions due to experimental noise.

(a) Possibility of a false random match.[17] A *false random match* between the back-computed NOE pattern and the experimental NOESY spectrum is defined as a match when $h_\tau(B,Y) \leq \delta$ occurs at random. We calculate the probability of a false random match and use it as a weighting factor for the similarity score in Equation (1). Let p be the probability for a back-computed NOE peak to randomly match to an experimental peak in Y_δ. Let θ be the probability of a false random match, which can be estimated using the following asymptotic approximation from Ref. 17:

$$\theta \approx \frac{1}{2}(\Phi(\frac{(1-p)m}{\rho}) - \Phi(\frac{(s-p)m}{\rho})),$$

where $\rho = \sqrt{2mp(1-p)}$, and $\Phi(\cdot)$ is the Gauss error function.

(b) Uncertainty from the NOE peak positions. Let $b_i = (\omega(a_1), \omega(a_2), \omega(a_3))$ denote the back-computed NOE peak for an NOE (a_1, a_2, a_3) in a 3D NOESY spectrum. The likelihood for a back-computed peak $b_i = (\omega(a_1), \omega(a_2), \omega(a_3))$ in the NOE pattern B to match an experimental NOESY peak within the distance δ in Y_δ can be defined as

$$\mathcal{N}_i(b_i) = \prod_{j=1}^{3} \mathcal{N}(|\omega(a_j) - p_j|, \sigma_j),$$

where (p_1, p_2, p_3) is the experimental NOESY peak matched to $(\omega(a_1), \omega(a_2), \omega(a_3))$ according to the Hausdorff distance measure, and $\mathcal{N}(|x-\mu|, \sigma)$ is the probability of observing the difference $|x-\mu|$ in a normal distribution with mean μ and standard deviation σ. Here we assume that the noise distribution of peak positions at each dimension is independent of each other. We note that the normal distribution and other similar distribution families have been widely and efficiently used to approximate the noise in the NMR data, e.g., see Refs. 29 and 22.

Then the expected number of peaks in $B \cap Y_\delta$ can be bounded by $|B \cap Y_\delta| = \sum_{i=1}^{m} \mathcal{N}_i(b_i)$. Thus, we have the following equation for the similarity score:

$$s = \frac{1}{m} \sum_{i=1}^{m} \mathcal{N}_i(b_i). \tag{2}$$

After considering both possibility from a false random match and uncertainty from the NOE peak positions, we obtain the following fitness score for a rotamer

$$s' = (1-\theta)s = \frac{1-\theta}{m} \sum_{i=1}^{m} \mathcal{N}_i(b_i). \tag{3}$$

For each rotamer, the computation of its similarity score s' can be computed in $O(mw)$ time, where m is the number of back-computed NOE peaks, and w is the total number of cross peaks in the experimental NOESY spectrum. The detailed pseudocodes for computing the similarity score and for HANA are provided in **SM Sections 3-4** available in Ref. 40.

5. ANALYSIS

5.1. Analysis of rotamer selection based on NOE patterns

Given a back-computed NOE peak $b_i = (\omega_{i1}, \omega_{i2}, \omega_{i3})$ in the NOE pattern of a rotamer, suppose that it finds a matched experimental peak in Y^δ with probability $g(\omega_{i1}, \omega_{i2}, \omega_{i3}, Y^\delta)$. Finding such a matched experimental NOESY peak for b_i can be regarded as a Poisson trial with success probability $g(\omega_{i1}, \omega_{i2}, \omega_{i3}, Y^\delta)$. We present the following result about the expected number of matched peaks for the back-computed NOE pattern of a rotamer.

Lemma 5.1. *Let X_i be an indicator random variable which is equal to 1 if the back-computed NOE peak b_i of a rotamer $\mathbf{r}$ finds a matched experimental peak; 0 otherwise. Let $X = \sum_{i=1}^{m} X_i$, where m is the total number of back-computed NOE cross-peaks for the rotamer $\mathbf{r}$. Then the expected number of back-computed NOE peaks that find matched experimental peaks is given by*

$$E(X) = \sum_{i=1}^{m} E(X_i) = \sum_{i=1}^{m} g(\omega_{i1}, \omega_{i2}, \omega_{i3}, Y^\delta).$$

Let $\mathbf{r_t}$ denote the rotamer closest to the real side-chain conformation for a residue, and let $\mathbf{r_f}$ denote another rotamer in the library for the same residue. We call $\mathbf{r_t}$ the *true* rotamer, and $\mathbf{r_f}$ the *false* rotamer. Let X_i and Y_i be indicator random variables as defined in Lemma 5.1 for each back-computed NOE peak in the true rotamer $\mathbf{r_t}$ and the false configuration $\mathbf{r_f}$ respectively. Let m_t and m_f denote the numbers of back-computed NOE peaks for the true rotamer $\mathbf{r_t}$ and the false rotamer $\mathbf{r_f}$. Let $X = \sum_{i=1}^{m_t} X_i$ and $Y = \sum_{i=1}^{m_f} Y_i$ denote the number of back-computed NOE peaks that find matched experiment peaks for rotamers $\mathbf{r_t}$ and $\mathbf{r_f}$ respectively. Let $\mu_t = E(X)$ and $\mu_f = E(Y)$ denote the expectations of X and Y. For simplicity of our theoretical analysis, we use Equation (1) to measure the fitness between the back-computed NOE pattern of a rotamer and the experimental spectrum in our theoretical model.

To measure the accuracy of the rotamer chosen based on our scoring function, we calculate the probability that the algorithm chooses the wrong rotamer $\mathbf{r_f}$ rather than the true rotamer $\mathbf{r_t}$, and show how it is bounded by certain threshold. The following theorem formally states this result. The proof of this theorem can be found in **SM**[40] **Section 5**.

Theorem 5.1. *Suppose that* $m_f\mu_t - m_t\mu_f \geq \max(m_f, \sqrt{m_f m_t}) \cdot 4\sqrt{\mu_t \ln m_t}$. *Then with probability at least* $1 - m_t^{-1}$, *our algorithm chooses the true rotamer* $\mathbf{r_t}$ *rather than the false rotamer* $\mathbf{r_f}$.

Theorem 5.1 indicates that if the difference between the expected numbers of matched NOE peaks for two roatmers is larger than certain threshold, we are able to distinguish these two roamters based on the Hausdorff distance measure with certain probability bound. By Theorem 5.1, we have the following result on the bound of the probability of picking the correct rotamer from the library based on the Hausdorff distance measure, if we select top k rotamers with highest similarity scores.

Theorem 5.2. *Let* t *denote the maximum number of rotamers for a residue. Suppose that* $m_f\mu_t - m_t\mu_f \geq 4\max(m_f, \sqrt{m_f m_t}) \cdot \sqrt{\mu_t \ln m_t}$ *and* $m_t > t - k$ *hold for the true rotamer* $\mathbf{r_t}$ *and every false rotamer* $\mathbf{r_f}$. *Then with probability at least* $1 - \frac{t-k}{m_t}$, *our algorithm chooses the correct rotamer.*

Proof. Since the total number of rotamers in a residue is t, by Theorem 5.1 the probability that the similarity score of the true rotamer is larger than that of at least $t - k$ rotamers is at least $(1 - \frac{1}{m_t})^{t-k}$. According to the fact $(1 + x)^a \geq 1 + ax$ for $x > -1$ and $a \geq 1$, we have $(1 - \frac{1}{m_t})^{t-k} \geq 1 - \frac{t-k}{m_t}$. Thus, the probability for the algorithm to choose the right rotamer is at least $1 - \frac{t-k}{m_t}$. □

Theorem 5.2 shows that if the discrepancy of the expected number of matched NOE peaks between the true rotamer and every other rotamer, and the number of back-computed NOE peaks are sufficiently large, the ensemble of top k rotamers with highest similarity scores will contain the true rotamer.

5.2. Time complexity analysis

The following theorem states that HANA runs in polynomial time.

Theorem 5.3. HANA *runs in* $O(tn^3 + tn\log t)$ *time, where* t *is the maximum number of rotamers at a residue and* n *is the total number of residues in the protein sequence.*

The detailed derivation of the time complexity can be found in **SM**[40] **Section 6**. We note that in practice, our NOE assignment algorithm HANA runs in 1-2 minutes on a 3 GHz single-processor Linux workstation.

6. RESULTS

HANA takes as input (a) protein sequence, (b) 3D NOESY-HSQC or 2D NOESY peak list, (c) assigned resonance list, (d) backbone computed by using the RDC-EXACT algorithm[37, 39] (Section 4.2), and (e) Xtalview rotamer library.[25] HANA was tested on experimental NMR data for human ubiquitin,[35, 9] zinc finger domain of the human DNA Y-polymerase Eta (pol η)[2] and human Set2-Rpb1 interacting domain (hSRI).[23] The high-resolution structures of these three proteins have been solved either by X-ray crystallography[35] or by traditional NMR approaches using both distance restraints from NOE data and orientational restraints from scalar and dipolar couplings.[9, 2, 23] We used these solved structures, which are also in the Protein Data Bank (PDB), as the reference structures to compare and check the quality of NMR structures determined from our NOE assignment tables. The NMR data for hSRI and pol η were recorded using Varian 600 and 800 MHz spectrometers at Duke University. Ubiquitin NMR data was obtained from Ref. 15 and from the PDB (ID: 1D3Z).

6.1. Robustness of Hausdorff distance and NOE assignment accuracy

To check the robustness of the Hausdorff distance measure for NOE pattern matching, we first computed a low-resolution structure of ubiquitin by combining the backbone determined from RDC-EXACT,[37, 36, 39] and rotamers selected based on the Hausdorff distance measure using patterns for backbone-sidechain NOEs. This low-resolution NMR structure is not the final structure, but is used to filter ambiguous NOE assignments (including backbone-backbone, backbone-sidechain and sidechain-sidechain NOE assignments). Our result shows that the low-resolution structure of ubiquitin obtained from our algorithm has a backbone RMSD 1.58 Å and an all-heavy-atom RMSD 2.85 Å from the corresponding X-ray structure (PDB ID: 1UBQ). Using this low-resolution structure, HANA was able to

Table 1. NOE assignment results for ubiquitin, pol η and hSRI.

Proteins	# of residues	# of NOESY peaks§	# of compatible assignments†	# of incompatible assignments†	Assignment accuracy
ubiquitin*	76	1580	901	93	90.6%
pol η**	39	1386	590	65	90.1%
hSRI***	112	5916	1429	119	92.3%

* The ubiquitin backbone calculated from the RDC data using RDC-EXACT has RMSD 1.58 Å from the X-ray reference structure (PDB ID: 1UBQ) (residues 2-71).

** The pol η backbone calculated from the RDC data using RDC-EXACT has RMSD 1.28 Å for the secondary structure regions and RMSD 2.71 Å for both secondary structure and loop regions (residues 8-36) from the NMR reference structure (PDB ID: 2I5O).

*** The hSRI backbone calculated from the RDC data using RDC-EXACT has RMSD 1.62 Å from the NMR reference structure (PDB ID: 2A7O) for the secondary structure regions (residues 15-34, 51-72, 82-97).

§ The NOESY peak list contains diagonal and symmetric cross peaks.

† Redundant symmetric NOE restraints have been removed from the final NOE assignment table.

resolve the NOE assignment ambiguity caused from the chemical shift degeneracy, and prune a sufficient number of ambiguous NOE assignments, as we will discuss next.

To measure the assignment accuracy of HANA, we define a *compatible* NOE assignment as one in which the distance between the assigned pair of NOE protons in the reference structure is within NOE distance bound of 6.0 Å. Otherwise, we call it an *incompatible* NOE assignment. The number of compatible NOE assignments can be larger than the number of total NOESY peaks, since it is possible that multiple compatible NOEs can be assigned to a single NOESY cross peak. Next, the *assignment accuracy* is defined as the fraction of compatible assignments in the final assignment table output by HANA.

As summarized in Table 1, our NOE assignment algorithm achieved above 90% assignment accuracy for all three proteins. We note that the fraction of assigned peaks of hSRI is less than the other two proteins. This is because we only used backbones in the secondary structure regions (residues 15-34, 51-72, 82-97) for pruning ambiguous NOE assignments for hSRI. Presently we are developing new algorithms to solve long loops. We believe that with more accurate loop backbone structures, we will be able to improve the accuracy of our NOE assignment algorithm, while assigning more NOE peaks. We note that the ubiquitin ^{13}C NOESY data from Ref. 15 are quite degenerate, thus we carefully picked a subset of NOESY peaks for assigning NOEs. Presently we are re-collecting a completely new set of ubiquitin NMR data including four-dimensional NOESY spectra for further testing of our algorithm.

Since the long-range NOEs, in which the spin-interacting protons are at least four residues away, play an important role in the structure determination, we also checked the fraction of incompatible long-range NOE assignments from our algorithm. We found that less than 3% of total assignments were from incompatible long-range NOEs in our computed assignments. As we will discuss next, such a small fraction of incompatible long-range NOE assignments can be easily resolved after one iteration of structure calculation.

6.2. Evaluation of structures from our NOE assignment tables

To test the quality of our NOE assignment results for structure determination, we fed the NOE assignment tables into the standard structure calculation program XPLOR.[3] The input files for the structure calculation include protein sequence, NOE assignment table, and dihedral restraints. Compared with Refs. 2 and 23, in which RDCs are incorporated along with NOE restraints into the final structure calculation, here we only used RDCs to compute the initial backbone fold. From an algorithmic point of view, our structure determination using only NOEs can be considered as a good "control" test of the quality of our NOE assignment. The structure calculation was performed in two rounds. After the first round of structure calculation, the NOE violations larger than 0.5 Å among top 10 structures with lowest energies were removed from the NOE assignment table. Then the refined NOE table was fed into the XPLOR program for the second-round structure calculation.

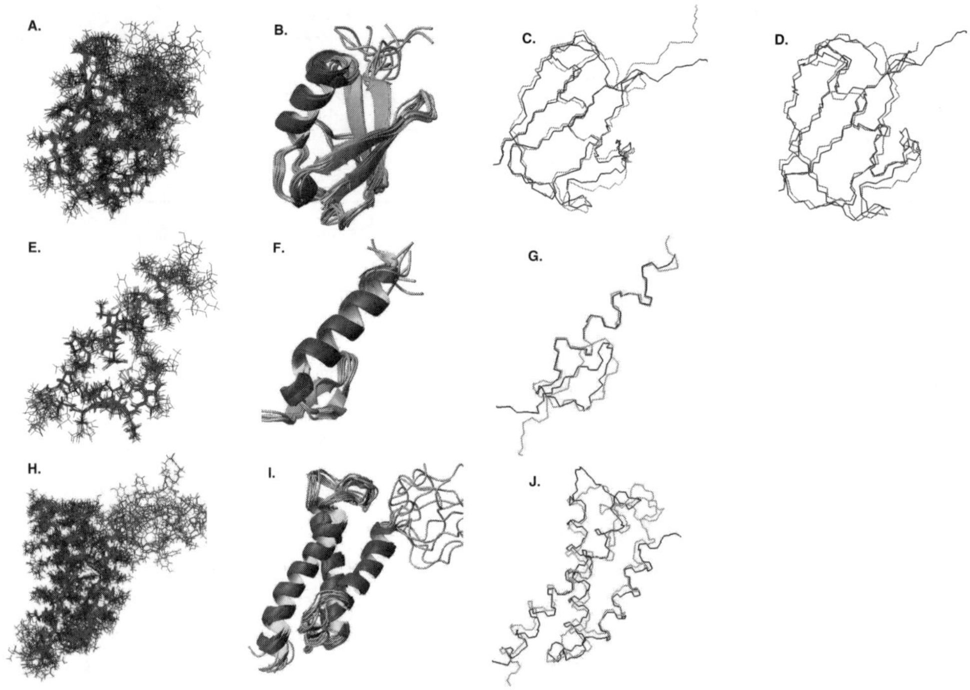

Fig. 2. The NMR structures of ubiquitin, pol η and hSRI computed from our automatically-assigned NOEs. Panels A, B, C and D in first row show the structures of ubiquitin, Panels E, F and G in the middle row show the structures of pol η, and Panels H, I and J in the bottom row show the structures of hSRI. Panels A, E and H show the ensemble of 10 best NMR structures with minimum energies. The backbones are shown in red while the side-chains are shown in blue. Panels B, F and I show the ribbon view of the ensemble of structures. Panel D shows the backbone overlay of the mean structures (in blue color) of ubiquitin with its X-ray reference structures[35] (in magenta color). The RMSD between the mean structure and the x-ray structure of ubiquitin is 1.23 Å for backbone atoms and 2.01 Å for all heavy atoms. Panels C, G and J show the backbone overlay of the mean structures (in blue color) with corresponding NMR reference structures (in green color) that have been deposited into the Protein Data Bank (PDB ID of ubiquitin[9]: 1D3Z; PDB ID of pol η [2]: 2I5O; PDB ID of hSRI[23]: 2A7O). The backbone RMSDs between the mean structures and the reference structures are 1.20 Å for ubiquitin, 1.38 Å for pol η, and 1.71 Å for hSRI. The all-heavy-atom RMSDs between the mean structures and the reference structures are 1.92 Å for ubiquitin, 2.39 Å for pol η, and 2.43 Å for hSRI.

Figure 2 illustrates final NMR structures of ubiquitin, pol η and hSRI calculated from XPLOR using our NOE restraint tables. For all three proteins, only a small number 18–60 (which is $1-4\%$ of the total number of NOE assignments) of NOE violations larger than 0.5 Å occurred after the first round of structure calculation. All final structures converged to an ensemble of low-energy structures with small RMSDs from the reference structure solved either by the X-ray crystallography or by traditional NMR approaches. For all three test cases, the mean structure of final top 10 structures with lowest energies had a backbone RMSD less than 1.7 Å and an all-heavy-atom RMSD less than 2.5 Å from the reference structure. This implies that our NOE assignment algorithm has provided a sufficient number of accurate distance restraints for protein structure determination. In particular, we examined the structure quality in secondary structure and loop regions. We found that the secondary structure regions have better RMSD from the reference structure than the loop regions. After the final structure calculated by XPLOR using our NOE assignment table output by HANA, the RMSD of secondary structure regions in pol η is 0.81 Å for backbone atoms and 1.74 Å for all heavy atoms, and the RMSD of secondary structure regions in ubiquitin is 0.93 Å for backbone atoms and 1.59 Å for all heavy atoms. These results show that the initial fold of secondary structure regions solved using the RDC-EXACT algorithm is accurate enough to combine with chosen rotamers from NOE patterns to resolve the NOE assignment ambiguities. In addition, we also found that the short loop regions of final structures can achieve about the same RMSD from the reference structure as the secondary structure regions. This indicates that the CCD algorithm with filtering of loops based on RDC fit can provide accurate short loops for our NOE assignment algorithm.

Our structure calculation protocol only requires one iteration, while other traditional NMR approaches in general take 7–10 iterations between NOE assignment and structure calculation. In addition, our NOE assignment algorithm only takes 1–2 minutes, versus hours to weeks for other methods. This efficiency is consistent with the proofs of correctness and time complexity of our algorithm. Therefore, the structure calculation framework based on our NOE assignment algorithm is more efficient than all other previous approaches in both theory and practice.

7. CONCLUSION

We have described a novel automated NOE assignment algorithm, HANA, that is combinatorially precise, and runs in polynomial time. To our knowledge, HANA is the first NOE assignment algorithm that simultaneously exploits the accurate algebraic geometry-based high-resolution backbone computation from RDC data,[37, 39] the statistical diversity of rotamers from a rotamer library,[25] and the robust Hausdorff measure[17, 19] for comparing the back-computed NOE patterns with the experimental NOE spectra and choosing accurate rotamers, to finally compute the NOE assignments with high accuracy. Owing to its simplicity, HANA runs extremely fast in practice. Furthermore, when applied to real biological NMR spectra for three proteins, our algorithm yields high assignment accuracy (> 90%) in each case suggesting its ability to play a role in high-throughput structure determination.

Although our current implementation of HANA uses 2D and 3D NOESY spectra, HANA is general and can be easily extended to use higher-dimensional (e.g., 4D) NOESY data.[6, 7] In addition, it would be interesting to extend the current version of HANA for NOE assignment with missing resonances. In general, acquisition of complete resonance assignment can require selective labeling of proteins, and is time-consuming. On the other hand, selection of correct rotamers can help the resonance assignment for side-chains. In principle, HANA can be extended to accommodate the NOE assignment with a partially assigned resonance list, as long as the back-computed NOE patterns with missing peaks are sufficient to identify accurate rotamers. Finally, it would be interesting to explore the use of side-chain rotamer packing algorithms[11] to choose rotamers that fit the data.

Acknowledgements

We thank Dr. L. Wang, Mr. A. Yan, Dr. S. Apaydin, Mr. J. Boyles, Prof. J. Richardson, Prof. D. Richardson, and all members of the Donald and Zhou Labs for helpful discussions and comments. We are grateful to Ms. M. Bomar for helping us with pol η NMR data.

References

1. C. Bailey-Kellogg, S. Chainraj, and G. Pandurangan. A random graph approach to nmr sequential assignment. *Journal of Computational Biology*, 12(6):569–583, 2005.
2. M. G. Bomar, M. Pai, S. Tzeng, S. Li, and P. Zhou. Structure of the ubiquitin-binding zinc finger domain of human DNA Y-polymerase η. *EMBO reports*, 8:247–251, 2007.
3. A. T. Brünger. X-PLOR, Version 3.1: a system for X-ray crystallography and NMR. *Journal of the American Chemical Society*, 1992.
4. A. A. Canutescu and R. L. Dunbrack Jr. Cyclic coordinate descent: A robotics algorithm for protein loop closure. *Protein Science*, 12:963–972, 2003.
5. G. M. Clore, J. G. Omichinski, K. Sakaguchi, N. Zambrano, H. Sakamoto, E. Appella, and A. M. Gronenborn. Interhelical angles in the solution structure of the oligomerization domain of the tumour suppressor p53. *Science*, 267:1515–1516, 1995.
6. B. E. Coggins, R. A. Venters, and P. Zhou. Filtered Back-projection for the Reconstruction of a High-Resolution (4,2)D CH3-NH NOESY Spectrum on a 29 kDa Protein. *J Am Chem Soc*, 127:11562–11563, 2005.
7. B. E. Coggins and P. Zhou. PR-CALC: A Program for the Reconstruction of NMR Spectra from Projections. *J Biomol NMR*, 34:179–95, 2006.
8. G. Cornilescu, F. Delaglio, and A. Bax. Protein backbone angle restraints from searching a database for chemical shift and sequence homology. *Journal of Biomolecular NMR*, 13:289–302, 1999.
9. G. Cornilescu, J. L. Marquardt, M. Ottiger, and A. Bax. Validation of Protein Structure from Anisotropic Carbonyl Chemical Shifts in a Dilute Liquid Crystalline

Phase. *Journal of the American Chemical Society*, 120:6836–6837, 1998.

10. G. M. Crippen and T. F. Havel. Distance Geometry and Molecular Conformations. *Wiley, New York.*, pages 635–642, 1988.
11. I. Georgiev, R. H. Lilien, and B. R. Donald. The minimized dead-end elimination criterion and its application to protein redesign in a hybrid scoring and search algorithm for computing partition functions over molecular ensembles. *Journal of Computational Chemistry*, [Epub ahead of print] PMID: 1829329, Feb 21, 2008.
12. P. Güntert. Automated NMR Protein Structure Determination. *Progress in Nuclear Magnetic Resonance Spectroscopy*, 43:105–125, 2003.
13. P. Güntert. Automated NMR protein structure calculation with CYANA. *Meth. Mol. Biol.*, 278:353–378, 2004.
14. T. Herrmann, P. Güntert, and K. Wüthrich. Protein NMR Structure Determination with Automated NOE Assignment Using the New Software CANDID and the Torsion Angle Dynamics Algorithm DYANA. *Journal of Molecular Biology*, 319(1):209–227, 2002.
15. R. Harris. The ubiquitin NMR resource page, BBSRC Bloomsbury Center for Structural Biology, http://www.biochem.ucl.ac.uk/bsm/nmr/ubq/. Jun 2007.
16. Y. J. Huang, R. Tejero, R. Powers, and G. T. Montelione. A topology-constrained distance network algorithm for protein structure determination from NOESY data. *Proteins: Structure Function and Bioinformatics*, 62(3):587–603, 2006.
17. D. P. Huttenlocher and E. W. Jaquith. Computing visual correspondence: Incorporating the probability of a false match. In *Procedings of the Fifth International Conference on Computer Vision (ICCV 95)*, pages 515–522, 1995.
18. D. P. Huttenlocher and K. Kedem. *Distance Metrics for Comparing Shapes in the Plane.* In B. R. Donald and D. Kapur and J. Mundy, editors, Symbolic and Numerical Computation for Artificial Intelligence, pages 201-219, Academic press, 1992.
19. D. P. Huttenlocher, G. A. Klanderman, and W. Rucklidge. Comparing Images Using the Hausdorff Distance. *IEEE Trans. Pattern Anal. Mach. Intell.*, 15(9):850–863, 1993.
20. H. Kamisetty, C. Bailey-Kellogg, and G. Pandurangan. An efficient randomized algorithm for contact-based nmr backbone resonance assignment. *Bioinformatics*, 22(2):172–180, 2006.
21. J. Kuszewski, C. D. Schwieters, D. S. Garrett, R. A. Byrd, N. Tjandra, and G. M. Clore. Completely automated, highly error-tolerant macromolecular structure determination from multidimensional nuclear overhauser enhancement spectra and chemical shift assignments. *J. Am. Chem. Soc.*, 126(20):6258–6273, 2004.
22. C. J. Langmead, A. K. Yan, R. H. Lilien, L. Wang, and B. R. Donald. A polynomial-time nuclear vector replacement algorithm for automated nmr resonance assignments. In *Proceedings of the seventh annual international conference on Research in computational molecular biology*, pages 176–187, 2003.
23. M. Li, H. P. Phatnani, Z. Guan, H. Sage, A. L. Greenleaf, and P. Zhou. Solution structure of the Set2-Rpb1 interacting domain of human Set2 and its interaction with the hyperphosphorylated C-terminal domain of Rpb1. *Proceedings of the National Academy of Sciences*, 102:17636–17641, 2005.
24. J. P. Linge, M. Habeck, W. Rieping, and M. Nilges. ARIA: Automated NOE assignment and NMR structure calculation. *Bioinformatics*, 19(2):315–316, 2003.
25. S. C. Lovell, J. M. Word, J. S. Richardson, and D. C. Richardson. The Penultimate Rotamer Library. *Proteins: Structure Function and Genetics*, 40:389–408, 2000.
26. G. T. Montelione and H. N. B. Moseley. Automated analysis of NMR assignments and structures for proteins. *Curr. Opin. Struct. Biol.*, 9:635–642, 1999.
27. C. Mumenthaler, P. Güntert, W. Braun, and K. Wüthrich. Automated combined assignment of NOESY spectra and three-dimensional protein structure determination. *J. Biomol. NMR*, 10(4):351–362, 1997.
28. M. Nilges, M. J. Macias, S. I. O'Donoghue, and H. Oschkinat. Automated NOESY interpretation with ambiguous distance restraints: the refined NMR solution structure of the pleckstrin homology domain from β-spectrin. *Journal of Molecular Biology*, 269(3):408–422, 1997.
29. W. Rieping, M. Habeck, and M. Nilges. Inferential Structure Determination . *Science*, 309:303 – 306, 2005.
30. J. B. Saxe. Embeddability of weighted graphs in k-space is strongly NP-hard. *Proc. 17th Alleron Conf. Commun. Control Comput.*, pages 480–489, 1979.
31. C. D. Schwieters, J. J. Kuszewski, N. Tjandra, and G. M. Clore. The Xplor-NIH NMR molecular structure determination package. *J Magn Reson*, 160:65–73, 2003.
32. A. Shehu, C. Clementi, and L. E. Kavraki. Modeling protein conformational ensembles: from missing loops to equilibrium fluctuations. *Proteins: Structure, Function, and Bioinformatics*, 65(1):164–79, 2006.
33. N. Tjandra and A. Bax. Direct measurement of distances and angles in biomolecules by NMR in a dilute liquid crystalline medium. *Science*, 278:1111–1114, 1997.
34. J. R. Tolman, J. M. Flanagan, M. A. Kennedy, and J. H. Prestegard. Nuclear magnetic dipole interactions in field-oriented proteins: Information for structure determination in solution. *Proc. Natl. Acad. Sci. USA*, 92:9279–9283, 1995.
35. S. Vijay-Kumar, C. E. Bugg, and W. J. Cook. Structure of ubiquitin refined at 1.8 A resolution. *Journal of Molecular Biology*, 194:531–44, 1987.
36. L. Wang and B. R. Donald. Analysis of a Systematic Search-Based Algorithm for Determining Protein Backbone Structure from a Minimal Number of Residual Dipolar Couplings. In *Proceedings of The IEEE Computational Systems Bioinformatics Conference (CSB), Stanford CA (August, 2004)*, 2004.
37. L. Wang and B. R. Donald. Exact solutions for internuclear vectors and backbone dihedral angles from NH residual dipolar couplings in two media, and their application in a systematic search algorithm for determining protein backbone structure. *Jour. Biomolecular NMR*, 29(3):223–242, 2004.
38. L. Wang and B. R. Donald. An Efficient and Accurate Algorithm for Assigning Nuclear Overhauser Effect Restraints Using a Rotamer Library Ensemble and Residual Dipolar Couplings. *The IEEE Computational Systems Bioinformatics Conference (CSB), Stanford CA. (August, 2005)*, pages 189–202, 2005.

39. L. Wang, R. Mettu, and B. R. Donald. A Polynomial-Time Algorithm for De Novo Protein Backbone Structure Determination from NMR Data. *Journal of Computational Biology*, 13(7):1276–1288, 2006.
40. J. Zeng, C. Tripathy, P. Zhou, and B. R. Donald. A Hausdorff-Based NOE Assignment Algorithm Using Protein Backbone Determined from Residual Dipolar Couplings and Rotamer Patterns – Supplementary Material. Department of Computer Science, Duke University, [online]. Available: http://www.cs.duke.edu/donaldlab/Supplementary/csb08/. May, 2008.

ITERATIVE NON-SEQUENTIAL PROTEIN STRUCTURAL ALIGNMENT

Saeed Salem and Mohammed J. Zaki*

Computer Science Department, Rensselaer Polytechnic Institute,
110 8th st. Troy, NY 12180, USA
*Email: {salems, *zaki}@cs.rpi.edu*

Structural similarity between proteins gives us insights on the evolutionary relationship between proteins which have low sequence similarity. In this paper, we present a novel approach called STSA for non-sequential pair-wise structural alignment. Starting from an initial alignment, our approach iterates over a two-step process, a superposition step and an alignment step, until convergence. Given two superposed structures, we propose a novel greedy algorithm to construct both sequential and non-sequential alignments. The quality of STSA alignments is evident in the high agreement it has with the reference alignments in the challenging-to-align RPIC set. Moreover, on a dataset of 4410 protein pairs selected from the CATH database, STSA has a high sensitivity and high specificity values and is competitive with state-of-the-art alignment methods and gives longer alignments with lower *rmsd*. The STSA software along with the data sets will be made available on line at `http://www.cs.rpi.edu/~zaki/software/STSA`.

Keywords: non-sequential alignment, CATH, PDB, RIPC set.

1. INTRODUCTION

Over the past years, the number of known protein structures has been increasing at a relatively fast pace, thanks to advancement in MR spectroscopy and X-ray crystallography. Recently (as of Oct 2007) the number of protein structures in the Protein Data Bank(PDB) [1] has reached 46377. Despite having the structural information about so many proteins, the function of a lot of these proteins is still unknown. Structural similarity highlights the functional relationship between proteins. Moreover, structural similarity between proteins allows us to study evolutionary relationship between remotely homologous proteins (with sequence similarity in the twilight-zone), thus allowing us to look farther in evolutionary time [2]. The goal of protein structural alignment is to find maximal substructures of proteins A and B, such that the similarity score is maximized. The two most commonly used similarity measures are: The coordinate distance-based root mean squared deviation (*rmsd*), which measures the spatial euclidean distance between aligned residues; and the distance matrix based measure that computes the similarity based on intra-molecular distances representing protein structures.

The complexity of protein structural alignment depends on how the similarity is assessed. Kolodny and Linial [3] showed that the problem is NP-hard if the similarity score is distance matrix based. Moreover, they presented an approximate polynomial time solution by discrediting the the rigid-body transformation space. In a more recent work, Xu et al. [4] proposed an approximate polynomial time solution, when the contact map based similarity score is used, using similar democratization techniques. Despite the polynomial time approximate algorithms and as the authors themselves noted, these methods are still too slow to be used in search tools.

There is no current algorithm that guarantees an optimal answer for the pair-wise structural alignment problem. Over the years, a number of heuristic approaches have been proposed, which can mainly be classified into two main categories.

1.1. Dynamic Programming Approach

Dynamic Programming (DP) is a general paradigm to solve problems that exhibit the optimal substructure property [5]. DP-based methods [6, 7, 8, 9, 10] construct a scoring matrix S, where each entry S_{OJ} corresponds to the score of matching the i-Th residue in protein A and the j-Th residue in protein B. Given a scoring scheme between residues in the two proteins, dynamic programming finds the global alignment that maximizes the score. Once the best equivalence is found, a superposition step is performed to find the transformation that minimizes the *rmsd* between the corresponding residues. In STRUCTAL [7], the structures are first superimposed onto each other using initial seeds (random or sequence-based). The similarity score S_{OJ} of match-

*Corresponding author.

ing the residues is a function of the spatial displacement between the residue pairs in the superimposed structures. DP is applied on the scoring matrix to get an alignment. The alignment obtained is an initial seed and the process of superposition and alignment is repeated till convergence.

Other methods employed local geometrical features to calculate the similarity score. CTSS [11] used a smooth spline with minimum curvature to define a feature vector of the protein backbone which is used to calculate the similarity score. Tyagi et al. [10] proposed a DP-based method where the similarity is the substitution value obtained from a substitution matrix for a set of 16 structural symbols. DP-based methods suffer from two main limitations: first, the alignment is sequential and thus non-topological similarity cannot be detected, and second, it is difficult to design a scoring function that is globally optimal [3].

1.2. Clustering Approach

Clustering-based methods [12, 13, 14, 15, 16, 17] seek to assemble the alignment out of smaller compatible (similar) element pairs such that the score of the alignment is as high as possible [18]. Two compatible element pairs are consistent (can be assembled together) if the substructures obtained by elements of the pairs are similar. The clustering problem is NP-hard [19], thus several heuristics have been proposed. The approaches differ in how the set of compatible element pairs is constructed and how the consistency is measured.

In [20], initial compatible triplets are found using geometric hashing. Two compatible triplets are consistent if they have similar transformations, where the transformation is defined such that it can transform one triplet onto the other with minimum distance. DALI [12] finds gapless fragment compatible pairs, which are similar hexapeptide fragments. It then uses a Monte Carlo procedure to combine consistent fragments into a larger set of pairs. The optimization starts from different seeds and the best alignment is reported. Compatible elements in SARF2 [13] are similar secondary structure elements (SSEs) which are obtained by sliding a typical α-helix or β-strand over the C_α trace of the protein. The set of the compatible pairs of the SSEs are filtered based on some distance and angle constraints; the final alignment is obtained by finding the largest set of mutually consistent fragment pairs. In an effort to reduce the search space in clustering methods, CE [14] starts with an initial fragment pair and the alignment is extended by the best fragment that satisfies a similarity criteria. In FATCAT [17], DP is used to chain the fragment pairs.

1.3. Our Contributions

We present STSA[a], an efficient non-sequential pairwise structural alignment algorithm. STSA is an iterative algorithm similar in spirit to the iterative Dynamic Programming(DP)-based methods, yet it employs a different technique in constructing the alignment. Specifically, we propose a greedy chaining approach to construct the alignment for a pair of superposed structures. One limitation of DP-based methods is that they only generate sequential alignments. Another limitation is the fact that we do not yet know how to design a scoring function that is globally optimal [3]. Our approach addresses these challenges by looking directly at the superposed structures and assembles the alignment from small closely superposed fragments. Unlike DP, this greedy approach allows for non-topological (non-sequential) similarity to be extracted.

We employ PSIST [21] to generate a list of similar substructures which serve as the initial alignment seeds. Our approach is decoupled such that we can use initial alignment seeds from other methods. In fact, we use SCALI seeds [16] for the RIPC results.

To assess the quality of the STSA alignment, we tested it on the recently published hard-to-align RIPC set [22]. STSA alignments have higher agreement (accuracy) with the reference alignment than state-of-the-art methods: CE, DALI, FATCAT, MATRAS, CA, SHEBA, and SARF. Moreover, we compiled a dataset of 4410 protein pairs from the CATH classification [23]. We measured the overall sensitivity and specificity of STSA to determine if two proteins have the same classification. Results from the CATH dataset indicate that STSA achieves high sensitivities at high specificity levels and is competitive to well established structure comparison methods like DALI, STRUCTAL, and FAST, as judged by the geometric match measure SAS_k [6].

[a]an acronym of **ST**ructural pair-wi**S**e **A**lignment

2. STSA ALIGNMENT

Our approach is based on finding an alignment based on initial seeds. We first discuss how to get the initial seeds and then explain our greedy chaining algorithm.

2.1. Alignment Seeding

The initial alignment seeds are similar substructures between protein A and protein B. An initial seed is an equivalence between a set of pairs. We obtain the seeds from our previous work PSIST [21]. PSIST converts each protein structure into a **S**tructure-**F**eature (**SF**) sequence and then uses suffix tree indexing to find the set of maximal matching segments (initial seeds)

Another source of seeds we use is the SCALI seeds [16]. The SCALI seeds are gapless local sequence-structure alignments obtained using HMM-STR [24], which is an HMM built on top of a library of local motifs.

An initial seed $s = F_i^A F_j^B(l)$ indicates that the fragment of protein A that starts at residue i matches the fragment from protein B that starts at residue j and both the fragments has equal length of l.

2.2. Iterative Superposition-Alignment Approach

Each alignment seed $(F_i^A F_j^B(l))$ is treated as an initial equivalence, E_0, between a set of residues from protein A and a set of residues from protein B. The correspondence between the residues in the equivalence is linear, i.e. $E = \{(a_i, b_j), \cdots, (a_{i+l-1}, b_{j+l-1})\}$. Given an equivalence E, we construct an alignment of the two structures as follows.

2.2.1. *Finding Optimal Transformation*

We first find a transformation matrix T_{opt} that optimally superposes the set of pairs of residues in the equivalence E such that the *rmsd* between the superposed substructures of A and B is minimized:

$$T_{opt} = arg_{min}(T) \quad RMSD_T(E) \ ,$$

where $RMSD_T(E) = \frac{1}{|E|} \sum_{(i,j) \in E} d(T[a_i], b_j)$. We find the optimal transformation T_{opt} using the Singular Value Decomposition [25, 26].

2.2.2. *Constructing Scoring Matrix*

We next apply the optimal transformation T_{opt} obtained in the previous step to protein A to obtain A^*. We then construct a $n \times m$ binary scoring matrix S, where n and m denote the number of residues in proteins A and B, respectively and $S_{ij} = score(dist(a_i^*, b_j))$; the score is 1 if the distance between corresponding elements, a_i^* and b_j is less than a threshold δ, and 0 otherwise.

2.2.3. *Finding an Alignment*

An alignment is a set of pair of residues $\{(a_i, b_j)\}$, a_i in A, and b_j in B. Based on the scoring matrix S we find the maximum correspondence by finding the maximum cardinality matching in the bipartite graph $G(U, V, E)$ where U is the set of residues in protein A, V is the set of residues in proteins B, and there is an edge $(a_i, b_j) \in E$ if $S_{ij} = 1$. However, the problem with the maximum matching approach is that it may yield several short, disjoint and even arbitrary matchings that may not be biologically very meaningful. Our goal is to find an alignment composed of a set of segments such that each segment has at least r residue pairs.

A run R_i is a set of consecutive diagonal 1's in the scoring matrix S which constitutes an equivalence, between a substructure in A and another in B, that can be aligned with a small *rmsd*. Specifically, a run R is a triplet (a_i, b_j, l), where a_i is the starting residue for the run in A (similarly b_j for B), and the the length of the run is l. The correspondence between residues in the run is as follows: $\{(a_i, b_j), \cdots, (a_{i+l-1}, b_{j+l-1})\}$.

The matrix S has a set of runs $R = \{R_1, R_2, \cdots, R_k\}$ such that $|R_i| \geq r$, where r is the minimum threshold length for a run. We are interested in finding a subset of runs $C \subseteq R$ such that all the runs in C are mutually non-overlapping and the length of the runs in C, $L(C) = \sum_{i \in C} |R_i|$ is as large as possible. The problem of finding the subset of runs with the largest length is essentially the same as finding the maximum weighted clique in a graph $G = (V, E)$ where V is the set of runs, with the weight for vertex i given as $w_i = |R_i|$, and there is an edge $(i, j) \in E$ if the runs R_i and R_j do not overlap. The problem of finding the maximum weighted clique is NP-hard [19], therefore we use greedy algorithms to find an approximate solution. Note that it is also possible to use a dynamic programming approach to align the proteins based on the scoring matrix S, however, this would yield only a sequential alignment. Since we are interested in non-sequential alignments, we adopt the greedy weighted clique fining approach.

The simplest greedy algorithm chooses the longest run $R_i \in R$ to be included in C, and then removes from R all the runs R_j that overlap with R_i. It then chooses the longest remaining run in R, and iterates this process until R is empty. We also implemented an enhanced greedy algorithm that differs in how it chooses the run to include in C. It chooses the run $R_i \in R$ that has the highest weight w where $w(R_i)$ is the length of R_i plus the lengths of all the remaining non-overlapping runs. In other words, this approach not only favors the longest run, but also favors those runs that do not preclude many other (long) runs.

Through our experiments, we found that the simple greedy algorithm gives similar alignments in terms of the length and *rmsd* as the enhanced one. Moreover, it is faster since we do not have to calculate the weights every time we choose a run to include to C. Therefore, we adopt the first heuristic as our basic approach. Note that it is also possible to use other recently proposed segment chaining algorithms [27]. The subset of runs in C makes up a new equivalence E_1 between residues in proteins A and B. The length of the alignment is the length of the equivalence $|E_1| = \sum_{i \in C} |R_i|$ and the *rmsd* of the alignment is the *rmsd* of the optimal superposition of the residue pairs in E_1.

2.2.4. *Refining the Alignment*

To further improve the structural alignment we treat the newly found equivalence E_1 as an initial alignment and repeat the previous steps all over again. The algorithm alternates between the superposition step and the alignment step until convergence (score does not improve) or until a maximum number of iterations has been reached. Figure 1 shows the pseudo-code for our iterative superposition-alignment structural alignment algorithm. The method accepts the set of maximal matching segments $M = \{F_i^A F_j^B(l)\}$ as initial seeds. It also uses three threshold values: δ for creating the scoring matrix, r for the minimum run length in S, and L for the maximum *rmsd* allowed for an equivalence. For every initial seed we find the optimal transformation (lines 4-5), create a scoring matrix (line 6), and derive a new alignment E_1 via chaining (line 7). If the *rmsd* of the alignment is above the threshold L we move on to the next seed, or else we repeat the steps (lines 3-10) until the score no longer improves or we exceed the maximum number of iterations. The best alignment found for each seed is stored in the set of potential alignments $\mathcal{E}$ (line 11). Once all seeds are processed, we output the best alignment found (line 13). We use the SAS_k [6] geometric match measure (explained in the next section) to score the alignments. We noticed that typically three iterations were enough for the convergence of the algorithm.

$M = \{F_i^A F_j^B(l)\}$, set of seed alignments
L, the *rmsd* threshold
r, the min threshold for the length of a run in S
δ, the max distance threshold for S

SEED-BASED ALIGNMENT (M, L, r, δ):
1. **for** every $F_i^A F_j^B(l) \in M$
2. $\quad$ E is the equivalence based on $F_i^A F_j^B(l)$
3. $\quad$ **repeat**
4. $\quad\quad$ $T_{opt} = RMSD_{opt}(E)$
5. $\quad\quad$ $A^* = T_{opt}A$
6. $\quad\quad$ $S_{ij} = 1$ if $d(a_i^*, b_j) < \delta$, 0 otherwise
7. $\quad\quad$ E_1= chain-segments(S, r)
8. $\quad\quad$ if $RMSD_{opt}(E_1) \geq L$ go to step 2
9. $\quad\quad$ $E \longleftarrow E_1$
10. $\quad$ **until** score does not improve
11. $\quad$ add E to the set of alignments $\mathcal{E}$
12. **end for**
13. Output best alignment from $\mathcal{E}$

Fig. 1. The STSA algorithm.

2.3. Scoring the Alignments

We assess the significance of STSA alignments by using the geometric match measure, SAS_k, introduced in [6], defined as follows:

$$SAS_k = rmsd(100/N_{mat})^k$$

where *rmsd* is the coordinate root mean square deviation, N_{mat} is the length of the alignment, and k is the degree to which the score favors longer alignments at the expense of *rmsd* values. In our implementation, we use $k = 1$, $k = 2$ and $k = 3$ to score the alignments, to study the effect of the scoring function. For each of the three scoring schemes SAS_1, SAS_2 and SAS_3, a lower score indicates a better alignment, since we desire lower *rmsd* and longer alignment lengths. Kolodny et al. [28] recently contended that scoring alignment methods by geometric measures yields better specificity and sensitivity; we observe consistent behavior in our results.

2.4. Initial Seeds Pruning

Since the quality of the alignment depends on the initial alignment (seed), we start with different initial seeds in an attempt to reach a global optimum alignment. This, however, results in a slow algorithm since we could potentially have a large number of initial seeds. Let the size of protein A be n and of B be m, respectively and $n \leq m$. The number of maximal matching segments can be as large as nm/l_{min}, where l_{min} is the length threshold. Most of these seeds do not constitute good initial seeds as judged by their final global alignments. In order to circumvent this problem, we only select heuristically the most promising seeds based on two heuristics: first, the length of the seed; second, the DALI rigid similarity score [12]. In the results section, we study the effect of these pruning heuristics on the quality of the alignments and the improvement in the running time that we gain.

2.5. Computational Complexity

The worst case complexity of finding the maximal matching segments using PSIST is $O(nm)$, where m and n denote the lengths of proteins A and B [21]. Assuming $m \leq n$, the complexity of constructing the full set of runs R is $O(nm)$, since we have to visit every entry of the scoring matrix. Since we use a threshold of $\delta = 5\mathring{A}$ to set $S_{ij} = 1$ in the scoring matrix, each residue, due to distance geometry, in A can be close to only a few residues in B (after superposition). Therefore, there are $O(n)$ 1's in the matrix S. And thus, we have d$O(n)$ diagonal runs, and sorting these runs takes $O(n \log n)$ time. In the greedy chaining approach, for every run we choose, we have to eliminate other overlapping runs, which can be done in $O(n)$ time per check, for a total time of $O(n^2)$. Over all the steps the complexity of our approach is therefore $O(n^2)$.

3. RESULTS

To assess the quality of STSA alignments compared to other structural alignment methods, we tested our method on the hard-to-align RIPC set [22]. Moreover, we evaluated the overall sensitivity and specificity of STSA compared to other alignment methods over 4410 alignment pairs using the CATH [23] classification as a gold standard.

The criteria on which we selected the other algorithms to compare with were: the availability of the program so that we could run it in-house, and the running time of the algorithm. We compared our approach against DALI [12], STRUCTAL [6], SARF2 [13], and FAST [15]. For the RIPC dataset, we used the published results for CE [14], FATCAT [17], CA [b], MATRAS [c], LGA [29], and SHEBA [30].

All the experiments were run on a 1.66 GHz Intel Core Duo machine with 1 GB of main memory running Ubuntu Linux. The default parameters for STSA were $r = 3$, $\delta = 5.5\mathring{A}$ and using top 100 initial seeds (see Section 3.3 for more details).

3.1. RIPC Set

The RIPC set contains 40 structurally related protein pairs which are problematic to align. Reference alignments for 23 (out of the 40) structure pairs have been derived based on sequence and function conservation. We measure the agreement of our alignments with the reference alignments provided in the RIPC set. As suggested in [22], we compute the percentage of the residues aligned identically to the reference alignment(I_s) relative to the reference alignment's length (L_{ref}).

As shown in Figure 2, while all the methods have mean agreements equal 60 percent or lower, the mean agreement of STSA alignments is 71%. As for the median, all the methods except FATCAT (63%) have median agreements less than 60%, while STSA alignments have a median agreement of 67% .

Some the alignments in the RIPC set are sequential. In these cases, most of the sequential alignment methods return a high agreement with the reference alignment. Thus, in few cases the sequential alignment of STSA gives a higher agreement than the non-sequential alignment. If we were to take the STSA alignment that gives a higher agreement with the reference alignment, then STSA alignments would have a mean and median agreement of 77% and 83% , respectively (STSABest in Figure 2).

As Mayr et al. [22] noted, there are seven chal-

[b] `http://bioinfo3d.cs.tau.ac.il/c_alpha_match/`
[c] `http://biunit.naist.jp/matras/`

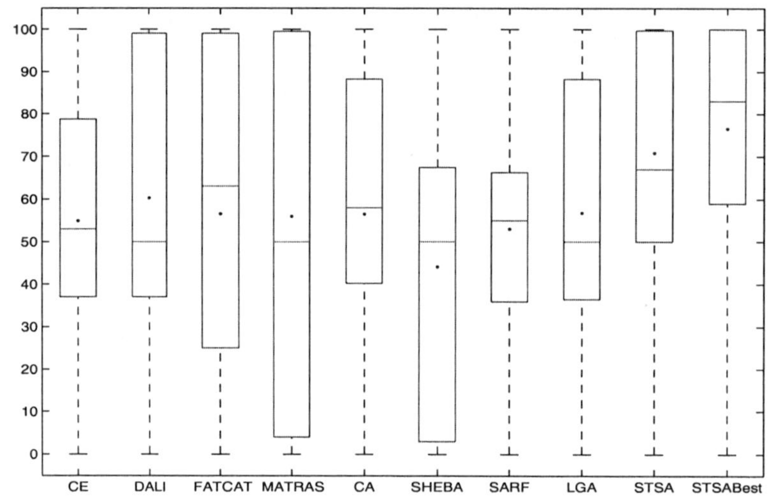

Fig. 2. Comparison of the alignments of 8 methods with the reference alignments from the RIPC set. Box-and-whisker plots for the distribution of agreements of the alignments produced by different methods as compared to the true reference alignments. The dark dots indicate the means, the horizontal lines inside the boxes indicate the medians, and the boxes show the range between the lower and the upper quartiles. Results for all the other methods (except SARF) are taken from [22].

lenging protein pairs which reveal how repetition, extensive indels, and circular permutation result in low agreement with the reference alignments. We found two protein pairs particularly problematic to align for all the sequential methods and sometimes the non-sequential ones, except STSA. First, for alignment of L-2-Haloacid dehalogenase (SCOP id: d1qq5a_, 245 residues) with CheY protein (d3chy__, 128 residues), all the methods (except SARF returned 33%) returned zero agreement with the reference alignment while STSA returned 100 percent agreement. The second problematic pair was of the alignment of NK-lysin (d1nkl__, 78 residues) with (Pro)phytepsin (d1qdma1, 77 residues) which has a circular permutation. For the second pair, all the methods (except CA returned 41%, and SARF returned 92%) returned zero agreement with the reference alignment while STSA returned 99 percent agreement. In this pair the N-terminal region of domain d1nkl__ has to be aligned with the C-terminal region of domain d1qdma1 to produce an alignment that matches the reference alignment (see Figure 3). By design, sequential alignment methods cannot produce such an alignment, and therefore fail to capture the true alignment. Among the non-sequential methods, the agreement of STSA alignments with the reference alignments are higher than the agreement of either CA and SARF.

As shown in Figure 3, all the last five methods (DALI, MATRAS, SHEBA, FATCAT, and LGA) have their alignment paths along the diagonal and do not agree with with the reference alignment (shown as circles). The CA method reports a non-sequential alignment that partially agrees with the reference alignment but it misses 59% of the reference alignment pairs. Both SARF and STSA alignments have excellent agreement with the reference alignment, 92%, 99%, respectively.

3.2. Measuring Sensitivity and Specificity Using CATH

Gerstein and Levitt [31] emphasized the importance of assessing the quality and significance of structural alignment methods using an objective approach. They used the SCOP database [32] as a gold standard to assess the sensitivity of the structural alignment program against a set of 2,107 pairs that have the same SCOP superfamily. In a more recent work, Kolodny et al. [28] presented a comprehensive comparison of six protein structural alignment methods. They used the CATH classification [23] as a gold standard to compare the rate of true and false positives of the methods. Moreover, they showed that the geometric match measures like SAS_k can better assess the quality of the structural alignment methods. We adopt a similar approach to assess the significance of our approach by comparing the

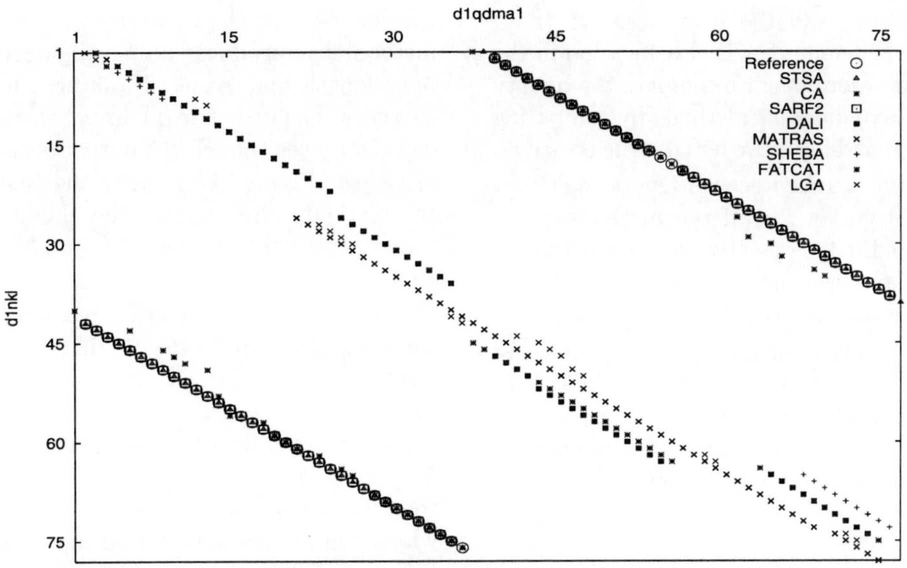

Fig. 3. Comparison of the agreement with the reference alignment of STSA alignment and 6 other alignment methods. Residue positions of d1qdma and d1nkl__ are plotted on the x-axis and y-axis, respectively. Note: The reference alignment pairs are shown in circles. The CA, SARF, and STSA plots overlap with the reference alignment. For this pair, we used the alignment's server of the corresponding method to get the alignment, except for DALI and SHEBA which we ran in-house.

true and false positive rates of STSA alignments to those of other three methods: DALI, STRUCTAL, and FAST. Since the other methods report only sequential alignments, for STSA we also used sequential alignments.

3.2.1. *The CATH Singleton Dataset*

CATH [23] is a hierarchical classification of protein domain clusters. The CATH database clusters structures using automatic and manual methods. The latest version (3.1.0; as for Jan'07) of the CATH database contains more than 93885 domains (63453 chains, from 30028 proteins) classified into 4 Classes, 40 Architectures, 1084 Topologies, and 2091 Homologous Superfamilies. The class level is determined according to the overall secondary structure content. The architecture level describes the shape of the domain structure. The topology (fold family) level groups protein domains depending on both the overall shape and connectivity of the secondary structures. Protein domains from the same homologous superfamily are thought to share a common ancestor and have high sequence identity or structure similarity.

We define protein domains that belong to homologous superfamilies which have only one member as *singletons*. There are 1141 singleton protein domains which belong to 648 different topologies in CATH. Since singleton domains are unique in their homologous subfamily, the structurally closest domains to the singleton domains are the domains in their neighboring H-levels in the same topology. We selected a set of 21 different topologies such that each topology has a singleton subfamily and at least ten other superfamilies. There are only 21 such topologies in CATH, and one domain for each homologous superfamily within a topology is randomly chosen as a representative. So, we have 21 singleton domains and 210 (10×21) domains selected from the different sibling superfamilies. Our final dataset thus has 4410 alignment pairs (21×210). The set of pairs which have the same CATH classification are labeled as positive examples, and as negative examples if they disagree. We have 210 positive pairs and 4200 negative pairs in our dataset.

3.2.2. *Alignment Results*

We ran all the methods on the 4410 structure pairs. The methods report the number of residues in the alignment, the *rmsd*, and the native alignment score: STRUCTAL reports a p-value for the alignment, FAST reports a normalized score, and DALI reports a z-score. For STSA, we score the alignments using the geometric matching score SAS_3. We sort the alignments by the methods' native score and calculate the true positives (TP), i.e., pairs with same

CATH classification, and the false positives (FP), i.e., pairs with a different CATH classification in the top scoring pairs. Moreover, we compare the quality of the alignments of different methods by comparing the average SAS matching score for the true positives.

Figure 4(a) shows the Receiver Operating Characteristic (ROC) curves for all the methods. The ROC graph plots the true positive rate (sensitivity), versus the false positive rate (1-specificity). Recall that the true positive rate is defined as $\frac{TP}{TP+FN}$, and the false positive rate is defined as $\frac{FP}{TN+FP}$, where TP and TN are the number of true positives and negatives, whereas FP and FN are the number of false positives and negatives. All the alignments were sorted by their native score (when applicable), or by the geometric score SAS_3.

Table 1. Comparison of the average alignment length.

TP	DALI	STRUCTAL	FAST	STSA
0.2	100.29/3.06	83.60/2.04	82.52/3.18	101.8 /3.1
	3.05/3.03	**2.44**/3.49	3.85/5.66	3.05/**2.94**
0.4	85.40/3.21	71.67/2.09	70.90/3.16	86.43/3.05
	3.76/5.15	**2.92**/5.68	4.46/8.87	3.53/**4.72**
0.6	75.77/3.36	65.97/2.20	63.90/3.22	76.66/3.03
	4.43/7.72	**3.33**/7.66	5.04/12.34	3.95/**6.73**
0.8	69.49/3.56	64.33/2.49	57.60/3.51	68.48/2.95
	5.12/10.61	**3.87**/9.35	6.09/18.37	4.31/**9.19**
1.0	63.03/3.76	62.09 /2.84	51.75/3.55	61.49/2.88
	5.97/15.02	**4.57/11.86**	6.86/25.62	4.68/12.39

The results are reported as follows: for each sensitivity value, the top row shows the average $N_{mat}/rmsd$, and the bottom row shows SAS_3/SAS_1, where the averages are calculated over the true positive alignments. The values in bold show the best SAS_3 and SAS_1 scores.

Having the best ROC curve does not imply the best alignments. [28] showed that the best methods, with respect to the ROC curves, do not necessarily have the best average geometric match score for the true positives pairs. Our results confirm this observation. Figure 4(b) shows the average SAS_3 measure of the true positives as we vary the number of top k scoring pairs. Clearly, STSA has the best average SAS score for the true positives. This can be explained by the fact that we use the SAS measure in our alignment algorithm. STRUCTAL comes second in the quality of the average SAS measure. Even though FAST was able to classify as many true positives as DALI and STSA, it still has the worst average SAS measure, indicating that it produces shorter alignments with higher *rmsd*. These results suggest is that if the goal is to simply discriminate between the classes, a method can score better than another method that produces better alignments in terms of both length and *rmsd*. However, since our goal is to assess the geometric quality of the alignments, we can clearly see that STSA outperforms the other approaches. Figure 4(c) shows the ROC curve of all the methods after sorting the alignments based on the geometric match score, SAS_3; STSA has the best ROC curve.

In fact, if we use different geometric scoring measures like SAS_2 and SAS_1, we find that STSA continues to give good alignments. Figures 5(a) and 5(c) show the average SAS_2 and SAS_1 scores, respectively, versus the true positive rates, and Figures 5(b) and 5(d) show the corresponding ROC curves. We find that for SAS_2, STSA is still the best. For SAS_1, which emphasizes lower *rmsd* more than length, we find that STRUCTAL is the best method, but is followed closely by STSA.

Table 1 summarizes these results in tabular form. It shows the average length and *rmsd* as well as the average SAS_3 and SAS_1 scores, for the true positive alignments for different sensitivities. At all sensitivities, the average STSA alignment length is longer than other methods. This gain in the alignment length comes at little or no cost in terms of the average *rmsd*. Compared to DALI and FAST, STSA is always superior in its alignment quality. Its SAS_3 score is much better (lower) than the other methods. On the other hand, if one prefers shorter, more accurate alignments, then STRUCTAL has the lowest SAS_1 scores, followed by STSA. If fact, by changing the parameters of STSA, we can explicitly bias, it to favor such shorter alignments if those are of more interest.

3.2.3. *Running Times*

Table 2 shows the total running time for the alignment methods on all the 4410 pairs in the singleton dataset. FAST is extremely fast but its alignments' quality is not so good. STSA is slightly slower than STRUCTAL, but is faster than DALI.

Table 2. Comparison of the running times on the CATH dataset.

Method	DALI	STRUCTAL	FAST	STSA
Time (s)	4932s	3179s	224s	3893s

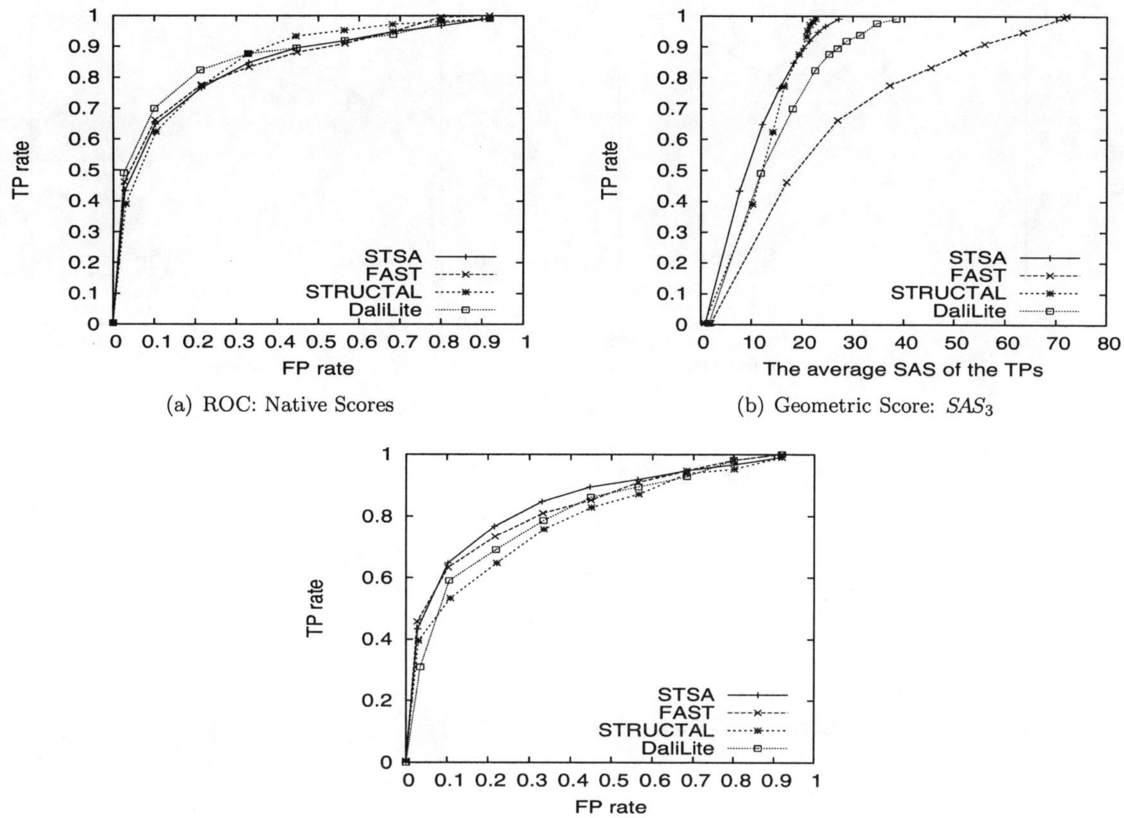

(a) ROC: Native Scores

(b) Geometric Score: SAS_3

(c) ROC: SAS_3

Fig. 4. Receiver Operating Characteristic (ROC) curves for the structural alignment methods measured over the 4410 pairs. (a) The alignments are sorting based on the native score or on the geometric match measure SAS, we tallied the number of true positives and false positives using CATH as a gold standard. (b) The average SAS_3 scores versus the true positive rate. (c) For all the methods, the alignments are sorted using SAS_3 scores and we plot the ROC curve showing the number of true and false positives.

3.3. Analysis of STSA

There are some parameters that affect the quality of the resulting alignment in STSA, namely the length of the smallest runs to consider r, and the threshold distance, δ which is used to populate the scoring matrix, and the number of initial seeds. The optimal values for $r = 3$ and $\delta = 5.5$ were found empirically such that they give the best ROC curve on the CATH data set. Here we investigate the effect of seed pruning on the sensitivity of STSA alignments, as well as the quality of the alignments. Figure 6 shows how the average SAS score changes when using different number of initial seeds for the two seed pruning heuristics. The first pruning approach sorts and selects the top k initial seeds based their length (in decreasing order), whereas the second approach uses the DALI rigid similarity scores [12]. Figure 6(a) shows that considering only the top $k = 100$ seeds, the average SAS scores for the true positives are almost as good as using all the seeds. Moreover, as seen in Figure 6(b), using the more sophisticated DALI rigid similarity score to sort the seeds performs the same as using the much simpler and cheaper length-based approach. As for the running time, pruning the seeds and using only the top 100 resulted in a drastic reduction in the running time. As reported in Table 2 STSA took $3893s$ when using the top 100 seeds, whereas it took $9511s$ seconds when using all the seeds.

3.4. Two Non-sequential Alignments

To demonstrate the quality of STSA in finding non-sequential alignments, we present the alignment on a pair of structures reported in SARF2 [13]. Figure 7 shows a non-sequential alignment between Leghemoglobin (2LH3:A) and Cytochrome P450 BM-3

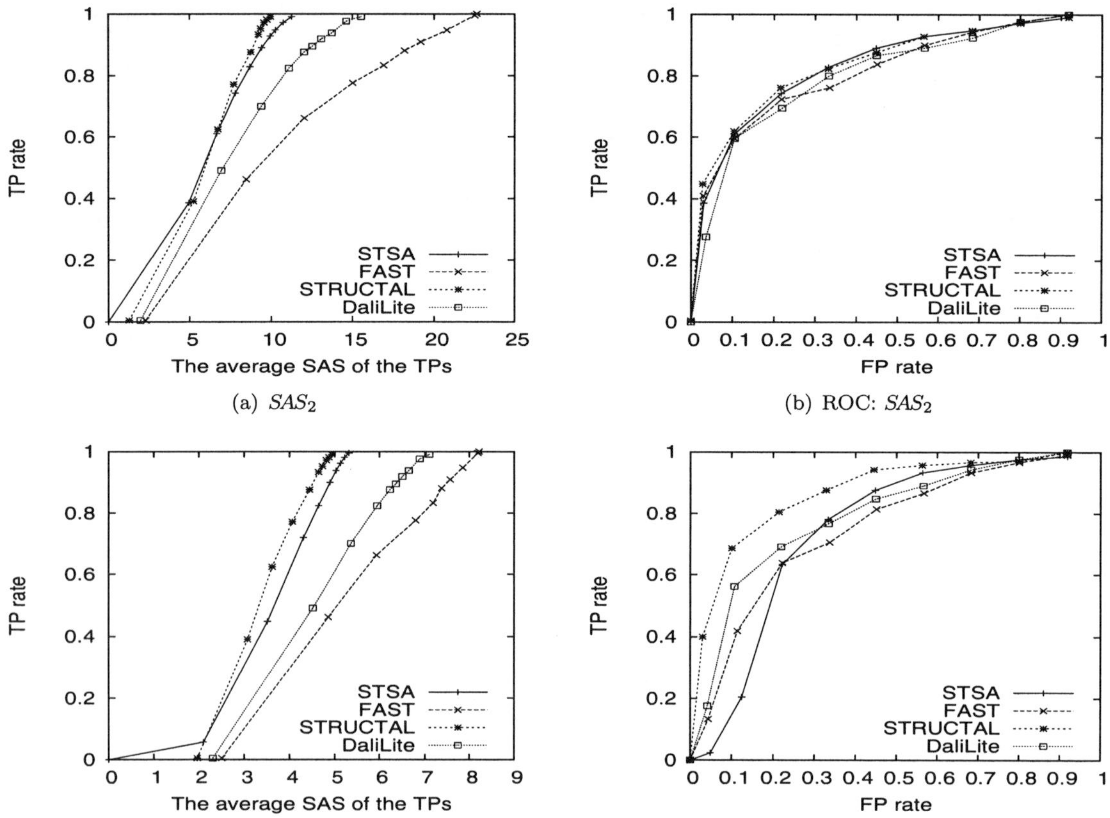

(a) SAS_2 (b) ROC: SAS_2 (c) SAS_1 (d) ROC: SAS_1

Fig. 5. Effect of different geometric matching scores, SAS_k, for $k = 2$ and $k = 1$. (a) The average SAS_2 for the true positive alignments. (b) ROC curve using SAS_2 score. (c) Average SAS_1 for true positives, and (d) ROC using SAS_1 score for sorting alignments.

(2HPD:A). STSA and SARF2 has some common aligned segments, but STSA yielded an alignment of length 114 and $rmsd = 3.37\text{Å}$, whereas SARF2 yielded an alignment with length 108 and $rmsd = 3.05\text{Å}$. The SAS_3 score of STSA is 2.27, which is better than SARF2's score of 3.84. On this example both SCALI and FAST failed to return an alignment. Also, as expected, this is a hard alignment for sequential alignment methods: STRUCTAL aligned 56 residues with $rmsd = 2.27$, DALI aligned 87 residues with $rmsd = 4.8$, and CE aligned 91 residues with $rmsd = 4.05$.

We took a second non-topological alignment pair from SCALI [16]. Figure 8 shows the non-topological alignment between 1FSF:A, and 1IG0:A. Our alignment had some common aligned segments with both SCALI and SARF2, but it returns a longer alignment. On the geometric SAS_3 measure STSA scored 1.27, SARF2 2.51 and SCALI 4.8. Among the sequential methods STRUCTAL was able to return a fairly good alignment for this pair, with a SAS_3 score of 1.6.

4. DISCUSSION

We presented STSA, an efficient algorithm for pairwise structural alignment. The STSA algorithm efficiently constructs an alignment from the superposed structures based on the spatial relationship between the residues. The algorithm assembles the alignment from closely superposed fragments, thus allowing for non-sequential alignments to be discovered.

Our approach follows a guided iterative search that starts from initial alignment seeds. We start the search from different initial seeds to explore different regions in the transformation search space.

On the challenging-to-align RIPC set [22], STSA alignments have higher agreement with the reference alignments than other methods: CE, DALI, FATCAT, MATRAS, CA, SHEBA, and SARF. The results on the RIPC set suggest that the STSA ap-

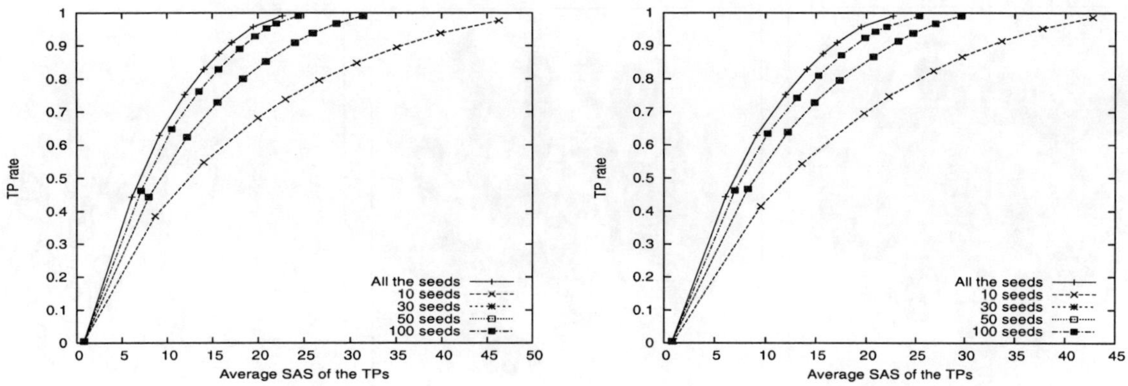

(a) Length score

(b) DALI rigid score

Fig. 6. Studying the effect of pruning on STSA. The average SAS score for the true positives as we consider different number of seeds is shown: (a) using length, (b) using DALI rigid score.

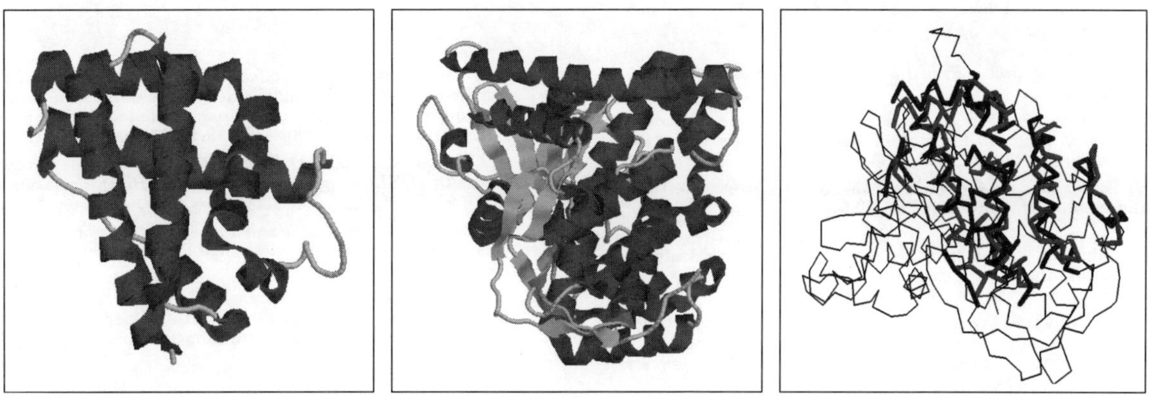

(a) (b) (c)

Fig. 7. A non-sequential alignment between (a) Leghemoglobin (2LH3:A, 153 residues) and (b) Cytochrome P450 BM-3 (2HPD:A, 471 residues). (c) STSA alignment: Leghemoglobin in black and Cytochrome in grey. The $N_{mat}/rmsd$ scores were 117/3.37Å for STSA, and 108/3.05Å for SARF2. For sequential methods, the scores were 56/2.27Å for STRUCTAL, 87/4.8Å for DALI and 91/4.05Å for CE.

proach is effective in finding non-sequential alignments, where the purely sequential (and in some cases non-sequential) approaches yield low agreement with the reference alignment.

Overall results on classifying the CATH singleton dataset show that STSA has high sensitivity for high specificity values. Moreover, the quality of STSA alignments, as judged by the SAS_3 geometric scores (longer alignments and lower *rmsd*), are better than the alignments of other methods: DALI, FAST, and STRUCTAL.

5. CONCLUSION & FUTURE WORK

Our experimental results on the RIPC set and the CATH dataset demonstrate that the STSA approach is efficient and competitive with state-of-the-art methods. Our next step is to extend our approach to address the multiple structure alignment problem. Moreover, we plan to add a functionality to handle flexible and reverse alignments.

References

1. H.M. Berman, J. Westbrook, Z. Feng, G. Gilliland, T.N. Bhat, H. Weissig, I.N. Shindyalov, and P.E. Bourne. The protein data bank. *Nucleic Acids Res*, 5(28):235–242, 2000.
2. J.F. Gibrat, T. Madej, and S.H. Bryant. Surprising similarities in structure comparison. *Curr Opin Struct Biol*, 6: 377–385, 1996.
3. R. Kolodny and N. Linial. Approximate protein structural alignment in polynomial time. *PNAS*, 101:12201–12206, 2004.
4. J. Xu, F. Jiao, and B. Berger. A parameterized algorithm for protein structure alignment. *J Comput Biol*, 5:564–77, 2007.
5. S.B. Needleman and C.D. Wunsch. A general method applicable to the search for similarities in the amino acid sequence of two proteins. *J Mol Biol*, 48:443–453, 1970.
6. S. Subbiah, D.V. Laurents, and M. Levitt. Structural similarity of dna-binding domains of bacteriophage repressors and the globin core,. *curr biol*, 3:141–148, 1993.
7. M. Gerstein and M. Levitt. Using iterative dynamic programming to obtain accurate pairwise and multiple alignments of

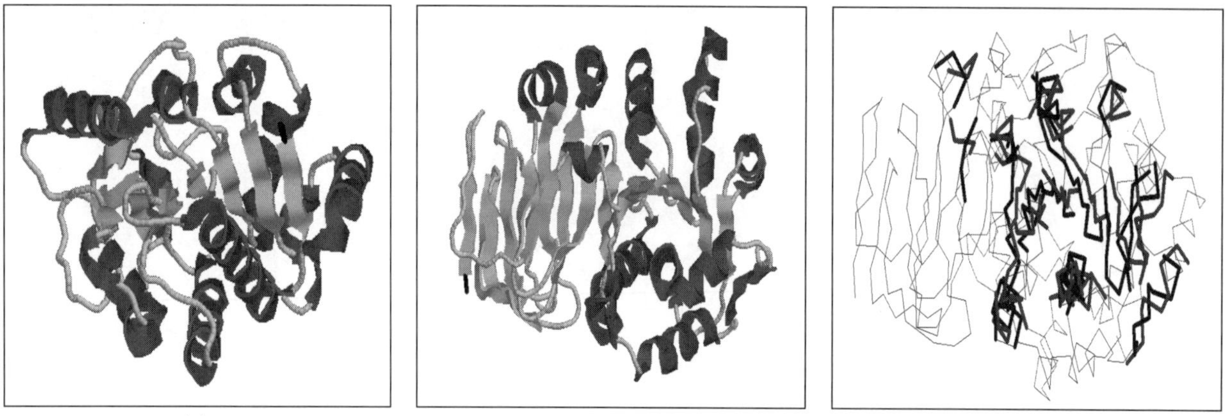

(a) (b) (c)

Fig. 8. A non-topological alignment between (a) Glucosamine-6-Phosphate Deaminase (1FSF:A, 266 residues) and (b) Thiamin Pyrophosphokinase (1IG0:A, 317 residues). (c) STSA alignment: 1FSF:A in black and 1IG0:A in grey. The $N_{mat}/rmsd$ scores were 139/3.42Å for STSA, 104/5.4Å for SCALI, and 105/2.9Å for SARF2. For the sequential methods the scores were 145/4.88Å for STRUCTAL, 106/4.9Å for DALI, and 111/5.1Å for CE.

protein structures. *Proc Int Conf Intell Syst Mol Biol*, 4: 59–67, 1996.

8. C.A. Orengo and W.R. Taylor. Ssap: sequential structure alignment program for protein structure comparison. *Methods Enzymol*, 266:617–35, 1996.
9. Y. Zhang and J. Skolnick. TM-align: A protein structure alignment algorithm based on TM-score.
10. M. Tyagi, V.S. Gowri, N. Srinivasan, A.G. Brevern, and B. Offmann. A substitution matrix for structural alphabet based on structural alignment of homologous proteins and its applications. *Proteins*, 65(1):32–39, 2006.
11. T. Can and T.F. Wang. Ctss:a robust and efficient method for protein structure alignment based on local geometrical and biological features. In *IEEE Computer Society Bioinformatics Conference (CSB)*, pages 169–179, 2003.
12. L. Holm and C. Sander. Protein structure comparison by alignment of distance matrices. *J Mol Biol*, 233(1):123–138, 1993.
13. N.N Alexandrov. Sarfing the pdb. *Protein Engineering*, 50 (9):727–732, 1996.
14. I.N. Shindyalov and P.E. Bourn. Protein structure alignment by incremental combinatorial extension (ce) of the optimal path. *Protein Eng*, 11:739–747, 1998.
15. J. Zhu and Z. Weng. Fast: A novel protein structure alignment algorithm. *Proteins:Structure, Function and Bioinformatics*, 14:417–423, 2005.
16. X. Yuan and C. Bystroff. Non-sequential structure-based alignments reveal topology-independent core packing arrangements in proteins. *Bioinformatics*, 21(7):1010–1019, 2003.
17. Y. Ye and A. Godzik. Flexible structure alignment by chaining aligned fragment pairs allowing twists. *Bioinformatics*, 19:II246–II255, 2003.
18. I. Eidhammer, I. Jonassen, and W.R. Taylor. Structure comparison and structure patterns. *J Comput Biol*, 7(5):685–716, 2000.
19. M.R. Garey and D.S. Johnson. Computers and intractability: A guide to the theory of np-completeness. In *W.H. Freeman, San Francisco, CA*, 1979.
20. R. Nussinov and H.J. Wolfson. Efficient detection of three-dimensional structural motifs in biological macromolecules by computer vision techniques. *proc. national academy of sciences of the usa (biophysics)*, 88:10495–10499, 1991.
21. F. Gao and M.J. Zaki. Indexing protein structures using suffix trees. In *IEEEComputational Systems Bioinformatics Conference, Palo Alto, CA*, 2005.
22. G. Mayr, F. Dominques, and P. Lackner. Comparative analysis of protein structure alignments. *BMC Structural Biol*, 7 (50):564–77, 2007.
23. C.A. Orengo, A.D. Michie, S. Jones, D.T. Jones, M.B. Swindells, and J.M. Thornton. Cath- a hierarchic classification of protein domain structures. *Structure*, 5(8):1093–1108, 1997.
24. C. Bystroff, V. Thorsson, and D. Baker. Hmmstr: a hidden markov model for local sequence-structure correlations in proteins. *J. Mol. Biol.*, 301:137–190, 2000.
25. W. Kabsch. A solution for the best rotation to relate two sets of vectors. *Acta Crystallogr*, A32:922–923, 1976.
26. G.H. Golub and C.F. Van Loan. Matrix computations. *Johns Hopkins University Press*, 3, 1996.
27. M.I. Abouelhoda and E. Ohlebusch. Chaining algorithms for multiple genome comparison. *Journal of Discrete Algorithms*, 50(3):321–341, 2005.
28. R. Kolodny, P. Koehl, and M. Levitt. Comprehensive evaluation of protein structure alignment methods: scoring by geometric measures. *J Mol Biol*, 346(4):1173–88, 2005.
29. A. Zemla. Lga - a method for finding 3d similarities in protein structures. *Nucleic Acids Research*, 31(13):3370–3374, 2003.
30. J. Jung and B. Lee. Protein structure alignment using environmental profiles. *Protein Engineering*, 13:535–543, 2000.
31. M. Gerstein and M. Levitt. Comprehensive assessment of automatic structural alignment against a manual standard, the scop classification of proteins. *it Protein Sci*, 7:445–456, 1998.
32. A. Murzin, S.E. Brenner, T. Hubbard, and C. Chothia. Scop: A structural classification of proteins for the investigation of sequences and structures,. *J Mol Biol*, 247:536–540, 1995.

COMBINING SEQUENCE AND STRUCTURAL PROFILES FOR PROTEIN SOLVENT ACCESSIBILITY PREDICTION

Rajkumar Bondugula†
Digital Biology Laboratory, 110 C.S. Bond Life Sciences Center, University of Missouri
Columbia, MO 65211, USA
Email: raj@bioanalysis.org

Dong Xu*
Digital Biology Laboratory, 201 Engineering Building West, University of Missouri
Columbia, MO 65211, USA
**Email: xudong@missouri.edu*

Solvent accessibility is an important structural feature for a protein. We propose a new method for solvent accessibility prediction that uses known structure and sequence information more efficiently. We first estimate the relative solvent accessibility of the query protein using fuzzy mean operator from the solvent accessibilities of known structure fragments that have similar sequences to the query protein. We then integrate the estimated solvent accessibility and the position specific scoring matrix of the query protein using a neural network. We tested our method on a large data set consisting of 3386 non-redundant proteins. The comparison with other methods show slightly improved prediction accuracies with our method. The resulting system does need not be re-trained when new data is available. We incorporated our method into the MUPRED system, which is available as a web server at *http://digbio.missouri.edu/mupred.*

1. INTRODUCTION

Predicting the three-dimensional structure of a protein from its sequence has been an open challenge in bioinformatics for more than three decades. In many cases, three-dimensional structures cannot be predicted accurately and researchers like to obtain structure features such as secondary structures and solvent accessibility (SA). While secondary structure captures some aspects of the protein structure, the SA characterizes different structural features. The concept of the SA was introduced by Lee and Richards[1] and can be defined as the extent to which water molecules can access the surface of a protein. The knowledge of SA helped to further the understanding of protein structure classification[2-4], protein interaction[5-7], etc.

A number of approaches such as information theory[8], support vector machines[9], neural networks[10-12], nearest-neighbor methods[13], and energy optimization[14] have been proposed for SA prediction. Almost all of these methods rely on protein position specific scoring matrix (PSSM)[15] from multiple sequence alignments. There are at least two drawbacks of these approaches. First, they predict the structural features of the proteins without using the structural information available in the Protein Data Bank[16] (PDB). Second, when proteins do not have close homologs in the database of known sequences (for example, *nr* at *http://www.ncbi.nlm.nih.gov*), the PSSM will not be well defined, making the predictions unreliable[17].

In our approach, both the structural information and the sequence profile information are used. We first build a structural profile by estimating the relative solvent accessibility of the query protein using a fuzzy mean operator (FMO) from the solvent accessibilities of proteins with known structures that have similar sequence fragments to the query protein. We then integrate the estimated solvent accessibility and the PSSM using a neural network (NN). We choose a NN as the appropproiate scheme for combining information from profiles and FMO is automatically learned by the network from the training data. The output of the NN is the predicted relative solvent accessibility of each residue. The user may either obtain real solvent accessibility values (in terms of Å^2) or classify solvent accessibility into multiple classes using any thresholds based on his/her specific needs. The proposed approach has the advantage of simplicity and transparency. Also,

---

*Corresponding author.
†Current address: Bldg 363 Miller Drive, Biotechnology HPC Software Applications Institute, US Army Medical Research and Materiel Command Ft. Detrick, MD 21702, USA

most of the existing methods were tested on small data sets containing up to a few hundred sequences. These results on small sets have significant variations in prediction accuracies. To overcome this problem, we tested our method on a large-scale data set of non-redundant proteins to obtain stable performance. The prediction program has been implemented into the MUPRED package as a public web server at *http://digbio.missouri.edu/mupred* along with the secondary structure prediction capacity.

2. METHOD AND MATERIALS

In our method, the relative solvent accessibility of each amino acid in the query protein is first estimated using the FMO. The calculated fuzzy means are used as the initial set of features. The second set of features is derived from the PSSM of the query protein. These two features are integrated using a neural network. In Section 2.1, we introduce the features and the data sets used in this work. The estimation of the relative solvent accessibilities using FMO is explained in Section 2.2. In Section 2.3, the process of deriving the second set of features and integrating these two feature sets using a neural network is described. In Section 2.4, the metrics used for performance assessment are presented.

2.1. Feature Inputs and Data Sets

The PSSM of the query protein is the starting point in generating input features. We use PSI-BLAST[15] and the *nr* database to generate the PSSM. We used the following parameters for generating the PSI-BLAST: *j* (number of iterations) = 3, *M* (substitution matrix) = BLOSUM90 with other parameters set at default values. We use the BLOSUM90 substitution matrix as we want only the hit fragments that are close subsequences of the query protein to contribute to the PSSM being generated. The parameters were experimentally determined on the training set. Similar results were obtained for a wide range of parameters (data not shown).

A database of representative protein set (RPS), whose three-dimensional structures (and hence, solvent accessibilities) are known is required to estimate the relative solvent accessibility of the query protein. We used the March 2006 release of the PDBSelect[18] database to prepare RPS. The PDBSelect database consists of representative proteins such that the sequence identity between any two proteins in the database is not more than 25%. Initially, the database had 3080 chains. We only selected the proteins whose structures are determined by X-ray crystallography method with a resolution of less than 3 Å and lengths of more than 50 residues. We further restricted our selection to proteins which have at least 90% of their residues composed of regular amino acids. The selection process has resulted in RPS that contains 1998 proteins with 310,114 residues.

First, we present the performance of our method on the RPS using a jack-knife procedure (query sequence eliminated from the RPS during prediction). We employed two widely used data sets (benchmark sets) to compare the performance of MUPRED with other methods. The first database used in reference [10] contains 126 representative proteins with 23,426 residues (hereafter referred as RS126). The second data set was introduced by Naderi-Manesh et al. in Reference [8]. The database consists of 215 representative proteins with 51,939 residues (hereafter referred as MN215). The proteins in RPS that are similar to any proteins in the benchmark sets are eliminated using the following procedure: each sequence in the RPS database was queried against proteins in the benchmark sets using the BLAST[19] program. If a hit with an e-value less than 0.01 is found, the query sequence was eliminated from the RPS. This procedure further reduced the number of proteins in RPS to 1657. In addition to testing our method on the RPS and the two standard benchmark sets, we employed a fourth data set derived from the Astral SCOP domain database[20] version 1.69. The original database with 25% maximum identity between any two sequences consists of 5457 protein domains. The proteins in the Astral SCOP data set that are similar to the proteins in the RPS are discarded using the same procedure outlined above (i.e., each sequence in the Astral SCOP database was queried against RPS using the BLAST program. If a hit with an e-value less than 0.01 is found, the sequence was eliminated from the Astral SCOP database). Similar to the procedure used to generate the RPS, domain sequences shorter than 40 residues were removed. If less than 90% of a domain sequence is composed of regular amino acids, it is discarded as well. The remaining 3386 domain sequences with 636,693 residues after the filtering make up the independent benchmark set.

2.2. Fuzzy Mean Operator

The profile of the query protein is used to search for the similar fragments in RPS by running the PSI-BLAST second time. The threshold value of e was set to 11,000 when searching the RPS. The higher the threshold, the larger the number of fragments returned by the PSI-BLAST. However, if the threshold is too high, the PSI-BLAST returns large number of informative hits as well as noises from the database. The best compromise was experimentally determined. The relative solvent accessibility (RSA) of each residue in the query protein is calculated using the hit fragments that have a residue aligned with the current residue using FMO. The process is explained in the following paragraphs.

The hit fragments returned by the PSI-BLAST program are scored using the following equation:

$$S = \max\{1, 7 + \log_{10}(e\text{-}value)\} \quad (1)$$

This score is formulated as a dissimilarity measure. For instance, the fragments of proteins in RPS that have high sequence similarity with the subsequences of the query protein have high statistical significance (or low e-value), therefore have low scores.

The RSA of each residue of the query protein is calculated from the RSAs of hits that have a residue aligned with the current residue. The SA of the hit fragments are calculated using the DSSP[21] program. For each residue, the absolute SA retuned by the DSSP program is transformed into RSA by dividing it with the maximum SA given in Reference [10]. The RSA of the query protein is calculated using the following expression for FMO:

$$RSA(r) = \frac{\sum_{j=1}^{K} RSA_j \left(\frac{1}{S^{\frac{2}{m-1}}} \right)}{\sum_{j=1}^{K} \left(\frac{1}{S^{\frac{2}{m-1}}} \right)} \quad (2)$$

where r is the current residue, K is number of hits that have residue aligned with the current residue, RSA_j is the relative solvent accessibility of the residue in the jth hit that is aligned with the current residue, S is the score defined in Equation (1), and m is a fuzzifier[22] that controls the weight of the dissimilarity measure S. The optimal value of fuzzifier was experimentally determined to be 1.5. Note that the Equation (2) is a special case of the fuzzy k-nearest neighbor algorithm[22].

2.3. Profile Feature Set and Integration of the Two Feature Sets

The second set of features is generated from the PSSM of the query protein. In the PSSM, each residue is represented by a 20 dimensional vector representing the likelihood of each of the 20 amino acids in that position. The profiles are first normalized and then rescaled into [-1 1] before converting them into vectors suitable for neural network training. We found that the maximum and minimum values in the profiles of all proteins in the RPS were -10 and 12, respectively. Therefore, the profiles were normalized and rescaled using the following expressions:

$$PSSM(i, j) \leftarrow 2x - 1,$$

$$\text{where } x \leftarrow \frac{(PSSM(i, j) + 10)}{22} \quad (3)$$

where $i \in [1, \ldots, n]$ (n is the length of the query protein) and $j \in$ {A,C,D,E,F,G,H,I,K,L,M,N,P,Q,R,S,T,V,W,Y}. An additional bit is used to represent if the current residue lies in the termini of the query protein. We arbitrarily choose 1 to represent the termini, while 0 is used for representing the interior of the protein. The transformed PSSM values, along with the additional bit are converted into vectors suitable for neural network training using a sliding window scheme, i.e., a vector representing the current residue is flanked by the vectors representing the neighbors on the both sides. This scheme allows us to capture that idea that a particular residue's solvent accessibility is dependent on the solvent accessibility states of its neighbors[28,36]. The number of neighbors on each side is determined by parameter W. We experimentally determined that the optimal number of neighbors on each side of the current residue to consider for this feature set is 7 and therefore, the total number of features in this set is (20+1)x15=315.

Similar to the features generated from the PSSM, the fuzzy means that originally lie in [0 1] are rescaled to lie in [-1 1] using the following transformation:

$$RSA(r) \leftarrow 2 \times RSA(r) - 1 \quad (4)$$

The rescaled fuzzy means are converted into vectors suitable for training the neural network using the sliding window scheme. Again, we use an extra bit to indicate the termini of the protein using the same encoding method as the PSSM feature set. We experimentally determined that the optimal window size is 13 and therefore, the total number of features in this feature set is 2x13=26. These two feature sets together (26+315 = 341 features/residue) are used to train the neural networks. The neural network used to integrate the fuzzy means and PSSM is a fully connected feed-forward network with one hidden layer, trained using standard back-propagation learning. We trained the networks with different number of nodes, starting at 170 and increased 10 units at a time. We found that 240 nodes resulted in an optimal performance. The output layer consists of a single neuron that produces the predicted RSA. The neural network has the following architecture 341×240×1 (input nodes × hidden nodes × output node). We randomly selected 50 of RPS proteins for generating the validation vectors and used the rest for training the neural networks. The networks were trained until the performance using the validation vectors started to decline. A total of 100 networks were trained using random initialization and the top six networks (networks with lowest re-substitution error the on the training data) were retained for prediction. Each query protein is simulated on all six networks and the average of the 6 networks is taken as the output of the prediction system. The block diagram of the MUPRED solvent accessibility prediction system is illustrated in Figure 1.

2.4. Prediction Accuracy Assessment

If the system is used as a classifier to group the residues into two classes (*buried* and *exposed*), the following two metrics are used to assess the performance:

Accuracy (Q_2):

$$Q_2 = \frac{p+n}{t} \tag{5}$$

Matthew's correlation coefficient[23] (MCC):

$$MCC = \frac{pn - uo}{\sqrt{(p+u)(p+o)(n+u)(n+o)}} \tag{6}$$

where p is the number of correctly classified *exposed* residues (true positives), n is the correctly classified *buried* residues (true negatives), o is the number of residues that were incorrectly classified as *exposed* residues (false positives), u is the number of residues that were incorrectly classified as the *buried* residues (false negatives), and $t = p + n + o + u$ (total number of residues).

To asses the performance of the RSA prediction ability of the system, the mean absolute error (MAE) as defined below is used:

$$MAE = \frac{1}{N}\sum \left|RSA_{observed} - RSA_{predicted}\right| \tag{7}$$

where $RSA_{observed}$ is the experimental RSA of a residue from the DSSP file divided by its maximum SA while $RSA_{observed}$ is the predicted RSA, and the summation is over all N residues in the protein.

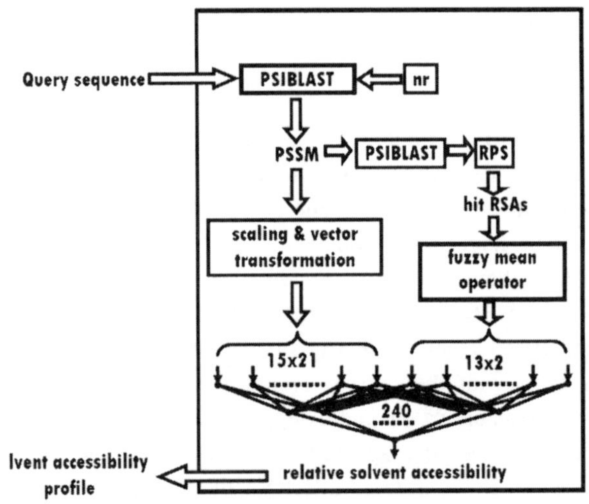

Fig. 1. MUPRED solvent accessibility prediction system. The profile of the query protein is first calculated and used to generate two feature sets. The first set consists of vectors derived from the normalized and rescaled PSSM using a sliding window scheme with window length (*W*) 15. This set consists of 15x21 =315 features/residue. The second feature set is generated by searching the local database of representative proteins based on profile-sequence alignment. The homologous fragments returned by the search process are used to estimate the relative solvent accessibility of each residue using the fuzzy mean operator. The vectors representing the second feature set are derived from the fuzzy means, using the sliding window of length (*W*) 13. Similar to the first feature set, an additional bit is used to represent the termini of the query protein. This feature set consists of 13x2 = 26 features, resulting in 341 features for each residue altogether. The neural network consists of 240 hidden units and a single output neuron that produces the predicted solvent accessibility.

3. RESULTS

In this section, we discuss the performance of the FMO alone, FMO with a neural network and finally,

MUPRED that uses both FMO and PSSM on the RPS and independent SCOP derived set. We then compare MUPRED with some existing methods for prediction accuracies on the two benchmarking data sets.

When we tested the SA profile generated by the FMO alone, we noticed that the trend of predicted SA profile often resembles the actual SA profile, except that the dynamic range of the predicted SA profile is consistently smaller. This may be due to the averaging effects over the neighboring residues when building the SA profile using Equation (2), although such an average reduces the noise for better prediction accuracy overall. Since the neural networks function well as the signal amplifiers, we trained a neural network using the sliding window scheme described in Section 2.2 with the window size 13. This network was not used in the final MUPRED as there appears to be no practical advantage in amplifying signals while integrating the feature sets. The performances of our systems as a two class-classifiers on the various data sets are given in Figure 2 (a-d). The plot on the left illustrates the distribution of the RSA in the corresponding data set, while the plot on the right contains the classification accuracies and the Matthew's correlation coefficients at various classification thresholds. The plots show that integrating FMO and PSSM using a neural network significantly improves the prediction accuracy over the FMO prediction alone or the FMO prediction with a neural network.

We compare MUPRED with existing methods on the two most widely used data sets. The comparison in terms of two-state accuracy on the RS126 data set is presented in Table 1, while the comparison on the MN215 is presented in Table 2. The MAEs of MUPRED on RPS, the SCOP derived independent set, RS126 and MN215 are 14.17%, 15.29%, 14.31% and 13.6%, respectively. The Pearson correlation coefficients of our method on RPS, the SCOP derived independent set, RS126 and MN215 are 0.72, 0.69, 0.71 and 0.72, respectively. Garg et al.[12] reported the Pearson correlation coefficient of 0.67 on the MN215 data set. In both the comparisons, MUPRED has the highest prediction accuracy in most cases. The MAE and the Pearson correlation coefficient on the RPS and the SCOP derived set indicate that the overtraining did not occur when we trained our neural networks.

The program is implemented in the ANSI compatible C programming language. The regression analysis performed on the computation time of our method on a Pentium-4, 3 GHz machine with 2 GB of RAM indicates that the prediction time is a linear function of the sequence length and requires 0.55 sec/residue, including the time required for calculating the profile using the PSI-BLAST. The peak memory requirement is under 20 MB.

Table 1. The comparison of MUPRED with existing methods on the RS126 data set. The performance reported is the two-state accuracy obtained by using different threshold values.

Threshold/Method	A	B	C	D	E
0	87	86	-	86	87
5	77	-	-	80	82
9	78	75	78	-	-
16	79	75	78	78	79
23	79	-	77	-	-
25	79	-	-	77	78

A- Current work; B-Rost and Sander, 1994; C-Manesh et al., 2001; D-Kim and Park, 2004;E-Sim et al., 2005. The '-' indicates that no information is available.

Table 2. The comparison of MUPRED with existing methods on the MN215 data set. The performace reported is the two-state accuracy obtained by using different threshold values.

Threshold/Method	A	C	F	G	H
4	77	75	-	-	-
5	77	-	75	77	75
9	78	76	-	-	-
10	78	-	71	78	77
16	79	76	-	-	-
20	79	-	-	78	78
25	79	74	70	78	-
30	79	-	-	-	78
36	80	74	-	-	-
40	80	-	-	-	78
49	81	80	-	-	-
50	2	-	76	-	81
60	86	-	-	-	85
64	88	97	-	-	-
70	91	-	-	-	91
80	95	-	-	-	95
81	96	81	-	-	-

A-current work; C- Manesh et al., 2001; F- Ahmed and Gromiha, 2002; G- Adamczak et al., 2004 ; H- Garg et al., 2005. The '-' indicates that no information is available.

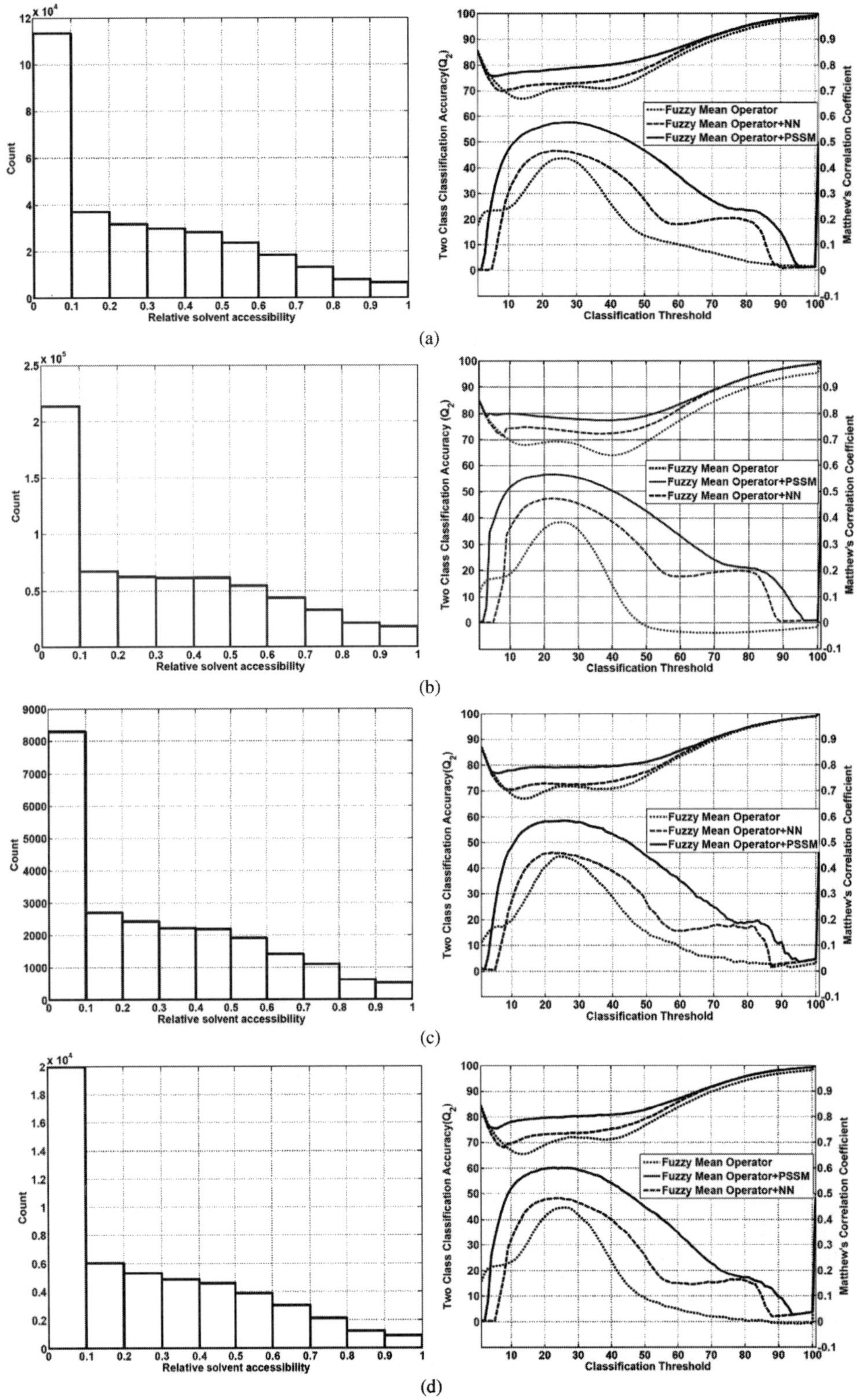

Fig. 2. The histograms showing the compositions of the RSAa in various data sets (Left) and performance of our methods on each of the data sets (Right). The classification threshold is varied along the x-axis, while the two-class classification accuracy (the top three curves) is plotted using the y-axis on the left, while the Matthew's correlation coefficient (the bottom three curves) is plotted using the y-axis on the right. (a) Training set of 1657 proteins; (b) SCOP data set with 3457 proteins; (c) Rost and Sander 126 protein set; (d) Manesh 215 protein set.

4. DISCUSSION

The proposed SA prediction system has some similarity to our secondary structure prediction system[24]. The key difference is that the former is a function approximation, while the later is a classification problem. Our method uses the structural information in the PDB more efficiently than the existing methods and therefore, reduces the dependence on availability of homologous sequences in a sequence database for building a well defined profile. At one extreme, the query sequence has many close homologs in the database of known sequences resulting in a well-defined PSSM. In such cases, our procedure uses profile-sequence alignment for finding similar fragments (exploiting both local and global similarities) in the RPS. Therefore, both PSSM and FMO contribute well for the final prediction. At other extreme where the sequence does not have close homologs, the PSSM is just the scoring matrix used in the alignment procedure. In such situations, our procedure is equivalent to searching for similar fragments in RPS using a sequence-sequence alignment. The homologous fragments (exploiting local similarities only) found by sequence-sequence alignment are effectively used by the FMO and therefore, has the protein structure contribution to the prediction with little or no help from PSSM. The latter case is emulated by the system with FMO followed by a neural network, which provides an estimate of the lower bound of accuracy. Since the output of the neural network is RSA (in [0 1]) of the protein, the system allows a user to choose the number of states and related thresholds, if a classification of residues is desired. The users can multiply the RSA by their maximum solvent accessible areas of respective amino acids to obtain the real solvent accessibility values in terms of $Å^2$. Unlike earlier methods, our system is transparent, weather it succeeds or fails. The predicted solvent accessibility for a given query protein can be traced back to proteins in the RPS that contributed for that prediction, giving additional insight to the users. One of the appealing features of our systems is that it need not be re-trained. As more and more representative structures are solved, their sequences just need to be added to the RPS and the algorithm will use the new information immediately. Over time, we expect our system increases the prediction accuracies automatically by having expanded *nr* and PDB databases, relieving the users or us from the burden of re-training the system in the future.

5. CONCLUSIONS

We developed a new and unique system for effective SA prediction. We use PSSM and fuzzy mean operator to seamlessly integrate sequence profile and structural information into one system, which has not been achieved before. This combination enables successful predictions for the sequences with or without homologs in the database of protein sequences. Our results prove that the additional, complementary information provided by using the structural information has slightly improved the prediction accuracy. Our system will have increased performance accuracy as more protein structures are added to PDB and the expansion of the *nr* databases.

Acknowledgements

This work was supported by a Research Board grant from University of Missouri and by an NIH grant (1R21GM078601). The authors would like to thank Travis McBee for his assistance in the project and Dr. James Keller for discussion on the fuzzy mean operator.

References

1. Lee B, Richards FM. The interpretation of protein structures: estimation of static accessibility. *J Mol Biol* 1971, **55(3)**:379–400.
2. Gromiha MM, Suwa M. Variation of amino acid properties in all-beta globular and outer membrane protein structures. *Int J Biol Macromol* 2003, **32(3-5)**:93-8.
3. Sujatha MS, Balaji PV. Identification of common structural features of binding sites in galactose-specific proteins. *Proteins* 2004, **55(1)**:44-65
4. Yu ZG, Anh VV, Lau KS, Zhou LQ. Clustering of protein structures using hydrophobic free energy and solvent accessibility of proteins. Phys Rev E Stat Nonlin Soft Matter Phys. *Physical Review* 2006, **E73(3.1)**:031920.
5. Ahmad S, Gromiha MM, Sarai A. Analysis and prediction of DNA-binding proteins and their binding residues based on composition, sequence and structural information. *Bioinformatics* 2004, **20(4)**:477-86.
6. Chen H, Zhou HX. Prediction of interface residues in protein-protein complexes by a consensus neural network method: test against NMR data. *Proteins* 2005, **61(1)**:21-35.

7. Hoskins J, Lovell S, Blundell TL. An algorithm for predicting protein-protein interaction sites: Abnormally exposed amino acid residues and secondary structure elements. *Protein Sci* 2006, **15(5)**:1017-29
8. Naderi-Manesh H, Sadeghi M, Arab S, Movahedi AAM. Prediction of protein surface accessibility with information theory. *Proteins* 2001, **42(4)**: 452-459.
9. Kim H, Park H. Prediction of protein relative solvent accessibility with support vector machines and long-range interaction 3D local descriptor. *Proteins* 2004, **54(3)**:557–562.
10. Rost B, Sander C. Conservation and prediction of solvent accessibility in protein families. *Proteins* 1994, **20(3)**:216–226.
11. Ahmad S, Gromiha MM. NETASA: neural network based prediction of solvent accessibility. *Bioinformatics* 2002, **18(6)**:819-24.
12. Garg A, Kaur H, Raghava G. Real value prediction of solvent accessibility in proteins using multiple sequence alignment and secondary structure. *Proteins* 2005, **61(2)**:318–324.
13. Sim J, Kim SY, Lee J. Prediction of protein solvent accessibility using fuzzy k-nearest neighbor method. *Bioinformatics* 2005, **21(12)**:2844-9.
14. Xu Z, Zhang C, Liu S, Zhou Y. QBES: predicting real values of solvent accessibility from sequences by efficient, constrained energy optimization. *Proteins* 2006, **63(4)**:961-6.
15. Altschul SF, Madden TL, Schäffer AA, Zhang J, Zhang Z, Miller W, Lipman DJ. Gapped BLAST and PSI-BLAST: a new generation of protein database search programs. *Nucleic Acids Res* 1997, **25**:3389-3402.
16. Berman HM, Westbrook J, Feng Z, Gilliland G, Bhat TN, Weissig H, Shindyalov IN, Bourne PE. The Protein Data Bank. *Nucleic Acids Res* 2000, **28**:235-242
17. Adamczak R, Porollo A, Meller J. Accurate prediction of solvent accessibility using neural networks-based regression. *Proteins* 2004, **56(4)**:753-67.
18. Hobohm U, Sander C. Enlarged representative set of protein structures. *Protein Science* 1994, 3(3):522-524.
19. Altschul SF, Gish W, Miller W, Myers EW, Lipman DJ. Basic local alignment search tool. *J Mol Biol* 1990, **215**:403-410.
20. Brenner SE, Koehl P, Levitt M. The ASTRAL compendium for sequence and structure analysis. *Nucleic Acids Res* 2000, **28**:254-256.
21. Kabsch W, Sander C. Dictionary of Protein Secondary Structure: Pattern Recognition of Hydrogen-bonded and Geometrical Features. *Biopolymers* 1983, **22**:2577-637.
22. Keller JM, Gray MR, Givens JA. A fuzzy K-Nearest Neighbor Algorithm. *IEEE Trans on SMC* 1985, **15**:580-585.
23. Mathews B. Comparison of the predicted and observed secondary structure of T4 phage lysozyme. *Biochim Biophys Acta* 1975, **405**:442-451.
24. Bondugula R, Xu D. MUPRED: A tool for bridging the gap between template based methods and sequence profile based methods for protein secondary structure prediction. *Proteins* 2007, **66(3)**:664-670.

EXTENSIVE EXPLORATION OF CONFORMATIONAL SPACE IMPROVES ROSETTA RESULTS FOR SHORT PROTEIN DOMAINS

Yaohang Li
Department of Computer Science, North Carolina A&T State University
Greensboro, NC 27411, USA
Email: yaohang@ncat.edu

Andrew J. Bordner, Yuan Tian, Xiuping Tao, and Andrey A. Gorin*
Computer Science and Mathematics Division, Oak Ridge National Laboratory,
Oak Ridge, TN, 37831, USA
**Email: agor@ornl.gov*

With some simplifications, computational protein folding can be understood as an optimization problem of a potential energy function on a variable space consisting of all conformation for a given protein molecule. It is well known that realistic energy potentials are very "rough" functions, when expressed in the standard variables, and the folding trajectories can be easily trapped in multiple local minima. We have integrated our variation of Parallel Tempering optimization into the protein folding program Rosetta in order to improve its capability to overcome energy barriers and estimate how such improvement will influence the quality of the folded protein domains. Here we report that (1) Parallel Tempering Rosetta (PTR) is significantly better in the exploration of protein structures than previous implementations of the program; (2) systematic improvements are observed across a large benchmark set in the parameters that are normally followed to estimate robustness of the folding; (3) these improvements are most dramatic in the subset of the shortest domains, where high-quality structures have been obtained for >75% of all tested sequences. Further analysis of the results will improve our understanding of protein conformational space and lead to new improvements in the protein folding methodology, while the current PTR implementation should be very efficient for short (up to ~80 a.a.) protein domains and therefore may find practical application in system biology studies.

1. INTRODUCTION

The Rosetta platform[1-4] is one of the most successful approaches in predicting overall backbone fold for the protein domains that lack any detectable structural analogs in Protein Data Bank (PDB). It has been ranked number one at the last three CASP competitions (Critical Assessment of Structure Prediction) among *ab initio* methods[5]. Unlike threading methods that rely on a known structure template, ab initio programs attempt to predict structure by generating polymer chain configurations from the whole conformational space and use scoring functions to estimate how good these conformations are.

The Rosetta approach combines many innovative ideas to overcome the enormous complexity of the protein chain conformational space. Two of the most important features are: (a) fragment libraries and (b) knowledge-based energy potentials derived from the statistical analysis of known conformations. The fragment libraries contain custom-made lists of conformers for 3-mer and 9-mer segments centered on each residue of the target chain. This arrangement replaces more traditional polymer chain representations (e.g. by dihedral angles or Cartesian coordinates of the atoms) with a set of discrete variables – numbers of the conformers from the fragment library – with each of them determining the structure of the whole short segment of the chain. The segment libraries reduce the dimensionality of the conformational space by many orders of magnitude, however, for a chain of 200 residues it is still ~200 dimensions to explore. The conformations are evaluated based on their backbone atoms, as all side groups are replaced with "elastic spheres" and not modeled explicitly.

Rosetta operates by starting 1,000 (in latest implementations sometimes 10,000 or even more) independent folding trajectories from random extended conformations and evolving them with a Monte-Carlo procedure, while gradually reducing the temperature. For each trajectory, the structure with the lowest observed energy is retained as the result of the folding, and the corresponding 1,000 (or more) results are further analyzed by various methods to determine the

* Corresponding author.

native fold. We will not be discussing the computational problem of finding the native fold, as our study is concerned with the folding trajectories and the quality of the ensemble of the resulting backbone conformations. We will demonstrate that introducing parallel tempering dramatically improves sampling properties of the method and leads to better final structures, but the same results suggest that there are other problems in the procedure preventing more complete success.

2. METHOD

The Parallel Tempering algorithm[6-8] (also known as the multiple Markov chains or replica-exchange method) allows multiple Markov chains to evolve at several temperature levels, forming a ladder, while replica exchanges are attempted between Markov chains at neighboring temperature levels. We have introduced a few modifications to the PT algorithm without changing its fundamentals[9].

A composite system is constructed with one molecule per temperature level and the Rosetta-style transitions take place in each Markov chain. However, instead of the Simulated Annealing[15] scheme used in Rosetta, we use an adaptable Metropolis[14] scheme that maintains a desired acceptance rate. The replica exchange transition takes place according to the Metropolis-Hastings criterion. The desired acceptance rate is decreased gradually to accelerate convergence of the composite system[10]. Moreover, in protein modeling, each replica configuration consists of a lot of information and thus the exchange of configurations is very costly. Alternatively, we exchange the temperatures of two neighbor levels instead to achieve a significant computational performance improvement[11]. The topic of the conformational sampling in protein folding is explored in many excellent stidues[16-25], our investigation was limited to specific issues of the Rosetta folding platform.

We have followed Rosetta methodology and generated an ensemble of 1000 structures for each of 50 domains that were included in this study and each folding experiment. Several types of folding experiments were conducted: the usual Rosetta folding (further referred to as a Rosetta run) with 32,000 Monte-Carlo steps, PTR folding (in the figures referred to as an MPI run as the MPI library was used for multiprocessor implementation) with the same 32,000 steps during the main simulation stage, as well as the PTR runs with 320,000 steps (LMPI - Long MPI), and the PTR runs with $1.5 \cdot 10^6$ steps (referred to as a VLMPI or Very Long MPI). Rosetta was outperformed in MPI runs without additional CPU costs, because the final structure was collected from each thread in the PTR simulations. Due to certain CPU time restrictions only the LMPI protocol was done for all 50 tested domains, and these are the best results that we currently have. Table 1 and Fig. 1 are based on the LMPI protocol.

All modifications made to the original Rosetta package were limited to the sampling procedure. Rosetta records all parameters of the conformation with the lowest energy and (if the native structure is provided) the Minimal Root Mean Square Deviation (MRMSD) distance to the native structure over all structures observed during the simulation. This distance is often smaller than the RMSD distance between the final lowest energy structure and native model, but it is a good measure of how close to the native structure we were able to "pass" during the simulation.

3. RESULTS

3.1. Capability of traversing a "rough" energy landscape

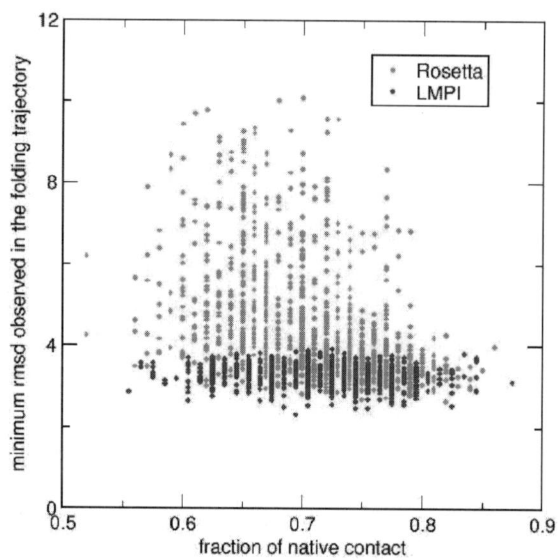

Fig. 1. Comparison of MRMSD to the fraction of native contacts in the final structure (Y-axis) for two ensembles of Rosetta and Parallel Tempering Rosetta simulations. All PTR trajectories pass within 4 Å RMSD of the native structure. Each point combines information from two different conformations, so there is no direct correlation between X and Y values.

All achieved improvements in the folding performance can be traced to the novel feature of Parallel Tempering Rosetta: the capability to traverse the rough energy landscape and get out from very deep local minima of the potential. In Fig. 1 two structure ensembles (each ~1000 structures) present results obtained for a Rosetta run (grey dots, wide area) and an LMPI run (darker dots, spread on much smaller area). The Y-axis represents the measure of the closest observed approximation to the native structure for a given trajectory — Minimal Root Mean Square Deviation (MRMSD) in Angstroms (Å). The X-axis displays the Fraction of Native residue Contacts (Cb-Cb under 8 Å) in the final structure for the corresponding trajectory. We know both of those quantities because we deal with a benchmark set, where the native structures are known.

There is a remarkable compression along the vertical axis. Only ~10% of all original trajectories have approached the native structure to the distance of 4 Å RMSD, but *all* 1000 trajectories in the PTR runs have passed below this limit. Actually, almost all of them have passed below 3 Å, with several trajectories reaching toward the 2 Å limit ("crystallographic" vicinity of native structure). It is important to note than any improvements in MRMSD is exponentially hard, as the conformational volume shrinks very fast when one considers smaller and smaller RMSD "volumes". For example, in the Cartesian coordinates representation the conformational volume of the structures within a 2 Å RMSD vicinity of the native one is at least 8 times smaller than those in a 4 Å RMSD vicinity. Table 1 confirms that the results were typical for almost all analyzed domains, as in almost all cases we observed dramatic improvements in MRMSD.

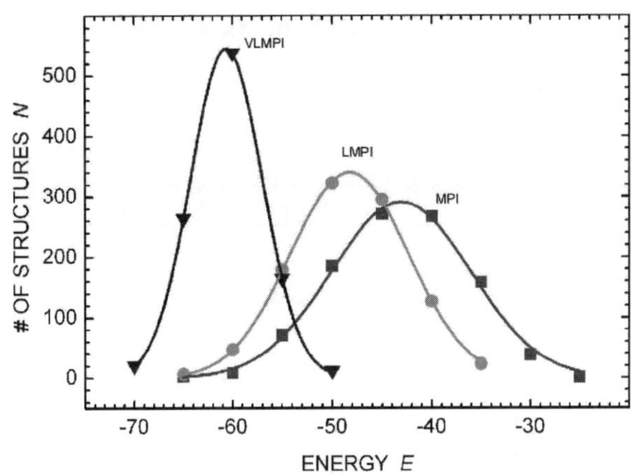

Fig. 2. Energy distribution of the final structures for 3 PTR runs: 32,000 step (MPI), 320,000 steps (LMPI) and $1.5*10^6$ steps (VLMPI). The effect of observing so many new conformations due to longer simulations has never been seen before for the Rosetta program.

Fig. 2 gives the most direct evidence that Parallel Tempering Rosetta reaches into new areas of the conformational space that could not be explored with standard simulated annealing Monte Carlo. The plot presents three energy distributions of the 1,000 final structures obtained in MPI, LMPI and VLMPI runs for the 1lev domain. The VLMPI run produces a much sharper distribution (twice as narrow), and it has little overlap with the MPI run. Here the lower energy is our "marker" that we indeed observe a novel conformation (Rosetta registers the lowest energy conformation seen).

As the distributions of the lowest energy visited by a particular trajectory show, it is clear that almost a half of the VLMPI runs have found conformations that were almost never visited by any of the MPI runs.

In the original Rosetta run, 32,000 steps used in the standard protocol were selected as the limit, after which there were no expectations of any improvements in the energy of the model. Here we observe an explosion in new conformations after extending the length of the run by 10 times (LMPI) and then by 5 times (VLMPI). In VLMPI case we even observe a semblance of convergence, as the width of the energy distribution starts to narrow. Interestingly, dramatic improvements in the final energy did not lead to equally dramatic improvements in the quality of the folded structures.

3.2. Results for the shortest domains

While improvements in the quality of the predictions have been seen across the whole benchmark, the simulations have reached a crucial "improvement threshold" for the shortest domains. The detailed results for the 16 shortest unique domains are presented in Table 1. In the original Rosetta run, the folding results are also systematically better for the shortest domains. With LMPI PTR simulations, several structures have been improved further, pushing the rate of good predictions to 75% of the total set in this size range (31 to 78 amino acids).

For 10 domains, the MRMSD parameter is under 2.5 Å (lines are shown in **bold** in Table 1). This means that at least one of the simulated trajectories passed within the *crystallographic quality vicinity* of the native structure (the corresponding numbers are underlined in the table). Excellent final structures were found for all of them. Out of the remaining 7, three had MRMSD in the range of 3 to 4 Å with relatively good quality final structures.

Only for 4 structures (the whole lines underlined in Table 1) did our platform fail to find structures with the percent of native contacts much above 40% (MRMSD was in the range of 5 to 6 Å). Yet those structures have shown some MRMSD improvements with longer simulation times. Between the MPI and LMPI runs the MinRMSD parameter has improved by 0.5 - 1 Å for four sequences. Actually in this whole set MRMSD did not improve for only four structures, which already had excellent prediction quality by the original Rosetta program. Overall, a higher rate of success than ours has never been reported, to our knowledge, in the literature before. Further experiments conducted in our group confirm this result on a much larger set of unique sequences. Initial results on homologous sequences (the idea was to fold with Rosetta homologous domains as well) have indicated further improvements in two of the four "hard" sequences, pushing the overall success rate even higher.

3.3. Insights into the protein folding process

The conducted simulations and significantly improved ability to search conformational space led to important insights into the obstacles that are faced in computational protein folding. Fig. 3 plots the dependence between the length of the folded domains and the maximum fraction of native contacts (100 means an ideal native structure) obtained in one of the accepted models for this domain. To iron out the structural differences, we used "sliding window" averages for both coordinates (each point represents averages over 10 structures close in length). The results for 50 folded domains produce 41 "sliding windows", and the corresponding 41 points are presented in Fig. 3.

The dependence is sharp and non-linear — for a domain length of around 110 the fraction of native contacts is projected to be only around 30%. At this level there are probably some correct elements of secondary structure, but likely no correct tertiary contacts. The good news is that the results are close to excellent for the domains <75 residues. Another encouraging point is that the problems, which rapidly escalate with increasing the length of the polymer chains, are probably tractable by applying more computer power. Indeed, we have observed the largest amplitude of improvements measured by the fraction of native contacts in the final structure in the longest considered domains (L>90), when we extended simulation from MPI to LMPI protocol.

Table 1. The results for 16 domains in range of 31 to 78 amino acids. The domains are shown by PDB is and chain identifier.

Structure ID	Best final RMSD (A)	Best MRMSD observed (A)	Best final FNC (%)
1tgz_B	**3.3**	**1.81**	**81**
1r0o_B	8.7	6.11	40
2bf8_B	**3.0**	**1.80**	**74**
1xt9_B	**3.6**	**2.07**	**69**
1r0o_A	5.8	5.25	41
1sv0_A	**3.0**	**2.01**	**74**
1le8_B	**6.1**	**2.31**	**87**
1dj7_B	7.9	5.95	40
1oey_A	5.6	5.29	43
1cf7_A	**2.8**	**1.87**	**78**
1bun_B	4.4	3.61	49
1le8_A	**1.4**	**0.82**	**96**
4sgb_I	4.5	4.32	41
1nql_B	4.3	2.60	54
1j2j_B	**1.4**	**0.61**	**99**
1mzw_B	**2.2**	**1.06**	**85**

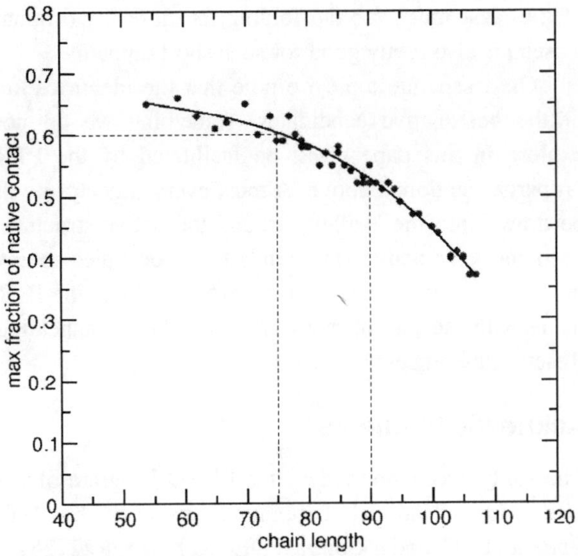

Fig. 3. The dependence between the length of the chain and the quality of the final structures. Below 75 amino acids the quality is very good, but it drops down sharply for longer domains.

The curve in Fig. 3 clearly spells trouble for Rosetta simulation of the domains longer than 105 residues. The average fraction of native contacts was only around 35% for domains in this range, and therefore correct folds can be expected only as a result of extraordinary luck.

4. DISCUSSION

In this study MRMSD measuremets have been used to access improvements in the capability to explore conformational space. Indeed, as the starting conformations are random elongated chains, during normal Rosetta simulations many of them never will fold successfully, and many simulation trajectories will never even pass in a close proximity of the native conformation. The fraction of the trajectories, which have conformation on a certain distance from the native conformation, is then an indirect indication of relative "freedom to travel" shown by the algorithm.

There are several important properties of the MRMSD that should be mentioned here. First, as we already mentioned above, the reduction of the "conformational volume" (defined in a reasonable metric it is simply a real volume in the space of conformational variables) is a power function of the reduction of the RMSD value. One can speculate that a reduction of the RMSD 2 times translates into 8 (or 16?) times reduction in the available conformational volume. Second, the MRMSD depends on the size of the protein chain in a complex way. For longer chains much smaller fraction of all configurations will satisfy the RMSD constraint of 2 Å than for shorter ones. Finally, even a very good MRMSD value does not guarantee that the folding will be successful. The structural trajectory will include a conformation with a 2 Å MRMSD value, but this conformation may have a high value of potential energy (due to some highly unfarobale interactions present in the overall correct model). As a result, the candidate conformation will not be saved, and in the following simulation the final conformation will be very different.

On other hand if a particular folding trajectory does not show a good value of the MRMSD than the simulation is bound to be unseccesful. Due to the definition an MRMSD value of, for example, 8 Å means that the best RMSD possible for the final structure will be greater or at best equal to 8 Å. This simple point explains our efforts to achieve a good MRMSD value for *all* folding trajectories. The trajectories with bad MRMSDs are essentially waste of the CPU time.

To access the quality of the resulting structures we have used (in addition to a standard RMSD) another measure: Fraction of Native residue Contacts (FNC). Two residues were consireded to be in contact if the distance between their Cb atoms (Ca for the glycin) was smaller than 8 Å. The "automatic" contacts (with neigbours -2, -1, +1, +2) were excluded. Many possible definitions of contacting residues are possible, for example, one can define differential contact cutoffs to take into account residue size differences.

By our experience almost all reasonable FNC definitions work well, and there is no clear advantages to prefer one definition to another. For some types of the analysis it seems to be useful to distingvish between short-range (local) and long-range contacts. The long-range contacts provide a more sensitive measure of the folding success, but then there is an additional uncertenty due to the noise effect, which is stronger on smaller sets of contacts. The FNC may provide a superior measure for the quality of the folded structures, but the questions about relative contributions of local and long-range contacts deserve a separate investigation. One possible way forward would be to use weights on all contacts derived, for example, from the separation between contacting residues in the primary sequence.

In the future we plan to conduct more comprehensive analysis of the folding trajectories.

Currently for each trajectory only two (most important) trajectory points are recorded: the conformation with the lowest energy (for this one we have the full set of data) and the lowest RMSD distance to the native fold (here we are limited to the distance value). Nevertheless several interesting and important conclusions both practical and theoretical can be drawn from the current work.

First, the Parallel Tempering dramatically improves sampling capabilities of the program. All local minima can be comprehensively explored. In the longest simulations we have observed an emerging Monte-Carlo convergence of the trajectories. Here we should note that these results obtained on relatively "soft" potentials. The real energy potentials (such as electrostatic and Van der Waals interactions) usually lead to rougher potential energy functions that the knowledge-based derived potentials. Yet there is no reason to believe that the Parallel Tempering algorithm cannot be adapted to such potentials with more temperature energy levels, etc.

Indeed, the role of the potential energy function constitutes a second lesson of our study. In a number of situations we observed that the current potential functions lead to a large "valley", where the native structure is located, but this valley does not have deep potential energy minimum located at the native conformation. While almost all folding trajectories cross the right "valley", only very few of them end up near the native conformation. There is no energy gradient leading through the remaining 2 Å of RMSD — and this process happens almost randomly, increasingly so for longer domains. Our approach will be useful for a more detailed exploration the conformational space and properties of the potentials. For example, we can produce structures with very low values of potential energy, which are really far from the native model, and in such way reveal shortcomings of the existing potentials.

The final (and helpful for applications) conclusion from our study is a sharp dependence between the probability to have a successful folding result and the length of the targeted domain (presented in Fig. 3). For short domains (75-90 residues long) the PTR implementation provides a significant improvement over the standard Rosetta, with high chances to have a structure with 80% of native contacts in the final ensemble. This improvement is something like making of the "last mile" for the folding, because the original Rosetta is also pretty good for such short domains.

On a separate topic we note that the identification of the best native candidates (something we do not explore in this paper) will be facilitated by the PTR property mentioned above. Almost every trajectory will be drawn into the "valley" around the native structure, so if the near native state tends to be occupied, many more near native decoys will be produced with the PTR than with usual Monte-Carlo simulated annealing Rosetta.References

Acknowledgements

The study was supported by the LDRD Program of the Oak Ridge National Laboratory managed by UT-Battelle, LLC, under Contract DE-AC05-00OR22725

References

1. Simons KT, Bonneau R, Ruczinski I, and Baker D. Ab initio Protein Structure Prediction of CASP III Targets Using ROSETTA, *Proteins: Structure, Function and Genetics*, 1999; **37**: 171-176.
2. Baker D. A surprising simplicity to protein folding, *Nature*, 2000; **405**: 39-42.
3. Bradley P, Chivian D, Meiler J, Misura KM, Rohl C, Schief W, Wedemeyer WJ, Schueler-Furman O, Murphy P, Schonbrun J, Strauss CE, Baker D. Rosetta predictions in CASP5: Successes, failures, and prospects for complete automation, *Proteins*, 2003; **53**: 457-468.
4. Rohl CA, Strauss CE, Misura KM, Baker D. Protein Structure Prediction using Rosetta, *Methods in Enzymology*, 2004; **383**: 66-93.
5. Moult J. A decade of CASP: progress, bottlenecks and prognosis in protein structure prediction, *Curr. Opin. Struct. Biol.*, 2005; **15**: 285–289.
6. Geyer CJ, Thompson EA. Annealing Markov Chain Monte Carlo with Applications to Ancestral Inference, *Journal of the American Statistical Association*, 1995; **90**: 909-920.
7. Hansmann U. Parallel Tempering Algorithm for Conformational Studies of Biological Molecules, *Chem. Phys. Letter*, 1997; **281**: 140-150.
8. Li Y, Strauss CE, Gorin A. Parallel Tempering in Rosetta Practice, *Proceedings of International Conference on*

Bioinformatics and its Applications, Fort Lauderdale, Florida, 2004.

9. Li Y, Protopopescu VA, Gorin A. Accelerated Simulated Tempering, *Physics Letters A*, 2004; **328**: 274-283.
10. Li Y, Strauss CE, Gorin A. Hybrid Parallel Tempering and Simulated Annealing Method – an Efficient Sampling Method in ab initio Protein Folding, *International Journal of Computational Science*, 2008; in print.
11. Li Y, Mascagni M, Gorin A. Decentralized Replica Exchange Parallel Tempering: An Efficient Implementation of Parallel Tempering using MPI and SPRNG, Proceedings of International Conference on Computational Science and Its Applications (ICCSA), Kuala Lumpur, 2007.
12. Schug A, Herges T, Verma A, Wenzel W. Investigation of the parallel tempering method for protein folding, *J. Phys: Condens. Matter*, 2005; **17**: 1641-1650.
13. Schug A, Wenzel W. Predictive in-silico all atom folding of a four helix protein with a free energy model, *J. Am. Chem. Soc.*, 2004; **126**: 16737.
14. Metropolis N, Rosenbluth AW, Rosenbluth MN, Teller TH, Teller E. Equation of State Calculations by Fast Computing Machines, *Journal of Chemical Physics*, 1953; **21**: 1087-1092.
15. Kirkpatrick S, Gelatt DDJ, Vecchi MP. Optimization by Simulated Annealing, *Science*, 1983; **220**: 671-680. Hansmann U. Parallel Tempering Algorithm for Conformational Studies of Biological Molecules, *Chem. Phys. Letter*, 1997; **281**: 140-150.
16. Hansmann U. Parallel Tempering Algorithm for Conformational Studies of Biological Molecules, Chem. Phys. Letter, 1997; 281: 140-150.
17. Lin C, Hu C. Hansmann UHE. Parallel Tempering Simulations of HP-36, *Proteins, Structure, Function, and Genetics*, 2003; **52**: 436-445.
18. Rabor AA, Scheraga. Improved Genetic Algorithm for the Protein Folding Problem by Use of a Cartesian Combination Operator, *Protein Science*, 1996; **5**(9): 1800-1815.
19. Pedersen JT, Moult J. Protein Folding Simulations with Genetic Algoithms and a Detailed Molecular Description, *J Mol Biol.*, 1997; **268**(2):240-259.
20. Damsbo M, Kinnear BS, Hartings MR, Ruhoff PT, Jarrold MF, Ratner MA, Application of evolutionary algorithm methods to polypeptide folding: Comparison with experimental results for unsolvated Ac-(Ala-Gly-Gly)5-LysH+, *Proceedings of the National Academy of Sciences*, 2004; **101**(19): 7215-7222.
21. Schulze-Kremer S. Application of Evolutionary Computation to Protein Folding, *Advances in evolutionary computing: Theory and Applications*, 2003; 915-940.
22. Kim JG, Fukunishi Y, Nakamura H. Average Energy Guided Simulated Tempering Implemented into Molecular Dynamics Algorithm for Protein Folding Simulation, *Chemical Physics Letters*, 2004; **392**: 34-39.
23. Okamoto Y. Generalized-ensemble algorithms: enhanced sampling techniques for Monte Carlo and molecular dynamics simulations, Journal of Molecular Graphics and Modelling, 2004; **22**: 425 - 439.
24. Mitsutake A, Sugita Y, Okamoto Y. Replica-exchange multicanonical and multicanonical replica-exchange Monte Carlo simulations of peptides, *Journal of Chemical Physics*, 2003; **118**: 6664 - 6675.
25. Sugita Y, Okamoto Y, Replica-exchange molecular dynamics method for protein folding, Chemical Physics Letters, 1999; 314: 141-151.

IMPROVING HOMOLOGY MODELS FOR PROTEIN-LIGAND BINDING SITES

Chris Kauffman, Huzefa Rangwala, and George Karypis

Department of Computer Science, University of Minnesota
117 Pleasant St SE, Room 464
Minneapolis, MN 55455
E-mail: {kauffman,rangwala,karypis}@cs.umn.edu

In order to improve the prediction of protein-ligand binding sites through homology modeling, we incorporate knowledge of the binding residues into the modeling framework. Residues are identified as binding or nonbinding based on their true labels as well as labels predicted from structure and sequence. The sequence predictions were made using a support vector machine framework which employs a sophisticated window-based kernel. Binding labels are used with a very sensitive sequence alignment method to align the target and template. Relevant parameters governing the alignment process are searched for optimal values. Based on our results, homology models of the binding site can be improved if a priori knowledge of the binding residues is available. For target-template pairs with low sequence identity and high structural diversity our sequence-based prediction method provided sufficient information to realize this improvement.

1. INTRODUCTION

Accurate modeling of protein-ligand interactions is an important step to understanding many biological processes. For example, many drug discovery frameworks include steps where a small molecule is docked with a protein to measure binding affinity[1]. A frequent approximation is to keep the protein rigid, necessitating a high-quality model of the binding site. Such models can be onerous to obtain experimentally.

Computational techniques for protein structure prediction provide an attractive alternative for this modeling task[2]. Protein structure prediction accuracy is greatly improved when the task reduces to homology modeling[3]. These are cases in which the unknown structure, the target, has a strong sequence relationship to another protein of known structure, referred to as the template. Such a template can be located through structure database searches. Once obtained, the target sequence is mapped onto the template structure and then refined.

A number of authors have studied the use of homology modeling to predict the structure of clefts and pockets, the most common interaction site for ligand binding[4–6]. Their consensus observation is that modeling a target with a high sequence similarity template is ideal for model quality while a low sequence similarity template can produce a good model provided alignment is done correctly. This sensitivity calls for special treatment of the interaction site during sequence alignment assuming ligand-binding residues can be discerned a priori.

Identifying structural properties of proteins from sequence has become a routine task exemplified by secondary structure prediction. Recent work has explored predicting interaction sites from sequence[7]. As a measure of how well these methods perform, they may be compared to methods that identify interaction sites from structure[8]. We employ both structure and sequence-based schemes to predict interaction sites but, even given perfect knowledge of which residues are involved in binding, it is not clear how best to utilize this knowledge to improve homology models.

In this work we incorporate knowledge of the residues involved in ligand binding into homology modeling to improve the quality of the predicted interaction site. Our contribution is to show that this knowledge does help and can be predicted from sequence alone with enough reliability to improve model quality in cases where target and template have low sequence identity. To our knowledge, this is the first attempt to explore the use of predicted interaction residues in a downstream application such as homology modeling. We explore a variety of parameters that govern the incorporation of binding residue knowledge, assess how much the best performing parameter sets improve model quality, and whether these these parameters generalize.

2. RELATED WORK

2.1. Prediction of ligand-binding residues

Small molecules interact with proteins in regions that are accessible and that provide energetically favorable contacts. Geometrically, these binding sites are generally deep, concave shaped regions on the protein surface, referred to alternately as clefts or pockets. We will refer to residues in clefts as ligand-binding residues.

Predicting ligand-binding site residues from sequence information is similar to several site interaction prediction problems involving DNA[9–11], RNA[12, 13], and other proteins[14–16]. Specifically, Soga and coworkers studied the prediction of ligand-binding site residues using conservation information in the form of profiles and solvent accessible properties of potentially interacting residues[7].

Several methods have been developed to identify putative ligand-binding sites by determining pockets on the protein's surface using its structure. A popular approach for this task is to place grid points at a small separation throughout the space of the protein. Potential binding sites are defined by all grid points, atoms, or residues within a fixed radius of a central grid point. This point is typically assigned based on burial criteria. Software packages such as AutoLigand[17], Ligsite^{csc}[18], VisGrid[19], and PocketPicker[8] utilize this paradigm.

2.2. Homology modeling of binding site

The factors involved in modeling protein interaction sites have received attention from a number of authors. These studies tend to focus on showing relationships between target-template sequence identity and the model quality of surface clefts/pockets.

DeWeese-Scott and Moult made a detailed study of CASP targets[a] that bind to ligands[4]. Their primary interest was in atom contacts between the model protein and its ligand. They measured deviations from true contact distances in the crystal structures of the protein-ligand complexes. Though the number of complexes they examined was small, they found that errors in the alignment of the functional region between target and template created problems in models, especially for low sequence identity pairs.

Chakravarty, Wang, and Sanchez did a broad study of various structural properties in a large number of homology models including surface pockets[5]. They noted in the case of pockets, side-chain conformations had a high degree of variance between predicted and true structures. Due to this noise, we will measure binding-site similarity using the α-carbons of backbone residues. They also found that using structure-induced sequence alignments improved number of identical pockets between model and true structures over sequenced-only alignments. This point underscores the need for a good alignment which is sensitive to the functional region. It also suggests using structure alignments as the baseline to measure the limits of homology modeling.

Finally, Piedra, Lois, and Cruz executed an excellent large-scale study of protein clefts in homology models[6]. To assess the difficulty of targets, the true structure was used as the template in their homology models and performance using other templates was normalized against these baseline models. Though a good way to measure the individual target difficulty, this approach does not represent the best performance achievable for a given target-template pair which led us to take a different approach for normalization. We follow their convention of assessing binding site quality using only the binding site residues rather than all residues in the predicted structure. As their predecessors noted, Piedra et al. point to the need for very good alignments between target and template when sequence identity is low.

The suggestions from these studies, that quality sequence alignments are essential, led us to employ sensitive alignment methods discussed in Section 4.3.

3. DATA

3.1. Primary structure and sequence data

Primary data for our experiments was taken from the RCSB Protein Data Bank (PDB)[20] in January of 2008. Protein sequences were derived directly from the structures using in-house software (Section 7). When nonstandard amino acids appeared in the se-

[a]`http://predictioncenter.org`

[b]`http://astral.berkeley.edu/seq.cgi?get=release-notes;ver=1.55`

quence, the three-letter to one-letter conversion table from Astral[21] version 1.55 was used to generate the sequence[b]. When multiple chains occurred in a PDB file, the chains were treated separately from one another. Identical sequences are removed by sequence clustering methods in later steps. Profiles for each sequence were generated using PSI-BLAST[22] with default options and the NCBI NR database (version 2.2.12 with 2.87 million sequence, downloaded August 2005). PSI-BLAST produces a position specific scoring matrix (PSSM) and position specific frequency matrix (PSFM) for a query protein, both of which are employed for our sequenced-based prediction and alignment methods.

3.2. Definition of binding residues

We considered ligands to be small molecules with at least 8 heavy atoms. Specifying a minimum number of atoms avoids single atom ligands such as calcium ions which are not of interest for this study. Protein residues involved in the binding were those with a distance less than 5Å between heavy atoms in protein and ligand. In-house software was developed to filter ligands, compute distances, and report ligand-binding residues (Section 7).

3.3. Ligand-binding residue prediction

The PDBBind database[23] provided the initial set of data used to train a support vector machine (SVM) classifier (Section 4.1). To remove redundant entries, sequences were extracted from the 'refined' set of PDBBind structures, 1300 total structures and 2392 sequences, and clustered at 40% identity using the CD-HIT software package.[24] This resulted in 400 independent sequences for which profiles were generated. This set had sequence independence at 40% identity from the evaluation set, described later.

3.4. Homology modeling data

Homology modeling requires target-template pairs with some sequence or structure relation. To construct such pairs, we started with the Binding MOAD database[25] which collects a large number of PDB entries with associated ligands. The database gives a family representative for related proteins. For each representative with a ligand of 8 atoms or more, we searched the DBAli database of structure alignments[26] for significant structurally related proteins, (DBAli structural significance score of 20 or better). Since our aim is to study the alignment of ligand binding residues, we eliminated templates which did not contain a ligand of at least 8 atoms. Targets which had no hits in the database which satisfied these criteria were also eliminated. Finally, in order to evaluate the performance of the binding-residue prediction, we eliminated any target which had greater than 40% sequence similarity to the prediction training set from Section 3.3 according to CD-HIT.

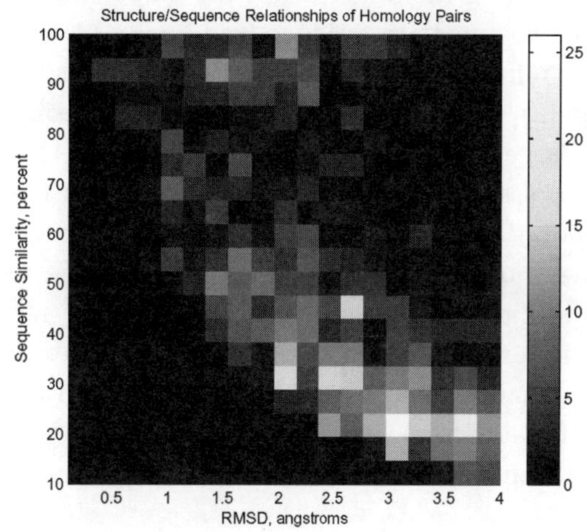

Fig. 1. The intensity of the heatmap indicates how many of the 1152 target-template pairs have the indicated RMSD-Sequence identity properties.

This left 409 unique targets, each having from one to twelve templates (average 2.8 templates per target) and 1,152 target-template pairs for the alignment. These pairs offer reasonable coverage of the sequence-structure relationship space according to their DBAli reports offering a range of easy (very similar sequences and structures) to hard homology modeling tasks (very different sequences and structures). DBAli is limited to structures related by less than a 4Å alignment and have at least 10% sequence identity which is reflected in our dataset. Figure 1 represents a distribution of the pairs over the RMSD-sequence identity landscape. The targets cumulatively represent 167,034 residues of which 9.1% are ligand-binding residues. This data was used for the evaluation of the ligand-binding residue prediction methods. An additional filtering step based on the generation of a quality baseline model was performed (see Section 5.2) which reduced the dataset to 1,000 target-template pairs for the statistical analysis of homology modeling results.

The identifiers for PDB entries used in our study may be obtained from the supplemental data (Section 7).

4. METHODS

The basis for most homology modeling approaches is to (1) obtain a structure template for a target sequence, (2) align the sequences of target and template, (3) let the target adopt the shape of the corresponding template residues, and finally (4) attempt some refinement of the shape. Our efforts center on step (2), properly aligning the binding residues of the target, assumed unknown, to those of the template, assumed known. Our hypothesis is that incorporating knowledge of these key residues will improve modeling of the binding site. In the following sections we describe how the binding residues of the target are predicted, how the target-template alignment is constructed, how baseline performance is generated from structure alignments, and the tools used to make a structure prediction.

4.1. Ligand residue prediction

4.1.1. *Structure-based prediction*

We chose to use PocketPicker for structure-based predictions of ligand-binding residues as it performed well in a recent benchmark by Weisel et al.[8]. It should be emphasized that in a true homology modeling situation, the target structure is unknown which precludes the use of structure-based predictors. They are employed here to benchmark whether binding residue prediction methods of any type are accurate enough to improve homology models.

PocketPicker reports the five largest pockets found in in the protein. Following the reasoning of Weisel et al., we defined binding residue prediction based on the single largest pocket (Pocket1) or on the largest three pockets (Pocket3) reported. These labels are evaluated for performance on the ligand-binding residue prediction task. For the homology modeling portion of the study, we used only the labels defined by the three largest pockets, Pocket3, to generate models.

4.1.2. *Sequence-based prediction*

Our predictions of ligand binding residues from sequence were made using a support vector machine (SVM) model[27]. In a previous work, we developed a generalized sequence annotation framework based on SVM learning which included prediction of ligand-binding residues[11,c]. In the present work we employed the same framework with a sliding window of size fifteen (seven to the left and right) around each residue to capture PSSM information on its neighbors. The framework is based off the SVM software package of Joachims[28] and eases the task of creating classification and regression models for sequence data.

A major advantage of SVM frameworks is their ability to exploit the so-called kernel trick which means roughly that similarity between data may be computed in a potentially high-dimensional, nonlinear space without greatly affecting efficiency. Thus, a kernel appropriate to a given type of data may be selected. In previous works, we have seen that the normalized second-order exponential kernel function (*nsoe*) is particularly useful in sequence prediction problems[11, 29, 30]. Details of the *nsoe* kernel and framework may be found in the references.

4.2. Predicted secondary structure

Incorporating aspects of predicted structure into sequence alignment scoring has been shown to improve alignment quality[31]. In our preliminary studies, we found that alignments which did not utilize secondary structure produced far inferior homology models. To that end, we predicted secondary structure using YASSPP, a SVM-based predictor[29]. YASSPP produces a vector of three scores, one for each of the three types of secondary structure, with high positive scores indicating confidence in that class. We would like to use true secondary structure for the templates but must be careful to use a score calibrated to the YASSPP outputs. In order to create these scores, we used knowledge of the true structures of targets to calculate the average SVM prediction values for true helices, strands, and coils.

[c]Available as a tech. report at `http://www.cs.umn.edu/research/technical_reports.php?page=report&report_id=07-023`

Template residues in a helix used the average helix vector for their secondary structure and similarly for template strands and coils. This approach follows from the observation of Przybylski and Rost[32] that scoring the predicted secondary structure between two sequences improves their alignment. However, we avoid the need to make predictions for the templates by using the averaged feature vector of the appropriate type of secondary structure.

4.3. Sequence alignment

Previous analyses of homology models for clefts have used alignment methods that employ global scoring matrices, for example the ALIGN command that MODELLER provides[5, 6]. We improve on these methods by employing sensitive profile-to-profile scoring and also explore special terms related specifically to binding residues.

4.3.1. *Alignment scoring*

The basic alignment algorithm we use is derived from the work on PICASSO by Mittleman[33] which was shown to be very sensitive in subsequent studies by others[34, 30]. The details of our modification are found in a previous work[35] but are briefly described as computing an optimal local alignment using an affine gap model with matching residues i and j in sequences X and Y, respectively, scored as

$$\begin{aligned} S_{P2P}(X_i, Y_j) = &\sum_{k=1}^{20} PSSM_X(i,k) \times PSFM_Y(j,k) \\ &+ \sum_{k=1}^{20} PSSM_Y(j,k) \times PSFM_X(i,k), \end{aligned} \tag{1}$$

where $PSSM$ is the position specific scoring of a sequence and $PSFM$ is the position specific frequency matrix of a sequence. This is known as profile-to-profile scoring (P2P).

In addition to the P2P scores, we included scoring between secondary structure elements in the target and template. This was computed as a dot product of the YASSPP descriptor vectors (Section 4.2) and is referred to hereafter as SSE.

The P2P and SSE scores were combined linearly with half the matching score coming from each. We used a subset of 48 target-template pairs, picked for sequence/structure diversity, to optimize our gap opening and extension penalties for our affine gap model. After a grid search, these were set to 3.0 and 1.5 which produced the best homology models on standard alignments.

4.3.2. *Modified alignments: using binding labels*

As we sought to give special attention to the ligand binding residues, we incorporated one additional term into matching residues to reflect this goal. Each residue was labelled either as ligand-binding or not. In the case of the targets, these labels were either the true labels, as described (Section 3.2), the structure predicted labels, or a sequence-predicted labels, (both in Section 4.1). Templates always used true labels. The contribution of matching and mismatching binding and nonbinding residues was controlled using a matrix of the form

$$M_{lig} = \begin{pmatrix} 0 & m_{nb} \\ m_{bn} & m_{bb} \end{pmatrix}. \tag{2}$$

The parameters relate to a target-template nonbinding-binding mismatch (m_{nb}), target-template binding-nonbinding mismatch (m_{bn}), and target-template binding-binding match (m_{bb}). In all cases we considered, m_{bn} and m_{nb} were negative, penalizing a mismatch, while m_{bb} was positive, rewarding a match. The parameter to score a nonbinding-nonbinding match would appear in the upper left entry of M_{lig} but this match was considered neutral and thus set to zero throughout the study. The ligand modification was not weighted when combining it with P2P and SSE scores. The final form of scoring between residue X_i of target and Y_j of template is

$$\begin{aligned} S(X_i, Y_j) = &\tfrac{1}{2} S_{P2P}(X_i, Y_j) + \tfrac{1}{2} S_{SSE}(X_i, Y_j) \\ &+ M_{lig}(X_i, Y_j), \end{aligned} \tag{3}$$

where S_{P2P} is the profile-to-profile score, S_{SSE} is the dot product of the secondary structure vectors, and $M_{lig}(X_i, Y_j)$ is the modification matrix score based on the whether the residues are considered binding or not.

We refer to alignments formed from $m_{nb} = m_{bn} = m_{bb} = 0$ as *standard* alignments as they do not incorporate knowledge of the ligand-binding residues in anyway. Nonzero modification parameters are termed *modified* alignments. Our hypothesis is that for some set of parameters, the modified alignment will produce better homology models than the standard alignment.

4.4. Structure alignments

The sequence alignment of target and template is intended to approximate a map of structurally related portions. Accordingly, one could expect a sequence alignment derived from a structure alignment to give a very good starting point for the homology modeling process. This is, of course, impossible when the target is unknown. However, in a benchmark study such as ours the structure induced sequence alignment will give a reasonable baseline for the best performance that can be expected of sequence alignment.

MUSTANG is a software package which aligns structures and produces their induced sequence alignment[36]. We used MUSTANG (version 0.3) to produce a baseline alignment for each target-template pair. Homology models were produced for the MUSTANG alignments and used to normalize scores, described in Section 4.6. These structure-induced alignments are referred to as *baseline* alignments as they use a true structure relation between target and template giving the homology model the best chance for success.

4.5. Homology modeling

Once a sequence alignment has been determined between target and template, we used MODELLER to predict the target structure[37]. We employed version 9.2 of the MODELLER package which is freely available. As input, MODELLER takes a target-template sequence alignment and the structure of the template. An optimization process ensues in which the predicted coordinates of the target are adjusted to violate, as little as possible, spatial constraints derived from the template.

Details of our use of MODELLER are as follows. The 'automodel' mechanism was used which, given the sequence alignment, performs all necessary steps to produce a target structure prediction. We chose to generate a single model as a brief preliminary exploration indicated little changes when multiple models are generated (data not shown). As mentioned earlier, some template structures contained nonstandard amino acids for which MODELLER will fail. To that end, we used a modified table of amino acid code to type conversions, taken from ASTRAL as in Section 3.1, to model nonstandard residues as an analogous standard residue. The mechanism for defining such a table is described in the MODELLER manual[d] and the specific table we used is available with the other supplementary data (Section 7).

4.6. Evaluation

4.6.1. *Ligand-binding residue predictions quality*

We evaluated the sequence-based prediction of ligand-binding residues using the receiver operating characteristic (ROC) curve[38]. This is obtained by varying the threshold at which residues are considered ligand-binding or not according to the SVM output of the predictor. For any binary predictor, the number of true positives (TP), false positives (FP), true negatives (TN), and false negatives (FN) determines standard classification statistics which we use for comparison between the structure-based and sequence-based predictors. These are

$$\text{Accuracy} = \frac{TP + TN}{TP + TN + FP + FN} \tag{4}$$

$$\text{Precision} = \frac{TP}{TP + FP} \tag{5}$$

$$\text{Recall} = \frac{TP}{TP + FN} \tag{6}$$

$$\text{Specificity} = \frac{TN}{TN + FP} \tag{7}$$

4.6.2. *Homology modeling quality*

We chose to evaluate predicted structures (models) based on their RMSD from the true structure of the protein in question. A low RMSD indicates similarity between two structures. Calculations were done using in-house software which implements the quaternion method of computing RMSD[39]. Only the α-carbon coordinates are used for the RMSD computation. Following the convention of Piedra et al.[6], we computed the RMSD between only the ligand-binding residues in the model and those in the true structure as these residues are most important to models of the binding site. For brevity, this will

[d] `http://www.salilab.org/modeller/manual/node105.html`

be called the ligRMSD for ligand-binding residues RMSD.

Difficult modeling tasks are not expected to achieve a low RMSD: there is not enough information present in the template to deduce a high quality target model. Evaluating purely on the above RMSD criteria would not account for this factor. We chose to normalize the RMSD in the following way. Using the baseline sequence alignment (generated from structure, Section 4.4), we produced a model for the target. The ligRMSD was calculated for this model against the true structure and is denoted $lig\text{RMSD}_{base}$. Sequence-only alignments were then used to generate homology models for the same target-template pairs. The ligRMSD for these models, denoted $lig\text{RMSD}_{seq}$, was divided by the ligand RMSD of the corresponding $lig\text{RMSD}_{base}$. The sequence alignments we produced were local while the baseline alignments were global. Using a local alignment means that some of the ligand-binding residues were potentially omitted from the alignment and subsequent model. For a given model, the total number of ligand binding residues is n_{tot} while the number of ligand-binding residues in the model is n_{mod}. We penalize the score of models by the ratio of total to missing residues. This gives a normalized homology score of

$$H = \frac{lig\text{RMSD}_{seq}}{lig\text{RMSD}_{base}} \times \frac{n_{tot}}{n_{mod}}. \tag{8}$$

Due to the ratio that is taken here, the scores should follow a log-normal distribution. When doing our statistical analysis, we convert into log-space to calculate significance but report results in the usual space.

To test whether knowledge of the ligand-binding residues improved or degraded binding site models, we performed Student's t-Test on the normalized scores of the standard alignment predictions paired with the corresponding normalized scores for modified alignments. The null hypothesis is that the two have equal mean while the alternative hypothesis is that the modified alignments produce models with a lower mean (a one-tailed test). We report p-values for the comparisons noting that a p-value below 0.05 is typically considered statistically insignificant. We also report the mean improvement (gain) from using modified alignments. If the mean of all normalized homology scores for the standard alignments is $\bar{H}_{stand}$ and that of a modified alignment is $\bar{H}_{mod}$, the percent gain is

$$\%\text{Gain} = \frac{\bar{H}_{stand} - \bar{H}_{mod}}{\bar{H}_{stand}}. \tag{9}$$

A positive gain indicates improvement through the use of the ligand-binding residue labels while a negative gain indicates label use degrades the homology models.

5. RESULTS

5.1. Ligand-bind residue prediction from sequence and structure

Figure 2 illustrates the receiver operating characteristic (ROC) for the sequence-based predictor on the evaluation set. To produce binary labels, a threshold was chosen so that the number of predicted positives was approximately equal to the number of true positives. The threshold point is shown in Figure 2 and statistics of the labels it induces are shown in Table 1. Also in Table 1 we show the performance of the structure-based predictor on the targets based on binding-residue definitions from the largest single and largest three pockets, labeled Pocket1 and Pocket3 (Section 4.1).

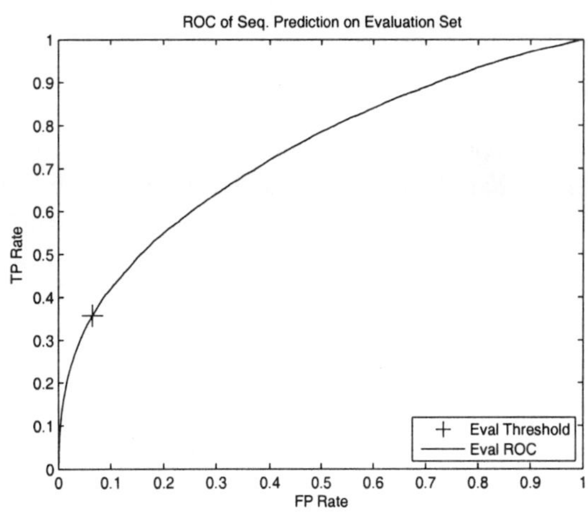

Fig. 2. ROC of sequence-only predictions of ligand-binding residues on evaluation set. The threshold position indicates the FPR and TPR for the predicted labels used in evaluation. The AUC is 0.7351 for the evaluation set.

In predicting ligand-binding residues, the sequence-only predictions are very comparable to those of the structure-based methods in terms of accuracy. As expected, the precision is worse than the best structure-derived labels method, but the two perform similarly when three of the largest pockets are used in the structure method.

Table 1. Performance statistics for predicting ligand-binding residues

Statistic	SeqPred	Pocket1	Pocket3
Accuracy	0.8813	0.8948	0.8302
Precision	0.3531	0.4430	0.3087
Recall	0.3572	0.5839	0.6907
Specificity	0.9341	0.9261	0.8443

A threshold of -0.91 was chosen for the sequence-based prediction as the cutoff for the positive class. Two variants of PocketPicker were used: positive residues generated from the single largest and three largest pockets, Pocket1 and Pocket3.

5.2. Homology modeling

Homology models were produced for the standard alignment procedure and for modified alignments that incorporated ligand labels derived from three sources: the true labels (Section 3.2), structure predicted labels, and sequence predicted labels (both in Section 4.1).

In some cases, the predicted structure that is produced by MODELLER is obviously wrong, for example when the model is in an extended rather than compact conformation. We removed structures for which the baseline alignment produced a model with greater than 10Å all-residue RMSD from the true structure. This left 1000 structures for the statistical analysis. Additional filtering was done on each target-template pair with failures being ignored for the analysis. Finally, we analyzed models in subgroups with specific sequence and structure properties and report the sample size of each group.

5.2.1. *Using true labels for binding residues*

The second section of Table 2 shows the improvement for alignments which used the true labels of ligand-binding residues. We found parameters $m_{bb} = 10$, $m_{nb} = m_{bn} = 0$ to provide the most improvement over standard alignments, though $m_{bb} \in \{7.5, 12.5\}$ with $m_{nb} = m_{bn} = 0$ produced only slightly inferior results. Also, $m_{bb} = 10, m_{nb} = -2.5, m_{bn} = 0$ performed well. We will discuss the issue of asymmetry in scoring later as it also pertains to the sequence and structure predicted labels.

The table shows sequence/structure subgroups along with the quality gained through the use of labels and whether the result is statistically significant (p-value ≤ 0.05). Improvement for the true labels occurs in low sequence identity groups with better gains in the higher structure diversity subgroup (2-4Å RMSD). At higher sequence identity, use of the labels improves performance only when the target and template are structurally diverse (0-50% identity and 2-4Å RMSD).

5.2.2. *Using structure-predicted labels*

We report the results of using structure predicted binding labels in the third section of Table 2. The best parameters we found in our grid search were $m_{bb} = 5, m_{nb} = 0$, and $m_{bn} = -2.5$, an assymetric scoring matrix. We see similar trends for the structure-predicted labels as were observed for the true labels with the largest gains appearing in the low sequence identity and high structural diversity areas of sequence-structure space. The magnitude of improvement for the structure-predicted labels appears greater in some cases than the true labels. We are still investigating the cause of this behavior.

5.2.3. *Using sequence-predicted labels*

The fourth section of Table 2 shows homology modeling results when sequence predicted labels are used. Again, asymmetric scoring parameters of $m_{bb} = 5, m_{nb} = 0, m_{bn} = -2.5$ provided the best performance. The significant gains are achieved only in the low sequence identity category and are greater in magnitude when the target-template structures are more diverse.

5.2.4. *Comparisons*

To compare the performance of true, structure-predicted, and sequence-predicted labels, we examine the first two rows of Table 2. These are the subgroups of pairs related by $\leq 30\%$ sequence identity and a DBAli structure alignment either between $0 \leq 4.0$Å or $2 \leq 4.0$Å. These two subgroups are where use of the ligand-binding labels appears to offer positive gains regardless of their source. The improvement given in these groups by the sequence-based labels are smaller than those for true and structure-

Table 2. Homology modeling results

SeqID	RMSD	True Labels N	True Labels $\frac{n_{mod}}{n_{tot}}$	True Labels %Gain	True Labels p-value	Structure Labels N	Structure Labels $\frac{n_{mod}}{n_{tot}}$	Structure Labels %Gain	Structure Labels p-value	Sequence Labels N	Sequence Labels $\frac{n_{mod}}{n_{tot}}$	Sequence Labels %Gain	Sequence Labels p-value
0≤ 30	2.0≤4.0	234	0.99	**3.51**	0.0009	234	0.98	**3.34**	0.0099	234	0.98	**2.03**	0.0276
0≤ 30	0.0≤4.0	254	0.99	**3.09**	0.0018	254	0.98	**3.95**	0.0037	254	0.98	**1.87**	0.0274
30≤ 60	0.0≤2.0	135	1.00	-0.02	0.5104	131	1.00	-0.93	0.7468	135	1.00	-0.50	0.7922
30≤ 60	2.0≤4.0	192	0.98	-1.40	0.9266	189	0.98	-1.10	0.7462	192	0.98	-2.33	0.9448
30≤ 60	0.0≤4.0	325	0.99	-0.83	0.9058	318	0.99	-1.04	0.8182	325	0.99	-1.58	0.9611
60≤100	0.0≤2.0	267	0.98	-0.34	0.9342	265	0.98	-0.72	0.8405	267	0.98	0.05	0.4334
60≤100	2.0≤4.0	121	0.99	-0.53	0.8451	120	0.99	-1.20	0.8492	121	0.99	-0.27	0.7274
60≤100	0.0≤4.0	388	0.98	-0.40	0.9626	385	0.98	-0.87	0.9217	388	0.98	-0.05	0.5838
0≤ 50	0.0≤2.0	116	1.00	-0.55	0.7718	114	1.00	1.28	0.2780	116	1.00	0.13	0.4109
0≤ 50	2.0≤4.0	395	0.98	1.73	0.0110	392	0.98	1.37	0.1204	395	0.98	0.03	0.4887
0≤ 50	0.0≤4.0	505	0.99	1.23	0.0230	500	0.99	1.38	0.0920	505	0.99	0.04	0.4769
50≤100	0.0≤2.0	312	0.98	-0.22	0.7796	308	0.98	-0.76	0.8812	312	0.98	-0.21	0.7647
50≤100	2.0≤4.0	152	0.99	-1.22	0.9072	151	0.99	-0.67	0.7519	152	0.99	-0.04	0.5167
50≤100	0.0≤4.0	464	0.98	-0.55	0.9374	459	0.98	-0.73	0.9123	464	0.98	-0.15	0.6701
0≤100	0.0≤2.0	426	0.99	-0.31	0.8587	420	0.99	-0.21	0.6091	426	0.99	-0.11	0.6688
0≤100	2.0≤4.0	546	0.99	0.92	0.0641	542	0.99	0.81	0.1817	546	0.99	0.01	0.4952
0≤100	0.0≤4.0	966	0.99	0.38	0.1469	956	0.99	0.37	0.2673	966	0.99	-0.05	0.5492

Columns one and two are the target-template sequence and RMSD ranges. The remaining columns relate specifically to each type of label. Columns three through six describe the sample size, ratio of modeled to total binding residues (Equation 8), percentage gain (Equation 9), and significance of results of models predicted using true labels. Columns six through eight describe the structure-predicted labels and columns nine through twelve the sequenced-predicted labels. The term $\frac{n_{ali}}{n_{tot}}$ is averaged over all models in the sample and, being close to one in all cases, indicates the majority of ligand binding residues are modeled.

based labels, but they are present and significant. It is also interesting to examine the last row of Table 2 and note that over the entire dataset, the true and structure-predicted labels offer positive though statistically insignificant gains while sequence-predicted labels slightly degrade model quality overall. This suggests use of labels only in the case when the only available templates are those with low sequence identity.

In many cases, the sequence-predicted labels did very well compared to the structure labels. An example of this is shown in Figure 3 for target 1h5q chain A produced by alignment to 1mxh chain D. In this case, the sequence-only method performs nearly identically to the structure-based method for deriving labels.

The magnitude of the ligand-ligand matching reward is different between the true and predicted label methods, 10 for true labels, 5.0 for the predicted labels. This is likely due to low precision for the predicted ligands.

The success of asymmetric scoring parameters for predicted labels still requires further investigation. It was expected that the true signal from template ligands to govern the success of the scoring parameters. This would lead to a negative m_{nb} to penalize 'missing' known ligand binding residue in the template. This appears to be the case for true labels which had good performance for $m_{bb} = 10, m_{nb} = -2.5, m_{bn} = 0$. However, the opposite has shown to be true for both the sequence and structure-based alignments, that m_{nb} is neutral while m_{bn} is used to penalize the alignment of a predicted binding residue to a nonbinder in the template.

5.2.5. *Generalization of model parameters*

When proposing a parameterized model that shows prediction improvements, care is needed to ensure that the chosen parameters are not highly dependant upon the data used for measurement. Since our modified alignments depend on a small number of parameters that affect the scoring binding residue matches, we want to ensure that these parameters will reproduce the reported performance on future data. To that end, we performed a permutation test to determine the validate the modified alignments.

For the sequence/structure subgroups of interest, we took random subsets and performed paired Student's t-Test on the standard and modified alignment normalized scores. We took the average p-value over 1000 random subsets and used it as an indication of how well the parameters are expected to perform on future data.

Models generated using the true labels and the parameters $m_{bb} = 10, m_{nb} = 0, m_{bn} = 0$ had better

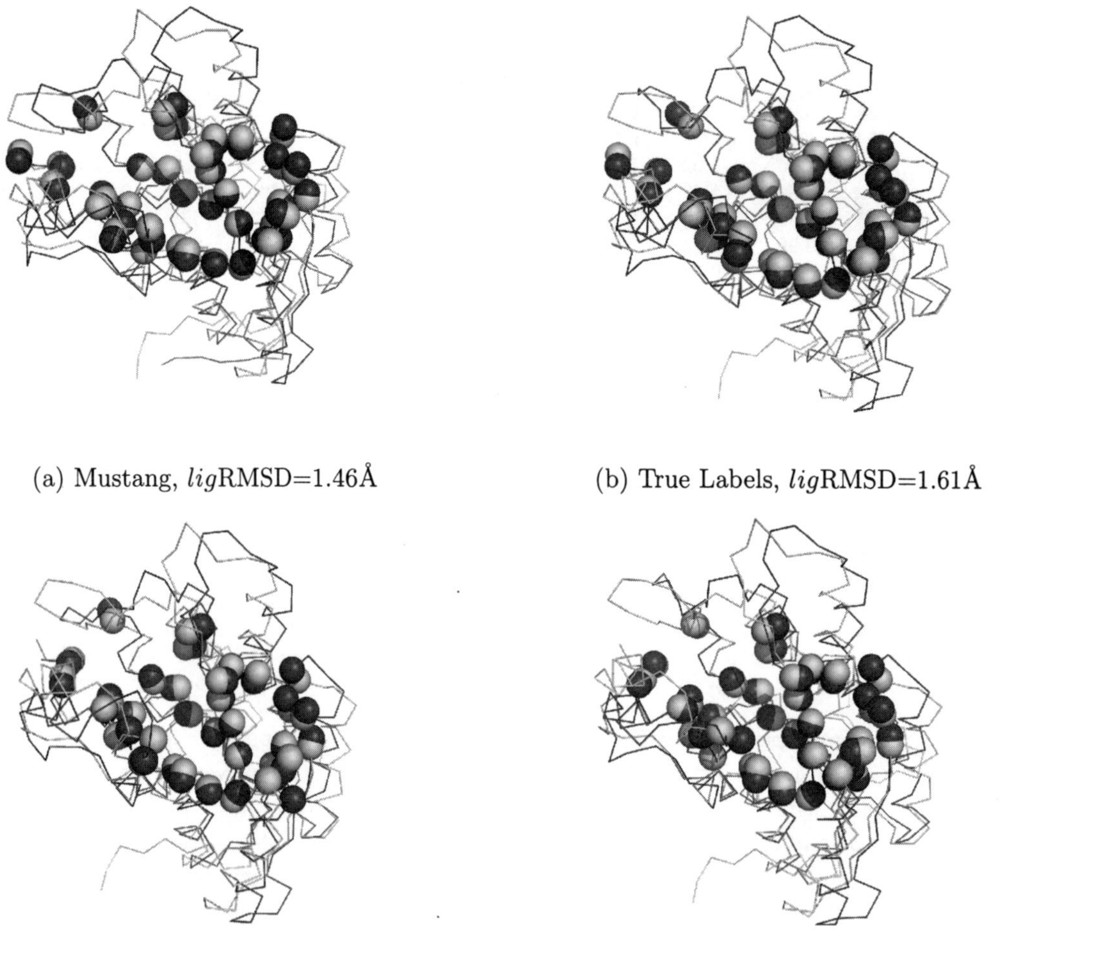

(a) Mustang, ligRMSD=1.46Å

(b) True Labels, ligRMSD=1.61Å

(c) SeqPred, ligRMSD=1.74Å

(d) PocketPicker, ligRMSD=1.75Å

Fig. 3. Homology models for target 1h5q chain A (template 1mxh chain D with 20% sequence identity and 2.48Å RMSD) produced by the 4 types of alignments. The protein has 260 residues with 35 ligand-binding residues. A backbone trace for the true model is shown in lightly colored, the predicted model in darkly colored, and the α-carbons of ligand-binding residues are shown as spheres. Images were produced with Pymol.

average p-values than other parameters in all the significant cases mentioned above indicating that they are likely to be applicable to future data.

Average p-values for the structure-based predicted labels and the parameters $m_{bb} = 5, m_{nb} = 0, m_{bn} = -2.5$ were better than other parameter sets. Again, significance was achieved in all the the cases above indicating good generalization.

Finally, the sequence predicted labels did not appear to have as good of generalization properties. At sequence identity $\leq$ 0-30% and RMSD $0 \leq 4$Å, the average p-values were between 0.08 and 0.11. An improved sequence predictions and a finer-grained grid-search will likely locate optimal parameters for the sequence-predicted labels generalize well.

6. CONCLUSIONS

We have explored the performance of a sequence-based and a structure-based ligand-binding residue predictor and have shown that making use of these predictions in a homology modeling framework can improve the overall quality of predicted structures. This effect is most pronounced when the sequence identity between the target and template is low.

Our prediction of ligand-binding residues from sequence was by no means perfect but the downstream application shows that even noisy predictions can benefit homology models.

It is unclear at this point why using the structure-predicted labels from PocketPicker outperform the true labels but this may be a moot point as in real homology modeling the structure of the target is unknown. This result may suggest that an alternate definition for ligand-binding residues should be used, one which accounts for the location of a residue in a pocket as well as being within contact distance of the ligand.

There are several relevant directions to pursue in order to expand on the current work. Improving ligand-binding residue prediction from sequence will no doubt boost the performance of models generated via this mechanism. Though the set of parameters we explored for alignment modification was sufficient to indicate improvement, it was by no means exhaustive enough to claim that the optimal parameters were located. The particular values used for modifications are highly dependent on other aspects of the alignment process such as P2P scoring function. This remains a general problem worth studying: what is the best way to incorporate diverse information (profiles, SEE, ligand labels) into the scoring scheme for alignments? Extending the notion of a 'label' for a residue to a continuous value, indicative of confidence, will increase the flexibility of this part of the scoring scheme and remove the need to derive a threshold separating positive and negative classes.

7. ACKNOWLEDGMENTS, SUPPLEMENTS

The authors gratefully acknowledge support from the NIH Training for Future Biotechnology Development grant, NIH T32GM008347, and NIH RLM008713A, NSF IIS-0431135, and the U of MN Digital Technology Center.

Supplementary materials for this work are available online at `http://bioinfo.cs.umn.edu/supplements/ligand-modeling/csb2008`. These include the MODELLER modified residue table, the cross-validation results of section 5.2.5 and the binary programs for extraction, sequence alignment, and structure alignment.

References

1. N Moitessier, P Englebienne, D Lee, J Lawandi, and C R Corbeil. Towards the development of universal, fast and highly accurate docking//scoring methods: a long way to go. *Br J Pharmacol*, 153(S1):S7–S26, November 2007.
2. Philippe Ferrara and Edgar Jacoby. Evaluation of the utility of homology models in high throughput docking. *Journal of Molecular Modeling*, 13:897–905, Aug 2007. 10.1007/s00894-007-0207-6.
3. D. Baker and A. Sali. Protein structure prediction and structural genomics. *Science*, 294(5540):93–96, Oct 2001.
4. Carol DeWeese-Scott and John Moult. Molecular modeling of protein function regions. *Proteins*, 55(4):942–961, Jun 2004.
5. Suvobrata Chakravarty, Lei Wang, and Roberto Sanchez. Accuracy of structure-derived properties in simple comparative models of protein structures. *Nucleic Acids Res*, 33(1):244–259, 2005.
6. David Piedra, Sergi Lois, and Xavier de la Cruz. Preservation of protein clefts in comparative models. *BMC Struct Biol*, 8(1):2, Jan 2008.
7. S. Soga, H. Shirai, M. Kobori, and N. Hirayama. Use of amino acid composition to predict ligand-binding sites. *Journal of Chemical Information and Modeling*, 47(2):400–406, 2007.
8. Martin Weisel, Ewgenij Proschak, and Gisbert Schneider. Pocketpicker: analysis of ligand binding-sites with shape descriptors. *Chemistry Central Journal*, 1(1):7, 2007.
9. Yanay Ofran, Venkatesh Mysore, and Burkhard Rost. Prediction of dna-binding residues from sequence. *Bioinformatics*, 23(13):i347–353, 2007.
10. Shandar Ahmad and Akinori Sarai. Pssm-based prediction of dna binding sites in proteins. *BMC Bioinformatics*, 6:33, 2005.
11. Huzefa Rangwala, Christopher Kauffman, and George Karypis. A generalized framework for protein sequence annotation. In *Proceedings of the NIPS Workshop on Machine Learning in Computational Biology*, 2007.
12. Michael Terribilini, Jae-Hyung Lee, Changhui Yan, Robert L. Jernigan, Vasant Honavar, and Drena Dobbs. Prediction of RNA binding sites in proteins from amino acid sequence. *RNA*, 12(8):1450–1462, 2006.
13. Manish Kumar, M. Michael Gromiha, and G. P S Raghava. Prediction of rna binding sites in a protein using svm and pssm profile. *Proteins*, 71(1):189–194, Apr 2008.
14. Yanay Ofran and Burkhard Rost. Predicted protein-protein interaction sites from local sequence information. *FEBS Lett*, 544(1-3):236–239, Jun 2003.
15. Ming-Hui Li, Lei Lin, Xiao-Long Wang, and Tao Liu. Protein protein interaction site prediction based on conditional random fields. *Bioinformatics*, 23(5):597–604, 2007.
16. Asako Koike and Toshihisa Takagi. Prediction of protein-protein interaction sites using support vector machines. *Protein Engineering, Design and Se-*

lection, 17(2):165–173, 2004.

17. Rodney Harris, Arthur J Olson, and David S Goodsell. Automated prediction of ligand-binding sites in proteins. *Proteins*, Oct 2007.
18. Bingding Huang and Michael Schroeder. Ligsitecsc: predicting ligand binding sites using the connolly surface and degree of conservation. *BMC Struct Biol*, 6:19, 2006.
19. Bin Li, Srinivasan Turuvekere, Manish Agrawal, David La, Karthik Ramani, and Daisuke Kihara. Characterization of local geometry of protein surfaces with the visibility criterion. *Proteins*, Nov 2007.
20. Helen M. Berman, John Westbrook, Zukang Feng, Gary Gilliland, T. N. Bhat, Helge Weissig, Ilya N. Shindyalov, and Philip E. Bourne. The protein data bank. *Nucl. Acids Res.*, 28(1):235–242, 2000.
21. John-Marc Chandonia, Nigel S Walker, Loredana Lo Conte, Patrice Koehl, Michael Levitt, and Steven E Brenner. Astral compendium enhancements. *Nucleic Acids Res*, 30(1):260–263, Jan 2002.
22. SF Altschul, TL Madden, AA Schaffer, J Zhang, Z Zhang, W Miller, and DJ Lipman. Gapped blast and psi-blast: a new generation of protein database search programs. *Nucl. Acids Res.*, 25(17):3389–3402, 1997.
23. Renxiao Wang, Xueliang Fang, Yipin Lu, and Shaomeng Wang. The pdbbind database: collection of binding affinities for protein-ligand complexes with known three-dimensional structures. *J Med Chem*, 47(12):2977–2980, Jun 2004.
24. Weizhong Li, Lukasz Jaroszewski, and Adam Godzik. Clustering of highly homologous sequences to reduce the size of large protein databases. *Bioinformatics*, 17(3):282–283, 2001.
25. Mark L Benson, Richard D Smith, Nickolay A Khazanov, Brandon Dimcheff, John Beaver, Peter Dresslar, Jason Nerothin, and Heather A Carlson. Binding moad, a high-quality protein-ligand database. *Nucleic Acids Res*, 36(Database issue):D674–D678, Jan 2008.
26. Marc A. Marti-Renom, Valentin A. Ilyin, and Andrej Sali. Dbali: a database of protein structure alignments. *Bioinformatics*, 17(8):746–747, 2001.
27. Vladimir N. Vapnik. *The Nature of Statistical Learning Theory*. Springer Verlag, 1995.
28. T. Joachims. *Advances in Kernel Methods: Support Vector Learning*, chapter Making large-Scale SVM Learning Practical. MIT-Press, 1999.
29. George Karypis. Yasspp: Better kernels and coding schemes lead to improvements in svm-based secondary structure prediction. *Proteins: Structure, Function and Bioinformatics*, 64(3):575–586, 2006.
30. Huzefa Rangwala and George Karypis. frmsdpred: predicting local rmsd between structural fragments using sequence information. *Comput Syst Bioinformatics Conf*, 6:311–322, 2007.
31. Jian Qiu and Ron Elber. Ssaln: an alignment algorithm using structure-dependent substitution matrices and gap penalties learned from structurally aligned protein pairs. *Proteins*, 62(4):881–891, Mar 2006.
32. Dariusz Przybylski and Burkhard Rost. Improving fold recognition without folds. *J Mol Biol*, 341(1):255–269, Jul 2004.
33. David Mittelman, Ruslan Sadreyev, and Nick Grishin. Probabilistic scoring measures for profile-profile comparison yield more accurate short seed alignments. *Bioinformatics*, 19(12):1531–1539, Aug 2003.
34. A. Heger and L. Holm. Picasso: generating a covering set of protein family profiles. *Bioinformatics*, 17(3):272–279, Mar 2001.
35. Huzefa Rangwala and George Karypis. Incremental window-based protein sequence alignment algorithms. *Bioinformatics*, 23(2):e17–e23, Jan 2007.
36. Arun S Konagurthu, James C Whisstock, Peter J Stuckey, and Arthur M Lesk. Mustang: a multiple structural alignment algorithm. *Proteins*, 64(3):559–574, Aug 2006.
37. A. Sali and T. L. Blundell. Comparative protein modelling by satisfaction of spatial restraints. *J Mol Biol*, 234(3):779–815, Dec 1993.
38. T. Fawcett. Roc graphs: Notes and practical considerations for researchers, 2004.
39. Chaok Seok Ken A. Dill Evangelos A. Coutsias. Using quaternions to calculate rmsd. *Journal of Computational Chemistry*, 25:1849–1857, 2004.

Computational Systems Bioinformatics 2008

Pathways, Networks, and Biological Systems

USING RELATIVE IMPORTANCE METHODS TO MODEL HIGH-THROUGHPUT GENE PERTURBATION SCREENS

Ying Jin*, Naren Ramakrishnan, and Lenwood S. Heath
Department of Computer Science, Virginia Tech, Blacksburg, VA 24061, U.S.A.
Email: {jiny,naren,heath}@cs.vt.edu.

Richard F. Helm
Department of Biochemistry, Virginia Tech, Blacksburg, VA 24061, U.S.A.
Email: helmrf@vt.edu.

With the advent of high-throughput gene perturbation screens (e.g., RNAi assays, genome-wide deletion mutants), modeling the complex relationship between genes and phenotypes has become a paramount problem. One broad class of methods uses 'guilt by association' methods to impute phenotypes to genes based on the interactions between the given gene and other genes with known phenotypes. But these methods are inadequate for genes that have no cataloged interactions but which nevertheless are known to result in important phenotypes. In this paper, we present an approach to first model relationships between phenotypes using the notion of 'relative importance' and subsequently use these derived relationships to make phenotype predictions. Besides improved accuracy on *S. cerevisiae* deletion mutants and *C. elegans* knock-down datasets, we show how our approach sheds insight into relations between phenotypes.

1. INTRODUCTION

There are now a variety of mechanisms to study loss of function phenotypes in specific cell types or at different stages of development in an organism. Genome wide deletion mutants, e.g., for *Saccharomyces cerevisiae* [1, 2], use homologous recombination to replace genes with targeted cassettes so that the resulting strain can be screened for specific phenotypes (or lack thereof). RNA interference methodologies, in organisms such as *Caenorhabditis elegans* [3, 4], use post-transcriptional gene silencing to degrade specific RNA molecules, thus causing a drastic attenuation of gene expression. Since RNAi may not completely deplete the expressed RNA molecules, its use is referred to as a 'knock-down', in contrast to a complete 'knockout' exhibited by a deletion mutant. Through the use of high-throughput screens, both these techniques now support large scale phenotypical studies.

A central goal of bioinformatics research is to model the phenotype effects of gene perturbations. The mapping between gene function and expressed phenotype is complex. A single gene perturbation (through deletion or RNAi interference) can lead to a cascade of changes in transcription or post-transcriptional pathways. It is impractical to make a comprehensive empirical analysis when there is a large number of candidate genes. An emerging area of interest therefore is to use diverse, highly heterogeneous, data (e.g., microarrays, RNAi studies, protein-protein interaction assays) to computationally model phenotype effects for mutations.

Previous studies have shown that by considering interactions between candidate genes and target genes (which have been known to result in a desired phenotype) the accuracy of phenotype prediction can be improved. Examples of interactions that have been considered by such works include physical interactions between proteins [5], interactions underlying protein complexes [6], and integrated gene networks constructed from multiple data sources [7]. Most of these methods can be classified as 'direct' methods since they require a direct interaction between a gene and another gene with the target phenotype in order to predict the phenotype for the given gene.

Statistical and computational methods to prioritizing genes by using combinations of gene expres-

*Corresponding author.

sion and protein interaction data have also been proposed, e.g., CGI [8] and GeneRank [9]. In addition to direct interactions, these methods take into account indirect interactions, i.e., links from genes to target genes through other intermediate genes. However, these approaches assume that there is at least one path from a candidate gene to some target gene(s). Since many genes do not have any catalogued interactions, this limits their applicability.

Markowetz *et al.* [10] proposed the NEM (nested effects models) approach to rank genes according to subset relations between phenotypes. NEM uses phenotype profiles only, i.e., it does not consider any protein-protein interactions. While this overcomes the limitations mentioned previously, NEM has shortcomings in scalability with respect to the number of phenotypes and to overcome the increased computational cost, NEM focuses on inference only from pairwise and triple relations.

Contributions: We propose a new graph theoretic approach to predicting phenotype effects of gene perturbation using phenotype relations (P^3). Our approach focuses on relative importance methods to infer relations between phenotypes and uses these relations to predict phenotype effects. We integrate phenotype profiles with the gene network to derive phenotype relations. It is assumed that genes tightly connected are likely to share the same phenotypes. We use a weighted directed graph to model the relations between phenotypes such that more complicated relations can be illustrated and interpreted instead of just subset relations. Since predictions are carried out purely based on the phenotype relations derived, there is no requirement for known interaction paths from candidate genes to target genes. Furthermore, once the relations between phenotypes are derived, they can be used repetitively in the prediction process. In particular, complete perturbation effects across all phenotypes can be predicted simultaneously from the relations between known phenotypes and others. Therefore, P^3 is more effective for large-scale phenotype prediction than previous methods that rank genes for each phenotype, one at a time. Experimental results on *S. cerevisiae* and *C. elegans* also show that our approach outperforms the direct and GeneRank methods consistently. In particular, for genes without any interactions in *S. cerevisiae*, we show that our method can predict 96% of their phenotypes with AUC (area under ROC curve) greater than 0.8, and 60% of the phenotypes in *C. elegans*.

2. WORKING EXAMPLE

Table 1 describes an example of phenotype profiles resulting from many gene perturbations. Each row represents a phenotype and each column a gene. The cell value indicates whether the gene perturbation exhibits the corresponding phenotype, e.g., g_1 gives rise to p_1 but not p_2 and p_3. A second form of data available is a gene network as shown in Figure 1 (left), that shows interactions between genes. For ease of interpretation, genes that result in the same phenotype as shown in Table 1 are also grouped in Figure 1 (left). Suppose that the only information about g_7 that we are given is that it results in phenotype p_3 and we desire to use computational methods to predict that it will also cause p_2 but not p_1 (see last column of Table 1).

Table 1. Example phenotype profiles.

	g_1	g_2	g_3	g_4	g_5	g_6	g_7
p_1	1	1	0	0	0	1	0
p_2	0	0	1	1	1	0	1
p_3	0	0	0	1	0	1	1

- **Using phenotype profiles:** If we were to use only Table 1 to make a prediction, it is not clear whether g_7 should result in p_1 or p_2. p_1 and p_2 involve three genes each, and p_3 has (exactly) one gene in common with both sets. Obviously, p_1 and p_2 have an equal chance to be predicted, no matter what association measure is used.
- **Using network information:** If we assume that all links in Figure 1 (left) have the same weight, then in fact the prediction result will be p_1. To see this, observe that g_7 has only one interaction partner g_2, and it is known that g_2 contributes to p_1 only. And there are no paths from g_7 to any genes resulting in phenotype p_2. Hence, no matter what graph theoretic methods are used, p_1 has a better chance of being predicted.

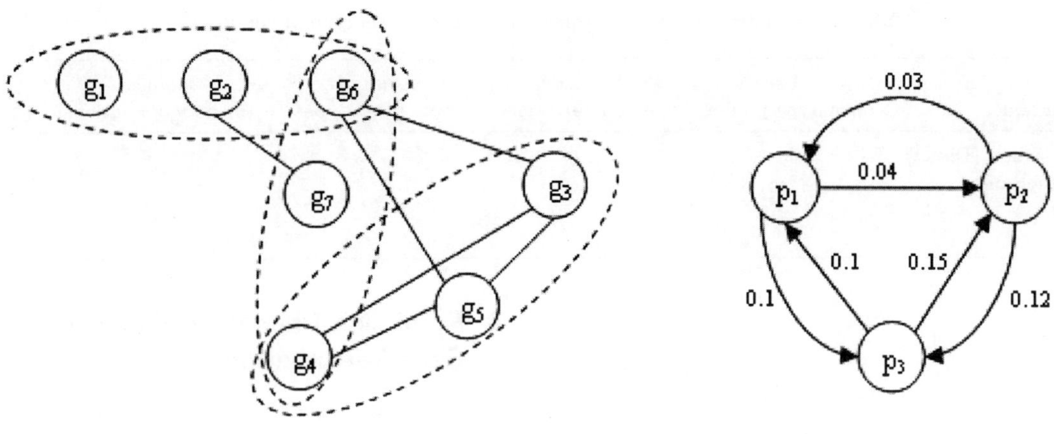

Fig. 1. (left) Example of gene network. (right) Induced relationships between phenotypes.

We propose to combine the selective superiorities of the two methods to model phenotypes. In this process, we develop a method that resembles a collaborative filtering algorithm [11] used in recommender systems research. First, we derive relationships between phenotypes from Table 1 and Figure 1 (left). Figure 1 (right) demonstrates the relationships between phenotypes obtained by applying our algorithm presented in the following section. The value on the arrow from phenotype p_i to phenotype p_j denotes the tendency that a gene perturbation causing p_i also causes p_j. From such a relation, we can predict that if a gene perturbation results in p_3, then it is more likely to result in p_2 rather than p_1. Some characteristics of existing methods and our approach are listed in Table 2.

3. METHODS

3.1. Inferring Relations Between Phenotypes

As stated earlier, inferring relations between phenotypes is a one-time cost and can be amortized over the prediction process. Our method is motivated by the study of relative importance in networks [12]. Original link analysis methods, such as PageRank [13] and HITS [14], rank nodes according to their "importance" or "authority" in the entire network. Relative importance analysis focuses on determining the importance of nodes with respect to a subset of the network, called the "root set." Multiple algorithms have been proposed for relative importance computation, such as k-short path, k-short node-disjoint paths, PageRank with priors, HITS with priors, and the k-step Markov approach, which are all surveyed by White and Smyth [12].

Suppose that there are n genes $G = \{g_i | 1 \leq i \leq n\}$, and m phenotypes $P = \{p_i | 1 \leq i \leq m\}$ in a study. Let $W_{n \times n}$ denote the connection matrix of the network, where $w_{i,j}$ denotes the weight of the connection between gene g_i and gene g_j. W is required to be a symmetric matrix whose diagonal is uniformly 0. For each phenotype p_j, there is a corresponding vector $\overline{p}_j = \langle v_1, v_2, ..., v_n \rangle$, where $v_i = 1$ indicates that gene g_i is known to result in p_j, otherwise $v_i = 0$. These vectors are grouped together to form a gene phenotype matrix $V_{m \times n}$, where rows are phenotypes and columns are genes. Given a phenotype p, genes resulting in this phenotype form a root set R. Similar to PageRank with priors, each gene is assigned a prior rank score, as shown in Equation 1. Observe that the sum of all initial rank scores is 1.

$$r^0_{g_i} = \begin{cases} \frac{1}{\|R\|} & \text{if } g_i \in R, \\ 0 & \text{otherwise.} \end{cases} \tag{1}$$

Let $N(g_i) = \{g_j | w_{i,j} > 0, i \neq j, \text{and } g_j \in G\}$ denote the set of all other genes that interact with g_i. Define parameter β, $0 \leq \beta \leq 1$, to be the relative weight between the original score of a gene and the score that results through the influence of its neighbors. The formula for iteratively computing gene rank scores is shown in Equation 2.

Table 2. Comparison of P[3] to other methods for phenotype prediction.

Method	Use phenotype profiles?	Use gene interactions?	Ability to rank phenotypes?	Induce phenotype relations?
Wormnet (Direct)		√		
GeneRank		√		
NEM	√		√	√
P^3	√	√	√	√

$$r_{g_i}^{k+1} = \beta r_{g_i}^0 + (1-\beta)\left(\sum_{g_j \in N(g_i)} \frac{w_{i,j}}{\pi_{g_i}} r_{g_j}^k\right). \quad (2)$$

Here, $\pi_{g_i} = \sum_{j=1}^{n} w_{i,j}$ is the total weight of interactions involving gene g_i. k in the equation indicates the number of iterations. After convergence, we obtain rank scores of all genes with respect to phenotype p. The above procedure can be repeated for every phenotype to obtain the corresponding list of rank scores of all genes. The list of rank scores of genes to a phenotype corresponds to a vector $\overline{\mathcal{R}_{p_i}} =< r_{g_1}, ..., r_{g_n} >$, where r_{g_k} is the rank score of g_k. Let $C_{m\times m}$ denote a "closeness" matrix of phenotypes, where both rows and columns are phenotypes, and each entry $c_{i,j}$ stores the closeness value from phenotype p_j to p_i. It is defined as the p_i's average rank scores of genes causing p_j. The formula is given in Equation 3, where $\overline{p_j}^T$ is the transpose of $\overline{p}_j$.

$$c_{i,j} = \overline{p}_j^T \times \overline{\mathcal{R}_{p_i}} \quad (3)$$

Note that this matrix is not necessarily symmetric, since the rank score of a gene to a phenotype depends on the scores of its neighbors, but for two phenotypes p and q, genes involved in phenotype p may not have the same neighbors as genes involved in phenotype q. For simplicity, the diagonal of the matrix is set to 0, because the closeness of a phenotype to itself is not of interest. This matrix thus maps to a weighted directed graph, such as seen in Figure 1 (right), where nodes are phenotypes, and the weight of the directed edge from phenotype p_i to phenotype p_j is $c_{i,j}$. After the whole matrix C is computed, prediction is carried out using this matrix.

3.2. Predicting Phenotype Effects of Gene Perturbations

Algorithms for ranking genes to a phenotype and ranking phenotypes for a gene using the phenotype graph are described below.

3.2.1. *Ranking Genes for a Phenotype*

Given a phenotype p, suppose that there is a gene g which is known to result in phenotypes $\{q_1, ...q_k\}$. The closeness of phenotype q_i, $1 \le i \le k$, to p is the weight of the edge from p to q_i in the phenotype graph. There are multiple ways to define the rank score of a gene g to the phenotype p, for example, we can utilize the maximum closeness from q_i, $1 \le i \le k$, to p. Here, we used the average closeness from known phenotypes of the gene to the target phenotype. The rank scores of all genes to all target phenotypes can be calculated simultaneously by a simple matrix computation, as shown in Equation 4.

$$RG = V' \times C \quad (4)$$

V', with entries $v'_{i,j} = \frac{v_{j,i}}{\sum_{k=1}^{m} v_{k,i}}$, is obtained by transposing the phenotype-gene matrix V and dividing each entry by the number of 1s in the corresponding row. RG is thus an $n \times m$ matrix, where rows are genes and columns are phenotypes, and the value of each cell is the rank score of the gene to the corresponding phenotype.

3.2.2. *Ranking Phenotypes for a Gene*

Given a gene g, assume that it is known to result in phenotypes $\{q_1, ..., q_k\}$. For any other phenotype p in the phenotype graph, the closeness from p to phenotype q_i, $1 \le i \le k$ is the weight of the edge from q_i to p. The method of ranking phenotypes to a gene is very similar to ranking genes for a phenotype, described above. In ranking genes, the weights on the edges incident on phenotypes $\{q_1, ..., q_k\}$ are used, but in ranking phenotypes, the edges outgoing from phenotypes $\{q_1, .., q_k\}$ are considered. The rank score of phenotype p to gene g is the average of the closeness values from p to phenotypes $\{q_1, ..., q_k\}$. Analogously as stated earlier, rank scores of all phenotypes to all genes can be computed at the same time. Equation 5 describes the method, where RG is the resulting rank score matrix.

$$RP = V' \times C^T \quad (5)$$

The only difference from the method to ranking genes is that the transpose of the closeness matrix is used here.

4. EXPERIMENTAL RESULTS

We illustrate the effectiveness of our methodology by comparing it to the Direct method (as used in Lee et al. [7]) and GeneRank [9] on two real datasets: deletion mutants on yeast and an RNAi study of early embryogenesis in the *C. elegans* nematode. We further analyze the phenotype graphs derived by clustering phenotypes with high closeness values and present a biological interpretation.

4.1. Data

Two datasets are used in this study: the dataset of *C. elegans* RNAi induced early embryo defects [4] and the yeast knockout dataset from the Munich Information Center for Protein sequences (MIPS) database [15].

We focus on 45 RNAi induced defect categories in the *C.elegans* early embryo (data available in [16]) and use an interaction network extracted from Wormnet [7]. The original core Wormnet contains 113,829 connections and 12,357 genes. To compare with the Direct and GeneRank methods, we select genes resulting in at least two early embryo defects and interacting with at least one other gene, and retain all interactions between them in Wormnet. To evaluate the applicability of P^3 on predicting phenotypes for genes without interactions, we prepare another gene set that retains genes without any interactions.

In the yeast data, the underlying network involves protein-protein interactions, and is built by combining the yeast protein interaction data from several sources (CYGD [17], SGD [18], and BioGrid [19]). Phenotypes and genes are selected according to the same criteria as above. The statistics of these datasets are listed in Table 3.

4.2. Experiment Setup

We implement the Direct method and use the log-likelihood value of each interaction published with Wormnet as the edge weights for the *C. elegans* network. For a given phenotype, genes known to result in that phenotype are considered as the seed set. The rank score of other genes are the sum of the log-likelihoods of interactions to the seed set. In the case of yeast, we simply set the same weight on all interactions.

Table 3. Statistics of datasets used in this work.

Organism	Genes	Interactions	Phenotypes
Caenorhabditis elegans	420	6677	45
Saccharomyces cerevisiae	1232	13228	72

In addition to the connectivity matrix of the network, GeneRank has another input, namely the expression changes vector, which is used to set initial ranks. In our case, we use a binary phenotype signature vector, where 1 means that the corresponding gene is known to show that phenotype, 0 otherwise. There is also a parameter d that determines relative weights of expression changes and connectivity information to the rank value. We tried multiple values on d from 0.1 to 0.9 with interval 0.1, and chose the one gives optimal prediction results in performance comparison (0.1). The implementation published with the original paper is used.

To compare with the above methods, the algorithm for ranking genes for a given phenotype is applied. Another algorithm to ranking phenotypes for a given gene is used to predict phenotypes for genes without any interactions. There is one parameter β in P^3 to derive relations between phenotypes. We studied different values on β from 0.1 to 0.9 with step 0.1, and found that 0.6 gives the best performance. We used 0.6 in all the experiments described below.

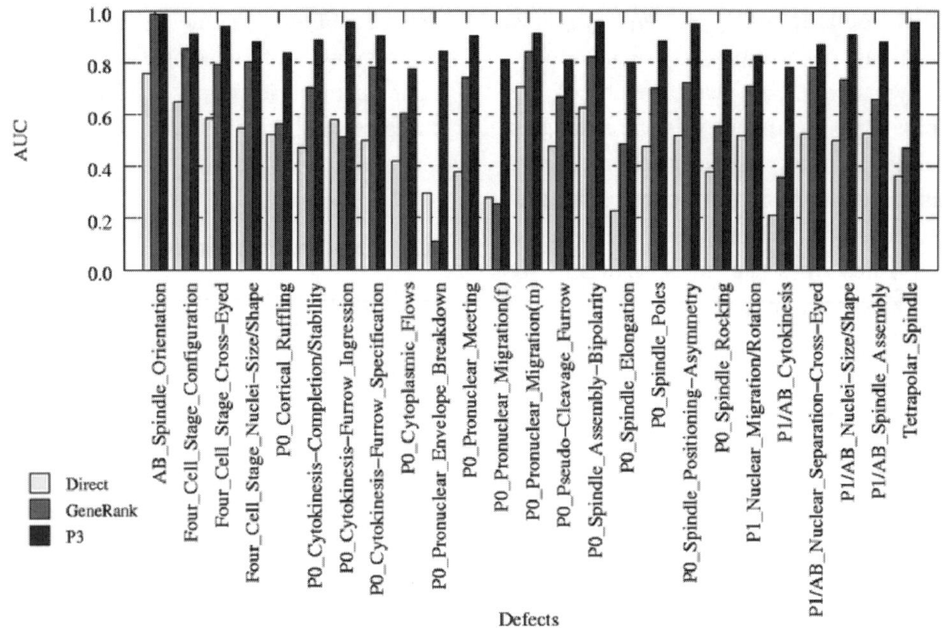

Fig. 2. Overall performance comparison on the *C. elegans* dataset. Direct : ranking genes using the interaction network only; GeneRank : $d = 0.1$; P^3 : $\beta = 0.6$

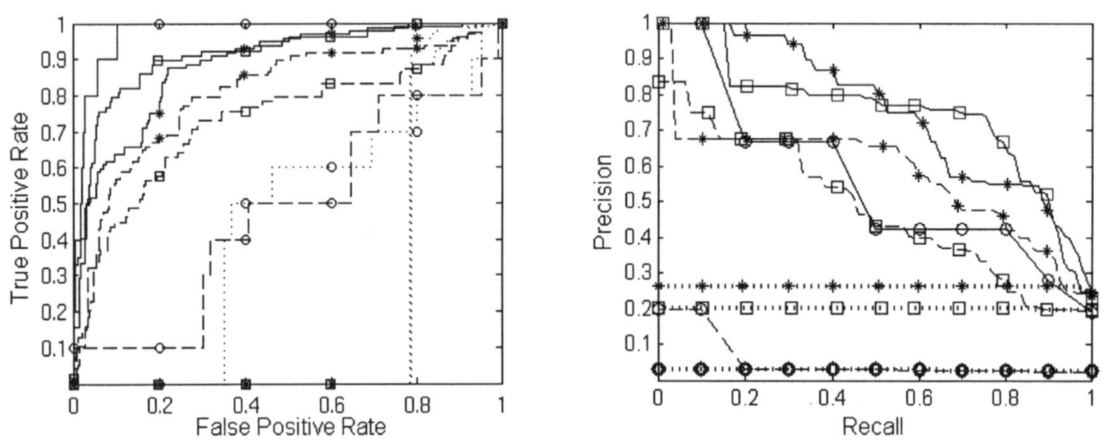

Fig. 3. (left) ROC curves on *C. elegans*. (right) Precision vs. Recall on *C. elegans*. Direct: points, GeneRank: dashed line, P^3: solid line; square: P1-AB Nuclear-Size-Shape, star: Four Cell Stage Nuclei-Size-Shape, circle: Tetrapolar Spindle.

4.3. Results

To evaluate the prediction performance for each phenotype, we used the leave-one-out and k-fold cross validation approaches. For the leave-one-out approach, one gene/phenotype pair is ignored from the original dataset each time, and the prediction algorithm is applied on the remaining dataset to see if that gene/phenotype pair is predicted correctly. Results show that our method outperforms the direct method and GeneRank method almost in all cases. We compared the Area Under the Receiver Operating Characteristic (AUC ROC) curve for each phenotype and plot the ROC curve and Precision-Recall curves for some phenotypes for further performance comparison. For k-fold cross validation, the original gene/phenotype pairs are separated into k groups, 10 in *C. elegans* and 5 in yeast, one of them is selected as test data and the remaining are used as training data. The distributions of AUC were compared. P^3 outperforms other methods in all cases. In the experiment of predicting phenotypes to genes without any interactions, results show that P^3 is able to predict a majority of these phenotypes with high accuracy.

4.3.1. *Leave-One-Out*

C. elegans: For each phenotype prediction, we computed true-positive rate versus false-negative rate to measure the recovery of genes with the given phenotype. The comparison of the area under the Receiver Operating Characteristic curve for each phenotype is shown in Figure 2. For visualization purpose, 20 defects are randomly selected for discussion here. The defect "AB Spindle Orientation" shows the highest AUC in the results of all three methods, with values of 0.99 in P^3 and GeneRank, and 0.76 in Direct method. P^3 is always better than the Direct method and outperforms the GeneRank method in most cases. The AUCs of P^3 are greater than those of Direct method and GeneRank by 0.37 and 0.2, in average respectively, and the maximum differences are 0.6 and 0.73, respectively. Only three defects, "Egg Size/Shape", "AB Spindle Orientation" and "P1/AB Cortical Activity" show that GeneRank method is slightly better than P^3, with the maximum difference of AUC as 0.028. Three phenotypes, "Tetrapolar Spindle", "Four Cell Stage Nuclei-Size-Shape", and "P1-AB Nuclear-Size-Shape", that have both high AUC and precision-recall for P^3 were chosen for further comparison. Figure 3 (left) shows their ROC curves, and the corresponding precision-recall curves are shown in Figure 3 (right).

Yeast: Similar to the study in *C. elegans*, we computed true-positive rate versus false-negative rate and precisions at certain recall levels. The comparison of the area under the Receiver Operating Characteristic curve for each phenotype is shown in Figure 4. For simplicity, we show the results for 28 phenotypes among the 72 examined phenotypes. The highest AUC in the selected results of P^3 is 0.98, from "Cell wall-Hygromycin B", that of the direct method is about 0.81, from "Peroxisomal mutants", and GeneRank has the highest AUC value about 0.88, from "Sensitivity to immunosuppressants". P^3 outperforms GeneRank and Direct method in most cases. The AUCs of P^3 are greater than those of Direct method and GeneRank by 0.4 and 0.2 in average respectively, and the maximum differences are 0.6 and 0.8 respectively. Three phenotypes that have both high AUC values and precisions among the result of P^3 method were chosen for further comparison. They are "Conditional phenotypes", "Carbon utilization", and "Cell morphology and organelle mutants". Figure 5 (left) shows their ROC curves, and the corresponding precision-recall curves are shown in Figure 5 (right).

4.3.2. *k-Fold Cross Validation*

C. elegans. 10-fold cross validation was carried out on *C. elegans* data. Figure 6 shows the distributions of AUC values of each method. The median, lower quantile and upper quantile of each group is plotted. As is evident, the performance is considerably improved by using P^3 for phenotype prediction.

Yeast. 5-fold cross validation was carried out on the yeast data. Figure 7 shows the comparison of distributions of AUC. The median, lower quantile and upper quantile of each group is plotted. P^3 outperforms the other two methods in all cases.

4.3.3. *Predicting Phenotypes to Genes Without Any Interactions*

To evaluate our approach in predicting phenotypes for genes without any interaction information, we identified those genes that have at least two phenotypes but without interactions in both datasets. We used the phenotype graphs obtained in the leave-one-out experiment, that were derived without any information about the test genes. The target gene/phenotype pairs are separated almost equally into two groups: one for training and another for testing. For example, for each gene, if it has two phenotypes then one is in the training group and another is in the test group. Results show that P^3 can predict most of the phenotypes successfully. Table 4 presents the characteristics of the data and results.

Table 4. Predicting phenotypes for genes without interactions.

Organism	Genes	Predicted with AUC $\geq$ 0.8
Caenorhabditis elegans	42	24
Saccharomyces cerevisiae	48	46

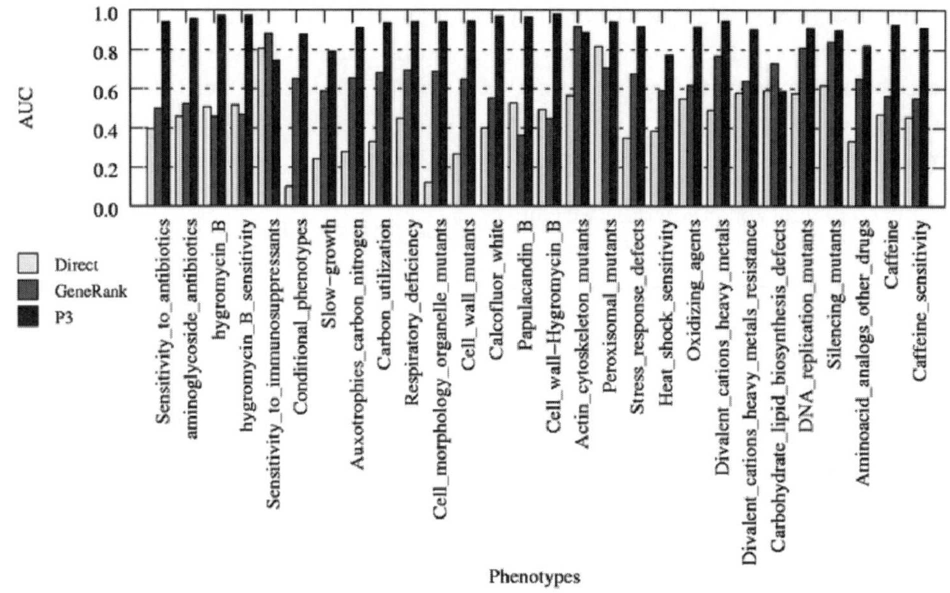

Fig. 4. Overall performance comparison on yeast phenotype dataset. Direct : ranking genes using the interaction network only; GeneRank : $d = 0.1$; P^3 : $\beta = 0.6$

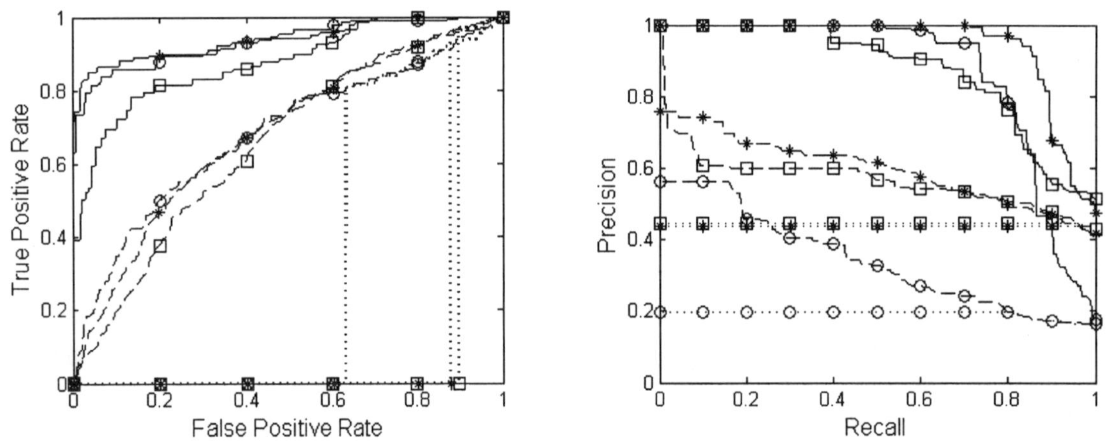

Fig. 5. (left) ROC curves on yeast. (right) Precision vs. Recall on yeast. Direct: points, GeneRank: dashed line, P^3: solid line; circle: Carbon utilization , square: Conditional phenotypes, star: Cell morphology and organelle mutants.

4.4. Phenotype Relations

The complete directed graph of phenotypes is too complex to describe in detail here. Therefore, we partition the graph into several highly connected subgraphs by using the CAST [20] algorithm. CAST is a heuristic approach for solving the 'corrupted cliques' problem. It transforms an undirected graph into a set of cliques or almost cliques by repetitively adding nodes having maximum average similarity to the current clique, as long as the similarity is above a threshold λ, and removing nodes with minimum average similarity to the clique, when the similarity is less than the threshold. The process stops when there are no more nodes to add or remove. First, directions are removed from the edges in the original phenotype graph. For each pair of phenotypes, two directed edges are merged into one undirected edge. Every new edge is assigned a new weight that is the average of weights of the original two edges. The graph is further adjusted by deletions of "weak" connections between phenotypes. For example, if the

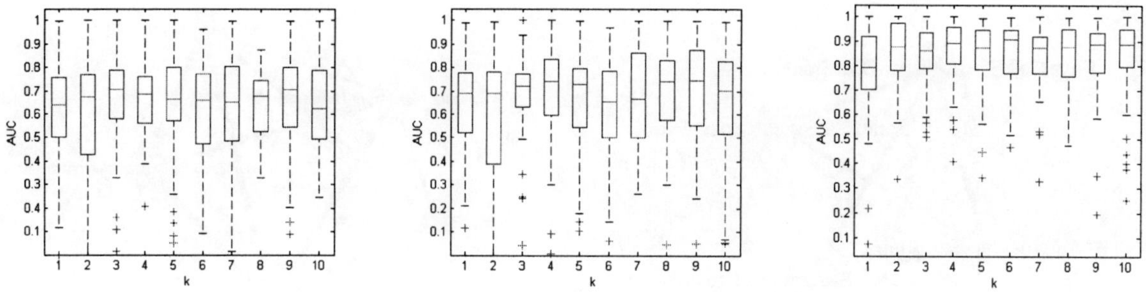

Fig. 6. AUC distributions on *C. elegans*. Direct method (left), GeneRank method (middle), and P^3 (right).

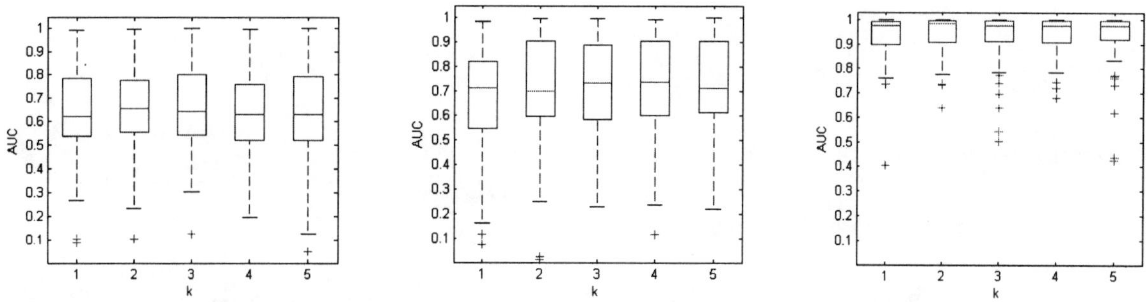

Fig. 7. AUC distributions on yeast. Direct method (left), GeneRank method (middle), and P^3 (right).

weight of the connection between phenotype p to q is less than a threshold t, then the corresponding edge is removed. We run the CAST algorithm on this simplified graph. A set of cliques and almost cliques are obtained. Each clique/almost clique is a cluster of single or a set of highly related phenotypes. Genes causing these phenotypes tend to interact or function together. Figure 8 and Figure 9 show some of the phenotype cliques obtained. The thickness of links represents the closeness between phenotypes. Multiple values are used for parameter t and λ. As t and λ decrease, the number of cliques decreases and the size of the maximum clique increases. We choose the parameter values that give small cliques so that they are relatively easy to interpret biologically. In *C. elegans*, there are 23 cliques/almost cliques, and the largest clique contains 11 nodes, one clique with 5 nodes, 4 nodes, and 3 nodes respectively, three cliques with 2 nodes, and the rest are singletons. In yeast, there are 41 cliques/almost cliques, and the largest clique contains 11 nodes, one clique with 4 nodes, six with 3 nodes, and six with 2 nodes, the remaining are singletons.

The *C. elegans* phenotypes identified in Figure 8 are all related to cell division. The edges suggest that there are distinct relationships between the formation and behavior of the nuclei, indicative of a functional role for structural proteins. The role of structural proteins, acting as conduits for macromolecular and organellar movement can also be seen in the largest clique where cytokinesis (splitting of the cytoplasm to form two cells) and furrow formation (where the cells are divided in half) are related.

The larger yeast clique in Figure 9 pertain to drug sensitivities, including antibiotics. Such associations could potentially be reflective of the role of the extracellular domain in resistance or non-resistance to select antibiotics. Inasmuch, caffeine sensitivity has been related to the synthesis of phospholipids (cell membrane components) and changes in calcium flux. Indeed, the smaller clique relates all of these concepts through sensitivity to immunosuppresants, a sensitivity that is related to phosphorylation-based signal transduction cascades.

5. DISCUSSION

In this paper, we have presented an approach to modeling phenotype relations and using these rela-

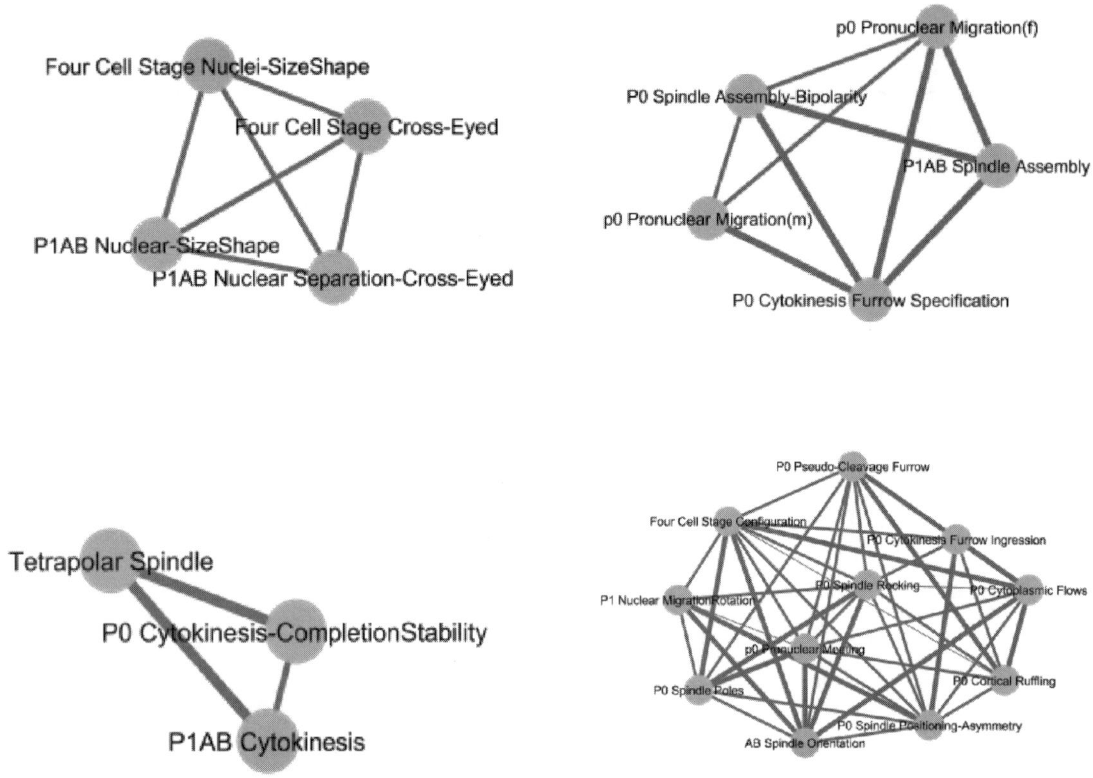

Fig. 8. Phenotype cliques in the *C. elegans* dataset derived from P^3.

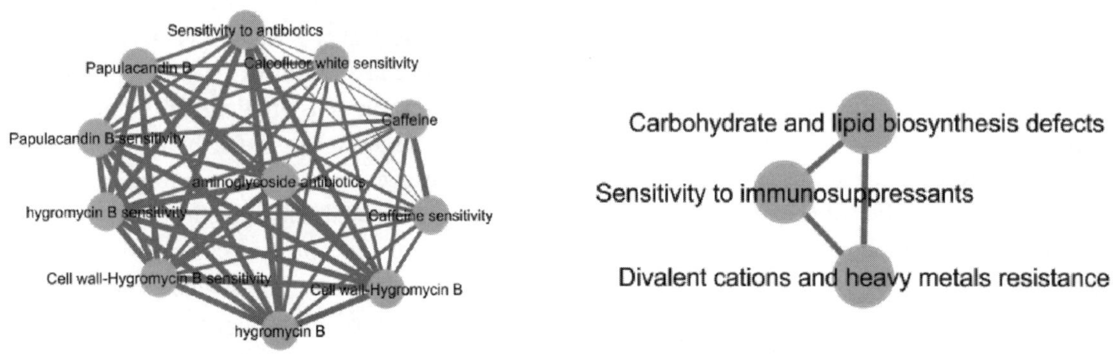

Fig. 9. Phenotype cliques in the *S. cerevisiae* dataset derived from P^3.

tionships to predict phenotypes for uncharacterized genes. The strong results indicate that the combination of gene networks and phenotype profiles provides a powerful synergy that is not obtainable with either method alone. One limitation is that to be able to make predictions, a gene should have at least one known phenotype. In future work, we seek to capture more complex many-many effects between genes and phenotypes and design new experiments to validate the predictions made.

Acknowledgments

This work is supported in part by US NSF grant ITR - 0428344.

References

1. Scherens, B., Goffeau, A.: The uses of genome-wide yeast mutant collections. Genome Biol 5(7) (2004)
2. Ohya, Y., *et al.*: High-dimensional and large-scale phenotyping of yeast mutants. PNAS 102(52) (December 2005) 19015–19020
3. Piano, F., *et al.*: Gene clustering based on RNAi phenotypes of ovary-enriched genes in C. elegans. Curr Biol 12(22) (November 2002) 1959–1964
4. Sonnichsen, B., *et al.*: Full-genome RNAi profiling of early embryogenesis in Caenorhabditis elegans. Nature 434(7032) (2005) 462–469
5. Oti, M., Snel, B., Huynen, M.A., Brunner, H.G.: Predicting disease genes using protein-protein interactions. J Med Genet 43(8) (2006) 691–698
6. Lage, K., *et al.*: A human phenome-interactome network of protein complexes implicated in genetic disorders. Nature Biotechnology 25(3) (2007) 309–316
7. Lee, I., *et al.*: A single gene network accurately predicts phenotypic effects of gene perturbation in Caenorhabditis elegans. Nature Genetics 40(2) (2008) 181–188
8. Ma, X., Lee, H., Sun, F.: CGI: a new approach for prioritizing genes by combining gene expression and proteinprotein interaction data. Bioinformatics 23(2) (2007) 215–221
9. Morrison, J., Breitling, R., Higham, D., Gilbert, D.: GeneRank: Using search engine technology for the analysis of microarray experiments. BMC Bioinformatics 6(1) (2005)
10. Markowetz, F., Kostka, D., Troyanskaya, O.G., Spang, R.: Nested effects models for high-dimensional phenotyping screens. Bioinformatics 23(13) (2007) i305–312
11. Herlocker, J.L., Konstan, J.A., Borchers, A., Riedl, J.: An algorithmic framework for performing collaborative filtering. In Proceedings of the 22nd annual international ACM SIGIR conference on Research and development in information retrieval, ACM Press (1999) 230–237
12. White, S., Smyth, P.: Algorithms for estimating relative importance in networks. In Proceedings of the 9th ACM SIGKDD international conference on Knowledge discovery and data mining, ACM Press (2003) 266–275
13. Page, L., Brin, S., Motwani, R., Winograd, T.: The PageRank Citation Ranking: Bringing Order to the Web. Technical report, Stanford Digital Library Technologies Project (1998)
14. Kleinberg, J.M.: Authoritative sources in a hyperlinked environment. JACM 46(5) (1999) 604–632
15. Mewes, H. W., *et al.*: MIPS: analysis and annotation of proteins from whole genomes in 2005. Nucleic Acids Res 34(Database issue) (2006)
16. Pati, A., Jin, Y., Klage, K., Helm, R.F., Heath, L.S., Ramakrishnan, N.: CMGSDB: integrating heterogeneous Caenorhabditis elegans data sources using compositional data mining. Nucleic Acids Res 36(Database issue) (2008)
17. Morrison, J., Breitling, R., Higham, D., Gilbert, D.: CYGD: Comprehensive Yeast Genome Database. BMC Bioinformatics 6(1) (2005)
18. Saccharomyces Genome Database, http://www.yeastgenome.org/
19. BioGrid, http://www.thebiogrid.org
20. Ben-Dor, A., Shamir, R., Yakhini, Z.: Clustering Gene Expression Patterns. J Comput Biol. 6(3/4) (1999) 281–297

CONSISTENT ALIGNMENT OF METABOLIC PATHWAYS WITHOUT ABSTRACTION

Ferhat Ay*[1], Tamer Kahveci[1], Valerie de Crécy-Lagard[2]

[1]*Department of Computer Science and Engineering, University of Florida,*
[2]*Department of Microbiology and Cell Science, University of Florida,*
Gainesville, FL 32611, USA
Email: {fay, tamer}@cise.ufl.edu, vcrecy@ufl.edu

Pathways show how different biochemical entities interact with each other to perform vital functions for the survival of organisms. Similarities between pathways indicate functional similarities that are difficult to identify by comparing the individual entities that make up those pathways. When interacting entities are of single type, the problem of identifying similarities reduces to graph isomorphism problem. However, for pathways with varying types of entities, such as metabolic pathways, alignment problem is more challenging. Existing methods, often, address the metabolic pathway alignment problem by ignoring all the entities except for one type. This kind of abstraction reduces the relevance of the alignment significantly as it causes losses in the information content. In this paper, we develop a method to solve the pairwise alignment problem for metabolic pathways. One distinguishing feature of our method is that it aligns reactions, compounds and enzymes without abstraction of pathways. We pursue the intuition that both pairwise similarities of entities (homology) and their organization (topology) are crucial for metabolic pathway alignment. In our algorithm, we account for both by creating an eigenvalue problem for each entity type. We enforce the consistency by considering the reachability sets of the aligned entities. Our experiments show that, our method finds biologically and statistically significant alignments in the order of seconds for pathways with $\sim$ 100 entities.
Keywords: metabolic pathway alignment, metabolic reconstruction, alternative enzyme identification

1. INTRODUCTION

One of the fundamental goals of biology is to understand the biological processes that are the driving forces behind organisms' functions. To achieve this goal, interactions between different components that build up metabolism should be examined in detail. These interactions can reveal significant information that is impossible to gather by analyzing individual entities. Recent advances in high throughput technology resulted in an explosion of different types of interaction data which is compiled in databases, such as KEGG[1] and EcoCyc[2]. Analyzing these databases is necessary to capture the valuable information carried by the pathways. An essential type of analysis is the comparative analysis which aims at identifying similarities between pathways of different organisms. Finding these similarities provides valuable insights for drug target identification[3], metabolic reconstruction of newly sequenced genome[4], and phylogenetic tree construction[5].

To identify similarities between two pathways, it is necessary to find a mapping of their entities. This problem is computationally interesting and challenging. Using a graph model for representing pathways, the graph/subgraph isomorphism problems can be reduced to global/local pathway alignment problems in polynomial time. However, since the graph and subgraph isomorphism problems are GI-complete and NP-complete respectively, global/local pathway alignment problems are GI/NP complete. Hence, efficient heuristics are needed to solve these problems in a reasonable time.

In order to reduce the time complexity of the alignment, some existing algorithms restrict the topology of query pathways[6, 7]. For instance, the method proposed by Tohsato *et al.*[7] works for only non-cyclic pathways, whereas the algorithm of Pinter *et al.*[8] restricts the query pathways to multi-source trees. However, those restrictions are far from the reality and they limit the applicability of the methods to only a small percentage of pathways.

A common delusion of existing algorithms for metabolic pathway alignment is to use a model that focuses on only one type of entity and ignores the others. This simplification converts metabolic pathways to the graphs with only compatible nodes. We use the word *compatible* for the entities that are of the same type. For metabolic pathways, two entities are compatible if they both are reactions or enzymes or compounds. We term the conversions that reduces the metabolic pathways to compatible entities as *abstraction*. Previously, reaction based[5], enzyme based[8, 9] and compound based[7] abstractions

*Corresponding author.

are used for representing metabolic pathways. Figure 1 illustrates the problems with the enzyme based abstraction used by Pinter *et al.*[8] and Koyutürk *et al.*[9]. In Figure 1(a), enzymes E_1 and E_2 interact on two different paths. Abstraction in Figure 1(b) loses this information and merges these two paths into a single interaction. After the abstraction, an alignment algorithm aligning the $E_1 \rightarrow E_2$ interactions in Figures 1(a) and 1(b) cannot realize through which path, out of two alternatives, the enzymes E_1 and E_2 are aligned. It is important to note that the amount of information lost due to abstraction grows exponentially with the number of branching entities.

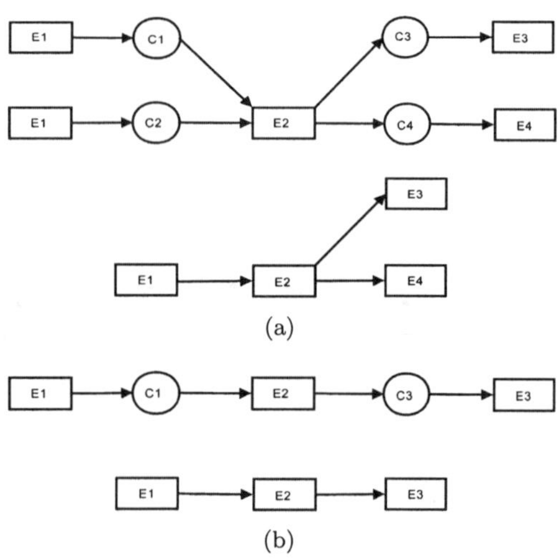

Fig. 1. Top figures in (a) and (b) illustrate two hypothetical metabolic pathways with enzymes and compounds represented by letters E and C, respectively. Bottom figures in (a) and (b) show the same pathways after abstraction when the compounds are ignored.

This paper addresses the pairwise alignment problem for metabolic pathways without any topology restriction or any abstraction. A distinguishing feature of our method is that reported alignments provide the individual mappings for reactions, compounds and enzymes. Furthermore, our algorithm can be extended to work for other pathway types that have entities from different compatibility classes.

In our method, we account for both pairwise and topological similarities of the entities since they both are very crucial for alignment. Singh *et al.*[10] combined homology and topology for protein interaction pathway alignment by creating an eigenvalue problem. A similar approach is previously used for discovery of authoritative information sources on the World Wide Web by Kleinberg[11]. In the case of protein interaction pathways, the alignment problem can be mapped to a single eigenvalue problem since all nodes are of the same type and interactions between them are assumed to be undirected. The algorithm proposed by Singh *et al.*, however, cannot be trivially extended to metabolic pathways as these pathways contain entities of varying types and the interactions are directed.

For metabolic pathway alignment, we first create three eigenvalue problems, one for compounds, one for reactions and one for enzymes. We, also, consider the directions of interactions. We solve these eigenvalue problems using power method. The principal eigenvectors of each of these problems define a weighted bipartite graph. We, then, extract reaction mappings using maximum weight bipartite matching on the corresponding bipartite graph. After that, to ensure consistency of the alignment, we prune the edges in the bipartite graphs of compounds and enzymes which lead to inconsistent alignments with respect to reaction mappings. Finally, we find the enzyme and the compound mappings using maximum weight bipartite matching. We report the extracted mappings of entities as an alignment together with a similarity score that we devise for measuring the similarity between the aligned pathways. Furthermore, we measure the unexpectedness of the resulting alignment by calculating its z-score.

Our experiments on KEGG Pathway database show that our algorithm successfully identifies functionally similar entities and sub-paths in pathways of different organisms. Our method produces biologically and statistically significant alignments of pathways very quickly.

Our Contributions:

- We introduce the *consistency* concept for alignment of pathways with different entity types by constructing reachability sets. We develop an algorithm that aligns pathways while enforcing consistency.
- We integrate the graph model that we devised earlier[3] into the context of pathway alignment. Using this model, we develop an algorithm to align pathways when there is no *abstraction*. Unlike existing graph models, this model is a nonredundant representation of pathways without any abstraction.

• We introduce a new scoring scheme for measuring the similarity of two reactions. We also devise a similarity score and a z-score for measuring similarities between two metabolic pathways.

The organization of the rest of this paper is as follows: Section 2 discusses the related work. Section 3 presents our graph model for representing pathways. Section 4 describes the proposed algorithm in detail. Section 5 illustrates the experimental results. Section 6 briefly concludes the paper.

2. BACKGROUND

Pathway alignment problem has been mostly considered for protein interaction networks (PPI). As a result, existing methods often can align two pathways only if all the interacting entities are of the same type[6, 10, 12, 13]. However, metabolic pathways are composed of enzymes, reactions, compounds and interactions between these three types of entities. Therefore, it is not trivial how PPI alignment methods can be extended to align metabolic pathways.

For solving the metabolic pathway alignment problem, existing methods model the pathways as interactions between entities of a single type. This abstraction causes significant information loss as seen in Figure 1. After this abstraction in modeling, a common approach for aligning metabolic pathways is to use graph theoretic techniques. Pinter *et al.*[8] mapped the metabolic pathway alignment problem to the subgraph homomorphism problem. However, they oversimplify the problem by restricting the topology of pathways to multi-source trees. By solely relying on Enzyme Commission (EC)[14] numbers, Tohsato *et al.*[15] proposed an alignment method for metabolic pathways in 2003. Due to the discrepancies in the EC hierarchy, the accuracy of this method is questionable. In 2007, they proposed another method[7], which only considers chemical structures of compounds for alignment. This idea, however, totally ignores the effect of other entities such as enzymes and reactions.

To overcome the above mentioned problems, in this paper, we refuse to use a model that is biased on one entity type. Equipped with a more comprehensive graph model without abstraction and an efficient iterative algorithm, our tool outperforms existing methods for metabolic pathway alignment.

3. MODEL

The first step in developing effective computational techniques to leverage metabolic pathways is to develop an accurate model to represent them. Existing graph models are not sufficient for representing all interactions between different entity types that are present in metabolic pathways. Figure 1 emphasizes the importance of the modeling scheme for pathway alignment. As discussed in Section 2, abstractions in modeling reduce the alignment accuracy dramatically.

In order to address the insufficiency of existing models, we developed a graph model for representation of metabolic pathways. Our model is a variation of boolean network model and is able to capture all interactions between all types of entities. We discuss this model in the rest of this section.

For the rest of this paper, we will use $\mathcal{P}$, $\mathcal{R}$, $\mathcal{C}$, $\mathcal{E}$ to denote the sets of all pathways, all reactions, all compounds and all enzymes, respectively. Let, $R \subseteq \mathcal{R}, C \subseteq \mathcal{C}, E \subseteq \mathcal{E}$ such that $R = \{R_1, R_2, \ldots, R_{|R|}\}$, $C = \{C_1, C_2, \ldots, C_{|C|}\}$ and $E = \{E_1, E_2, \ldots, E_{|E|}\}$ denote the reactions, compounds and enzymes of the pathway P, respectively. The definition below formalizes our graph model:

Definition 1. A directed graph, $G(V, I)$ for representing the metabolic pathway $P \in \mathcal{P}$, is constructed as follows: The node set, $V = [R, C, E]$, is the union of reactions, compounds and enzymes of P. The edge set, I, is the set of interactions between the nodes. An interaction is represented by a directed edge that is drawn from a node x to another node y, if and only if one of the following three conditions holds:

1) x is an enzyme that catalyzes reaction y.
2) x is an input compound of reaction y.
3) x is a reaction that produces compound y.

Figure 2 illustrates the conversion of a KEGG metabolic pathway to our graph model. As suggested, our model is capable of representing metabolic pathways without losing any type of entities or interactions between these entities. We avoid any kind of abstraction in alignment by using this model. Besides, our model is a nonredundant representation of pathways since it represents each entity using a single node.

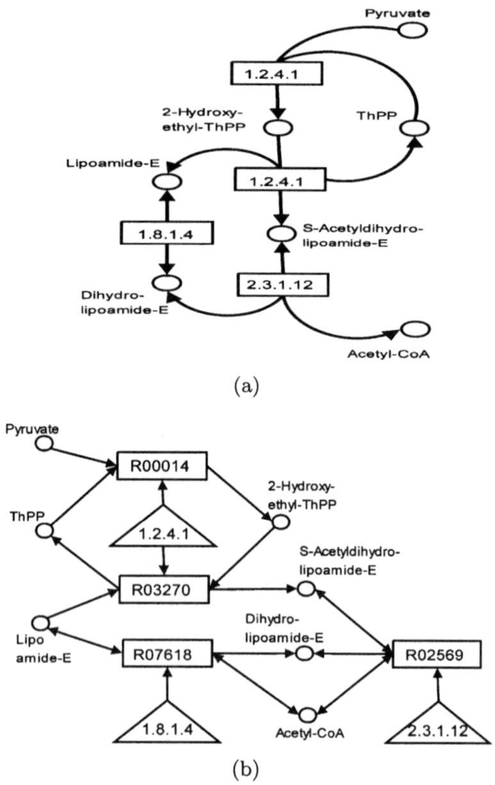

Fig. 2. Graph representation of metabolic pathways: (a) A portion of the reference pathway of Alanine and aspartate metabolism from KEGG database (b) Our graph representation corresponding to this portion. Reactions are shown by rectangles, compounds are shown by circles and enzymes are shown by triangles.

4. ALGORITHM

Motivated by previous research on alignment of pathways and growing demand on fast and accurate tools for analyzing biological pathways, in this section we describe our algorithm for pairwise alignment of metabolic pathways. Before going into the details of the algorithm, it is better to formally state the alignment problem. To do this we first need to define an alignment and the consistency of an alignment.

Let, $P, \bar{P} \in \mathcal{P}$ stand for the two query metabolic pathways which are represented by graphs $G(V, I)$ and $\bar{G}(\bar{V}, \bar{I})$, respectively. Using our graph formalization V can be replaced by $[R, C, E]$, where R denotes the set of reactions, C denotes the set of compounds and E denotes the set of enzymes of P. Similarly, $\bar{V}$ is replaced by $[\bar{R}, \bar{C}, \bar{E}]$.

Definition 2. An *alignment* of two metabolic pathways $P = G(V, I)$ and $\bar{P} = \bar{G}(\bar{V}, \bar{I})$, is a *mapping* $\varphi : V' \rightarrow \bar{V}'$ where $V' \subseteq V$ and $\bar{V}' \subseteq \bar{V}$.

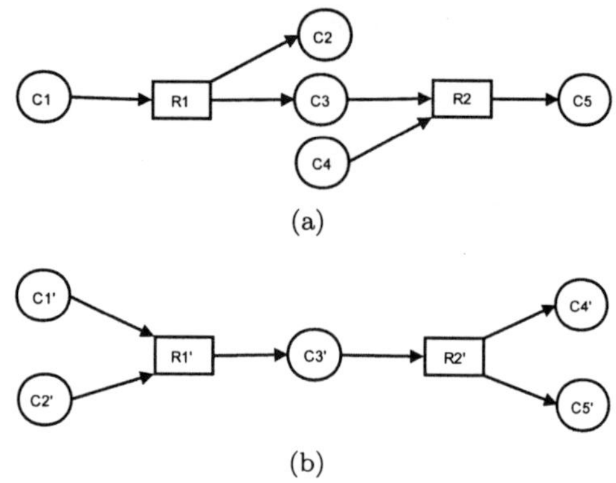

Fig. 3. *Consistency of an alignment and an example nonsensical matching:* Figures in (a) and (b) are graph representations of two query pathways. Enzymes are not displayed for simplicity. Suppose that our alignment algorithm mapped the reactions R1 to R1' and R2 to R2'. In this scenario, a trivial consistent matching is C1-C1'. An example for a nonsensical matching that cause inconsistency is C2' - C5. When C1 is matched to C1', a consistent matching might be C2' - C4 since they are inputs of two neighbor reactions.

Before arguing the consistency of an alignment, we discuss the reachability concept for entities. Given two entities $v_i, v_j \in V$ which are compatible, v_j is *reachable* from v_i if there is a directed path from v_i to v_j in graph G. As a shorthand notation, $v_i \Rightarrow v_j$ denotes that v_j is reachable from v_i.

Using the definition and the notation above, we define a consistent alignment as follows:

Definition 3. An alignment of two pathways $P = G(V, I)$ and $\bar{P} = \bar{G}(\bar{V}, \bar{I})$ defined by the mapping $\varphi : V' \rightarrow \bar{V}'$ is *consistent* if and only if all the conditions below are satisfied:

- For all $\varphi(v) = \bar{v}$ where $v \in V$ and $\bar{v} \in \bar{V}$, v and $\bar{v}$ are compatible.
- $\varphi(v)$ is one-to-one.
- For all $\varphi(v_i) = \bar{v_i}$, there exists $\varphi(v_j) = \bar{v_j}$ where $v_i, v_j \in V$ and $\bar{v_i}, \bar{v_j} \in \bar{V}$, such that $v_i \Rightarrow v_j$ and $\bar{v_i} \Rightarrow \bar{v_j}$, or $v_j \Rightarrow v_i$ and $\bar{v_j} \Rightarrow \bar{v_i}$.

The first condition filters out matchings of different entity types. The second condition ensures that none of the entities are mapped to more than one entity. The last condition restricts the mappings to the ones which are supported by at least one other mapping. Additionally, it eliminates nonsensical matchings that cause inconsistency as described in Figure 3.

Now, let, $SimP_{\varphi} : \mathcal{P} \times \mathcal{P} \rightarrow \Re \cap [0, 1]$ be a pairwise pathway similarity function, induced by the mapping φ. The maximum score, $SimP_{\varphi} = 1$, is achieved when two pathways are identical. In Section 4.5, we will describe in detail how $SimP_{\varphi}$ is computed after φ is created. In order to restate our problem, it is only necessary to know that there exists such a similarity function for pathways.

Under the light of the above definitions and formalizations, here is the *problem statement* for pairwise metabolic pathway alignment:

Definition 4. Given two metabolic pathways, $P = G(V, I)$ and $\bar{P} = \bar{G}(\bar{V}, \bar{I})$, the alignment problem is to find a consistent mapping $\varphi : V \rightarrow \bar{V}$ that maximizes $SimP_{\varphi}(P, \bar{P})$.

In the following sections, we describe our algorithm for metabolic pathway alignment.

4.1. Pairwise Similarity of Entities

Metabolic pathways are composed of entities which are either enzymes, compounds or reactions. The degree of similarity between pairs of entities of two pathways is a good indicator of the similarity between these pathways.

A number of similarity measures have been devised for each type of entity in the literature. In the rest of this section, we describe the similarity functions we used for enzyme and compound pairs. We also discuss the similarity function we developed for reaction pairs. All pairwise similarity scores are normalized to the interval of $[0, 1]$ to ensure compatibility between similarity scores of different entities.

Enzymes: An enzyme similarity function is of the form $SimE : \mathcal{E} \times \mathcal{E} \rightarrow \Re \cap [0, 1]$. In our implementation, the two options we provide the user for enzyme similarity scoring are:

• *Hierarchical enzyme similarity score*[15] depends only on Enzyme Commission (EC)[14] numbers of enzymes.

• *Information content enzyme similarity score*[8] uses EC numbers of enzymes together with the information content of this numbering scheme.

Compounds: Two different methods we use for compound similarity are:

• *A trivial compound similarity score* returns 1 if two compounds are identical and 0 otherwise.

• *SIMCOMP compound similarity score* for compounds is defined by Hattori *et al.*[16]. This score is assessed by mapping chemical structures of compounds to graphs and then measuring the similarity between these graphs.

Reactions: Our similarity score for reactions depends on the similarities of the components that take place in the reaction process such as enzymes, input compounds and output compounds. It is of the form $SimR : \mathcal{R} \times \mathcal{R} \rightarrow \Re \cap [0, 1]$. Our reaction similarity score employs the maximum weight bipartite matching technique. The following is a brief description of the maximum weight bipartite matching problem:

Definition 5. Let, U and V be two disjoint node sets and S be a $|U| \times |V|$ matrix representing edge weights between all possible pairs with one element from U and one element from V, where existing edges correspond to a nonzero entry in S. Maximum Weight Bipartite Matching problem is to find a list of node pairs, such that the sum of edge weights between the elements of these pairs is maximum. We denote this sum of edge weights by $MBS(U, V, S)$.

Let, R_i and R_j be two reactions from $\mathcal{R}$. Define R_i as a combination of input compounds, output compounds and enzymes and denote it by $[C_{in}^{R_i}, C_{out}^{R_i}, E^{R_i}]$, where $C_{in}^{R_i}, C_{out}^{R_i} \subseteq C$ and $E^{R_i} \subseteq E$. Similarly, define R_j as $[C_{in}^{R_j}, C_{out}^{R_j}, E^{R_j}]$. Additionally, compute the edge weight matrices $S_{C_{out}}$ and $S_{C_{in}}$ using the selected compound similarity score and S_E using the selected enzyme similarity.

The similarity score of (R_i, R_j) is computed as:

$$\begin{aligned} SimR(R_i, R_j) = {} & \gamma_{C_{in}} MBS(C_{in}^{R_i}, C_{in}^{R_j}, S_{C_{in}}) \\ & + \gamma_{C_{out}} MBS(C_{out}^{R_i}, C_{out}^{R_j}, S_{C_{out}}) \\ & + \gamma_E MBS(E^{R_i}, E^{R_j}, S_E) \qquad (1) \end{aligned}$$

Here, $\gamma_{C_{in}}, \gamma_{C_{out}}, \gamma_E$ denote the relative weights of input compounds, output compounds and enzymes on reaction similarity, respectively. Typical values for these parameters are $\gamma_{C_{in}} = \gamma_{C_{out}} = 0.3$ and $\gamma_E = 0.4$. These values are empirically determined after a number of experiments. One more factor that defines reaction similarity is the choice of $SimE$ and $SimC$ functions. Since we have two options for each, we end up having four different options for reaction similarity depending on the choices of $SimE$ and $SimC$.

Now, we can create the pairwise similarity vectors $\overrightarrow{H_R}^0$, $\overrightarrow{H_C}^0$, $\overrightarrow{H_E}^0$ for reactions, compounds and enzymes, respectively. Since, calculation of these vectors is very similar for each entity type we just describe the one for reactions.

The entry ${H_R}^0((i-1)|R|+j)$ of $\overrightarrow{{H_R}^0}$ vector stands for the similarity score between $R_i \in R$ and $\bar{R_j} \in \bar{R}$, where $1 \leq i \leq |R|$ and $1 \leq j \leq |\bar{R}|$. We will use the notation ${H_R}^0(i,j)$ for this entry since $\overrightarrow{{H_R}^0}$ can, also, be viewed as a $|R| \times |\bar{R}|$ matrix. One thing we need to be careful about is that $\overrightarrow{{H_R}^0}, \overrightarrow{{H_C}^0}, \overrightarrow{{H_E}^0}$ vectors should be of the unit norm. This normalization is crucial for stability and convergence of our alignment algorithm, as we will clarify in Section 4.2. We, therefore, compute an entry of $\overrightarrow{{H_R}^0}$ as:

$$ {H_R}^0(i,j) = \frac{SimR(R_i, \bar{R_j})}{||\overrightarrow{{H_R}^0}||_1} \tag{2} $$

In a similar fashion, we compute all entries of $\overrightarrow{{H_C}^0}, \overrightarrow{{H_E}^0}$ by using $SimC$ and $SimE$ functions, respectively. We use these three vectors to carry the homology information throughout the algorithm. In Section 4.3, we will describe how they are combined with topology information to produce an alignment.

4.2. Similarity of Topologies

Previously, we discussed why and how we use pairwise similarities of entities. However, although pairwise similarities are necessary, they are not sufficient. The induced topologies of the aligned entities should also be similar. In order to account for topological similarity, in this section, we define the notion of neighborhood for each compatibility class. After that, we create *support matrices* that allow us to exploit this neighborhood information.

To be consistent with our reachability definition, we define our neighborhood relations according to directions of interactions. In other words, we distinguish between *backward neighbors* and *forward neighbors* of an entity.

Let, $BN(x)$ and $FN(x)$ denote the backward and forward neighbor sets of an entity x. We need to show how to construct these sets for each entity type. We start by defining neighborhood of reactions to build backbones for topologies of the pathways. Then, using that backbone we define neighborhood concepts for compounds and enzymes.

Consider two reactions R_i and R_j of the pathway P. If an output compound of R_i is an input compound for R_j, then R_i is a backward neighbor of R_j and R_j is a forward neighbor of R_i. We construct the forward and backward neighbor sets of each reaction in this manner. For instance, in Figure 2(b), R02569 is a forward neighbor of R03270, and R03270 is a backward neighbor of R02569.

A more generalized version of neighborhood definition can be given to include not only instant neighbors but also neighbors of neighbors, and so on. However, it complicates the algorithm unnecessarily, since our method already considers the support of indirect neighbors as described in Section 4.3.

As stated before, neighborhood definitions of compounds and enzymes depend on the topology of reactions. Let, C_i and C_j be two compounds, R_s and R_t be two reactions of the pathway P. If $R_s \in BN(R_t)$ and C_i is an input (output) compound of R_s and C_j is an input (output) compound of R_t then $C_i \in BN(C_j)$ and $C_j \in FN(C_i)$. For example, in Figure 2(b), Lipoamide-E and Dihydro-lipoamide-E are neighbors since they are inputs of two neighbor reactions R02569 and R03270, respectively. For enzymes the construction is similar.

In the light of the above definitions, we create *support matrices* for each compatibility class. These matrices contain the information about topological similarities of pathways. In here, we only describe how to calculate the support matrix for reactions. The calculations for enzymes and compounds is done similarly.

Definition 6. Let, $P = G([R,C,E],I)$ and $\bar{P} = \bar{G}([\bar{R},\bar{C},\bar{E}],\bar{I})$ be two metabolic pathways. The support matrix for reactions of P and $\bar{P}$ is a $|R||\bar{R}| \times |R||\bar{R}|$ matrix denoted by A_R. An entry of the form $A_R[(i-1)|R|+j][(u-1)|R|+v]$ identifies the fraction of the total support provided by $R_u, \bar{R_v}$ matching to $R_i, \bar{R_j}$ matching. Let, $N(u,v) = |BN(R_u)||BN(\bar{R_v})|+|FN(R_u)||FN(\bar{R_v})|$ denote the number of possible neighbor matchings of R_u and $\bar{R_v}$.

Each entry of A_R is computed as:

$$ A_R[(i-1)|R|+j][(u-1)|R|+v] = $$

$$ \begin{cases} \frac{1}{N(u,v)} & \text{if } (R_i \in BN(R_u) \text{ and } \bar{R_j} \in BN(\bar{R_v})) \\ & \text{or } (R_i \in FN(R_u) \text{ and } \bar{R_j} \in FN(\bar{R_v})) \\ 0 & \text{otherwise} \end{cases} $$

After filling all entries, we replace the zero columns of A_R with $|R||\bar{R}| \times 1$ vector $[\frac{1}{|R||\bar{R}|}, \frac{1}{|R||\bar{R}|}, \ldots, \frac{1}{|R||\bar{R}|}]^T$. This way support of the matching indicated by the zero column is uniformly distributed to all other matchings.

For example, in Figure 1(a), $|BN(E2)| = 1$ and $|FN(E2)| = 2$ and in Figure 1(b), $|BN(E2)| = 1$ and $|FN(E2)| = 1$. Hence, the support of matching E2 of Figure 1(a) with E2 of Figure 1(b) should be equally distributed to its 3 (i.e. $1\times1+2\times1$) possible neighbor matching combinations by assigning 1/3 to the corresponding entries of A_E matrix.

We use the terms A_R, A_C and A_E to represent the support matrices for reactions, compounds and enzymes, respectively. Power of these support matrices is that, they allow us to distribute the support of a matching to other matchings according to the distances between them. This distribution is crucial for favoring the matchings whose neighbors can also be matched as well. The method for distributing the matching scores appropriately is described in the next section.

4.3. Combining Homology and Topology

Both the pairwise similarities of entities and the organization of these entities together with the interactions between them provide us precious information about the functional correspondence and evolutionary similarity of metabolic pathways. Hence, an accurate alignment strategy needs to combine these factors cautiously. In this subsection we describe our strategy for achieving this combination.

From the previous sections, we have $\overrightarrow{H_R}^0$, $\overrightarrow{H_C}^0$, $\overrightarrow{H_E}^0$ vectors containing pairwise similarities of entities and A_R, A_C, A_E matrices containing topological similarities of pathways. Using these vectors and matrices together with a weight parameter $\alpha \in [0,1]$, for adjusting the relative effect of topology and homology, we transform our problem into three eigenvalue problems as follows:

$$\overrightarrow{H_R}^{k+1} = \alpha A_R \overrightarrow{H_R}^k + (1-\alpha)\overrightarrow{H_R}^0 \quad (3)$$

$$\overrightarrow{H_C}^{k+1} = \alpha A_C \overrightarrow{H_C}^k + (1-\alpha)\overrightarrow{H_C}^0 \quad (4)$$

$$\overrightarrow{H_E}^{k+1} = \alpha A_E \overrightarrow{H_E}^k + (1-\alpha)\overrightarrow{H_E}^0 \quad (5)$$

for $k \geq 0$.

For stability purposes $\overrightarrow{H_R}^k, \overrightarrow{H_C}^k$ and $\overrightarrow{H_E}^k$ are normalized before each iteration.

Lemma 4.1. *A_R, A_C and A_E are column stochastic matrices.*

Proof. Each entry of A_R, A_C and A_E are nonnegative by Definition 6. By construction, entries of each column of these matrices sums up to one. □

Lemma 4.2. *Let, A be an $N \times N$ column stochastic matrix and E be an $N \times N$ matrix such that $E = \vec{H}e^T$, where $\vec{H}$ is an N-vector with $||\vec{H}||_1 = 1$ and e is an N-vector with all entries equal to 1. For any $\alpha \in [0,1]$ define the matrix M as:*

$$M = \alpha A + (1-\alpha)E \quad (6)$$

The maximal eigenvalue of M is $|\lambda_1| = 1$. The second largest eigenvalue of M satisfies $|\lambda_2| \leq \alpha$.

Proof. Omitted, see Haveliwala *et al.*[17] □

Using an iterative technique called power method, our aim is to find the stable state vectors of the Equations (3), (4) and (5). We know by Lemma 4.1 that A_R, A_C and A_E are column stochastic matrices. By construction of $\overrightarrow{H_R}^0, \overrightarrow{H_C}^0, \overrightarrow{H_E}^0$, we have $||\overrightarrow{H_R}^0||_1 = 1, ||\overrightarrow{H_C}^0||_1 = 1, ||\overrightarrow{H_E}^0||_1 = 1$. Now, by the following theorem, we show that the stable state vectors for Equations (3), (4) and (5) exist and they are unique.

Theorem 4.1. *Let, A be an $N \times N$ column stochastic matrix and H^0 be an N-vector with $||H^0||_1 = 1$. For any $\alpha \in [0,1]$, there exists a stable state vector H^s, which satisfies the equation:*

$$H = \alpha AH + (1-\alpha)H^0 \quad (7)$$

Furthermore, if $\alpha \in [0,1)$, then H^s is unique.

Proof. *Existence:* Let, e be the n-vector with all entries equal to 1. Then, $e^T H = 1$ since $||H||_1 = 1$ after normalizing H. Now, Equation 7 can be rewritten as:

$$\begin{aligned} H &= \alpha AH + (1-\alpha)H^0 = \alpha AH + (1-\alpha)H^0 e^T H \\ &= (\alpha A + (1-\alpha)H^0 e^T)H = MH \end{aligned}$$

where $M = \alpha A + (1-\alpha)H^0 e^T$. $H^0 e^T$ is a column stochastic matrix, since its columns are all equal to

H^0 and $||H^0||_1 = 1$. Created as a weighted combination of two column stochastic matrices, M is also column stochastic. Then, by Lemma 4.2, $\lambda_1 = 1$ is an eigenvalue of M. Hence, there exists an eigenvector H^s corresponding to the eigenvalue λ_1, which satisfies the equation $\lambda_1 H^s = MH^s$.
Uniqueness: Applying Lemma 4.2 to the M matrix defined in the existence part, we have $|\lambda_1| = 1$ and $|\lambda_2| \leq \alpha$. If $\alpha \in [0, 1)$, then $|\lambda_1| > |\lambda_2|$. This implies that, λ_1 is *the* principal eigenvalue of M and H^s is the unique eigenvector corresponding to it. □

Convergence rate of power method for Equations (3), (4) and (5) are determined by the eigenvalues of the M matrices (as defined in Equation 6) of each equation. Convergence rate is proportional to $O(\frac{|\lambda_2|}{|\lambda_1|})$, which is $O(\alpha)$, for each equation. Therefore, choice of α not only adjusts the relative importance of homology and topology, but it also affects running time of our algorithm. Our experiments showed that our algorithm performs well and converges quickly with $\alpha = 0.7$.

In Equations (3), (4) and (5), before the first iteration of power method we only have initial pairwise similarity scores. In the kth iteration, the weight of pairwise similarity score stays to be $(1 - \alpha)$, whereas weight of total support given by $(k - t)^{th}$ degree neighbors of $R_i, \bar{R_j}$ is $\alpha^{k-t}(1 - \alpha)$. That way, neighborhood topologies of matchings are thoroughly utilized without ignoring the effect of initial pairwise similarity scores. As a result, stable state vectors calculated in this manner, are convenient candidates for extracting the entity mappings to create the overall alignment for the query pathways.

4.4. Extracting the Mapping of Entities

Having combined homological and topological similarities of query metabolic pathways, now, it only remains to extract the mapping, φ, of entities. However, since we restrict our consideration to consistent mappings, this extraction by itself is still challenging. Figure 3 points out the importance of maintaining consistency of an alignment.

An alignment is described by the mapping φ, that gives the individual matchings of entities. Lets denote φ as $\varphi = [\varphi_R, \varphi_C, \varphi_E]$, where φ_R, φ_C and φ_E are consistent mappings for reactions, compounds and enzymes, respectively.

If we go back to definition of consistency, there are three conditions that φ should satisfy. The first one is trivially satisfied for any φ of the form $[\varphi_R, \varphi_C, \varphi_E]$, since we beforehand distinguished each entity type. For the second condition, it is sufficient to create one-to-one mappings for each entity type. By using maximum weight bipartite matching we get one-to-one mappings $\varphi_R, \varphi_C, \varphi_E$, which in turn implies φ is one-to-one since intersections of compatibility classes are empty.

The difficult part of finding a consistent mapping is combining mappings of reactions, enzymes and compounds without violating the third condition. For that purpose, we choose a specific ordering between extraction of reaction, enzyme and compound mappings. We create the mapping φ_R first. We extract this mapping by using maximum weight bipartite matching on the bipartite graph constructed by the edge weights in $\overrightarrow{H_R}^S$ vector. Then, using the aligned reactions and the reachability concept, we prune the edges from the bipartite graph of compounds (enzymes) for which the corresponding compound (enzyme) pairs are inconsistent with the reaction mapping. In other words, we prune the edge between two compounds (enzymes), $x, \bar{x}$, if there does not exist any other compound (enzyme) pair $y, \bar{y}$ such that, x is reachable from $\bar{x}$ and y is reachable from $\bar{y}$, or $\bar{x}$ is reachable from x and $\bar{y}$ is reachable from y. By pruning these edges we make sure that for any φ_C and φ_E extracted from the pruned bipartite graphs, $\varphi = [\varphi_R, \varphi_C, \varphi_E]$ is consistent.

Recall that, our aim is to find a consistent alignment which maximizes the similarity score $SimP_\varphi$. The φ defined above satisfies the consistency criteria. For the maximality of similarity score, in the next section, we define $SimP_\varphi$ and then discuss that φ is the mapping that maximizes this score.

4.5. Similarity Score of Pathways

As we present in the previous section, our algorithm guarantees to find a consistent alignment represented by the mappings of entities. One can discuss the accuracy and biological significance of our alignment by looking at the individual matchings that we reported. However, this requires a solid background of the specific metabolism of different organisms. To computationally evaluate the degree of similarity between pathways we devise a similarity score.

We use pairwise similarities of aligned entities to calculate the overall similarity between two query pathways. The definition of similarity function $SimP_{\varphi}$, is as follows:

Definition 7. Let, $P = G([R, C, E], I)$ and $\bar{P} = \bar{G}([\bar{R}, \bar{C}, \bar{E}], \bar{I})$ be two metabolic pathways. Given a mapping $\varphi = [\varphi_R, \varphi_C, \varphi_E]$ between entities of P and $\bar{P}$, similarity of P and $\bar{P}$ is calculated as:

$$SimP_{\varphi}(P, \bar{P}) = \frac{\beta}{|\varphi_C|} \sum_{\forall (C_i, \bar{C_j}) \in \varphi_C} SimC(C_i, \bar{C_j}) + \frac{(1-\beta)}{|\varphi_E|} \sum_{\forall (E_i, \bar{E_j}) \in \varphi_E} SimE(E_i, \bar{E_j})$$

where $|\varphi_C|$ and $|\varphi_E|$ denote the cardinality of corresponding mappings and $\beta \in [0, 1]$ is a parameter that adjusts the relative influence of compounds and enzymes on the alignment score.

Calculated as above, $SimP_{\varphi}$ gives a score between 0 and 1, such that a bigger score implies a better alignment between pathways. We use $\beta = 0.5$ in our experiments, since we do not want to bias our score towards enzymes or compounds. The user can choose $\beta = 0$ to have an enzyme based similarity score or $\beta = 1$ to have a compound based similarity score. Reactions are not considered while calculating this score since reaction similarity scores are already determined by enzyme and compound similarity scores.

Now, having defined the pathway similarity score, we need to show that the consistent mapping $\varphi = [\varphi_R, \varphi_C, \varphi_E]$ found in the previous section, is the one that maximizes this score. But, this follows from the fact that we used maximum weight bipartite matching on the pruned bipartite graphs of enzymes and compounds. In other words, since maximality of the total edge weights of φ_C and φ_E are beforehand assured by the extraction technique, their summation is guaranteed to give the maximum $SimP_{\varphi}$ value for a fixed β.

Complexity Analysis

Let, $P = G([R, C, E], I)$ and $\bar{P} = \bar{G}([\bar{R}, \bar{C}, \bar{E}], \bar{I})$ be two query pathways. The overall time complexity of our algorithm, which is dominated by the power method iterations, is:
$O(|R|^2|\bar{R}|^2 + |C|^2|\bar{C}|^2 + |E|^2|\bar{E}|^2)$.

5. EXPERIMENTS

In order to evaluate the performance of our algorithm we conduct various experiments.

Datasets: We use KEGG Pathway database, which currently has 72,628 pathways which are generated from 360 reference pathways. We convert the pathway data into our graph model.

Parameters: We allow users to change a set of parameters in our implementation. This flexibility is important in some scenarios. For instance, if a user is interested only in enzyme similarities or compound similarities between pathways, then it would be enough to set the parameters accordingly. Due to space limitations, we report the results with only one parameter setting.

α is the parameter that adjusts the relative weight of topology and homology. As we discussed, $\alpha = 0.7$ works well for our method. There is no significant difference between different $SimE$ and $SimC$ scores. We use the information content enzyme similarity score for $SimE$ and the SIMCOMP similarity score for $SimC$ in our experiments. $\gamma_{C_{in}}, \gamma_{C_{out}}, \gamma_E$ are relative weights of each component in reaction similarity calculation. We set $\gamma_{C_{in}} = 0.3, \gamma_{C_{out}} = 0.3, \gamma_E = 0.4$ to balance the effect of compounds and enzymes on reaction similarity. β_C, β_E are relative weights of compounds and enzymes in overall similarity score and they are set to $\beta_C = 0.5, \beta_E = 0.5$.

5.1. Biological Significance

Our first experiment focuses on the biological significance of the found alignments. An alignment should reveal functionally similar entities or sub-paths between different pathways. More specifically, it is desirable to match the entities that can substitute each other or the sub-paths that serve similar functions. We use pathway pairs which are known to contain not identical but functionally similar entities or subpaths in this experiment.

Alternative Enzymes: Two enzymes are called *alternative enzymes*, if they catalyze two reactions with different input compounds that produce a specific target compound. Similarly, we name these reactions as *alternative reactions* and their inputs as *alternative compounds*. Identifying alternative entities is important and useful for various applica-

Table 1. Alternative enzymes that catalyze the formation of a common product using different compounds. [1]Pathways: 00620-Pyruvate metabolism, 00252-Alanine and aspartate metabolism, 00860-Porphyrin and chlorophyll metabolism. [2]Organism pairs that are compared. [3]KEGG numbers of aligned reaction pairs. [4]EC numbers of aligned enzyme pairs. [5] Aligned compounds pairs are put in the same column. Common target products are underlined. Alternative input compounds are shown in **bold**. Abbreviations of compounds: MAL, malate; FAD, Flavin adenine dinucleotide; OAA, oxaloacetate; NAD, Nicotinamide adenine dinucleotide; Pi, Orthophosphate; PEP, phosphoenolpyruvate; Asp, L-Aspartate; Asn, L-Aspargine; Gln, L-Glutamine; PPi, Pyrophosphate; Glu, L-Glutamate; AMP, Adenosine 5-monophosphate; CPP, coproporphyrinogen III; PPHG, protoporphyrinogen; SAM, S-adenosylmethionine; Met, L-Methionine.

P. Id[1]	Organism[2]	Reaction								
		R. Id[3]	Enzyme[4]	Compounds[5]						
00620	*S. aureus*	R01257	EC:1.1.1.96	MAL	+ ***FAD***		→	OAA	+ $FADH_2$	
	H. sapiens	R00342	EC:1.1.1.37	MAL	+ ***NAD***		→	OAA	+ NADH	
00620	*A. thaliana*	R00345	EC:4.1.1.31	OAA	+ ***Pi***		→	PEP	+ CO_2	
	S. aureus	R00341	EC:4.1.1.49	OAA	+ ***ATP***		→	PEP	+ CO_2	+ ADP
00252	*C. hydro.*	R00578	EC:6.3.5.4	Asp	+ ATP	+ ***Gln***	→	Asn	+ AMP	+ PPi
	C. parvum	R00483	EC:6.3.1.1	Asp	+ ATP	+ ***NH_3***	→	Asn	+ AMP	+ Glu
00860	*S. aureus*	R06895	EC:1.3.99.22	CPP	+ ***O_2***		→	PPHG	+ CO_2	
	H. sapiens	R03220	EC:1.3.3.3	CPP	+ ***SAM***		→	PPHG	+ CO_2	+ Met

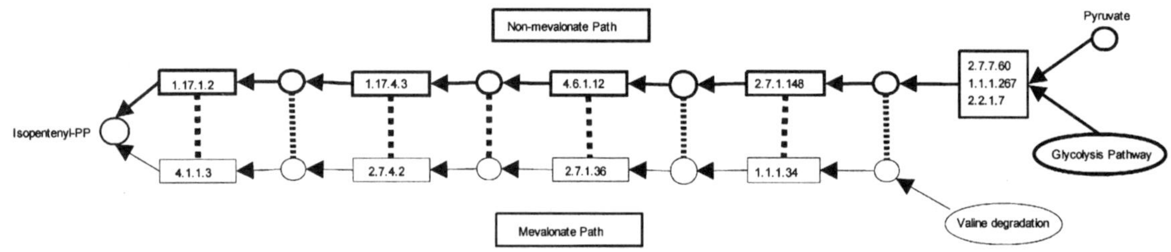

Fig. 4. Identification of alternative sub-paths: A portion of the metabolic pathway of steroid biosynthesis from KEGG. *H.sapiens* produces Isopentenyl-PP via the lower path which is called Mevalonate Path. However, *E.coli* uses a totally different path called Non-mevalonate Path, for producing Isopentenyl-PP which is shown in bold. Using our algorithm, we align the Steroid biosynthesis pathways of *H.sapiens* and *E.coli.* We illustrate the resulting matchings of entities by dashed lines. Compound names are omitted for simplicity.

tions. Some examples are, metabolic reconstruction of newly sequenced organisms[4] and identification of drug targets[3, 18, 19].

We test our tool to search for well-known alternative enzymes presented in Kim *et al.*[20] Table 1 illustrates four cases in which our algorithm successfully identifies alternative enzymes, with the corresponding reaction mappings. Furthermore, resulting compound matchings are consistent with the alternative compounds proposed in Kim *et al.* For instance, there are two different reactions that generate Asparagine (Asn) from Aspartate (Asp) as seen in Table 1. One is catalyzed by aspartate ammonia ligase (EC:6.3.1.1) and uses Ammonium (NH_3) directly, whereas the other is catalyzed by transaminase (EC:6.3.5.4) that transfers the amino group from Glutamine (Gln). We compare the Alanine and aspartate pathways (00252) of two organisms that use the two different routes. Our algorithm aligns the alternate reactions, enzymes and compounds correctly. Our alignment results for the other 3 examples in Table 1 are also consistent with the experimental results, see 20.

Alternative Paths: As metabolic pathways are experimentally analyzed, it is discovered that different organisms may produce the same compounds by totally different paths. Experimental identification provide us well documented examples of such alternative paths. We use our algorithm to identify these known alternative paths in metabolic pathways.

It is shown that, two alternative paths for Isopentenyl-PP production in different organisms exist[21]. Figure 4 illustrates these paths and the entity mappings found by our algorithm. Despite the fact that EC numbers of aligned enzymes are totally different, which indicates that their initial pairwise similarity scores are 0, our algorithm aligns these functionally similar paths since it also accounts for the topological similarities of pathways.

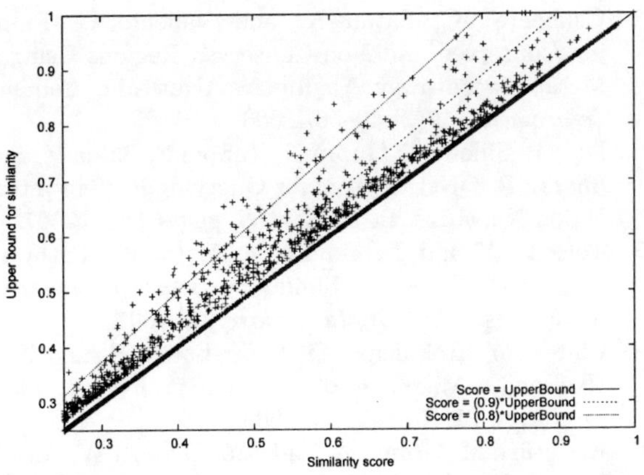

Fig. 5. Effect of consistency restriction on alignment scores: Similarity scores of alignments with consistency restriction and upper bound on the similarity of corresponding pathways without any restriction are shown for pairs of 15,000 randomly selected pathways. Scores below 0.25 are discarded as they indicate dissimilar pathways.

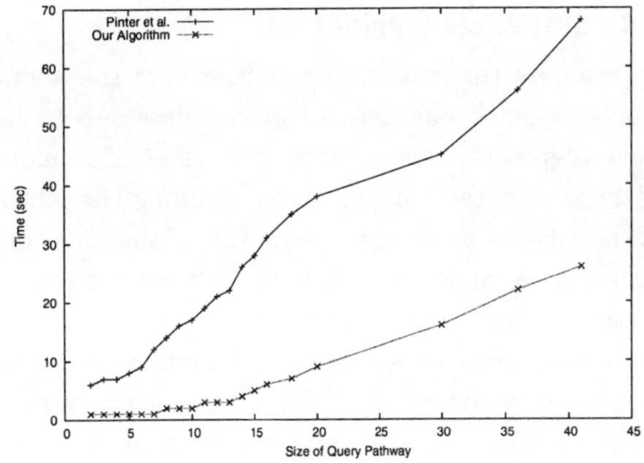

Fig. 6. Running time comparison of our method and the method of Pinter *et al.*: Pathways of varying size are queried against a database of pathways. Total time for each query including IO operations and unexpectedness calculations are plotted for each pathway size. Pathway size is measured as the number of enzymes in the pathway.

Since our method finds one-to-one mappings, only four of seven enzymes in the Non-mevalonate path are mapped to four enzymes of the Mevalonate path. A future work would be to relax the restriction that mappings should be one-to-one. That way alternative paths with different numbers of entities would be aligned without individual entity mappings.

5.2. Effect of Consistency

In order to output meaningful alignments, we report the alignments that are induced by consistent mappings. We ensure the consistency of an alignment by restricting entity mappings to reachable entities. This restriction is necessary for filtering out nonsensical mappings that degrade the accuracy of the alignment. We compute an upper bound to the loss of similarity score due to consistency restriction. We find upper bounds on similarities for each alignment by removing the consistency restriction. This is done by ignoring the pruning phase, which is described in Section 4.4.

Figure 5 demonstrates the effect of consistency restriction on similarity score. For 91 % of the alignments the similarity score found by consistency restriction is not less than 90 % of the upper bound score. Alignments with similarity scores not less than 80 % of the upper bound score constitute 98.5 % of all pathways. Hence, the loss of similarity score due to consistency restriction is not significant.

5.3. Running Time

As discussed theoretically in Section 4.3, our algorithm is guaranteed to find entity mappings with a high convergence rate. We implement the proposed algorithm in C programming language and compare its performance with an existing metabolic pathway alignment tool designed by Pinter *et al.*[8].

The graph model of Pinter *et al.* oversimplifies the metabolic pathways in two ways. First, they totally discard the compounds and reactions from the pathway and use only enzymes. Second, they ignore some interactions between enzymes to have acyclic graphs. Generally, they map a pathway with n enzymes to a graph with n nodes and $n-1$ edges. Since we refuse to have any kind of abstraction, the graph size for the same pathway is considerably larger in our model. For example, Folate biosynthesis pathway of *E.coli* has 12 enzymes. Their simplified model represent this pathway as a graph with 12 nodes and 11 edges, whereas in our graph model the *same* pathway is represented by 55 nodes (22 reactions, 12 enzymes, 21 compounds) and 84 edges. Since we measure the pathway size by the number of enzymes in this experiment, these two pathways are considered to be of the *same size*. Although our algorithm builds a larger graph, Figure 6 shows that our algorithm still runs significantly faster for all pathway sizes. Our method is at least three times faster than the method of Pinter *et al.* for all test cases.

5.4. Statistical Significance

To evaluate the statistical significance of the alignments found by our method, we calculate z-score for each alignment. We generate a number of random pathways for each alignment by shuffling the labels of the entities of query pathways. Label shuffling corresponds to randomly switching the rows of support matrices of each entity type.

Our experiments show that alignment of same metabolic pathways in different organisms create higher z-scores than different pathways in the same or different organisms. In a specific organism pathways that have similar functions, such as different amino acid metabolisms, give higher z-scores than pathways that belong to different functional groups. Due to space constraints we do not present any results for this part.

6. CONCLUSION

In this paper, we considered the pairwise alignment problem for metabolic pathways. We developed a method that aligns reactions, compounds and enzymes. In our algorithm, we considered both the homology and the topology of pathways. We enforced the consistency of the alignment by considering the reachability sets of the aligned entities. Using maximum weight bipartite matching, we first extracted reaction mappings. Then, we enforced the consistency by applying a pruning technique and we extract the mappings for enzymes and compounds. Our experiments showed that, our method is capable of finding biologically and statistically significant alignments very quickly.

References

1. Ogata H, Goto S, Sato K, Fujibuchi W, Bono H, and Kanehisa M. KEGG: Kyoto Encyclopedia of Genes and Genomes. *NAR*, 27(1):29–34, 1999.
2. Keseler IM, Collado-Vides J, Gama-Castro S, Ingraham J, Paley S, Paulsen IT, Peralta-Gil M, and Karp PD. EcoCyc: a comprehensive database resource for Escherichia coli. *NAR*, 33:334–337, 2005.
3. Sridhar P, Kahveci T, and Ranka S. An iterative algorithm for metabolic network-based drug target identification. In *PSB*, volume 12, pages 88–99, 2007.
4. Francke C, Siezen RJ, and Teusink B. Reconstructing the metabolic network of a bacterium from its genome. *Trends in Microbiology*, 13(11):550–558, November 2005.
5. Clemente JC, Satou K, and Valiente G. Finding Conserved and Non-Conserved Regions Using a Metabolic Pathway Alignment Algorithm. *Genome Informatics*, 17(2):46–56, 2006.
6. Dost B, Shlomi T, Gupta N, Ruppin E, Bafna V, and Sharan R. QNet: A Tool for Querying Protein Interaction Networks. In *RECOMB*, pages 1–15, 2007.
7. Tohsato Y and Nishimura Y. Metabolic Pathway Alignment Based on Similarity between Chemical Structures. *IPSJ Digital Courier*, 3, 2007.
8. Pinter RY, Rokhlenko O, Yeger-Lotem E, and Ziv-Ukelson M. Alignment of metabolic pathways. *Bioinformatics*, 21(16):3401–8, 2005.
9. Koyutürk M, Grama A, and Szpankowski W. An efficient algorithm for detecting frequent subgraphs in biological networks. In *ECCB*, pages 200–207, 2004.
10. Singh R, Xu J, and Berger B. Pairwise Global Alignment of Protein Interaction Networks by Matching Neighborhood Topology. In *RECOMB*, pages 16–31, 2007.
11. Kleinberg JM. Authoritative sources in a hyperlinked environment. *Journal of the ACM*, 46(5):604.
12. Dutkowski J and Tiuryn J. Identification of functional modules from conserved ancestral protein interactions. *Bioinformatics*, 23(13):i149–158, 2007.
13. Koyutürk M, Grama A, and Szpankowski W. Pairwise Local Alignment of Protein Interaction Networks Guided by Models of Evolution. In *RECOMB*, pages 48–65, 2005.
14. Webb EC. *Enzyme nomenclature 1992* . Academic Press, 1992.
15. Tohsato Y, Matsuda H, and Hashimoto A. A Multiple Alignment Algorithm for Metabolic Pathway Analysis Using Enzyme Hierarchy. In *ISMB*, pages 376–383, 2000.
16. Hattori M, Okuno Y, Goto S, and Kanehisa M. Development of a chemical structure comparison method for integrated analysis of chemical and genomic information in the metabolic pathways. *J Am Chem Soc*, 125(39):11853–11865, 2003.
17. Haveliwala TH and Kamvar SD. The Second Eigenvalue of the Google Matrix. *Stanford University Technical Report*, March 2003.
18. Sridhar P, Song B, Kahveci T, and Ranka S. Mining metabolic networks for optimal drug targets. pages 291–302, 2008.
19. Song B, Sridhar P, Kahveci T, and Ranka S. Double Iterative Optimization for Metabolic Network-Based Drug Target Identification. *International Journal of Data Mining and Bioinformatics*, 2007.
20. Kim J and Copley SD. Why Metabolic Enzymes Are Essential or Nonessential for Growth of Escherichia coli K12 on Glucose. *Biochemistry*, 46(44), 2007.
21. McCoy AJ, Adams NE, Hudson AO, Gilvarg C, Leustek T, and Maurelli AT. L,L-diaminopimelate aminotransferase, a trans-kingdom enzyme shared by Chlamydia and plants for synthesis of diaminopimelate/lysine. *PNAS*, November 2006.

DETECTING PATHWAYS TRANSCRIPTIONALLY CORRELATED WITH CLINICAL PARAMETERS

Igor Ulitsky and Ron Shamir

School of Computer Science, Tel Aviv University, Tel Aviv, Israel

Email: {ulitskyi,rshamir}@post.tau.ac.il

The recent explosion in the number of clinical studies involving microarray data calls for novel computational methods for their dissection. Human protein interaction networks are rapidly growing and can assist in the extraction of functional modules from microarray data. We describe a novel methodology for extraction of connected network modules with coherent gene expression patterns that are correlated with a specific clinical parameter. Our approach suits both numerical (e.g., age or tumor size) and logical parameters (e.g., gender or mutation status). We demonstrate the method on a large breast cancer dataset, where we identify biologically-relevant modules related to nine clinical parameters including patient age, tumor size, and metastasis-free survival. Our method is capable of detecting disease-relevant pathways that could not be found using other methods. Our results support some previous hypotheses regarding the molecular pathways underlying diversity of breast tumors and suggest novel ones.

1. INTRODUCTION

Systems biology has the potential to improve the diagnosis and management of complex diseases by offering a comprehensive view of the molecular basis behind the clinical pathology. To achieve this, a computational analysis extracting mechanistic understanding from the available data is required. Such data include many thousands of genome-wide expression profiles obtained using the microarray technology. A wide variety of approaches have been suggested for reverse engineering of mechanistic molecular networks from expression data[1-3]. However, most of these methods are effective only when using expression profiles obtained under diverse conditions and perturbations, while the bulk of data currently available on human clinical studies are expression profiles of groups of individuals sampled from the natural population. The standard methodologies for analysis of such datasets usually include: (a) unsupervised clustering of the samples to reveal the basic correlation structure, and (b) focus on a specific clinical parameter and the application of statistical methods for identification of a gene signature that best predicts it. While these methods are successful in identifying potent signatures for classification purposes[4,5], the insights that can be obtained from examining the gene lists they produce are frequently limited.

It is thus desirable to develop novel computational tools that will utilize additional information in order to extract more knowledge from gene expression studies. Various parameters are commonly recorded in such studies, and they can be classified into two types: (a) logical parameters (e.g., gender or tumor subtype) and (b) numerical parameters (e.g., patient age or tumor grade). A key question is how to identify genes significantly related to a specific clinical parameter. As it is frequently difficult to suggest novel hypotheses based on individual genes, it is desirable to identify the *pathways* that are correlated with a clinical parameter. By considering together the whole pathway, correlations that would have been missed if we tested each gene separately can be revealed. One approach to this problem uses predefined gene sets describing pathways and quantifies the change in their expression levels[6-8]. The drawback of this approach is that pathway boundaries are often difficult to assign, and in many cases only part of the pathway is altered during disease. Moreover, unknown pathways are harder to find in this approach. To overcome these problems, the use of gene networks was suggested. Several approaches for integrating microarray measurements with network knowledge have been

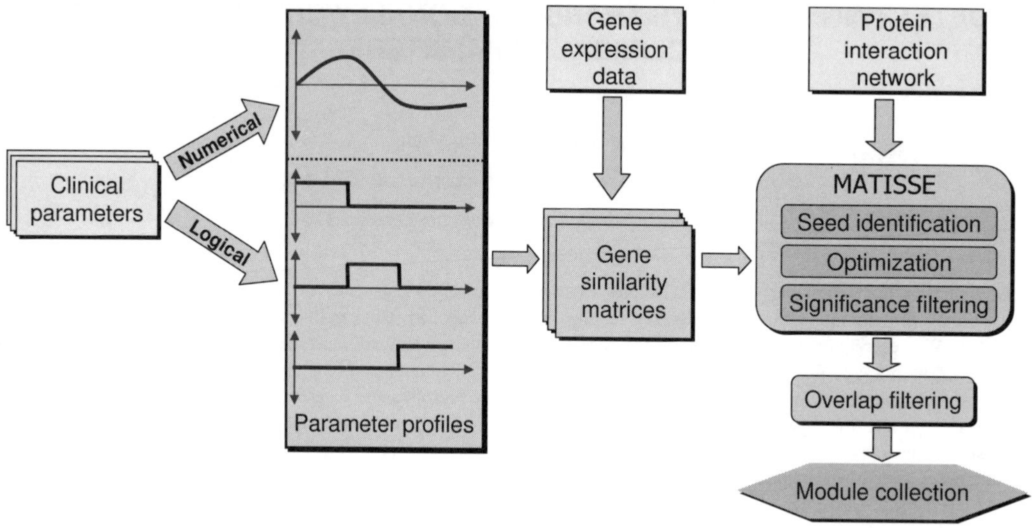

Fig. 1. Study outline. Clinical parameters are used to generate a collection of parameter profiles. The parameter profiles are used, together with gene expression data, to generate gene similarity scores. These scores, together with a protein interaction network serve as an input to MATISSE, which identifies a set of modules for each parameter. The modules are then filtered and a collection of non-redundant modules is produced.

proposed, some of which can be applied also for binary clinical parameters. Some proposed computational methods for detection of subnetworks that show correlated expression[9-11]. A successful method for detection of active subnetworks' was proposed by Ideker *et al.*[12] and extended by other groups[13-16]. These methods are based on assigning a significance score to every gene in every sample and looking for subnetworks with statistically significant combined scores. Breitling *et al.*[17] proposed a simple method named GiGA which receives a list of genes ordered by expression relevance and extracts subnetworks corresponding to the most relevant genes. Other tools use network and expression information together, but for sample classification[18,19].

The most basic parameter in clinical studies is the binary disease status (case vs. control). Other studies provide more clinical information in the form of additional parameters. For example, in the breast cancer expression data published by Minn *et al.*[20], each sample was accompanied by up to 10 different parameters (Table 1). These parameters include general characteristics of the patients (e.g., age), pathological status of the tumor and follow-up information. Given such data, we wish to identify pathways whose transcription is dysregulated in a manner that is consistent with a particular clinical parameter. This information can then be used both for predictive purposes and for improving our understanding of the biology underlying the disease progression. This requires identifying subnetworks with expression patterns correlated to numerical or multi-valued logical parameters with more than two possible values.

We have previously developed the MATISSE algorithm for extraction of functional modules from expression and network data[9]. It receives as input a protein interaction (PI) network alongside a collection genome-wide mRNA expression profiles. The output of MATISSE is a collection of modules: connected subnetworks in the PI graph, whose corresponding mRNAs exhibit significantly correlated expression patterns. Here we describe an extension of the MATISSE algorithm aimed at extraction of modules of genes whose expression profiles are not only correlated to one another, but also correlated with one of the clinical parameters. These two requirements aim to identify subnetworks that constitute functional modules in the cell and are involved with a specific clinical phenotype.

We used a human PI network consisting of 10,033 nodes and 41,633 interactions (see Methods) and applied our algorithm to 99 breast cancer samples (BC dataset[20]) in conjunction with 10 numerical and logical parameters (Figure 1). This analysis identified several modules significantly correlated with various parameters such as patient age, tumor size, Her2 status and metastases-free survival period length.

Table 1. Parameters from the breast cancer dataset that were used in this study.

Parameter	Samples*	Type	Distribution
Age at diagnosis	99	Numerical	55.80±13.6
Tumor Size (cm)	99	Numerical	3.62±1.7
Positive Lymph Nodes	99	Numerical	3.59±6.3
Estrogen receptor (ER) status	99	Logical	
Progesterone receptor (PR) status	98	Logical	
Her2 staining (grade)	88	Numerical	0.53±0.98
Metastasis after 5 years?	68	Logical	
Metastasis free survival (years)	82	Numerical	5.17±2.3
Lung metastasis free survival (years)	82	Numerical	5.50±2.3
Bone metastasis free survival (years)	82	Numerical	5.34±2.3

* Number of samples for which the parameter was available

Importantly, our results provide support for the correlation between the expression levels of several pathways, such as the ribosomal proteins and the patient prognosis. However, this is not always the case, as we did not find support for the correlation between survival and the levels of the unfolded protein response pathway genes. Finally, we show that the specific disease-related insights suggested by our method can not be picked up using existing alternative methods.

2. METHODS

2.1. The basic methodology

Our approach builds on the MATISSE methodology for identifying co-expressed subnetworks[9]. We first outline that methodology here. The input to MATISSE includes an undirected *constraint graph* $G^C = (V, E)$, a subset $V_{sim} \subseteq V$ and a symmetric matrix S where S_{ij} is the similarity between $v_i, v_j \in V_{sim}$. The goal is to find disjoint subsets $U_1, U_2, ..., U_k \subseteq V$ called *modules*, so that each subset induces a connected subgraph in G^C and contains elements that share high similarity values. We call the nodes in V_{sim} *front nodes* and nodes in $V \backslash V_{sim}$ *back nodes*.

In the biological context, V represents genes or gene products (we shall use the term 'gene' for brevity), and E represents interactions between them. S_{ij} measures the similarity between genes i and j. Originally, we used the Pearson correlation between gene expression patterns as a similarity metric[9]. The set V_{sim} is smaller than V in several cases. For example, when using mRNA microarrays, some of the genes may be absent from the array, and others may be excluded due to insignificant expression changes across the tested conditions. Hence, a module aims to capture a set of genes that have highly similar behavior, and are also topologically connected, and thus may belong to a single complex or pathway. The quantification of gene similarity is obtained by formulating the problem as a hypothesis testing question. In this approach statistically significant modules correspond to heavy subnetworks in a similarity graph, with nodes inducing a connected subgraph in G^C. A three-stage heuristic is used to obtain high-scoring modules.

2.2. Identifying modules correlated with clinical parameters

Here, we are interested in extracting groups of genes that are not only similar across the experimental conditions, but also exhibit significant correlation with one of the clinical parameters. To this end we devised a hybrid similarity score that reflects these two phenomena. Importantly, our scheme can handle both numerical and logical parameters. The advantage of a uniform scheme is that the modules identified for different parameters are directly comparable, and in case of overlaps, the more significant module can be picked.

Formally, we are given a set of parameters $P_1,...,P_m$ (numerical and logical) and we wish to quantify, for each gene pair (i,j), the extent to which the genes are correlated to P_k and to each other. For each parameter we first discard the samples for which the value of the parameter is not available. Let m be the number of samples that survived this filter. Then, we compute one or more *parameter profiles* $p_{ij} = (p_{ij}^1, p_{ij}^2, ..., p_{ij}^m)$. If P_i is a numeric parameter, it is assigned a single parameter profile vector p_{i1}, *and* p_{i1}^{k} equals the value of P_i in sample k. If P_i is a logical parameter that attains with k different values $c_i^1, c_i^2, ..., c_i^l$, then for each $1 \le j \le l$ we compute a 0/1 parameter profile vector $p_{ij} = (p_{ij}^1, p_{ij}^2, ..., p_{ij}^m)$ where $p_{ij}^k = 1$ if the value of P_i in sample k is c_j and 0 otherwise.

We denote the expression pattern of gene k by $x_k = (x_k^1, x_k^2, ..., x_k^m)$. We are interested in quantifying the similarity between p_{ij} and x_k. Let r_{ijk} be the Pearson correlation coefficient between p_{ij} and x_k. If P is numerical, then r_{j1k} is close to 1 if the transcript and the parameter are strongly correlated. If P is logical, r_{ijk} is close to 1 if the transcript levels are high when the value of P_i is c_j and low otherwise. Transcript correlation to such 0/1 profiles was previously used successfully as a differential gene expression score[21].

Recall that we are interested in gene pairs a,b that are: (i) correlated with p_{ij} and (ii) correlated with each other. To address (i) we would like the similarity score of genes a and b to be high only if both a and b are correlated with the parameter. We thus first set $S_{diff}(i,j) = \min\{r_{ija}, r_{ijb}\}$. To address (ii) we use the partial correlation coefficient between the gene patterns conditioned on p_{ij}. Formally:

$$S_{corr}(a,b \mid p_{ij}) = \frac{r_{a,b} - r_{ija} r_{ijb}}{\sqrt{(1-r_{ija}^2)(1-r_{ijb}^2)}}$$

where $r_{a,b}$ is the Pearson correlation coefficient between the profiles of genes a and b. Intuitively, S_{corr} conveys the information about how similar a and b are, given their correlation to p_{ij}. Finally, we use the similarity score:

$$S = \frac{S_{diff} + \lambda \cdot S_{corr}}{1+\lambda}$$

where λ is a tradeoff parameter setting the relative importance of the correlation with the clinical parameter. For each parameter profile S scores were computed for both positive and negative correlations with the parameter. Note that the values of S are always between -1 and 1. Note that standard Pearson correlation can also be used as S_{corr}. We chose to use partial correlation in this work, as it allows us to penalize gene pairs for which most of the correlation can be explained by their separate correlations with the clinical parameter. The S scores are then modeled using the probabilistic model described previously[9].

2.3. Finding high-scoring modules

MATISSE uses a three-step heuristic to identify high-scoring modules. The heuristic consists of (a) identification of small high-scoring seeds; (b) seed optimization using a greedy algorithm; (c) significance filtering. The seed finding step was performed as described previously[9]. The greedy algorithm was improved in this study. To allow improvement of modules that reached the maximum size limit, we added two additional operation types: (a) a "replace" operation in which a node is added to a module replacing the node that contributes least to the module score; (b) a "swap" operation, in which module assignments of two nodes are swapped. Both operations are performed only if they improve the total solution weight jeopardizing the connectivity of the modules.

In order to evaluate the statistical significance of the modules found in a dataset, we randomly shuffled the expression pattern of each gene and re-executed the algorithm. This process was repeated 100 times and the best score of a module in each run was recorded. These scores were then used to compute an empirical p-value for modules found in the real data. Only modules with $p<0.1$ were retained.

2.4. Filtering overlapping modules

We removed modules that overlapped by >50% with another module that was more significantly correlated with a clinical parameter.

2.5. MATISSE parameters

We used $\lambda=4$ for the analysis described in this paper. The upper bound on module size was set to 120. The rest of the parameters were set as described previously[9].

2.6. Network and expression data

A human PI network was compiled from the HPRD[22], BIND[23], BioGrid[24], HDBase (http://hdbase.org/), and SPIKE[25] databases. The resulting network consisted of 10,033 proteins (mapped to Entrez Gene entries) and 41,633 interactions.

The expression dataset was obtained from GEO (Accession GSE2603). We used the normalized expression values available in the respective GEO records. Affymetrix probe identifiers were mapped to

Entrez Gene. If several probes mapped to the same Entrez Gene, the highest intensity was used in every sample. Values <20 were set to 20 and values >20,000 were set to 20,000. 2,000 genes that showed the highest gene pattern variance were used as front nodes.

2.7. Module annotation

We annotated the modules using Gene Ontology (http://www.geneontology.org/) and MSigDB (http://www.broad.mit.edu/gsea/, "curated gene sets" collection[6]). Gene Ontology enrichment p-values were computed using TANGO[26], which uses resampling to correct for multiple testing and annotation overlap. All other p-values were Bonferroni corrected for multiple testing.

3. RESULTS

3.1. Breast cancer dataset

The breast cancer (BC) dataset contained 99 expression profiles of tumor samples from the MSKCC cohort[20]. 15 different parameters were available for each sample, some of which were not sufficiently clear or redundant. The 10 parameters we used are listed in Table 1. For 9 parameters at least one significant module was identified. After filtering module overlaps (see Methods) we identified 10 significant non-redundant modules, with sizes ranging from 84 to 118 (Table 2).

Using GO and MSigDB annotations (see Methods) we found that 6 modules (60%) were significantly enriched with at least one GO biological process and all 10 modules (100%) were enriched with at least one MSigDB category (Table 2). Seven modules (70%) were enriched with at least one of the 16 MSigDB gene sets related to breast cancer. Overall, eight of the breast cancer related gene sets were enriched in at least one module.

Module BC-1 was positively correlated with the age of the woman at the time of breast cancer diagnosis. Inspection of the expression data revealed that the module was particularly up-regulated in women above age 72 (Figure 2). The module did not show significant GO enrichment categories. When examining 27 MSigDB gene sets related to aging, we found a significant between BC-1 and the MSigDB category "AGED_RHESUS_UP" (8 genes, p=0.002), which contains genes identified as up-regulated in the muscles of aged rhesus monkeys when compared to young ones[27]. One of these eight genes is RELA, a transcription factor component of the NFκB complex. BC-1 contained two additional genes from the PKC pathway which activates NFκB – NFKBIA and PKCA (MSigDB gene set PKCPATHWAY, p=0.04). Increased activity of the NFκB pathway has been recently implicated in aging in a study utilizing diverse expression data and transcription factor binding motifs[28]. Adler *et al.* have also shown that blocking of this pathway can reverse the age-related transcriptional program. Note that our methodology connecting NFκB to aging is completely different: Adler *et al.* sought motifs over-represented in age-dependent genes in various microarray datasets, whereas we looked for connected PI subnetworks that are correlated with age on the expression level. Our results thus provide further support for the relationship between NFκB and age-dependent transcriptional changes.

BC-2 is an apoptosis-related module that is positively correlated to the size of the tumor. This module is also significantly enriched with genes related to unfolded protein response (UPR) and the TNF pathway. Accordingly, this module also significantly enriched with heat shock factor (HSF) targets (p=0.03) and genes localized to the ER (from GO, $p=6.81*10^{-9}$). Interestingly, heat shock protein level has been traditionally associated with poor breast cancer prognosis and higher metastasis likelihood[29]. However, BC-2 was only weakly correlated with metastases-free survival period in our dataset (r=0.038).

Two modules, BC-3 and BC-4, were identified as negatively correlated with tumor size. Both modules were enriched with genes previously associated with ER-positive tumors. However, the correlation of the module profiles with ER status was very weak in our dataset (r=0.001 and r=0.008, for BC-3 and BC-4, respectively). However, we did find a significant overlap between genes in BC-3 and the recently mapped targets of the estrogen receptor[30] ($p=1.13*10^{-4}$). Finally, estrogen receptors Esr1 and Esr2 both appeared in BC-3. This suggests that increased ER transcription factor activity could result in smaller tumors.

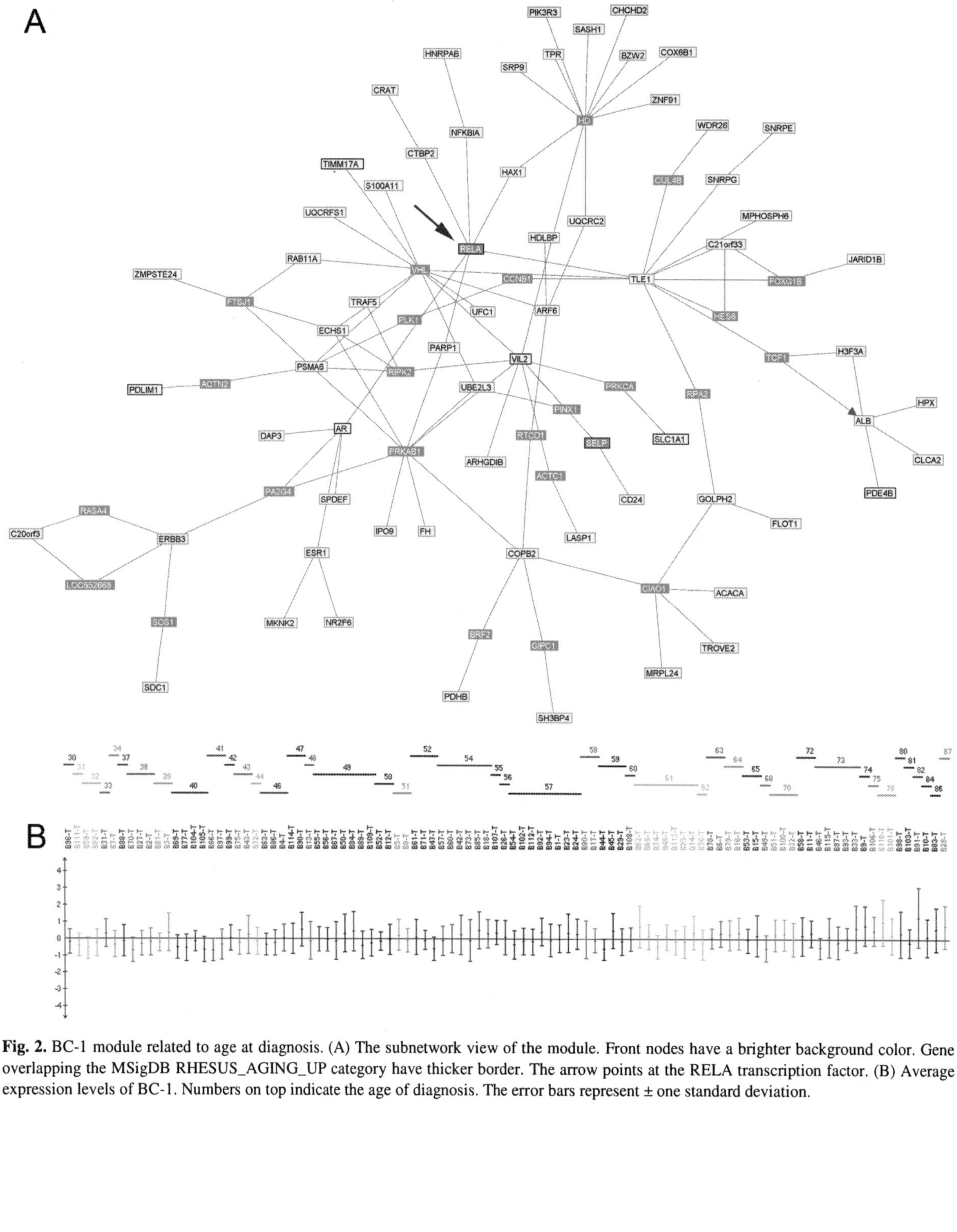

Fig. 2. BC-1 module related to age at diagnosis. (A) The subnetwork view of the module. Front nodes have a brighter background color. Gene overlapping the MSigDB RHESUS_AGING_UP category have thicker border. The arrow points at the RELA transcription factor. (B) Average expression levels of BC-1. Numbers on top indicate the age of diagnosis. The error bars represent ± one standard deviation.

Table 2. Modules identified in the breast cancer dataset of Minn *et al.* Front nodes are nodes for which expression data are used (see Methods). GO enrichment p-values were computed using TANGO. MSigDB enrichment p-values are Bonferroni corrected. For MSigDB, up to 5 most significantly enriched gene sets are shown.

Module	Parameter	Average correlation	Total nodes	Front nodes	Score FDR	GO biological process	p-value	MSigDB gene set	p-value
BC-1	Age at diagnosis	0.196	90	64	0.08			HUMAN_MITODB_6_2002	0.016
								MITOCHONDRIA	0.022
								BRCA_ER_POS	0.026
								PKCPATHWAY	0.04
BC-2	Tumor Size	0.188	118	82	<0.01	response to unfolded protein	<0.001	ST_TUMOR_NECROSIS_FACTOR_PATHWAY	9.36E-10
								BRCA_ER_NEG	8.76E-08
								STEMCELL_NEURAL_UP	9.11E-08
								APOPTOSIS	3.79E-07
								APOPTOSIS_GENMAPP	1.68E-06
BC-3	Tumor Size	-0.175	115	86	<0.01			BRCA_ER_POS	2.13E-09
								ALZHEIMERS_DISEASE_DN	1.92E-05
								BREASTCA_TWO_CLASSES	3.05E-04
								DRUG_RESISTANCE_AND_METABOLISM	9.96E-04
								CARM_ERPATHWAY	0.034
BC-4	Tumor Size	-0.157	97	60	0.09			BRCA_ER_POS	0.002
								AKAPCENTROSOMEPATHWAY	0.009
								P53PATHWAY	0.023
BC-5	Positive lymph nodes	-0.143	84	66	<0.01			BRCA_ER_NEG	1.32E-09
								STEMCELL_NEURAL_UP	1.41E-05
								TARTE_PLASMA_BLASTIC	7.84E-05
								PENG_GLUTAMINE_DN	8.87E-04
								ALZHEIMERS_DISEASE_DN	0.004
BC-6	Her2 staining	0.204	107	80	0.01	positive regulation of I-kappaB kinase/NF-kappaB cascade	0.009	ALZHEIMERS_DISEASE_DN	2.74E-08
								HUMAN_MITODB_6_2002	9.84E-05
								FLECHNER_KIDNEY_TRANSPLANT_REJECTION_DN	2.83E-04
								PGC	3.67E-04
								MITOCHONDRIA	9.48E-04
BC-7	Metastasis after 5 years?	-0.203	96	74	0.04	translation	0.004	RIBOSOMAL_PROTEINS	9.23E-33
								JISON_SICKLECELL_DIFF	3.86E-08
								FLOTHO_CASP8AP2_MRD_DIFF	3.32E-07
								HCC_SURVIVAL_GOOD_VS_POOR_DN	3.43E-04
								TRANSLATION_FACTORS	0.009
BC-8	Metastasis after 5 years?	0.224	116	86	0.02	antigen processing	<0.001	WIELAND_HEPATITIS_B_INDUCED	1.09E-11
						antigen presentation	<0.001	PROTEASOME	9.97E-11
						modification-dependent protein catabolism	<0.001	FLECHNER_KIDNEY_TRANSPLANT_WELL_UP	5.12E-08
								PROTEASOMEPATHWAY	7.40E-08
								TCRAPATHWAY	3.04E-06
BC-9	Mestassis free survival	0.191	118	91	<0.01	translation	0.02	RIBOSOMAL_PROTEINS	1.40E-33
								JISON_SICKLECELL_DIFF	4.30E-11
								FLOTHO_CASP8AP2_MRD_DIFF	2.22E-10
								MYC_TARGETS	6.95E-04
								HCC_SURVIVAL_GOOD_VS_POOR_DN	0.003
BC-10	Lung metastatis free survival	0.195	102	74	0.01	positive regulation of I-kappaB kinase/NF-kappaB cascade	<0.001	RIBOSOMAL_PROTEINS	7.08E-11
								NFKBPATHWAY	3.23E-06
								JISON_SICKLECELL_DIFF	7.28E-06
								ST_TUMOR_NECROSIS_FACTOR_PATHWAY	1.96E-05
								APOPTOSIS_GENMAPP	3.04E-04

Three modules (BC-7, BC-9 and BC-10) were significantly enriched with ribosomal proteins (RPs). Expression levels of these modules were correlated with Her2- and ER-positive longer metastases-free survival in the lungs and in the bone marrow. High expression of RPs is indicative of a higher metabolic rate within malignant cells. High levels of RP expression have been previously associated with Her2 overexpression in BC cell lines[31]. RP over-expression was also associated with less aggressive ovarian tumors[32]. Our results provide additional support for the notion that RP expression is positively correlated with longer survival. Surprisingly, two of the modules enriched for ribosomal proteins (BC-7 and BC-9) were enriched with the MSigDB class "HCC_SURVIVAL_GOOD_VS_POOR_DN" described as representing genes associated with poor survival in hepatocellular carcinoma. However, this class is not associated with any publication and BC-7 and BC-9 were not enriched with other gene sets related to survival in MSigDB, so further corroboration is required here.

BC-8 was significantly enriched with proteasomal genes and associated with shorter metastases-free survival periods. This module contained 16 different proteasomal subunits, all as front nodes. It also contained multiple genes associated with antigen representation and the immune response. Interestingly, this module was also significantly enriched with genes located on chromosome 6 (p=1.29*10^{-6}, the most significant module-chromosome association). Therefore, it is possible that the up-regulation results from aberrations of this chromosome in a subset of the tumors.

3.2. Comparison with other methods

We first compared the parameter-correlated modules (PCMs) to the modules obtained using the standard MATISSE algorithm with the same parameters. MATISSE identified 19 modules covering 996 genes. 8 of the modules (42%) were significantly enriched for a GO category and 11 (58%) were enriched for an MSigDB category (all 11 were enriched with at least one breast-cancer related category), indicating that a larger percentage of PCMs are functionally relevant compared to MATISSE modules. However, 18 GO annotations were enriched in the MATISSE solution only, compared to 5 in the parameter-correlated solution only (195 vs. 47 for MSigDB gene sets), indicating a trade-off between specificity and selectivity in functional module selection. As expected, the MATISSE module genes were more strongly correlated on the expression level (average r=0.3 vs. 0.14), whereas PCMs were more strongly correlated with clinical parameters (average maximum correlation of 0.14 per PCM, compared to 0.12 for MATISSE modules).

Some of the insights described above could not be revealed using MATISSE: only two small modules (9 genes each) were slightly correlated with age and they did not overlap the rhesus aging signature; (b) the MATISSE modules that were slightly correlated with tumor size were not enriched for the UPR pathway; (c) no MATISSE modules were enriched for ribosomal or other translation-related proteins; (d) the maximum enrichment for same-chromosome genes was significantly lower (p=0.002 vs. p=1.29*10^{-6}). Thus we conclude that while using expression correlation alone can lead to more diverse functional modules, using clinical parameter correlation enables detection of more specific disease-relevant modules that are missed otherwise.

The insights also could not be based on parameter correlation alone. When taking the 200 genes with the highest enrichment with the parameters: (a) the genes correlated with age at prognosis were not enriched with the rhesus gene set and did not contain RELA; (b) the genes correlated with tumor size were not enriched with UPR pathway genes; (c) the genes negatively correlated with tumor size were not enriched with ER targets; (c) the genes correlated with metastases-free survival were not enriched with ribosomal proteins.

Finally, logical parameters can be analyzed using GSEA[6]. GSEA found 130 (9) gene sets associated with poor (good) prognosis at FDR<0.1. 31 (3) were associated with negative (positive) ER status, none of them breast cancer related. No gene sets were significantly associated with PR status. Similar to our analysis, GSEA identified the correlation between survival and the levels of the ribosomal proteins and the proteasome. However, only one breast cancer related gene set appeared in the GSEA results (BRCA_ER_POS), and none of the pathways we identified using continuous parameters could be found using GSEA.

4. DISCUSSION

The increasing availability of network and expression data in multiple species led to development of several methods for detecting modular structures through joint analysis of network and expression data[9,11-17]. As the coverage and quality of the interaction networks improve, we expect that these tools will play a central part in the analysis of microarray data. A prominent current challenge is to enable these tools to use as much additional information as possible in order to produce more accurate and biologically relevant results. Clinical parameters of the profiled tissue can help in association of genes and pathways with clinical phenotypes.

To the best of our knowledge, the method we described here is the first capable of jointly analyzing interaction data, expression profiles and continuous numerical clinical parameters. A simple alternative for joint analysis of the three sources is to first apply a module finding algorithm to network and expression data, and then associate modules with parameters. As our results show, module finding algorithms are indeed successful at identifying multiple functional modules. However, clinically important pathways are missed if the clinical data are used only in the post-processing of the modules.

While the results we present are encouraging, there is certainly room for improvement. In particular, it would help to incorporate confidence levels for individual interactions[33] and to further improve our optimization algorithm. Our methodology for integrating parameter data currently analyzes each parameter in isolation, ignoring correlations between parameters. Another important frontier is to associate modules with combinations of different parameter values, e.g., up-regulation in poor prognosis and in ER-negative tumors.

Finally, we are currently developing a user-friendly interface to the methods described here that will allow analysis through the MATISSE software (http://acgt.cs.tau.ac.il/matisse).

Acknowledgements

IU is a fellow of the Edmond J Safra Bioinformatics Program at Tel-Aviv University. This research was supported in part by the "GENEPARK: GENomic Biomarkers for PARKinson's disease" project that is funded by the European Commission within its FP6 Programme (contract number EU-LSHB-CT-2006-037544), and by the Israel Science Foundation (grant no. 385/06).

References

1. Gat-Viks, I. & Shamir, R. Refinement and expansion of signaling pathways: the osmotic response network in yeast. *Genome Res* **17**, 358-67 (2007).
2. Bansal, M., Belcastro, V., Ambesi-Impiombato, A. & di Bernardo, D. How to infer gene networks from expression profiles. *Mol Syst Biol* **3**, 78 (2007).
3. Sprinzak, D. & Elowitz, M.B. Reconstruction of genetic circuits. *Nature* **438**, 443-8 (2005).
4. van de Vijver, M.J. et al. A gene-expression signature as a predictor of survival in breast cancer. *N Engl J Med* **347**, 1999-2009 (2002).
5. Ben-Dor, A. et al. Tissue classification with gene expression profiles. *J Comput Biol* **7**, 559-83 (2000).
6. Subramanian, A. et al. Gene set enrichment analysis: a knowledge-based approach for interpreting genome-wide expression profiles. *Proc Natl Acad Sci U S A* **102**, 15545-50 (2005).
7. Kim, S.Y. & Volsky, D.J. PAGE: parametric analysis of gene set enrichment. *BMC Bioinformatics* **6**, 144 (2005).
8. Jiang, Z. & Gentleman, R. Extensions to gene set enrichment. *Bioinformatics* **23**, 306-13 (2007).
9. Ulitsky, I. & Shamir, R. Identification of functional modules using network topology and high-throughput data. *BMC Syst Biol* **1**, 8 (2007).
10. Segal, E., Wang, H. & Koller, D. Discovering molecular pathways from protein interaction and gene expression data. *Bioinformatics* **19 Suppl 1**, i264-71 (2003).
11. Hanisch, D., Zien, A., Zimmer, R. & Lengauer, T. Co-clustering of biological networks and gene expression data. *Bioinformatics* **18 Suppl 1**, S145-54 (2002).
12. Ideker, T., Ozier, O., Schwikowski, B. & Siegel, A.F. Discovering regulatory and signalling circuits in molecular interaction networks. *Bioinformatics* **18 Suppl 1**, S233-40 (2002).
13. Rajagopalan, D. & Agarwal, P. Inferring pathways from gene lists using a literature-derived network of biological relationships. *Bioinformatics* **21**, 788-93 (2005).

14. Cabusora, L., Sutton, E., Fulmer, A. & Forst, C.V. Differential network expression during drug and stress response. *Bioinformatics* **21**, 2898-905 (2005).
15. Nacu, S., Critchley-Thorne, R., Lee, P. & Holmes, S. Gene expression network analysis and applications to immunology. *Bioinformatics* **23**, 850-8 (2007).
16. Liu, M. et al. Network-based analysis of affected biological processes in type 2 diabetes models. *PLoS Genet* **3**, e96 (2007).
17. Breitling, R., Amtmann, A. & Herzyk, P. Graph-based iterative Group Analysis enhances microarray interpretation. *BMC Bioinformatics* **5**, 100 (2004).
18. Chuang, H.Y., Lee, E., Liu, Y.T., Lee, D. & Ideker, T. Network-based classification of breast cancer metastasis. *Mol Syst Biol* **3**, 140 (2007).
19. Rapaport, F., Zinovyev, A., Dutreix, M., Barillot, E. & Vert, J.P. Classification of microarray data using gene networks. *BMC Bioinformatics* **8**, 35 (2007).
20. Minn, A.J. et al. Genes that mediate breast cancer metastasis to lung. *Nature* **436**, 518-24 (2005).
21. Troyanskaya, O.G., Garber, M.E., Brown, P.O., Botstein, D. & Altman, R.B. Nonparametric methods for identifying differentially expressed genes in microarray data. *Bioinformatics* **18**, 1454-61 (2002).
22. Peri, S. et al. Human protein reference database as a discovery resource for proteomics. *Nucleic Acids Res* **32**, D497-501 (2004).
23. Bader, G.D. et al. BIND--The Biomolecular Interaction Network Database. *Nucleic Acids Res* **29**, 242-5 (2001).
24. Stark, C. et al. BioGRID: a general repository for interaction datasets. *Nucleic Acids Res* **34**, D535-9 (2006).
25. Elkon, R. et al. SPIKE - a database, visualization and analysis tool of cellular signaling pathways. *BMC Bioinformatics* **9**, 110 (2008).
26. Shamir, R. et al. EXPANDER--an integrative program suite for microarray data analysis. *BMC Bioinformatics* **6**, 232 (2005).
27. Kayo, T., Allison, D.B., Weindruch, R. & Prolla, T.A. Influences of aging and caloric restriction on the transcriptional profile of skeletal muscle from rhesus monkeys. *Proc Natl Acad Sci U S A* **98**, 5093-8 (2001).
28. Adler, A.S. et al. Motif module map reveals enforcement of aging by continual NF-kappaB activity. *Genes Dev* **21**, 3244-57 (2007).
29. Calderwood, S.K., Khaleque, M.A., Sawyer, D.B. & Ciocca, D.R. Heat shock proteins in cancer: chaperones of tumorigenesis. *Trends Biochem Sci* **31**, 164-72 (2006).
30. Kwon, Y.S. et al. Sensitive ChIP-DSL technology reveals an extensive estrogen receptor alpha-binding program on human gene promoters. *Proc Natl Acad Sci U S A* **104**, 4852-7 (2007).
31. Oh, J.J., Grosshans, D.R., Wong, S.G. & Slamon, D.J. Identification of differentially expressed genes associated with HER-2/neu overexpression in human breast cancer cells. *Nucleic Acids Res* **27**, 4008-17 (1999).
32. Welsh, J.B. et al. Analysis of gene expression profiles in normal and neoplastic ovarian tissue samples identifies candidate molecular markers of epithelial ovarian cancer. *Proc Natl Acad Sci U S A* **98**, 1176-81 (2001).
33. Suthram, S., Shlomi, T., Ruppin, E., Sharan, R. & Ideker, T. A direct comparison of protein interaction confidence assignment schemes. *BMC Bioinformatics* **7**, 360 (2006).

Computational Systems Bioinformatics 2008

Computational Genomics

THE EFFECT OF MASSIVE GENE LOSS FOLLOWING WHOLE GENOME DUPLICATION ON THE ALGORITHMIC RECONSTRUCTION OF THE ANCESTRAL POPULUS DIPLOID

Chunfang Zheng

Department of Biology, University of Ottawa,
Ottawa, Ontario K1N 6N5, Canada
Email: czh033@uottawa.ca

P. Kerr Wall

Biology Department, Penn State University,
University Park, PA 16802, USA
Email: pkerrwall@psu.edu

Jim Leebens-Mack

Department of Plant Biology, University of Georgia,
Athens, GA 30602, USA
Email: jleebensmack@plantbio.uga.edu

Victor A. Albert

Joint Centre for Bioinformatics, University of Oslo,
NO-0316 Oslo, Norway
Department of Biological Sciences, SUNY Buffalo,
Buffalo, NY 14260, USA
Email: victor.albert@nhm.uio.no

Claude dePamphilis

Biology Department, Penn State University,
University Park, PA 16802, USA
Email: cwd3@psu.edu

David Sankoff*

Department of Mathematics and Statistics, University of Ottawa,
Ottawa, Ontario K1N 6N5, Canada
**Email: sankoff@uottawa.ca*

We improve on guided genome halving algorithms so that several thousand gene sets, each containing two paralogs in the descendant T of the doubling event and their single ortholog from an undoubled reference genome R, can be analyzed to reconstruct the ancestor A of T at the time of doubling. At the same time, large numbers of defective gene sets, either missing one paralog from T or missing their ortholog in R, may be incorporated into the analysis in a consistent way. We apply this genomic rearrangement distance-based approach to the recently sequenced poplar (*Populus trichocarpa*) and grapevine (*Vitis vinifera*) genomes, as T and R respectively.

1. INTRODUCTION

Following an episode of whole genome doubling, intra- and interchromosomal rearrangement processes over evolutionary time redistribute segments both large and small across the genome. The present-day genome can be largely decomposed into a set of duplicated DNA segments dispersed among the chromosomes, with all the duplicate pairs exhibiting a similar degree a sequence divergence. A linear-time "genome halving" algorithm, based only

*Corresponding author.

on the chromosomal segment order, can find an ancestral genome that minimizes the genomic distance to the present-day genome[1, 2]. Unfortunately, the output of the combinatorial optimization method does not really suffice as a solution to the reconstruction problem, since there may be a large number of very different, equally optimal solutions. Our guided genome halving (GGH) strategy to oversome this non-uniqueness problem is to guide the reconstruction of the ancestor by one or more reference, or outgroup, genomes. This strategy does not imply sacrificing optimality of the halving solution.

The flowering plants are well-known for numerous historical events of genome doubling[3]. The recently sequenced poplar genome (*Populus trichocarpa*)[4], which shows very clear evidence of recent genome duplication, and the grapevine genome (*Vitis vinifera*)[5, 6], whose ancestor diverged before the aforementioned duplication, provide a pair of analytical incentives to the GGH strategy. On the one hand, the poplar data has an order of magnitude more duplicated elements than has previously been analyzed, straining computational resources. On the other hand, the richness of the data allows us to assess the effects on ancestral genome reconstruction of the apparently massive loss of duplicate genes, as suggested by the modest proportion of paralogous pairs we can detect, as the poplar genome discarded most of its duplications.

This paper thus contributes two advances on the methodological level: first, the scaling up, by more than an order of magnitude, of the amount of data amenable to our analysis, and second, the incorporation of data from gene duplicate pairs that have lost one member, making use of chromosomal context in both the genome that can be traced to the doubling event and in the outgroup.

1.1. Background

Algorithms for guided genome halving (GGH), or reconstruction of the pre-doubling genome with the help of an outgroup, were first used for the ancestral doubled genome of the maize (*Zea mays*) genome, with the rice (*Oryza sativa*) and sorghum (*Sorghum bicolor*) genomes as outgroups[7]. We generated all the 1.5×10^6 solutions to the genome halving problem for the maize genome, and then identified the subset, containing only a handful of relatively similar solutions that have a minimum rearrangement distance with the rice (or sorghum) genome.

This approach was feasible with the small number (34) of doubled blocks identified in maize that were also present in one copy in each outgroup, but in a subsequent analysis[8], when we attempted to reconstruct the ancient doubled yeast genome from which *Saccharomyces cerevisiae* is descended, guided simultaneously by both of the undoubled outgroup genomes *Ashbya gossypii* and *Kluyveromyces waltii*, the number of doubled genes we could use as evidence was an order of magnitude greater than the number of blocks in the cereals data, and the number of solutions to the halving problem astronomical. It was no longer feasible to exhaustively search the halving solutions to find those that are closest to the outgroups. Instead we took a random sample of several thousand solutions in the hope that the best one might be optimal, or close to it. It was not clear, however, how large the sample should be, or how to validate the results, since the local optima found in that study remained fairly far apart, as measured by genomic rearrangement distance.

In our current use of GGH, on yeast[9] and on the flowering plants studied in the present article, we seek to replace the brute force approach of generating all (or a random sample of) halving solutions first, i.e., before taking into consideration the outgroup genome. Instead, we inject all pertinent information derivable from the outgroup into the halving algorithm, influencing hitherto arbitrary choices in that algorithm so that the halving solution is guided towards the outgroup.

1.2. Outline

In the next section, we sketch the necessary background about genomic rearrangement distance and the genome halving and GGH algorithms. In Section 3, we describe the sources for our data and how we processed them to obtain the gene sets for the GGH analysis. In Section 4 we present the GGH algorithm incorporating both full and defective gene sets. We apply this method to the full gene sets in combination with one or both of two defective gene sets from *Populus* and *Vitis* in Section 5. We present the reconstructed undoubled *Populus* ancestor based on over 6000 gene sets and evaluate the evolutionary signal versus noise (a) in the ancestor-*Populus* and ancestor-*Vitis* comparisons, (b) in the full and defective gene sets, and (c) in genes with two or three common adjacencies in the data and those with weaker positional evidence.

2. FORMAL PRELIMINARIES AND PREVIOUS WORK

2.1. Genomes, rearrangement operations and genomic distance

A genome G is represented by a set of strings (called *chromosomes*) of form $\{g_{11} \cdots g_{1n_1}, ..., g_{\chi 1} \cdots g_{\chi n_\chi}\}$, where $n = n_1 + \cdots + n_\chi$ and $\{|g_{..}|\} = \{1, \cdots, n\}$; i.e., each integer $i \in \{1, \cdots, n\}$ appears exactly once in the genome and may have either positive or negative polarity. The biologically-motivated operations of reversal or inversion, reciprocal translocation, chromosome fission or fusion, and transposition, can all be represented by an operation (called double-cut and join, or DCJ) of cutting the genome twice, each time between two elements on one of the chromosomes and rejoining the four resulting cut ends differently[10, 11]. Whether the two cuts are on the same chromosome or not, and how the endpoints are rejoined, determine which rearrangement operation pertains.

The genome rearrangement distance $d(G, H)$ is defined to be the minimum number of DCJ operations required to convert one of the genomes, G, into the other, H.

Rearrangement algorithms[12, 13, 10] can be formulated in terms of the bi-coloured "breakpoint graph", where each end (either $5'$ or $3'$) of a gene in genome G is represented by a vertex joined by a black edge to the vertex for adjoining end of the adjacent gene, and these same ends, represented by the same $2n$ vertices in the graph, are joined by gray edges determined by the adjacencies in genome H. In addition, each vertex representing a first or last term of some chromosome in G or in H is connected by a edge of the appropriate colour to an individual "cap" vertex, and there are specific rules for adding caps to the genome with fewer chromosomes and for joining the caps among themselves. if G has χ chromosomes and H has no more than χ, there are $2n + 4\chi$ vertices in all. The breakpoint graphs necessarily consist of disjoint alternating colour cycles, and it can be shown that, in the DCJ formulation, $d(G, H) = n + \chi - c$, where c is the number of cycles in the breakpoint graph. Calculating the distance can be done in time linear in n.

2.2. Genome halving

Let T be a genome consisting of ψ chromosomes and $2n$ genes $a_1^{(1)} \cdots, a_n^{(1)}; a_1^{(2)}, \cdots, a_n^{(2)}$, dispersed in any order on the chromosomes. For each i, we call $a_i^{(1)}$ and $a_i^{(2)}$ "duplicates", but there is no particular property distinguishing all elements of the set of $a_i^{(1)}$ in common from all those in the set of $a_i^{(2)}$. A potential "doubled ancestor" of T is written $A' \oplus A''$, and consists of 2χ chromosomes, where some half (χ) of the chromosomes, symbolized by the A', contains exactly one of $a_i^{(1)}$ or $a_i^{(2)}$ for each $i = 1, \cdots, n$. The remaining χ chromosomes, symbolized by the A'', are each identical to one in the first half, in that where $a_i^{(1)}$ appears on a chromosome in the A', $a_i^{(2)}$ appears on the corresponding chromosome in A'', and where $a_i^{(2)}$ appears in A', $a_i^{(1)}$ appears in A''. We define A to be either of the two halves of $A' \oplus A''$, where the superscript (1) or (2) is suppressed from each $a_i^{(1)}$ or $a_i^{(2)}$. *The genome halving problem for T is to find an A for which some $d(A' \oplus A'', T)$ is minimal.*

In the rearrangement distance algorithm, construction of the breakpoint graph is an easy step. The genome halving algorithms [2] also make use of the breakpoint graph, but the problem here is the more difficult one of building the breakpoint graph where one of the genomes (the doubled ancestor $A' \oplus A''$) is unknown. This is done by segregating the vertices of the graph in a natural way into subsets, such that all the vertices of each cycles must fall within a single subset, and then constructing these cycles in an optimal way within each subset so that the black edges correspond to the structure of the known genome T and the gray edges define the adjacencies of $A' \oplus A''$.

As a first step each gene a in a doubled descendant is replaced by a pair of vertices (a_t, a_h) or (a_h, a_t) depending if the DNA is read from left to right or right to left. The duplicate of gene $a = (a_t, a_h)$ is written $\bar{a} = (\bar{a_t}, \bar{a_h})$.

Following this, for each pair of neighbouring genes, say (a_t, a_h) and (b_h, b_t), the two adjacent vertices a_h and b_h are linked by a black edge, denoted $\{a_h, b_h\}$ in the notation of Ref. 11. For a vertex at the end of a chromosome, say b_t, it generates a virtual edge of form $\{b_t, \text{end}\}$. Note that the use of "end" instead of "cap" reflects a somewhat different bookkeeping for the beginnings and ends of chromosome in the halving algorithm compared to the distance algorithm in Section 2.1.

The edges thus constructed are then partitioned into *natural graphs* according to the following principle: If an edge $\{x, y\}$ belongs to a natural graph, then so does some edge of form $\{\bar{x}, z\}$ and some edge of form $\{\bar{y}, w\}$. If a natural graph has an even number of edges, it can be shown that in all optimal ancestral doubled genomes, the edges coloured gray, say, representing adjacent vertices in the ancestor, and incident to one of the vertices in this natural graph, necessarily have as their other endpoint another vertex within the same natural graph.

For all other natural graphs, there are one or more ways of grouping them pairwise into *supernatural graphs* so that an optimal doubled ancestor exists such that the edges coloured gray incident to any of the vertices in a supernatural graph have as their other endpoint another vertex within the same supernatural graph. Thus the supernatural graph may be completed one at a time.

An important detail in this construction is that before a gray edge is added during the completion of a supernatural graph, it must be checked to see that it would not inadvertently result in a circular chromosome. Key to the linear worst-case complexity of the halving algorithm is that this check may be made in constant time.

Along with the multiplicity of solutions caused by different possible constructions of supernatural graphs, within such graphs and within the natural graphs, there may be many ways of drawing the gray edges. Without repeating here the lengthy details of the halving algorithm, it suffices to note that these alternate ways can be generated by choosing one of the vertices within each supernatural graph as a starting point.

2.3. Genome halving with outgroups

Let T be a genome consisting of ψ chromosomes and $2n$ genes $a_1^{(1)} \cdots, a_n^{(1)}; a_1^{(2)}, \cdots, a_n^{(2)}$, dispersed in any order on the chromosomes, where for each i, genes $a_i^{(1)}$ and $a_i^{(2)}$ are duplicates. Any genome R is a reference or outgroup genome for T if it contains the n genes $a_1, \cdots, a_n$.

Let R be a reference genome for T. The GGH problem with one outgroup is to find a potential ancestral genome A such that some $d(R, A) + d(A' \oplus A'', T)$ is minimal. In practice, A is either one of the solutions to the unconstrained halving problem, or it is close to such a solution, so little is lost in restricting our search to the set of solutions of the genome halving problem for T.

One strategy, suitable for small data sets, as in Ref. 7, is to generate the entire set $\mathbf{S}$ of genome halving solutions of T, then to evaluate each $A \in \mathbf{S}$ to find the one that minimizes $d(R, A)$.

When $\mathbf{S}$ is so large that it is not feasible to generate all of $\mathbf{S}$ in order to find the best A, we may resort to sampling $\mathbf{S}$, as in Ref. 8. In defining the gray edges in the supernatural graphs of Section 2.2, we generally have several choices at some of the steps. By randomizing this choice, we are effectively choosing a random sample of $X \in \mathbf{S}$.

3. THE POPULUS-VITIS COMPARISON

Annotations for the *Populus* and *Vitis* genomes were obtained from databases maintained by the U.S. Department of Energy's Joint Genome Institute[4] and the French National Sequencing Center, Genoscope[6], respectively. An all-by-all BLASTP search was run on a data set including all *Populus* and *Vitis* protein coding genes, and orthoMCL[14] was used to construct 2104 full and 4040 defective gene sets, in the first case containing two poplar paralogs (genome T) and one grape ortholog (genome R), and in the second case missing a copy from either T or R. The chromosomal location and orientation of these paralogs and orthologs was used to construct our database of gene orders for these genomes, and the input to the GGH algorithm.

4. THE GGH ALGORITHM

The key idea in our improvement over brute force algorithms is to incorporate information from R during the halving process. It is important to take advantage of the common structure in T and R as early as possible, before it can be destroyed in the course of construction. To this end, we drop the practice of completing all the gray edges in one supernatural graph before starting another. We simply look for elements of common structure and add gray edges accordingly, always making sure that no circular chromosomes are inadvertently created.

Missing homologs The halving algorithm requires full gene sets at several steps in reconstructing the ancestor, so we algorithmically restore the missing homologs to appropriate positions in T and R at the outset.

Paths We define a path to be any connected fragment of a breakpoint graph, namely any connected fragment of a cycle. We represent each path by an unordered pair $(u, v) = (v, u)$ consisting of its current endpoints, though we keep track of all its vertices and edges. Initially, each black edge in T is a path, and each black edge in R is a path.

Pathgroups A pathgroup Γ is an ordered triple of paths, two in T and one in R, where one endpoint of one of the paths in T is the duplicate of one endpoint of the other path in T and both are orthologous to one of the endpoints of the path in R. The other endpoints may be duplicates or orthologs to each other, or not.

4.1. The algorithms

In adding pairs of gray edges to connect duplicate pairs of terms in the breakpoint graph of T versus $A' \oplus A''$, (which is being constructed), our approach is basically greedy, but with a sophisticated look-ahead. We can distinguish five different levels of desirability, or priority, among potential gray edges, i.e., potential adjacencies in the ancestor.

Recall that in constructing the ancestor A to be close to the outgroup R, such that $A' \oplus A''$ is simultaneously close to T, we must create as many cycles as possible in the breakpoint graphs between A and R and in the breakpoint graph of $A' \oplus A''$ versus T.

(1) Adding two gray edges would create two cycles in the breakpoint graph defined by T and $A' \oplus A''$, by closing two paths. When this possibility exists, it must be realized, since it is an obligatory choice in any genome halving algorithm. It may or may not also create cycles in the breakpoint graph comparison of X with the outgroup, but this does not affect its priority.

(2) Adding two gray edges would create two cycles, one for T and one for the outgroup.

(3) Adding two gray edges would create a cycle in the T versus $A' \oplus A''$ comparison, but none for the outgroup. It would, however, create a higher priority pathgroup.

(4) Adding two gray edges would create a cycle in the T versus $A' \oplus A''$ comparison, but none for the outgroup, nor would it create any higher priority pathgroup.

(5) Each remaining path terminates in duplicate terms, which cannot be connected to form a cycle, since in $A' \oplus A''$ these must be on different (and identical) chromosomes. In supernatural graphs containing such paths, there is always another path and adding two gray edges between the endpoints of the two paths can create a cycle.

In not completing each supernatural graph before moving on to another, we lose the advantage in Ref. 2 of a constant time check against creating circular chromosomes. The worst case becomes a linear time check. This is a small liability, because the worst case scenario is seldom realized, the check almost always requiring only one or two steps.

Algorithm GGH:
Guided Genome Halving with Full and Defective Gene Sets

Input. Two genomes: duplication descendant T', outgroup genome R', where each gene is has three homologs (full set) or two homologs (defective set), in the patterns TTR, TT or TR.
Output. Augmented genomes T, and R, where all gene sets are full, and
Genome A, a halving solution of T, minimizing $d(A' \oplus A'', T) + d(A, R)$.
insertMH
Initialize paths (black edges) in T and R.
Construct supernatural graphs.
Construct two pathgroups for each gene g in R, one based on g_t, the other on g_h.
If number of chromosomes in T is odd,
add pathgroup with two paths of form (end, end).
While there remains at least one pathgroup
For each pathgroup $((x,y),(\bar{x},z),(x,m))$
classify it by case and priority, and find a pathgroup Γ that has the highest priority. To choose among Priority 2 pathgroups, find one that maximizes the number of "real" black edges, i.e., edges in T' and R', not just edges created by **insertMH**. Similarly for Priority 3 pathgoups.
Case 1: $\bar{x} \neq y$, and adding xy and $\bar{x}\bar{y}$ would not create a circular chromosome.
Priority 1: $z = \bar{y}$.
Priority 2: $y = m$.
Priority 3: adding xy and $\bar{x}\bar{y}$ would create a pathgroup with priority 2.
Priority 4: None of 1, 2 or 3.
Case 2: $\bar{x} \neq y$, and adding $x\bar{z}$ and $\bar{x}z$ would not create a circular chromosome.
Priority 2: $z = m$.
Priority 3: adding $x\bar{z}$ and $\bar{x}z$ would create a pathgroup with priority 2.
Priority 4: Neither of 2 or 3.
Case 3:$\bar{x} = y$.
Priority 5:
If Γ is Case 1, **addGrayEdge**$(xy, \bar{x}\bar{y})$.
If Γ is Case 2, **addGrayEdge**$(x\bar{z}, \bar{x}z)$.
If Γ is Case 3, find some
$W = ((w,\bar{w}),(\bar{w},w),(w,s))$ in the same supernatural graph and **addGrayEdge**$(xw, \bar{x}\bar{w})$.

Algorithm: addGrayEdge($rt, \bar{r}\bar{t}$)

Add gray edges $rt, \bar{r}\bar{t}$ to partially completed genome $X" \oplus X''$.
Add gray edge rt to partially completed genome X.
Update paths in pathgroups that are affected by the new gray edges.
Remove pathgroups that start with r and t.

Algorithm: insertMH:
Insert Missing Homologs in Chromosomes

Input. Two genomes: duplication descendant T', outgroup R', where each gene is has two or three homologs, in the patterns TTR, TT, TR.
Output. Augmented genomes T and R containing exactly three homologs for each gene, in the pattern TTR, maximizing the number of common edges of form $\{a_1, b_1\}$, $\{a_2, b_2\}$ in T and $\{a, b\}$ in R.
(Or $\{a_1, b_2\}$, $\{a_2, b_1\}$ in T and $\{a, b\}$ in R.)
While there are genes that have only two copies, **count edgeDiff** for each such, which simultaneously finds the BestPosition.
Insert the gene with the minimum edgeDiff value into the BestPosition of this gene.

Algorithm: count edgeDiff

If a gene g just has one copy (g_1) in T' and one copy (g) in R', then we must insert another copy (g_2) into T'.
If a gene g just has two copies (g_1, g_2) in T', then we must insert g into R'.

(The details are omitted here. This is essentially a greedy heuristic to add adjacencies reflecting, as if possible, adjacencies already existing in R' and T'.)

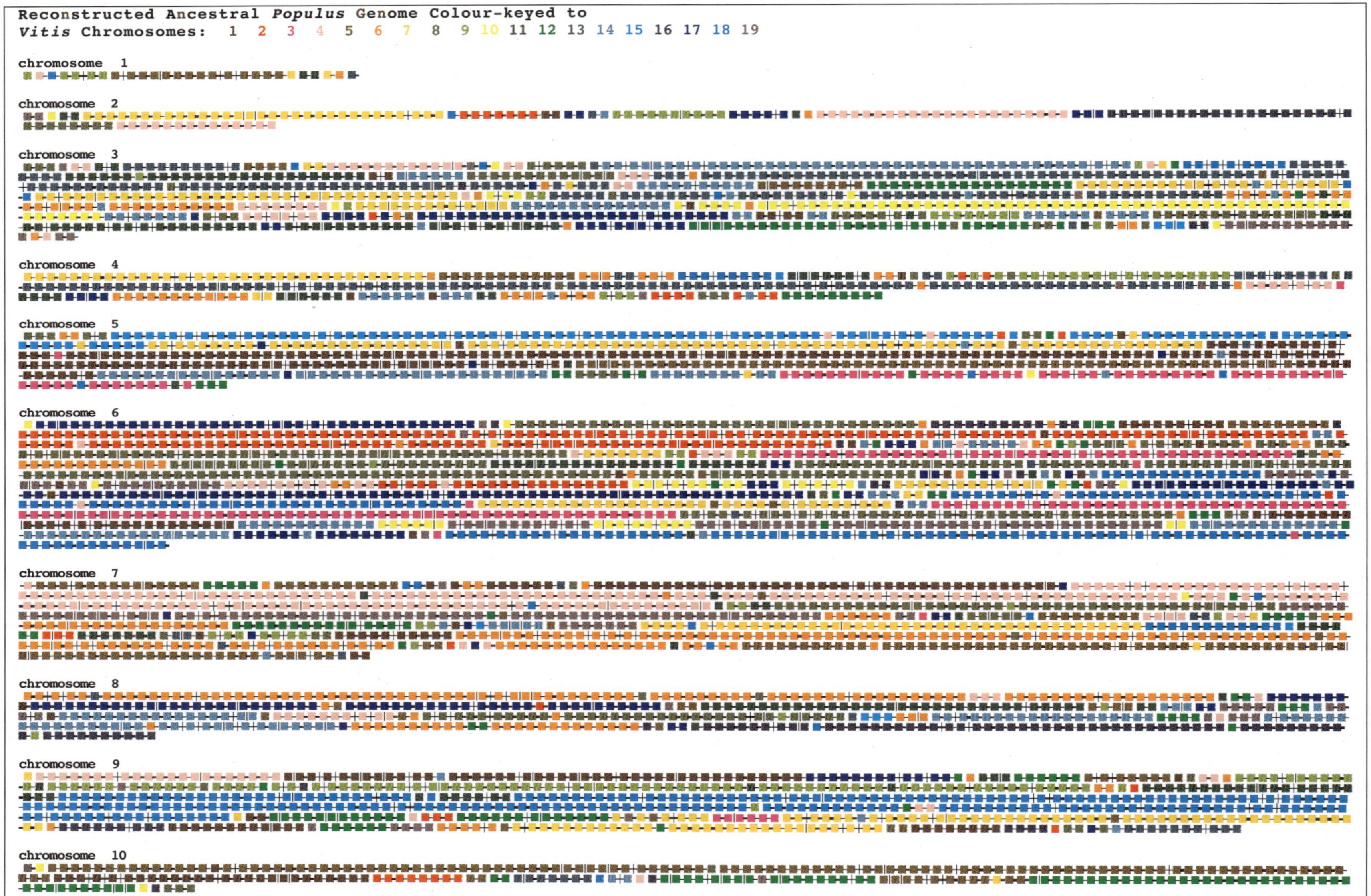

Fig. 1. Ten chromosomes (wrapped) of ancestral poplar genome reconstructed by GGH algorithm from 6144 full and defective gene sets. Only genes with grape orthologs are indicated. Adjacencies present three times, i.e., twice in poplar and once in grape are indicated by $\equiv$, those present twice by $=$ and those present once by $-$. Intrachromosomal breakpoints within segments indicated by $|$.

5. RESULTS AND DISCUSSION

Our data consisted of 6144 gene sets, of which only 2104 were full sets. There were only 836 defective sets by virtue of a missing ortholog in V, while 3204 genes lacked one paralog in T.

Table 1. Comparisons of the reconstructed immediate pre–doubling ancestor A with the *Vitis* genome and of the immediate doubled ancestor $A \oplus A$ with *Populus*. PPV: full gene sets, PP: defective, missing grape ortholog, PV: defective, missing one poplar paralog. Projected: genes not in PPV ancestor deleted from solution A, d: genomic distance, b:,number of breakpoints, $r = 2d/b$: the re-use statistic.

data sets	genes in A	$d(A, Vitis)$ d	b	r	$d(A\oplus A, Populus)$ d	b	r
PPV	2104	638	751	1.70	454	690	1.32
PPV,PP	2940	649	757	1.71	737	1090	1.35
projected	2104	649	757	1.71	581	823	1.41
PPV,PV	5308	1180	1331	1.77	1083	1457	1.49
projected	2104	663	758	1.75	670	833	1.61
PPV,PP, PV	6144	1208	1363	1.77	1337	1812	1.48
projected	2104	664	757	1.75	750	926	1.62
without singletons							
PPV	2020	560	661	1.69	346	541	1.28
PPV,PP	2729	594	690	1.72	453	714	1.27
projected	2006	571	664	1.72	416	628	1.32
PPV,PV	4203	573	686	1.67	751	1031	1.46
projected	1955	489	580	1.69	490	644	1.52
PPV,PP, PV	4710	675	797	1.69	856	1211	1.41
projected	1986	528	622	1.70	558	744	1.50

Table 1 shows the results of the analysis on the full gene sets only, on combinations of the full sets with one kind of defective sets, and all three sets. For each case we study not only the reconstructed ancestor but also a "projected" version where genes from the defective sets are simply erased, in order to assess the changes in gene order due to the defective gene sets. Whereas the distance between each T and its reconstructed ancestor A is given by GGH, the distance between projected ancestor and T required a heuristic, not detailed here, for attributing each paralog in T to one of the two copies of the ancestral genome.

Figure 1 depicts the result of analyzing all the 6144 gene sets with GGH, although the 836 genes with no grape ortholog are not visible. The large number of singleton genes disrupting otherwise homogeneous synteny blocks suggests that "noise" due to uncertainties inherent in homology identification and especially orthology identification may be artifactually inflating genomic distance d and the number of breakpoints b. Since the rigorous noise elimination techniques of Refs. 15 and 16 have not yet been extended in the context of genome doubling, we simply identified singletons as gene sets lacking two real (i.e., not inferred from **insertMH**) common adjacencies out of six possible in the original genomes, and ran all the analyses again without these genes.

In each case, we counted the breakpoints and calculated the appropriate genomic distance d, i.e., from the doubled ancestor to *Populus* and from the undoubled version of the same ancestor to *Vitis*. This enabled us to calculate the "breakpoint re-use" statistic $r = 2d/b$, which is a measure of how much signal about conserved order (among segments, not within segments) remains in the comparison of two genomes after a period of evolutionary rearrangements. When $r = 1$, we can have high confidence in the rearrangement distance and history. When r approaches two, the segment order in the two genomes being compared are essentially random with respect to each other, i.e., calculating r for random genomes gives a value approaching 2[a]. In Table 1, we see both from changes in d and changes in r that

- most of the signal contained in the order among conserved chromosomal segments has been lost between the ancestor and *Vitis*, but is retained to a great degree between the ancestor and *Populus*, probably reflecting the difference in divergence time but also possible biases towards T in the GGH algorithm,
- the addition of the defective PV gene sets degrades the analysis, more than the addition of PP sets, though this may due to the four times greater number of gene sets in the former,
- the elimination of singletons improves all the analyses, but where PV is present, this comes about largely by discarding most of the sets, which turn out to be singletons.

With the application of our method to the more than

[a]If breakpoints are frequently re-used during evolution, then r will also be close to 2; unfortunately there is no internal way of testing the breakpoint re-use hypothesis against the null hypothesis of complete loss of signal about segment order[17].

6000 gene sets, we have shown that any realistic case of genome doubling should be amenable, even if all the gene paralogs remain in the sequenced descendant. The analysis with 6144 gene sets required almost 48 hours on a MacBook, but this was anomalously large, since those with 4000 or 5000 required less than 5 hours and those with 2000 about 1 hour. Much of the running time is due to the check on the number of real edges in a pathgroup to choose among Priority 2 or among Priority 3 options. This could be reduced by optimizing data structures in our software.

The inclusion of defective PV gene sets would appear to add little more than noise to the analysis, but the PP sets would seem to add significant information, especially to the ancestor-*Populus* comparison.

The elimination of singletons proves to be a meaningful way of drastically decreasing the number of segments (as measured by b) and the genomic distance to credible levels, though this still does not result in a detectible signal in the ancestor-*Vitis* comparison. The recently sequenced and asembled *Carica papaya* genome, which is phylogenetically more closely related to *Populus*, but like *Vitis* diverged before the *Populus* doubling event, should be better able play the outgroup role in our analysis, once it is published and we have been able to identify orthologs.

Acknowledgments

Research supported in part by grants from the Natural Sciences and Engineering Research Council of Canada (NSERC). DS holds the Canada Research Chair in Mathematical Genomics.

References

1. El-Mabrouk N, Bryant D, Sankoff D. Reconstructing the pre-doubling genome. In: Istrail S, Pevzner P, Waterman M (eds.), Third Annual International Conference on Computational Molecular Biology (RECOMB 99). ACM Press, New York. 1999: 154–163.
2. El-Mabrouk N, Sankoff D. The reconstruction of doubled genomes. *SIAM Journal on Computing* 2003; **32**: 754–792.
3. Cui L, Wall PK, Leebens-Mack JH, Lindsay BG, Soltis DE, Doyle JJ, Soltis PS, Carlson JE, Arumuganathan K, Barakat A, Albert VA, Ma H, dePamphilis CW. Widespread genome duplications throughout the history of flowering plants. *Genome Research* 2006; **16**: 738–749.
4. Tuskan GA, Difazio S, Jansson S, Bohlmann J, Grigoriev I, Hellsten U, Putnam N, Ralph S, Rombauts S, Salamov A, Schein J, Sterck L, Aerts A, Bhalerao RR, Bhalerao RP, Blaudez D, Boerjan W, Brun A, Brunner A, Busov V, Campbell M, Carlson J, Chalot M, Chapman J, Chen GL, Cooper D, Coutinho PM, Couturier J, Covert S, Cronk Q, Cunningham R, Davis J, Degroeve S, Djardin A, Depamphilis C, Detter J, Dirks B, Dubchak I, Duplessis S, Ehlting J, Ellis B, Gendler K, Goodstein D, Gribskov M, Grimwood J, Groover A, Gunter L, Hamberger B, Heinze B, Helariutta Y, Henrissat B, Holligan D, Holt R, Huang W, Islam-Faridi N, Jones S, Jones-Rhoades M, Jorgensen R, Joshi C, Kangasjrvi J, Karlsson J, Kelleher C, Kirkpatrick R, Kirst M, Kohler A, Kalluri U, Larimer F, Leebens-Mack J, Lepl JC, Locascio P, Lou Y, Lucas S, Martin F, Montanini B, Napoli C, Nelson DR, Nelson C, Nieminen K, Nilsson O, Pereda V, Peter G, Philippe R, Pilate G, Poliakov A, Razumovskaya J, Richardson P, Rinaldi C, Ritland K, Rouz P, Ryaboy D, Schmutz J, Schrader J, Segerman B, Shin H, Siddiqui A, Sterky F, Terry A, Tsai CJ, Uberbacher E, Unneberg P, Vahala J, Wall K, Wessler S, Yang G, Yin T, Douglas C, Marra M, Sandberg G, Van de Peer Y, Rokhsar D. The genome of black cottonwood, Populus trichocarpa (Torr. & Gray). *Science* 2006; **313**: 1596–1604. http://genome.jgi-psf.org/Poptr1/Poptr1.download.html
5. Velasco R, Zharkikh A, Troggio M, Cartwright DA, Cestaro A, Pruss D, Pindo M, Fitzgerald LM, Vezzulli S, Reid J, Malacarne G, Iliev D, Coppola G, Wardell B, Micheletti D, Macalma T, Facci M, Mitchell JT, Perazzolli M, Eldredge G, Gatto P, Oyzerski R, Moretto M, Gutin N, Stefanini M, Chen Y, Segala C, Davenport C, Dematt L, Mraz A, Battilana J, Stormo K, Costa F, Tao Q, Si-Ammour A, Harkins T, Lackey A, Perbost C, Taillon B, Stella A, Solovyev V, Fawcett JA, Sterck L, Vandepoele K, Grando SM, Toppo S, Moser C, Lanchbury J, Bogden R, Skolnick M, Sgaramella V, Bhatnagar SK, Fontana P, Gutin A, Van de Peer Y, Salamini F, Viola R. A high quality draft consensus sequence of the genome of a heterozygous grapevine variety. *PLoS ONE* 2007; **2**: e1326.
6. Jaillon O, Aury JM, Noel B, Policriti A, Clepet C, Casagrande A, Choisne N, Aubourg S, Vitulo N, Jubin C, Vezzi A, Legeai F, Hugueney P, Dasilva C, Horner D, Mica E, Jublot D, Poulain J, Bruyre C, Billault A, Segurens B, Gouyvenoux M, Ugarte E, Cattonaro F, Anthouard V, Vico V, Del Fabbro C, Alaux M, Di Gaspero G, Dumas V, Felice N, Paillard S, Juman I, Moroldo M, Scalabrin S, Canaguier A, Le Clainche I, Malacrida G, Durand E, Pesole G,

Laucou V, Chatelet P, Merdinoglu D, Delledonne M, Pezzotti M, Lecharny A, Scarpelli C, Artiguenave F, P ME, Valle G, Morgante M, Caboche M, Adam-Blondon AF, Weissenbach J, Qutier F, Wincker P; French-Italian Public Consortium for Grapevine Genome Characterization. The grapevine genome sequence suggests ancestral hexaploidization in major angiosperm phyla. *Nature* 2007; **449**: 463–467. http://www.genoscope.cns.fr/externe/English/Projets/Projet_ML/data/annotation/

7. Zheng C, Zhu Q, Sankoff D. Genome halving with an outgroup. *Evolutionary Bioinformatics* 2006; **2**: 319–326.
8. Sankoff D, Zheng C, Zhu Q. Polyploids, genome halving and phylogeny. *Bioinformatics* 2007; **23**: i433–i439.
9. Zheng C, Zhu Q, Adam Z, Sankoff D. Guided genome halving: hardness, heuristics and the history of the Hemiascomycetes. *Bioinformatics* 2008; **24**.
10. Yancopoulos S., Attie O, Friedberg R. Efficient sorting of genomic permutations by translocation, inversion and block interchange. *Bioinformatics* 2005; **21**: 3340–3346
11. Bergeron A, Mixtacki J, Stoye J. A unifying view of genome rearrangements. In: Bücher P, Moret BME (eds.), Workshop on Algorithms in Bioinformatics (WABI 2006). Lecture Notes in Computer Science **4175**, 2006:163–173.
12. Bafna V, Pevzner P. Genome rearrangements and sorting by reversals. *SIAM Journal of Computing* 1996; **25**: 272–289.
13. Tesler G. Efficient algorithms for multichromosomal genome rearrangements. *Journal of Computer and System Sciences* 2002; **65**: 587–609.
14. Li L, Stoeckert CJ Jr, Roos DS. OrthoMCL: identification of ortholog groups for eukaryotic genomes. *Genome Research* 2003; **13**: 2178–2189.
15. Zheng C, Zhu Q, Sankoff D. Removing noise and ambiguities from comparative maps in rearrangement analysis. *Transactions on Computational Biology and Bioinformatics* 2007; **4**: 515–522.
16. Choi V, Zheng C, Zhu Q, Sankoff D. Algorithms for the extraction of synteny blocks from comparative maps. In: Giancarlo R, Hannenhalli S. (eds.), Workshop on Algorithms in Bioinformatics (WABI 2007). Lecture Notes in Bioinformatics **4645**, 2007: 277–288.
17. Sankoff D. The signal in the genomes. *PLoS Computational Biology* 2006; **2**: e35.

ERROR TOLERANT SIBSHIP RECONSTRUCTION IN WILD POPULATIONS

Saad I. Sheikh and Tanya Y. Berger-Wolf
Department of Computer Science, University of Illinois at Chicago,
851 S. Morgan (M/C 152), Room 1120 SEO, Chicago, IL 60607
E-mail: ssheikh,tanyabw@cs.uic.edu

Mary V. Ashley and Isabel C. Caballero
Department of Biological Sciences, University of Illinois at Chicago,
840 West Taylor Street, SEL 1031 M/C 067, Chicago, IL 60607
E-mail: ashley@uic.edu,icabal2@uic.edu

Wanpracha Chaovalitwongse
Department of Industrial and Systems Engineering, Rutgers University,
CoRE Building, 96 Frelinghuysen Rd., Piscataway, NJ 08854

Bhaskar DasGupta
Department of Computer Science, University of Illinois at Chicago,
851 S. Morgan (M/C 152) Room 1120 SEO, Chicago, IL 60607
E-mail: dasgupta@cs.uic.edu

Kinship analysis using genetic data is important for many biological applications, including many in conservation biology. Wide availability of microsatellites has boosted studies in wild populations that rely on the knowledge of kinship, particularly sibling relationships (sibship). While there exist many methods for reconstructing sibling relationships, almost none account for errors and mutations in microsatellite data, which are prevalent and affect the quality of reconstruction. We present an error-tolerant method for reconstructing sibling relationships based on the concept of consensus methods. We test our approach on both real and simulated data, with both pre-existing and introduced errors. Our method is highly accurate on almost all simulations, giving over 90% accuracy in most cases. Ours is the first method designed to tolerate errors while making no assumptions about the population or the sampling.

Keywords: Sibship Reconstruction, Kinship Analysis, Consensus, Combinatorial Optimization.

1. INTRODUCTION

Kinship analysis of wild populations is an important and necessary component of studying mating systems, dispersal patterns and kin selection. In wild populations, kinship relationships (lower order pedigree) are typically inferred from microsatellite markers, rather than *Single Nucleotide Polymorphisms* (SNPs) which are more commonly used in model organisms (see Ref. 6 for discussion). There are two main approaches to kinship inference from microsatellite data: using genetic distance estimates and statistical likelihood methods [1,8,23–25], and enumeration of feasible relationships based on Mendelian constraints [2,5,6,10,22]. However, with the exception of COLONY [25], none of the existing kinship reconstruction methods is designed to tolerate genotyping errors or mutation. Yet, both errors and mutations cannot be avoided in practice and identifying these errors without any prior kinship information is a challenging task.

In Refs. 5, 6, 10, 22 we have presented a method for reconstructing sibling relationships from single generation microsatellite data that optimally identifies the most parsimonious set of sibling groups subject only to Mendelian inheritance constraints. We have shown that our method performs comparably or better than other sibling reconstruction approaches on both biological and simulated data. While our method was not designed for data with genotyping errors, it did perform relatively well on data that contained a limited number of errors. In this paper we present a new approach for reconstructing sibling relationships from microsatellite data designed explicitly to tolerate genotyping errors and mutations in data.

1.1. Microsatellite Markers

While there are several molecular markers used in population genetics such as allozymes, AFLPs, RFLPs, microsatellites (also known as SSRs, STRs, SSLPs, and VNTRs) are the most commonly used in population biology for non-model organisms. Microsatellites are repeats of short DNA sequences distributed throughout the genome. These are co-dominant, unlinked, multi-allelic markers that offer numerous advantages for population studies. Generally, phase or haplotype information is not available for microsatellite loci in non-model organisms.

1.2. Sibling Reconstruction Problem Statement

The main focus of our paper is to design a method that accurately reconstructs sibling groups from microsatellite data of a single generation in presence of genotyping errors and mutations. We have formally defined the problem of sibling reconstruction in Ref. 6 and we restate it here. Let $U = \{X_1, ...X_n\}$ be a population U of n diploid individuals of the same generation. Each individual is represented by a genetic (microsatellite) sample at l loci. That is, $X_i = (\langle a_{i1}, b_{i1}\rangle, ..., \langle a_{il}, b_{il}\rangle)$ and a_{ij} and b_{ij} are the two alleles of the individual i at locus j represented as some identifying string. We assume that the same string in the same locus corresponds to the same allele, however alleles from different loci are independent. The goal is to reconstruct the full sibling groups which is a partition of individuals into $P_1, ...P_m$ where individuals in the same partition P_i have the same parents. We assume no knowledge of parental information.

1.3. 2-Allele Algorithm

In Ref. 6 we presented a combinatorial 2-ALLELE MIN SET COVER algorithm for the siblings reconstruction problem. We rely on Mendelian inheritance constraints that dictate that full siblings must share their parents' alleles at all loci. We formalize this rule as the 2-ALLELE PROPERTY in Ref. 5 as follows: for a set of individuals there exists a reordering of individuals' alleles within a locus such that the total number of distinct alleles on each side at this locus is at most 2. Note, that the 2-allele property is a necessary constraint for a group of individuals to be siblings but not sufficient. Notice, also, that *any* two individuals necessarily satisfy the 2-allele property since by default the number of alleles on each side of any locus is at most two.

The 2-ALLELE MIN SET COVER algorithm works by first generating all maximal sibling groups that obey the 2-ALLELE PROPERTY. The algorithm then uses set cover[19] to find the minimum number of sibling groups necessary to explain the data.

1.4. Errors in Microsatellite Data

Errors and mutation cannot be verifiably avoided when genotyping wild populations. While there may be several sources and types of errors (see Refs. 13, 17), here we are concerned primarily with how they affect the sibling reconstruction problem. We now discuss errors typically present in microsatellite data.

Allelic Dropout occurs when one or both alleles are not amplified during polymerase chain reaction (PCR) and is one of the most common errors [13]. If one of the alleles is not amplified, the result mimics a homozygote. The case when both alleles are missing is easily identifiable as no amplification has occured and is handled by a simple extension of the 2-ALLELE ALGORITHM (see section 1.5).

Heterozygous Mistype occurs when two alleles are amplified by PCR but one or both of them, for a variety of reasons, are not recorded as present. In the context of sibling reconstruction any allele that was not present in either one of the parents is a mistype.

Homozygous Mistype occurs when only one allele is amplified by PCR, and does not match any of the parental alleles.

Genetic Mutation is the actual variation in the alleles, also called *polymorphism*. This arises from mistakes made during DNA replication. A mutation may also be classified as Mistypes when reconstructing sibling relationships.

Allele Combination Error occurs when one or both alleles at a locus are present in the parents (or sibling group) but Mendelian inheritance rules are still violated.

Null Alleles is the lack of any amplification. When no allele is amplified it can be explicitly marked as a missing allele.

1.5. Accommodating Missing Alleles

To accommodate known missing alleles in the data we denote them by a special symbol, *e.g.*, a wildcard (*). The 2-ALLELE MIN SET COVER algorithm then proceeds to construct feasible sibling sets treating the wildcard as any possible allele.

1.6. Consensus Methods

We base our idea of error-tolerant sibling reconstruction on the consensus-based approach. The idea behind consensus methods is to combine different solutions to the same problem into one solution, *i.e.*, group decision making. The formal theory of voting and group decision making dates back to the eighteenth century [11,12] and modernized by Kenneth J. Arrow in 1951 [3]. Recently mathematical and computational group choice and consensus techniques have been applied to biological problems, mostly in the context of phylogenetic reconstruction [9]. Our solution is based on using such methods to tolerate genotyping errors. In Section 2.1 we define consensus in the context of siblings reconstruction problem, and discuss some approaches and their feasibility.

2. CONSENSUS BASED APPROACH FOR ERROR-TOLERANT SIBLINGS RECONSTRUCTION

We now describe our approach to reconstructing sibling relationships in presence of genotyping errors. Consider an individual X_i which has some genotyping error(s). Any error that is affecting siblings reconstruction must be preventing X_i's sibling relationship with at least one other individual X_j, who in reality is its sibling. It is unlikely that an error would cause two unrelated individuals to be paired up as siblings, unless all error-free loci do not contain enough information. It is possible that an individual has more than one error (albeit extremely rarely [13,17]), yet it is unlikely that all the errors bias the solution in the same direction.

Thus, we can discard one locus at a time, assuming it to be erroneous, and obtain a sibling reconstruction solution based on the remaining loci. If all such solutions put the individuals X_i and X_j in the same sibling group (*i.e.*, there is a consensus among those solutions), we consider them to be siblings. The bulk of our error-tolerant approach is concerned with pairs of individuals that do not consistently end up in the same sibling group during this process, that is, there is no consensus about their sibling relationship.

We now present a formal definition of consensus in the context of sibling reconstruction and describe our consensus-based algorithm for error- tolerant sibling reconstruction.

2.1. Consensus Methods for Siblings Reconstruction

Recall that for a population of individuals $U = \{X_1 \ldots X_n\}$ the goal of a siblings reconstruction problem is to find a partition of the population into sibling groups $S = \{P_1 \ldots P_m\}$ where all individuals are covered with no overlap:

$$\bigcup_{1 \le j \le m} P_j = U \qquad \text{and} \qquad \forall j, k \; P_j \cap P_k = \emptyset$$

A partition defines an equivalence relationship. Two individuals are equivalent if and only if they are in the same partition of the solution S.

$$X_i \equiv_S X_j \iff \exists P_k \in S \text{ s.t. } X_i \in P_k \wedge X_j \in P_k$$

We are now ready to give the definition of a consensus method.

Definition 2.1. A *consensus* method for sibling groups is a computable function f that takes k solutions $\mathcal{S} = \{S_1, ..., S_k\}$ as input and computes *one* final solution.

Definition 2.2. A *strict consensus* [21] $\mathcal{C}$ is a partitioning of sibling groups where two individuals are together only if they were in the same partition for all solutions:

$$\mathcal{C} = \{P_{\mathcal{C},1} \ldots P_{\mathcal{C},m}\} \text{ where}$$
$$X_j \equiv_{\mathcal{C}} X_k \iff \forall S_i \in \mathcal{S} \;\; X_j \equiv_{S_i} X_k$$

The strict consensus defines a true equivalence relation and, thus, is a transitive function:

$$(X_i \equiv_{\mathcal{C}} X_j \quad \wedge \quad X_j \equiv_{\mathcal{C}} X_k) \quad \Rightarrow \quad X_i \equiv_{\mathcal{C}} X_k$$

Any individual that is not consistently placed into a partition in all solutions will be added as a singleton. Such a consensus solution is reliable for the individuals that have been placed together in a group, but there may be many singleton groups.[a]

2.2. Distance-based Consensus

The original 2-Allele Algorithm finds the most parsimonious solution with the fewest number of sibling groups. While the algorithm performs well in absence of errors, it is not designed to handle errors. Moreover, the resulting sibling groups returned by the algorithm may overlap. The strict consensus, on the other hand, conservatively identifies reliable sibling relationships and puts the rest back into singleton groups. In order to combine the best aspects of both methods we present a distance based consensus method. We start with a strict consensus of the "leave-one-locus-out" solutions and search for the *nearest good parsimonious solution.* In order to search for such a solution we need quantitative measures to 1) assess the quality of a solution, f_q, and 2) calculate the pairwise distance between solutions, f_d. Assume that we have the two functions f_q and f_d.

$$f_q : S \to \mathbf{R} \qquad \text{and} \qquad f_d : S \times S \to \mathbf{R}$$

Since we start with a strict consensus $\mathcal{C}$ the partitions in the solution cannot be refined any further. Therefore to improve the solution, we use the operations of merging

[a] If we relax the above constraint to require not all but most of the solutions to agree on the equivalence relationship it gives us a "majority consensus". While it performs well in other applications, such as phylogenetic reconstruction [7], it is too biased towards loci with errors in the context of sibling reconstruction.

two sets. The following monotonic property must be obeyed by any improved solution $\mathcal{C}'$:

$$\forall X_i, X_j \in U \quad X_i \equiv_{\mathcal{C}} X_j \implies X_i \equiv_{\mathcal{C}'} X_j. \quad (1)$$

Thus, given a solution $\mathcal{C}$, we look for an improved solution $\mathcal{C}'$ that minimizes $f_d(\mathcal{C}, \mathcal{C}')$ and maximizes $f_q(\mathcal{C}')$. To combine the two objectives we can formulate the following optimization problems:

(1) Maximize f_q with an upper bound on f_d
(2) Minimize f_d with a lower bound on f_q
(3) Maximize/Minimize some (linear) combination of f_d and f_q

We prove all of these problems to be NP-Hard in general for arbitrary f_q and f_d.

Theorem 2.1. *Let $\mathcal{C}$ be a collection of sibling groups and $k \in \mathbf{R}$. Let $\mathcal{S}$ be the set of all solutions that are an improvement of $\mathcal{C}$ and are obtainable from $\mathcal{C}$ by merging sibling sets. The problem of finding an improved solution $\mathcal{C}' \in \mathcal{S}$ such that*

$$f_q(\mathcal{C}') = \max_{\substack{S \in \mathcal{S} \\ f_d(\mathcal{C}, S) \leq k}} f_q(S)$$

is NP-hard.

Proof. We show that this problem is NP-hard by reducing from the 2-ALLELE MIN SET COVER problem, which we have proven to be MAX SNP-hard [4]. We start with a collection $\mathcal{C}$ of singleton only sets and aim to minimize the number of sibling groups.

Formally, for an input $U = \{X_1, ...X_n\}$ to the 2-ALLELE MIN SET COVER, the corresponding input to the distance-based consensus problem is $\mathcal{C} = \{\{X_1\}, ..., \{X_n\}\}$ and $k = 0$. We define the distance function f_d to be

$$f_d(\mathcal{C}, \mathcal{C}') = \begin{cases} 0 & \text{groups in } \mathcal{C} \text{ can be merged to form } \mathcal{C}' \\ 1 & \text{otherwise} \end{cases}$$

Finally, we define the quality function $f_q(\mathcal{C}') = |U| - |\mathcal{C}'|$. This ensures the minimum number of sets to maximize the objective function since $|\mathcal{C}'| < |U|$.

The bound on f_d guarantees that any merged sibgroups obey the 2-allele property and the quality maximization objective ensures that the solution is a minimum set cover. □

Thus, finding improved solutions subject to the first objective is NP-hard. We now show that the second objective is NP-hard as well.

Theorem 2.2. *Let $\mathcal{C}$ be a collection of sibling groups and $k \in \mathbf{R}$ be the lower bound on f_q. Let $\mathcal{S}$ be the set of all solutions that are an improvement of $\mathcal{C}$ and are obtainable from $\mathcal{C}$ by merging sibling sets. The problem of finding an improved solution $\mathcal{C}' \in \mathcal{S}$ such that*

$$f_d(\mathcal{C}, \mathcal{C}') = \min_{\substack{S \in \mathcal{S} \\ f_q(S) \geq k}} f_d(\mathcal{C}, S)$$

is NP-hard.

Proof. Similar to the proof of Theorem 2.1 above, we again reduce from the 2-ALLELE MIN SET COVER problem. Given an input $U = \{X_1, ...X_n\}$ to the 2-ALLELE MIN SET COVER, the corresponding input to the distance-based consensus problem is $\mathcal{C} = \{\{X_1\}, ..., \{X_n\}\}$ and $k = n$. We define the distance function f_d as follows:

$$f_d(\mathcal{C}, \mathcal{C}') = \begin{cases} \infty & \text{groups in } \mathcal{C} \text{ can be merged to form } \mathcal{C}' \\ 1 & \text{otherwise} \end{cases}$$

We define the quality function f_q as the sum of the distance from strict consensus and the number of sets:

$$f_q(\mathcal{C}') = f_d(\mathcal{C}, \mathcal{C}') + |\mathcal{C}'|.$$

A reduction from the 2-ALLELE MIN SET COVER problem follows. □

Lastly, for an arbitrary combination of f_q and f_d, Objective 3 is unattainable as well.

Theorem 2.3. *Let $\mathcal{C}$ be a collection of sibling groups. Let $\mathcal{S}$ be the set of all solutions that are an improvement of $\mathcal{C}$ and are obtainable from $\mathcal{C}$ by merging sibling sets and* *let $g(f_q, f_d)$ be a (linear) combination of the functions f_q and f_d. The problem of finding an improved solution $\mathcal{C}' \in \mathcal{S}$ such that*

$$g(f_d(\mathcal{C}, \mathcal{C}'), f_q(\mathcal{C}')) = \underset{S \in \mathcal{S}}{OPT}\{g(f_d(\mathcal{C}, S), f_q(S))\}$$

is NP-hard.

Proof. This theorem follows from the Theorem 2.1 (OPT is max) and Theorem 2.2 (OPT is min). The objective with only one function is a special case of a linear combination. Hence both the minimization and the maximization objectives are NP-Hard. □

We have shown that the three versions of the problem of finding the closest best solution to a given solution of the sibling reconstruction problem are NP-hard. In the next section we present a heuristic approach that efficiently finds good solutions.

3. GREEDY DISTANCE-BASED CONSENSUS

We now present a greedy algorithm that given a collection of sibling reconstructions attempts to find a good solution with few sibling groups while allowing for a small number of errors in the data. The GREEDY CONSENSUS ALGORITHM uses costs associated with errors in data to define a merging cost and to find and merge the pair of sibling groups with the minimum (merging) cost. The sibling groups to be merged are selected by exhaustively examining all pairs of groups and identifying the merge that results in the *lowest total merging cost* for the *merged* group. Our quality function is based on the parsimony assumption: we try to find the minimum number of sibling groups and errors that explain the data. Therefore, to get the minimum number of sibling groups our quality function is defined as $f_q = |U| - |\mathcal{C}|$.

3.1. Distance Function

We define two functions necessary to calculate the distance f_d : the cost and the benefit of assigning an individual to a sibling group. The cost of an assignment is used when an individual cannot be assigned to a group without violating the 2-allele property. The total cost of tolerating errors is computed using user-defined costs for each type of possible error in data. These costs are provided by the user depending on the expected error rates and number of loci. By default, these may be uniform. The *benefit* of an assignment is determined by the shared alleles and allele pairs of the new individual, which can be added *without violating 2-allele property.*

More formally, we assume that we are given as an input the relative costs of the four distinct error types and the upper bounds on the number of errors per individual, per sibling group, and per individual in a sibling group. [b] The cost and the benefit of assigning an individual X to a sibling group P_i is defined as:

$$f_{assign}(P_i, X) = \begin{cases} benefit & \text{If } X \text{ can be added to } P_i \\ cost & \text{otherwise.} \end{cases}$$

Suppose $\mathcal{C} = \{P_1, ..., P_m\}$ is a collection of sibling groups and $\mathcal{C}'$ is a collection of groups obtained from $\mathcal{C}$ by merging groups P_i and P_j. Then we define the distance between $\mathcal{C}$ and $\mathcal{C}'$ as follows:

$$f_d(\mathcal{C}, \mathcal{C}') = \min\left\{ \sum_{X \in P_i} f_{assign}(P_j, X), \sum_{X \in P_j} f_{assign}(P_i, X) \right\}$$

3.2. Greedy Algorithm

Given an upper bound on the number of errors and the relative error costs, GREEDY CONSENSUS ALGORITHM searches for the solution with the fewest number of sibling

[b] Note that COLONY [25], the only other sibling reconstruction method that explicitly tolerates errors in data, requires considerably more detailed information about the types, costs, and frequencies of errors.

groups and errors necessary to explain the data. We denote by $U_{|i}$ the set of individuals U with the ith locus being omitted. The GREEDY CONSENSUS ALGORITHM has three phases:

(1) Calculate the 2-ALLELE MIN SET COVER solutions for $U_{|1} \dots U_{|l}$,
(2) Calculate the strict consensus C of the above solutions,
(3) Merge sibling groups greedily as allowed by the parameters.

Phase 1 runs the 2-ALLELE MIN SET COVER algorithm to obtain solutions for dropping one locus. Any technique for siblings reconstruction may be used here. We use the 2-ALLELE MIN SET COVER algorithm as the basis since it performs as well or better than other available methods and makes fewest assumptions [6]. Phase 2 works by examining all the solutions from phase 1 and placing two individuals in the same sibling group only if all the solutions agree. Unpaired individuals are placed in singleton groups. Finally, phase 3 works iteratively by merging the *closest* pair of sibling groups. This is done by calculating the f_d distance for all pairs of sibling groups at every iteration. The pair that gives the smallest distance is merged. This continues until the minimum distance is greater than either the *maximum editing cost per sibling group* or the average edit cost exceeds *maximum average editing cost per sibling group*. Both of these costs are input parameters.

To analyze the computational time complexity of the GREEDY CONSENSUS ALGORITHM, we consider each phase separately. Computing the 2-ALLELE MIN SET COVER for each subset of the input is the most expensive part of the algorithm. The 2-ALLELE MIN SET COVER problem is MAX SNP-hard [4], which means that it cannot be approximated within some constant factor in polynomial time, unless $P = NP$. We use the commercial mixed integer program solver CPLEX[c] to solve the problem to optimality. GREEDY CONSENSUS ALGORITHM executes l runs of 2-ALLELE MIN SET COVER to compute the l solutions for consensus method.

The consensus part (Phases 2-3) of GREEDY CONSENSUS ALGORITHM is polynomial: The total time for $O(n)$ iterations is $O(n^3 l)$. Note that our approach is not exclusive to the 2-ALLELE MIN SET COVER but may be used with a faster algorithm for a base solution.

4. EXPERIMENTAL METHODOLOGY

We tested our approaches on random datasets generated by coupling the random simulations used in Refs. 5, 6, 10 and adding random errors to the dataset. We compared results of our GREEDY CONSENSUS ALGORITHM with the original 2-ALLELE MIN SET COVER [6] as well as those of the FAMILY FINDER software [8] and a limited comparison to COLONY software [25]. The comparison to COLONY was limited by the computational resources since as a maximum likelihood-based method it is computationally intensive.

We also tested our approach on biological datasets with known sibling groups: Tiger Shrimp (*Penaeus monodon*[18]), ants (*Leptothorax acervorum* [15]), and Atlantic Salmon (*Salmo salar* [16]). Only the shrimp dataset had original errors in it. We introduced errors into other datasets to test our approach.

4.1. Random Simulations

We validate our approach using random simulations. We first create random diploid parents (male and female) and then generate complete genetic data for offspring. We use a range of values for the parameters of the varying the number of males, females, alleles,

[c]CPLEX is a registered trademark of ILOG

loci, number of offsprings and sibling groups (families). We then introduce errors into the data and use various methods to reconstruct the sibling groups. We compare our results to the actual known sibling groups in the data to assess accuracy. We measure the error rates using the Gusfield's partition distance [14]. The base population is generated using uniform distribution as described in Ref. 6.

We used the following ranges of parameter settings for the fixed error rate of 10% of individuals:

- The number of adult females F and the number of adult males M were equally set to $5, 20$.
- The number of loci sampled $l = 4, 6, 8, 10$.
- The number of alleles per locus (with uniform allele frequency distribution) $a = 10, 15$.
- The the number of true sibling groups $j = 5, 10, 20$.
- The maximum number of offspring per couple (with uniform family size distribution) $o = 5, 10$.

4.2. Random Errors

Errors were introduced uniformly at random with a probability of 0.1 of an individual having an error (which is higher than typical for real data). Once an individual to have an error is chosen, we choose the locus to be introduce an error uniformly at random. Then the type of the error to be introduced is chosen by generating a random number between 0 and 1, and choosing the corresponding error from Table 1(a). While the probability of 0.1 for having an individual with an error may seem large compared to Ref. 13, it is meant to exhibit how robust our method is to genotyping errors. It also is affected by the number of loci since we introduce only one erroneous locus for an individual. We further test our approach by varying the error rate for selected parameters.

Table 1. Random errors and associated costs.

(a) Error ranges for the different error types

Type of Error	Random Number Range
Allelic Dropout	$[0, 0.5]$
Heterozygous Mistype	$(0.5, 0.7]$
Homozygous Mistype	$(0.7, 0.95]$
Genetic Mutation	otherwise

(b) Costs and relative thresholds used for Greedy Algorithm Simulations

Cost	value
Allelic Dropout	0.34
Heterozygous Mistype	0.7
Homozygous Mistype	1
Allele Combination Error	0.4
Maximum Editing per individual	2.0
Maximum Editing per group	∞
Maximum Avg Edit Ind in Group	0.45

The error rates in Table 1(a) have been derived from biological data as well as Ref. 13. The values used in our experiments are shown in Table 1(b).

4.3. Evaluation

We measure the accuracy of the solution by comparing the known sibling sets with those generated by our algorithm, and calculating the minimum partition distance [14]. The solution error is the percentage of individuals that would need to be removed to make the reconstructed sibling set equal to the true sibling sets. Note that the 2-ALLELE MIN SET COVER does not return a partitioning of the individuals, whereas the GREEDY CONSENSUS ALGORITHM partitions them.

The experiments were run on Intel^{TM} Quad Core Xeon Processor (2.66GHz) with 24 GB RAM memory.

4.4. Sibling Group Reconstruction Methods

We compare the performance of the GREEDY CONSENSUS ALGORITHM to the 2-ALLELE MIN SET COVER algorithm and two other sibship reconstruction methods. While there are other sibling reconstruction methods available, in our evaluations, partially presented in Ref. 6, Family Finder and COLONY, together with the 2-ALLELE MIN SET COVER, were the best.

Family Finder. The approach proposed in Ref. 8 is a mixture of likelihood and combinatorial techniques. The algorithm constructs a graph with individuals as nodes and the edges weighted by the pairwise likelihood ratio that the individuals are siblings versus being unrelated. Very light edges are ignored. Sibling groups are then dense areas of the graph.

COLONY. Wang [25] has proposed the only other known error tolerant approach. The method uses a simulated annealing algorithm that works by starting with known sibling groups. Similar to the consensus approach, individuals whose sibling groups are not known are placed into singleton sibling groups. Iteratively alternate solutions are created by randomly changing group memberships of individuals. It uses a "cool-down" approach to reduce exploration after a large number of iterations. The method assumes that at least one gender in the population is monogamous.

5. RESULTS

We have compared the accuracy of reconstruction of sibling groups by the new error-tolerant approach to the best existing sibling reconstruction methods. We use simulated data with a wide range of parameters. On simulated data our GREEDY CONSENSUS algorithm performs better than all other methods on almost all parameters. When the number of loci is small the 2-ALLELE MIN SET COVER performs better in some cases, but overall the consensus method performs best on simulated data. Both Family Finder and COLONY are very inaccurate when the number of loci is small, thus making them expensive for wild populations. For all simulations with 6 or more loci, our approach was 95% or more accurate, even if the number of erroneous individuals went up to 20%. Family Finder and COLONY showed considerable improvement with increase in the number of loci and alleles per locus.

We present the results on simulated data in Figure 1. We show the accuracy as a function of the number of sampled loci, number of alleles per locus, number of families, and the size of a family.

On real biological data all methods performed comparably well with slight variation around the 90% accuracy. The consensus approach achieved over 90% accuracy for all the biological datasets, which was slightly better than the 2-ALLELE MIN SET COVER.

6. CONCLUSIONS

We have proposed an error-tolerant approach for reconstructing sibling relationships from microsatellite data. Our method is based on the idea of taking a consensus of partial solutions obtained by omitting one locus at a time and then locally improving the resulting combined solution. We proved the intractability of any general formulation of distance based consensus methods. We proposed a new combinatorial algorithm for the problem of reconstructing sibling relationships from single generation microsatellite genetic data in presence of genotyping errors. We have implemented and tested our approach on both simulated and real data. We have provided a framework for distance based consensus methods which may be used with any combination of distance and quality functions, possibly yielding better results.

Consensus methods give a partition, unlike a set cover, and the proposed GREEDY CONSENSUS ALGORITHM acheived over 95% accuracy for most datasets and performs comparably or better than other approaches in most cases.

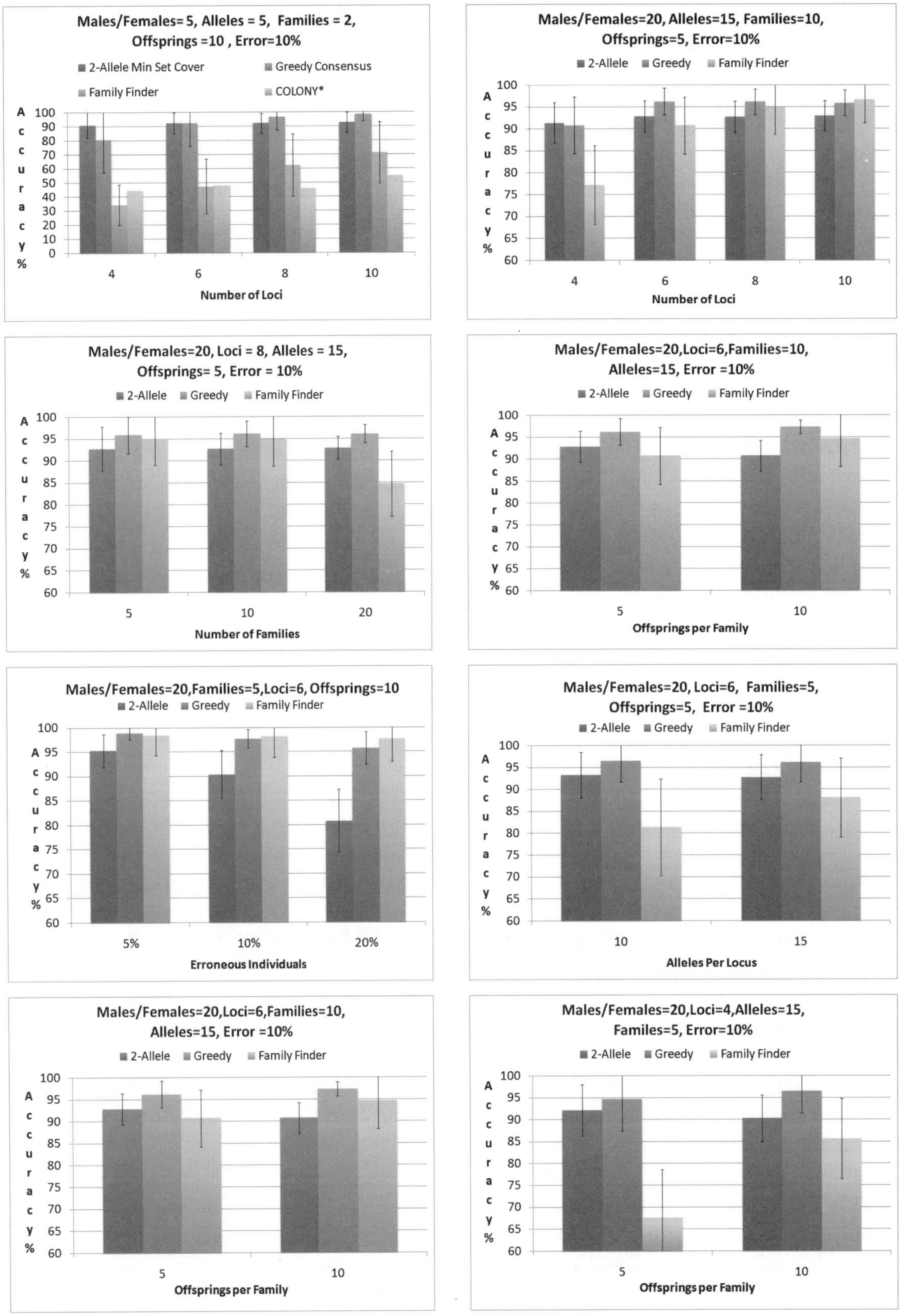

Fig. 1. Results on simulated datasets. Only 50 iterations were used for the COLONY algorithm due to its computational inefficiency and time constraints.

While Family Finder and COLONY performed comparably well in some scenarios, our method requires considerably less input, makes fewer assumptions, and was consistently over 90% accurate. Moreover, our method was considerably faster than COLONY which performed an almost an entire exhaustive search for a global minimum of the likelihood function. COLONY also requires one of the parents to be monogamous which is an unrealistic assumption for many species. Family Finder did not perform well for large families, especially when the allele frequency was high.

6.1. Future Work

Our approach can be combined with a variety of methods for both generating the input solutions, and developing a consensus among them. In the future we intend to explore other than greedy optimization objectives to avoid local minima in the distance function.

Our technique can be extended to solve other problems in kinship analysis. Since our approach is not restricted to the methods used for generating input solutions, it can be used as a general consensus between different methods of sibling reconstruction. For example, a tree-based consensus method can be used to merge pedigrees.

Acknowledgments

This research is supported by the following grants: NSF IIS-0612044 (Berger-Wolf, Ashley, Chaovalitwongse, DasGupta), Fulbright Scholarship (Saad Sheikh), NSF CCF-0546574 (Chaovalitwongse). We are grateful to the people who have shared their data with us: Jeff Connor, Atlantic Salmon Federation, Dean Jerry, and Stuart Barker. We would also like to thank Anthony Almudevar, Bernie May, and Dmitri Konovalov. We are also grateful to our collaborators: Ashfaq Khokhar and Satya Lahari.

References

1. A. Almudevar. A simulated annealing algorithm for maximum likelihood pedigree reconstruction. *Theoretical Population Biology*, 63, 2003.
2. A. Almudevar and C. Field. Estimation of single generation sibling relationships based on DNA markers. *Journal of Agricultural, Biological, and Environmental Statistics*, 4:136–165, 1999.
3. K. J. Arrow. *Social Choice and Individual Values.* John Wiley, New York, second edition, 1963.
4. M. Ashley, T. Y. Berger-Wolf, P. Berman, W. Chaovalitwongse, B. DasGupta, and M.-Y. Kao. On approximating four covering/packing problems with applications to bioinformatics. Technical report, DIMACS, 2007.
5. T. Y. Berger-Wolf, B. DasGupta, W. Chaovalitwongse, and M. V. Ashley. Combinatorial reconstruction of sibling relationships. In *Proceedings of the 6th International Symposium on Computational Biology and Genome Informatics (CBGI 05)*, pages 1252–1255, Utah, July 2005.
6. T. Y. Berger-Wolf, S. I. Sheikh, B. DasGupta, M. V. Ashley, I. C. Caballero, W. Chaovalitwongse, S. P. Lahari. Reconstructing sibling relationships in wild populations. *Bioinformatics*, 23(13), i49-i56, 2007..
7. T. Y. Berger-Wolf, T. L. Williams, B. E. Moret, and T. J. Warnow. An experimental evaluation of phylogenetic consensus methods. Technical Report TR-CS-2003-19, Department of Computer Science, University of New Mexico, 2003.
8. Jen Beyer and B. May. A graph-theoretic approach to the partition of individuals into full-sib families. *Molecular Ecology*, 12:2243–2250, 2003.
9. M. Janowitz, F. J. Lapointe, F. McMorris, B. Mirkin, and F. Roberts, editors. Bioconsensus. DIMACS Series in Discrete Mathematics and Theoretical Computer Science. DIMACS-AMS, 2001.
10. W. Chaovalitwongse, T. Y. Berger-Wolf, B. Dasgupta, and M. V. Ashley. Set covering approach for reconstruction of sibling relationships. *Optim, Methods and Soft.*, 22(1):11 – 24, Feb 2007.
11. J. C. de Borda. Mémoire sur les élections au scrutin. *Histoire de l'Académie Royale des Sci.*, 1784.
12. Marie Jean Antoine Nicolas de Caritat marquis de Condorcet. Essay on the application of analysis to the probability of majority decisions, 1785.
13. P. Gagneux, C. Boesch, and D. S. Woodruff. Microsatellite scoring errors associated with noninvasive genotyping based on nuclear DNA amplified from shed hair. *Mol.Eco.*, 6(9):861-8,Sep 1997.
14. D. Gusfield. Partition-distance: A problem and class of perfect graphs arising in clustering. *Information Processing Letters*, 82(3):159–164, May 2002.
15. R. L. Hammond, A.F. G. Bourke, and M. W. Bruford. Mating frequency and mating system of the polygynous ant, *Leptothorax acervorum. Molecular Ecology*, 10(11):2719–2728, 1999.
16. C. M. Herbinger, P. T. O'Reilly, R. W. Doyle, J. M. Wright, and F. O'Flynn. Early growth performance of atlantic salmon full-sib families reared in single family tanks or in mixed family tanks. *Aquaculture*, 173(1–4):105–116, 1999.
17. J. I. Hoffman and W. Amos. Microsatellite genotyping errors: detection approaches, common sources and consequences for paternal exclusion. *Molecular Ecology*, 14(2):599–612, 2005.

18. D. R. Jerry, B. S. Evans, M. Kenway, and K. Wilson. Development of a microsatellite dna parentage marker suite for black tiger shrimp *Penaeus monodon*. *Aquaculture*, 255(1-4):542-547, 2006.
19. R. M. Karp. Reducibility among combinatorial problems. In R. E. Miller and J. W. Thatcher, editors, *Complexity of Computer Computations*, 85–103. Plenum Press, 1972.
20. D. A. Konovalov, C. Manning, and M. T. Henshaw. KINGROUP: a program for pedigree relationship reconstruction and kin group assignments using genetic markers. *Molecular Ecology Notes*,4:779-782, 2004.
21. F. R. McMorris, D. B. Meronik, and D. A. Neumann. A view of some consensus methods for trees. In J. Felsenstein, editor, *Numerical Taxonomy*, pages 122–125. Springer-Verlag, 1983.
22. S. I. Sheikh, T. Y. Berger-Wolf, W. Chaovalitwongse, and M. V. Ashley. Reconstructing sibling relationships from microsatellite data. In *Proceedings of the European Conference on Computational Biology (ECCB)*, January 07.
23. B. R. Smith, C. M. Herbinger, and H R. Merry. Accurate partition of individuals into full-sib families from genetic data without parental information. *Genetics*, 158(3):1329–1338, July 2001.
24. S. C. Thomas and W. G. Hill. Sibship reconstruction in hierarchical population structures using Markov chain monte carlo techniques. *Genet. Res., Camb.*, 79:227–234, 2002.
25. J. Wang. Sibship reconstruction from genetic data with typing errors. *Genetics*, 166:1968–1979, April 2004.

ON THE ACCURATE CONSTRUCTION OF CONSENSUS GENETIC MAPS

[1]Yonghui Wu, [2]Timothy J. Close, and [1]Stefano Lonardi*

[1]*Department of Computer Science and Engineering, University of California, Riverside, CA 92521, USA*
[2]*Department of Botany and Plant Sciences, University of California, Riverside, CA 92521, USA*
**Email: stelo@cs.ucr.edu*

We study the problem of merging genetic maps, when the individual genetic maps are given as directed acyclic graphs. The problem is to build a *consensus map*, which includes and is consistent with all (or, the vast majority of) the markers in the individual maps. When markers in the input maps have ordering conflicts, the resulting consensus map will contain cycles. We formulate the problem of resolving cycles in a combinatorial optimization framework, which in turn is expressed as an integer linear program. A faster approximation algorithm is proposed, and an additional speed-up heuristic is developed. According to an extensive set of experimental results, our tool is consistently better than JOINMAP, both in terms of accuracy and running time.

Keywords: genetic map, consensus, genotype, population genetics

1. INTRODUCTION

Genetic linkage maps are arguably the cornerstone of a variety of biological applications including map-assisted breeding, association genetics and map-assisted gene cloning, just to name a few. Traditionally scientists have focused on building genetic maps for a single mapping population, task for which a wide variety of software tools are available and have satisfactory performance, e.g., JOINMAP[1], CARTHAGENE[2], ANTMAP[3], RECORD[4] TMAP[5] and MSTMAP[6].

In recent years, the rapid adoption of high-throughput genotyping technologies has been paralleled not only by an increase in the map density but also by a variety of marker types. Today it is increasingly common to find several genetic maps available for the same organism, usually for different sets of genetic markers and obtained with a variety of genotyping technologies. Notable examples are genetic linkage maps based on microsatellites in human[7] and in cattle[8], and maps based on sequence length polymorphism in mouse[9] and rat[10], just to name a few. In the case of maize, for instance, seven distinct mapping populations of *Zea mays* have been used[11]. When multiple maps are available, one could envision to construct a bigger single map (hereafter called *consensus map*) that includes all the markers.

A consensus map provides a higher density of markers and therefore a greater genome coverage than the individual maps. As the name suggests, the consensus map should be consistent with the order of the markers from the individual maps. However, this may not always be possible since the presence of errors is very likely to introduce conflicts between the individual maps. Due to the way individual genetic maps are assembled, two types of errors are observed, namely local reshuffles and global displacements. Local reshuffles refer to inaccuracies in the order of nearby markers, whereas global displacements refer to the cases where a few markers are placed at positions far from the correct ones. When addressing conflicts to build the consensus maps, one should take into account both types of errors.

1.1. Related works

Several systematic approaches have been proposed to construct consensus maps[12; 13; 1; 14; 11]. The method adopted by Beavis *et al.*[12] for the integration of maize maps is to pool together the genotyping data from the individual mapping populations, and then rely on traditional mapping algorithms to build the consensus map. Although this pooling strategy is commonly used, it has several shortcomings. First, it cannot be used in all circumstances. For example, when the data are obtained from different populations (e.g., one dataset obtained from a double haploid population and another from a recombinant inbred lines population), then they cannot be merged and treated equivalently afterward. Second, the pooling method results in a large number of missing observations. A large amount of missing observations combined with the limited tolerance to missing data by exist-

*Corresponding author.

ing mapping algorithms inevitably deteriorates the quality of the consensus map.

An alternative approach, like the one used in the tool JOINMAP[13; 1], is to first obtain the consensus estimates of pairwise genetic distances by weighting for population structure and size. Then, the tool searches for a map that minimizes an objective function that measures the fit of the map to the distance estimates and the overall quality of the map. The drawbacks of this approach are twofold. First, it is well-known that distance estimates are not very accurate when based on a small sample of recombination events. Construction of genetic maps based on approximate estimates will result in inaccuracies in the ordering between markers on the consensus map. Second, the computational problem of searching for an optimal map with respect to the objective function being used is very time consuming. For instance, the most recent version of JOINMAP took *three months* of computation to construct a consensus map from three individual maps of barley containing a total of 1,800 markers (the markers are divided into 7 linkage groups of roughly equal sizes). Despite these drawbacks, JOINMAP is still the only off-the-shelf software package available to build consensus maps.

The most recent approach to the problem relies on graph theory and was initially proposed by Yapa *et al.*[14] and later extended by Jackson *et al.*[11] Yapa *et al.*[14] use directed acyclic graphs (DAG) to represent maps from individual populations. The set of DAGs are then merged into a consensus graph on the basis of their shared vertices. A directed cycle in the resulting graph indicates an inconsistency among the individual maps with regard to the order of the markers involved. In order to resolve the inconsistencies, Jackson *et al.*[11] propose to break cycles by removing a minimum weight set of feedback edges. This objective function is reasonable when dealing with local reshuffles. However in the presence of global relocations, it is not appropriate because too many edges need to be deleted in order for all the cycles to be broken. A similar approach is to remove a minimum weight feedback vertex set from the graph. The obvious drawback of this method is that the markers corresponding to the deleted vertices will be excluded from the consensus map.

1.2. Our contribution

We follow the graph theoretical paradigm outlined in [11, 14] and represent individual genetic maps as DAGs. The individual maps are combined into a single directed graph according to their shared vertices. Any ordering conflict among the individual maps will result in cycles in the combined graph. Here, we propose to resolve the cycles by removing the smallest set of (feedback) marker occurrences. Note that we are not deleting markers but marker occurrences. A marker may occur in multiple individual maps. A marker occurrence refers to the appearance of a marker in a particular individual map. The deletion of a marker occurrence will not affect the occurrences of the same marker in other maps.

Trying to identify and eliminate a small number of marker occurrences from some of the maps is a better strategy than the one proposed in [11] because it more accurately reflects the type of errors that may be present in the individual maps. We formulate the optimization problem resulting from this strategy via integer linear programming (Section 3), and we propose an approximation algorithm to solve it. We also devise an heuristic to decompose the original problem, in the case the size of the instance to be solved is too large.

As soon as all cycles in the consensus map are resolved, we process the resulting graph with another novel algorithm whose objective is to simplify the DAG to help geneticists to be able to visualize and make use of the consensus map (Section 4). This step involves removing redundant edges and merging nodes on the consensus map without introducing conflicts. In the last step, a final algorithm produces a linear order of the markers which is consistent with the consensus graph (Section 5).

The last two steps of our approach, i.e., condensing the markers and linearizing the DAG, further distinguishes our approach from those in [14, 11]. The final output of our workflow is a linear order of marker bins[a] which is a format geneticists are used to work with. The output of the methods by Yapa *et al.*[14] and Jackson *et al.*[11] is however a DAG, which is often too complex and convoluted to make much sense out of it. For the same reasons, we did not compare our experimental results against those methods.

In Section 6 we carry out an extensive evaluation of our algorithms on synthetic data. We compare the perfor-

[a]A *bin* is a set of markers for which the relative orders are undetermined.

mance of our method with JOINMAP. Our approach produces consistently better results than JOINMAP, both in terms of accuracy and running time. Our method is also superior to the method of pooling together genotyping data from individual maps. We have also employed our software on the genotyping data we collected for three mapping populations (about 1,800 markers) for barley, but we had to omit those results from this manuscript due to lack of space.

2. PRELIMINARIES AND NOTATIONS

A *genetic linkage map* represents the linear order and the pairwise distance of markers on a chromosome. The distance between two adjacent markers, expressed in *centimorgans* (cM), is determined by the frequency of genetic recombination occurring in the region between them. Two markers are one centimorgan apart if one observes an average of 0.01 crossovers per meiosis in the region enclosed by the two markers. The set of markers for which no recombination is detected is called a *bin*. For markers in the same bin, their relative orders are undetermined. From this point forward, a genetic map is composed of a sequence of bins (of marker) and the distance between them.

Some notations are in order. Let Π denote a genetic linkage map, and let M_Π denote the set of markers included in Π. Given a set of maps $\Omega = \{\Pi_1, \Pi_2, \ldots, \Pi_K\}$, we define M_Ω to be the *universe* of all the markers, i.e., $M_\Omega = \cup_{i=1}^{K} M_{\Pi_i}$.

Given a map Π we define $G_\Pi = (M_\Pi, E_\Pi)$ to be the directed weighted graph *induced* by the map, where the set of edges E_Π is defined as $E_\Pi = \{(m_i, m_j) |\, m_i$ is in the bin immediately preceding the bin of $m_j\}$ and the weight of an edge (m_i, m_j) is set to the distance between the corresponding bins. The notion of induced graph can be extended to a set of maps. Let $G_\Omega = (M_\Omega, E_\Omega)$ be the directed weighted graph induced by Ω, where $E_\Omega = \cup_{i=1}^{K} E_{\Pi_i}$. The weight of an edge in G_Ω is set to be the average of the weights of the corresponding edges in the original maps.

We use m_i to refer to a generic marker, and m_i^j to refer to the occurrence of marker m_i in map Π_j. We further define N_Ω to be the set containing all the marker occurrences. If we select a set $R \subseteq N_\Omega$, a *submap* $\Pi(R)$ of Π with respect to R is defined by deleting the occurrences of all markers not in R from the map Π. The subproblem $\Omega(R)$ of the original problem Ω restricted to R is defined as $\Omega(R) = \{\Pi_i(R) | \Pi_i \in \Omega\}$.

Figure 1 illustrates the notations Π, Ω, G_Π, G_Ω, M_Ω, N_Ω, $\Pi(R)$ and Ω_R for a small example.

3. RESOLVING ORDERING CONFLICTS

Let $\Omega = \{\Pi_1, \Pi_2, \ldots, \Pi_K\}$ be the set of input maps for which we want to build a consensus map. Merging maps $\Pi_1, \Pi_2, \ldots, \Pi_K$ into a consensus DAG is straightforward when there are no conflicts. If some of the markers have conflicting orders among the input maps, then G_Ω contains cycles. In order to resolve cycles, we propose to delete the smallest set of marker occurrences. More precisely, if we first assign weights to the individual maps to represent their quality (i.e., high weight is associated with high quality), the problem is to delete the minimum-weight set of marker occurrences so that the resulting subproblem is conflict-free. The optimization problem that emerges from this strategy is the following.

Minimum-Weight Feedback Marker Occurrence Set (MWFMOS)

Input: Ω and w, where Ω is a set of individual maps from which one would like to build a consensus map, and w is the associated weight function on N_Ω where $w(m_i^j)$ is the weight of marker occurrence m_i^j. Without loss of generality, we assume that $w(m_i^j) > 1$ for all $m_i^j \in N_\Omega$.

Objective: identify a set $\mathcal{D}$ of minimum total weight so that the subproblem restricted to $N_\Omega - \mathcal{D}$ is conflict-free (i.e., the graph induced by the subproblem, $G_{\Omega(N_\Omega - D)}$, is acyclic).

It is relatively easy to prove that MWFMOS is NP-complete when the number of maps is unbounded. The proof uses a reduction from the minimum feedback edge set problem (not shown here due to space restrictions). We still do not know whether MWFMOS is still NP-complete when the number of maps is bounded by a constant, but we suspect it is.

The solution to the MWFMOS problem with input (Ω, w) can be obtained by solving MWFMOS for the non-overlapping subproblems corresponding to the strongly connected components in G_Ω. The optimal solution to the original problem is simply the concatenation of the optimal solutions to the subproblems. In the following, we will be focusing on solving MWFMOS for a subproblem only.

$$
\begin{aligned}
&\Omega = \{\Pi_1, \Pi_2\}\\
&\Pi_1 = [(m_2)\ 2\ (m_3, m_4)\ 1\ (m_5)\ 2\ (m_6, m_7)]\\
&\Pi_2 = [(m_1)\ 1\ (m_2, m_3)\ 2\ (m_5)\ 1\ (m_4, m_7)]\\
&M_\Omega = \{m_1, m_2, m_3, m_4, m_5, m_6, m_7\}\\
&N_\Omega = \left\{ \begin{matrix} m_2^1, m_3^1, m_4^1, m_5^1, m_6^1, m_7^1, \\ m_1^2, m_2^2, m_3^2, m_4^2, m_5^2, m_7^2 \end{matrix} \right\}\\
&R = \left\{ \begin{matrix} m_2^1, m_3^1, m_4^1, m_5^1, m_6^1, m_7^1, \\ m_1^2, m_2^2, m_3^2, m_4^2, m_7^2 \end{matrix} \right\}\\
&\Pi_1(R) = \Pi_1\\
&\Pi_2(R) = [(m_1)\ 1\ (m_2, m_3)\ 3\ (m_4, m_7)]\\
&\Omega(R) = \{\Pi_1(R), \Pi_2(R)\}
\end{aligned}
$$

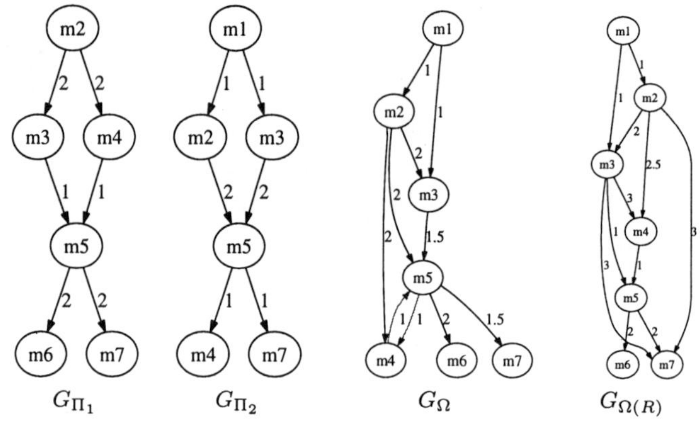

Fig. 1. Two simple genetic linkage maps, along with the corresponding notations used in this paper. Maps Π_1 and Π_2 both consist of four bins (enclosed in parentheses). The numbers in between adjacent bins indicate the distances between them. Maps Π_1 and Π_2 are not consistent with each other because there is a cycle in G_Ω between m_4 and m_5. Removing m_5 from Π_2 resolves the conflict.

3.1. An LP-based algorithm

Let $\mathcal{I} = \{F_1, F_2, \ldots, F_K\}$ be a subproblem of Ω corresponding to a strongly connected component in G_Ω. A submap F_i is hereafter called a *fragment* since it is a contiguous piece of an individual map from Ω. Each fragment F_i has the same format as Π_i. Throughout this paper, we use Ω to denote the original problem, and $\mathcal{I}$ to denote a subproblem of Ω.

A conflict in $\mathcal{I}$ is characterized by a path $m_{i_1}^{j_1} \to m_{i_2}^{j_1}, m_{i_2}^{j_2} \to m_{i_3}^{j_2}, \ldots, m_{i_k}^{j_n} \to m_{i_1}^{j_n}$ (not to confused with a path in $G_\mathcal{I}$), wherein $m_{i_1}^{j_1} \to m_{i_2}^{j_1}$ means that marker m_{i_1} precedes marker m_{i_2} in fragment F_{j_1} (markers m_{i_1} and m_{i_2} do not have to be in adjacent bins). Note that the path starts and ends with the same marker in two different fragments. Let P be the set of such paths.

Given P, we formulate MWFMOS as an Integer Linear Program (ILP) as follows.

$$
\begin{aligned}
&Min\ \ \textstyle\sum x_i^j w(m_i^j)\\
&S.T.\ \textstyle\sum_{m_i^j \in p} x_i^j \geq 1\ \forall p \in P\\
&\qquad x_i^j \in \{0, 1\}
\end{aligned}
\tag{1}
$$

where x_i^j is the binary variable associated with the marker occurrence m_i^j which is set to 1 if m_i^j is to be deleted and set to 0 otherwise. The LP relaxation of the above ILP is straightforward. The number of constraints in (1) is $|P|$, which is at most $O(K!|M_\mathcal{I}|^K)$, where K is number of fragments in $\mathcal{I}$ and $|M_\mathcal{I}|$ is the total number of distinct markers in $\mathcal{I}$. The upper bound is a polynomial in the size of the input if K is fixed. The dual of the LP relaxation of (1) is the following program.

$$
\begin{aligned}
&Max \qquad \textstyle\sum y_p\\
&S.T.\ \textstyle\sum_{p \ni m_i^j} y_p \leq w(m_i^j)\ \forall m_i^j \in N_\mathcal{I}\\
&\qquad y_p \geq 0 \qquad \forall p \in P
\end{aligned}
\tag{2}
$$

where y_p is the associated variable with path $p \in P$, and $N_\mathcal{I}$ is the set containing all the marker occurrences in $\mathcal{I}$. The following LP is equivalent to (2).

$$
\begin{aligned}
&Min \qquad \lambda\\
&S.T.\ \textstyle\sum_{p \ni m_i^j} y_p \leq \lambda w(m_i^j)\ \forall m_i^j \in N_\mathcal{I}\\
&\qquad \textstyle\sum_{p \in P} y_p = 1\\
&\qquad y_p \geq 0 \qquad \forall p \in P
\end{aligned}
\tag{3}
$$

The optimal solution to (3) is the reciprocal of the solution of (2). To simplify the notation, we can rewrite (3) in the matrix representation.

$$
\begin{aligned}
&Min \qquad \lambda\\
&S.T. \qquad A\vec{y} \leq \lambda\vec{w}\\
&\qquad \vec{y} = 1 \text{ and } \vec{y} \geq 0
\end{aligned}
\tag{4}
$$

Each row of A corresponds to a marker occurrence in $N_\mathcal{I}$ and each column of A refers to a path in P. We have $A[r, c] = 1$ if and only if $m_i^j \in N_\mathcal{I}$ corresponding to the rth row of A is on the path corresponding to the cth column of A. With $\vec{y} = 1$, we mean $\sum_{p \in P} y_p = 1$.

Due to the large number of variables, solving optimally (4) can be very time consuming. In the following, we show how to achieve an $(1 + \epsilon)$-optimal (or simply

[b] A solution λ is said to be $(1 + \epsilon)$ optimal if $\lambda < (1 + \epsilon)\lambda_{opt}$, where λ_{opt} is the optimal solution. An $(1 + \epsilon)$ optimal solution is sometimes simply called an ϵ-optimal solution.

ϵ-optimal) solution[b]. To find such an approximate solution, we follow the method proposed by Plotkin *et al.*[15] Let $\vec{z}$ be the dual variables associated with (4), and let us define $C(\vec{z}) = \min_{\vec{y}|\vec{y}=1} \vec{z}^t A\vec{y}$.

Consider an error parameter $0 < \epsilon < 1/6$, a feasible primal solution $(\vec{y}, \lambda)$, and a feasible dual solution $\vec{z}$. λ is 6ϵ optimal if the following two relaxed optimality conditions are met:

$$(1-\epsilon)\lambda \vec{z}^t \vec{w} \leq \vec{z}^t A\vec{y} \qquad (5)$$

$$\vec{z}^t A\vec{y} - C(\vec{z}) \leq \epsilon(\vec{z}^t A\vec{y} + \lambda \vec{z}^t \vec{w}) \qquad (6)$$

A sketch of the algorithm to find a 6ϵ optimal solution is presented in Figure 2. The performance guarantee of our algorithm APPROXSOLVE is presented as Theorem 1, and the time complexity of the algorithm is presented as Theorem 2.

Lemma 1. *Let $(\vec{y}, \lambda)$ and $\vec{z}$ be feasible primal and dual solutions that satisfy both condition (5) and (6). Then, $(\vec{y}, \lambda)$ is an $(1+6\epsilon)$ optimal solution.*

Proof. This Lemma corresponds as Lemma 2.1 in [15]. To be self-contained, we present the proof here.
From (5) and (6), we have: $C(\vec{z}) \geq (1-\epsilon)\vec{z}^t A\vec{y} - \epsilon\lambda\vec{z}^t\vec{w} \geq (1-\epsilon)^2\lambda\vec{z}^t\vec{w} - \epsilon\lambda\vec{z}^t\vec{w} \geq (1-3\epsilon)\lambda\vec{z}^t\vec{w}$.
Hence, $\lambda \leq (1-3\epsilon)^{-1}C(\vec{z})/\vec{z}^t\vec{w} \leq (1-3\epsilon)^{-1}\lambda^* \leq (1+6\epsilon)\lambda^*$. □

Theorem 1. *Algorithm* APPROXSOLVE *returns an $(1+6\epsilon)$ optimal solution to (4).*

Proof. The theorem follows from Lemma 2.2 in [15]. To be self-contained, we present the proof here.

According to Lemma 1, in order to prove Theorem 1, we only have to show that condition (5) and (6) are both satisfied when Algorithm APPROXSOLVE stops. Since condition (6) is met by the while loop at line 4, we only have to show that (5) is satisfied when the algorithm stops.

We first show that when $\alpha \geq \alpha_0 = \frac{2\ln(2|N_{\mathcal{I}}|\epsilon^{-1})}{\lambda\epsilon}$, $\vec{z}$ as assigned by the "for" loops at line 3 and 10 in algorithm APPROXSOLVE will satisfy condition (5).

Let $I = \{i : (1-\epsilon/2)\lambda w_i \geq \vec{a_i}^t\vec{y}\}$. Let $j \in I$. $\lambda z_j w_j = \lambda e^{\alpha\vec{a_j}^t\vec{y}/w_j} \leq \lambda e^{\alpha(1-\epsilon/2)\lambda} = \lambda e^{\alpha\lambda}e^{-\alpha\epsilon\lambda/2} \leq \lambda e^{\alpha\lambda}e^{-ln(2|N_{\mathcal{I}}|\epsilon^{-1})} \leq \frac{\epsilon}{2|N_{\mathcal{I}}|}\lambda e^{\alpha\lambda} \leq \frac{\epsilon}{2|N_{\mathcal{I}}|}[\lambda\vec{z}^t\vec{w}]$. Consequently, $\lambda\vec{z}^t\vec{w} = \sum_{i\in I}\lambda z_i w_i + \sum_{i\notin I}\lambda z_i w_i \leq \sum_{i\in I}\lambda z_i w_i + \sum_{i\notin I}\frac{1}{1-\epsilon/2}z_i\vec{a_i}^t\vec{y} \leq \sum_{i\in I}\lambda z_i w_i + \frac{1}{1-\epsilon/2}\vec{z}^t A\vec{y} \leq \frac{\epsilon}{2}\lambda\vec{z}^t\vec{w} + \frac{1}{1-\epsilon/2}\vec{z}^t A\vec{y}$. Therefore, we have $(1-\epsilon)\lambda\vec{z}^t\vec{w} \leq \vec{z}^t A\vec{y}$

Notice that α is initialized to be $2\alpha_0$ and whenever $\max_r \vec{a_r}^t\vec{y}/w_r \leq \lambda/2$, α gets recomputed. Therefore condition (5) is satisfied throughout the execution of Algorithm APPROXSOLVE. □

Lemma 2. *Let $(\vec{y_1}, \lambda)$ and $\vec{z_1}$, where $\vec{z_1} = \{\frac{1}{w_r}e^{\alpha\vec{a_r}^t\vec{y_1}/w_r}\}_{r=1}^{|N_{\mathcal{I}}|}$, be primal and dual solutions that do not satisfy (6). Let $(\vec{y_2}, \lambda)$ and $\vec{z_2}$ be the solutions in the next iteration, i.e. $\vec{y_2} = (1-\delta)\vec{y_1} + \delta\tilde{y}$, and let α, δ and ϵ be defined in Algorithm* APPROXSOLVE. *Let $\Phi_1 = \vec{z_1}^t\vec{w}$, $\Phi_2 = \vec{z_2}^t\vec{w}$. $\Phi_1 - \Phi_2 > \lambda\epsilon^2\Phi_1/4$.*

Proof. $\Phi_2 = \vec{z_2}^t\vec{w} = \sum_i e^{\alpha\vec{a_i}^t\vec{y_2}/w_i} = \sum_i e^{\alpha\vec{a_i}^t((1-\delta)\vec{y_1}+\delta\tilde{y})/w_i} = \sum_i e^{\alpha\vec{a_i}^t\vec{y_1}/w_i}e^{\alpha\delta\vec{a_i}^t(\tilde{y}-\vec{y_1})/w_i}$.
Since $w_i > 1, \vec{y_1} = 1$ and $\tilde{y} = 1 \implies |\vec{a_i}^t(\tilde{y}-\vec{y_1})/w_i| < 1$. Since $\delta = \frac{\epsilon}{4\alpha} \implies |\alpha\delta\vec{a_i}^t(\tilde{y}-\vec{y_1})/w_i| < \epsilon/4 < 1/4$.
According to Taylor's expansion, $e^x < 1 + x + 2x^2$ for $|x| < 1/4$. By plugging in $x = \alpha\delta\vec{a_i}^t(\tilde{y}-\vec{y_1})/w_i$ we get $e^{\alpha\delta\vec{a_i}^t(\tilde{y}-\vec{y_1})/w_i} < 1 + (\alpha\delta\vec{a_i}^t(\tilde{y}-\vec{y_1})/w_i) + 2(\alpha\delta\vec{a_i}^t(\tilde{y}-\vec{y_1})/w_i)^2$
$< 1 + (\alpha\delta\vec{a_i}^t(\tilde{y}-\vec{y_1})/w_i) + \frac{\epsilon}{2}(\alpha\delta\vec{a_i}^t(\tilde{y}+\vec{y_1})/w_i)$
Therefore, $\Phi_2 = \sum_i e^{\alpha\vec{a_i}^t\vec{y_1}/w_i}e^{\alpha\delta\vec{a_i}^t(\tilde{y}-\vec{y_1})/w_i} < \sum_i e^{\alpha\vec{a_i}^t\vec{y_1}/w_i} + \alpha\delta(C(\vec{z_1}) - \vec{z_1}A\vec{y_1}) + \frac{\epsilon}{2}\alpha\delta(C(\vec{z_1}) + \vec{z_1}A\vec{y_1})$
$\Phi_2 < \Phi_1 + \alpha\delta(C(\vec{z_1}) - \vec{z_1}A\vec{y_1}) + \frac{\epsilon}{2}\alpha\delta(C(\vec{z_1}) + \vec{z_1}A\vec{y_1})$
$\Phi_1 - \Phi_2 > \alpha\delta(\vec{z_1}A\vec{y_1} - C(\vec{z_1})) - \epsilon\alpha\delta\vec{z_1}A\vec{y_1}$
Due to the fact that $(\vec{y_1}, \lambda)$ and $\vec{z_1}$ do not satisfy (6), we have $\Phi_1 - \Phi_2 > \lambda\epsilon\alpha\delta\vec{z_1}^t\vec{w}$.
According to the choice of δ, $\Phi_1 - \Phi_2 > \lambda\epsilon^2\Phi_1/4$ □

Theorem 2. *Algorithm* APPROXSOLVE *converges in $O(\frac{1}{\epsilon^3\lambda^*}\log(|N_{\mathcal{I}}|\epsilon^{-1}))$ iterations, where λ^* is the optimal solution to (4).*

Proof. Notice that during the execution of Algorithm APPROXSOLVE, λ is a monotonically decreasing sequence with $\lambda_i > 2\lambda_{i+1}$. Let the sequence of λ be $\lambda_0, \lambda_1, \lambda_2, \ldots, \lambda_n$, where $\lambda_n > \lambda^*$ is the final output. When $\lambda = \lambda_k$, then $e^{\alpha\lambda_k/2} \leq \Phi \leq |N_{\mathcal{I}}|e^{\alpha\lambda_k}$

Due to Lemma 2, it takes at most $O(\frac{1}{\epsilon^3\lambda_k}\log(|N_{\mathcal{I}}|\epsilon^{-1}))$ iterations to cut λ from λ_k to λ_{k+1}. Since $\lambda_i > 2\lambda_{i+1}$, the overall time complexity is determined by the last step. Hence the overall running time is $O(\frac{1}{\epsilon^3\lambda^*}\log(|N_{\mathcal{I}}|\epsilon^{-1}))$. □

Step 5 in algorithm APPROXSOLVE can be solved by running all pairs shortest path algorithm (details not

APPROXSOLVE($\vec{y_0}, \epsilon$)

$\vec{y} \leftarrow \vec{y_0}$; $\lambda \leftarrow \max_r \vec{a_r}^t \vec{y}/w_r$; $\alpha \leftarrow 4\ln(2|N_\mathcal{I}|\epsilon^{-1})/(\lambda\epsilon)$; $\sigma \leftarrow \epsilon/(4\alpha)$;
{$\vec{a_r}$ is the transpose of the rth row vector of matrix A. $|N_\mathcal{I}|$ is the number of rows in A}
for $r = 1, \ldots, |N_\mathcal{I}|$ **do**
  $z_r \leftarrow e^{\alpha \vec{a_r}^t \vec{y}/w_r}/w_r$
while $(\vec{y}, \lambda, \vec{z})$ does not satisfy (6) **do**
  $\widetilde{y} \leftarrow argmin_{\vec{y}|\vec{y}=1} \vec{z}^t A\vec{y}$
  $\vec{y} \leftarrow (1-\sigma)\vec{y} + \sigma\widetilde{y}$
  if $\max_r \vec{a_r}^t \vec{y}/w_r \leq \lambda/2$ **then**
    $\lambda \leftarrow \max_r \vec{a_r}^t \vec{y}/w_r$; $\alpha \leftarrow 4\ln(2|N_\mathcal{I}|\epsilon^{-1})/(\lambda\epsilon)$; $\sigma \leftarrow \epsilon/(4\alpha)$;
  for $r = 1, \ldots, |N_\mathcal{I}|$ **do**
    $z_r \leftarrow e^{\alpha \vec{a_r}^t \vec{y}/w_r}/w_r$
$\lambda \leftarrow \max_r \vec{a_r}^t \vec{y}/w_r$;
return $\vec{y}, \lambda, \vec{z}$

Fig. 2. A sketch of our LP-based algorithm

shown here), which takes time $O(|N_\mathcal{I}|^3 \log |N_\mathcal{I}|)$. The vector $\vec{y}$ does not have to be stored in memory explicitly since all we need is $A\vec{y}$ which takes space $O(|N_\mathcal{I}|)$. Combining the running time for each iteration with the upper bound on the number of iterations, the overall time complexity of APPROXSOLVE is $O(\frac{1}{\epsilon^3\lambda*} \log(|N_\mathcal{I}|\epsilon^{-1})|N_\mathcal{I}|^3 \log |N_\mathcal{I}|)$. Note that the time complexity does not depend on $|P|$.

Given the near optimal solution $\vec{z}$ to the dual of (4), the near optimal solution to the LP relaxation of (1) is $\vec{x} \leftarrow \vec{z}/C(\vec{z})$. In our algorithm we apply two types of rounding to convert the fractional solution $\vec{x}$ to an integral solution, and then choose the better one among the two.

The first method is randomized. The randomized rounding algorithm progressively deletes marker occurrences until all the conflicts are resolved. In each step, the method samples a marker to be deleted according to a probability distribution proportional to $\vec{x}$. The solution obtained is further reduced to a minimal solution by removing redundant marker occurrences. The second rounding method employs a greedy strategy. The markers occurrences in $N_\mathcal{I}$ are sorted into the descending order according to their associated probabilities. We delete just enough marker occurrences to resolve all the conflicts. Again, the solution is further reduced to a minimal solution by removing redundancies.

3.2. A speed-up heuristic

The LP-based algorithm works well when either the size of the subproblem is small (i.e., $|N_\mathcal{I}|$ is small) or the number of markers to be deleted is small (i.e., $1/\lambda*$ is small), the latter of which is usually the case in practice. However if both $|N_\mathcal{I}|$ and $1/\lambda*$ are large, the LP-based algorithm can be still too slow. In this case, we advise to employ an heuristic algorithm which breaks a large subproblem $\mathcal{I}$ into even smaller sub-subproblems.

Our heuristic algorithm uses the notion of node betweenness originally proposed by Girvan and Newman[16]. Recall that the *betweenness centrality* of a node in a graph is equal to the number of shortest paths that go through it. The intuition is that nodes with high betweenness usually correspond to hubs, and their deletion will likely break the graph into disconnected components.

Now let $m_i^{j_1}$ and $m_i^{j_2}$ be a pair of occurrences of the same marker in two individual maps. A path between $m_i^{j_1}$ and $m_i^{j_2}$ is the shortest if it traverses the smallest number of marker occurrences. Let Q be the set of all such pairwise shortest paths. If there are multiple shortest paths between a pair, we arbitrarily choose one to be included in Q. Observe that Q is a subset of P in the ILP (1).

We define the *weighted betweenness centrality* of a marker occurrence m_i^j as the number of shortest paths in Q that go through node m_i^j divided by its weight $w(m_i^j)$. The higher the centrality, the more likely it is the "true bad" marker occurrence that should be deleted. Our heuristic algorithm works by computing the centrality for every marker occurrences and then iteratively deleting the ones with the highest value. The step is repeated until the sizes of the sub-problems are all small enough to be handled by our LP-based algorithm. The pseudo-code for our algorithm is presented in Figure 3.

4. CONDENSING THE MARKERS

Having resolved the conflicts in Ω, the graph G_Ω is now a DAG. In practice however, the graph G_Ω is overly complex. For example, G_Ω contains a large number of redundant edges. A directed edge (m_i, m_j) is said to be *redundant* if there is an alternative (distinct) path from m_i to m_j in G_Ω. For reasonably large individual maps, the resulting consensus graphs obtained after the removal of redundant edges can still be very complex.

Recall that a bin represents a set of nearby markers for which the relative orders are undetermined. In this step, we aim to simplify G_Ω by condensing markers into bins. In order to clearly differentiate the bins constructed in this step from the bins in the original maps, we refer to the former ones as *super-markers.*

The rationale for combining markers into super-markers is the following. If two markers always appear paired in the same bin of the original individual maps, then there is no way to determine the relative order between them and they should be drawn as a single super-marker. Generalizing this observation, we define the notion of *co-segregating* markers as follows. Given a set of maps $\Omega = \{\Pi_1, \Pi_2, \ldots, \Pi_K\}$, two markers (m_i, m_j) are said to be *co-segregating* if they satisfy the following two conditions (A) m_i and m_j belong to the same bin in at least one of the maps in Ω, and (B) there is no path from m_i to m_j or from m_j to m_i in G_Ω. The first condition is intended to ensure that the markers to be condensed into a super-marker are indeed close. The second condition is intended to ensure that the relative order between the markers to be condensed into a super-marker is undetermined.

The co-segregation relation does not define an equivalent relation, because it does not satisfy the transitivity property. Furthermore, when we group markers into super-markers, we must be careful not to introduce new conflicts. In order to address these issues, we employ a greedy iterative algorithm to carry out a maximal decomposition of the markers into super-markers. In each step, we condense one pair of co-segregating markers into a super-marker. The original problem Ω is being transformed into a new problem Ω' (which has one less marker than Ω). We keep repeating this iterative process until no co-segregating markers can be found. Let Ω^f be the final set of maps and G_{Ω^f} be the corresponding induced DAG. We further remove redundant edges from G_{Ω^f}, and let the final graph obtained by this procedure be DAG_Ω. DAG_Ω is guaranteed to have the following property.

Theorem 3. *The in-degree and out-degree of the vertices in DAG_Ω are at most K, where K is the number of maps.*

Proof. Let $\Pi^f \in \Omega^f$ be one of the final individual maps. Let M_{Π^f} be the set of super-markers contained in Π^f. Consider any two super-markers m_i and m_j from M_{Π^f}. If m_i and m_j belong to different bins in Π^f, then m_i and m_j are ordered (meaning either m_i is before m_j or the reverse). On the other-hand, if m_i and m_j belong to the same bin, since m_i and m_j do not form a co-segregating pair (due to the greediness of our algorithm), there must be a path from either m_i to m_j or from m_j to m_i in DAG_Ω. Therefore, if we restrict our attention to a single map $\Pi^f \in \Omega^f$, DAG_Ω defines a total order on the set of super-markers M_{Π^f}. As a result, each super-marker can have at most one immediate predecessor and one immediate successor from one individual map. Since each super-marker can appear in at most K maps, the theorem follows. □

5. LINEARIZING THE DAG

In this last step, we process DAG_Ω to produce a linear order of the bins (super-markers). The linear order must be consistent with the partial order of the bins, i.e., if there is a path from bin b_i to bin b_j in DAG_Ω, then b_i should precede b_j in the linear order. In the case when there is no path between a pair of bins, we have to impute the order of the two bins as well as the distance between them.

Let us define $D[b_i, b_j]$ to be the distance from a bin b_i to another bin b_j in DAG_Ω. If there is only one path from b_i to b_j, then $D[b_i, b_j]$ is trivially assigned the length of that path. If there are multiple paths from b_i to b_j, we set $D[b_i, b_j]$ to be the average length of all paths from b_i to b_j, which can be efficiently computed by dynamic programming.

Now, let b_i and b_j be two bins that are not ordered in DAG_Ω. Our algorithm determines the relative order between b_i and b_j as follows. There are three cases.

```
FASTDELETE(Ω, δ)
S ← ∅;
done ← false
while not done do
  done ← true;
  for each connected components in G_{Ω(N_Ω − S)} do
    I ← the corresponding sub-problem
    if |N_I| > δ then
      done ← false;
      Q ← ∅;
      for m ∈ M_I do
        A ← the set of occurrences of marker m in I
        for m^i ∈ A do
          for m^j ∈ A do
            p ← a shortest path from m^i to m^j if there exists one
            Q ← Q + {p}
      calculate the node centrality for each maker occurrence in N_I based on set Q and the associated weights
      v ← the marker occurrence with the highest centrality
      S ← S + {v}
return S
```

Fig. 3. A sketch of our heuristic-based algorithm

- If b_i and b_j have common ancestors and common successors. Let A be the set of common ancestors and S be the set of common successors. Let $p \in A$ be one of the ancestors and $s \in S$ be one of the successors. We define the distance from bin b_i to bin b_j with respect to the common ancestor and successor pair (p, s) as $\Delta^{(p,s)}[b_i, b_j] = D[p, s]\left(\frac{D[p,b_j]}{D[p,b_j]+D[b_j,s]} - \frac{D[p,b_i]}{D[p,b_i]+D[b_i,s]}\right)$. The final distance $\Delta[b_i, b_j]$ is averaged over all (p, s) pairs, i.e. $\Delta[b_i, b_j] = \sum_{p\in A, s\in S} \Delta^{(p,s)}[b_i, b_j]/(|A|\,|S|)$. If $\Delta[b_i, b_j]$ is positive, then the preferred order is b_i before b_j. Otherwise, the preferred order is b_j before b_i.
- If b_i and b_j have only common successors. Let S be the set of successors and let $s \in S$ be one of the successor. The distance from bin b_i to bin b_j with respect to s is defined as: $\Delta^s[b_i, b_j] = D[b_i, s] - D[b_j, s]$. The final distance $\Delta[b_i, b_j]$ is again averaged over all successors, i.e. $\Delta[b_i, b_j] = \frac{\sum_{s\in S} \Delta^s[b_i,b_j]}{|S|}$.
- If b_i and b_j have only common ancestors. $D[b_i, b_j]$ is similarly computed as in the previous case.

The algorithm we propose to linearize DAG_Ω is similar to the topological sorting algorithm. Let T be the list of ordered bins ($T = \emptyset$ initially). At each iteration, our algorithm determines the next marker s to be ordered. If s is uniquely determined under the partial order given by DAG_Ω, then we simply append s to the end of T. Otherwise, if S is the set of multiple choices, s is chosen so that $\sum_{t\in S, t\neq s} \Delta_{s,t}$ is maximized.

6. EXPERIMENTAL RESULTS

We implemented our algorithms in C++ and carried out extensive evaluations on both real data sets and synthetic data sets. Due to lack of space we will present only the results for synthetic data. Our software tool, called MERGEMAP, is available upon request from the authors.

6.1. Evaluation of the conflict-resolution

The purpose of this set of experiments is to assess the effectiveness and efficiency of our conflict-resolving algorithm. Each data set of this experiment consists of six individual maps, which are all noisy copies of one single *true* map. The *true* map of a data set is simply a permutation of m markers, where the parameter m ranges from 100 to 500 (representing a spectrum of maps from medium sizes to extra large sizes). The distances between adjacent markers are fixed to be 1 cM. To generate an individual map from the *true* map, we first swap η randomly chosen adjacent pairs, and then we relocate γ randomly chosen markers to a random position. The η swaps are intended to mimic local reshuffles while the γ relocations are intended to mimic global displacements, the two types of errors that may present in a genetic map. In our experiments, η ranges from 10 to 30 and γ ranges from 2 to 6.

For each data set, a consensus map was constructed by MERGEMAP by running the conflict resolution module, followed by the bin condensation and the final linearization. The consensus map was compared with the *true* map and the number of erroneous marker pairs were counted. We call a pair of markers *erroneous* when their order in consensus map differs from the order in the true map. When the consensus map is identical to the true map, the number of erroneous marker pairs is zero. On the other hand, when the consensus map is the reverse of the *true* map, the number of erroneous markers is equal to $m(m-1)/2$. For each choice of m, η and γ, ten independent random data sets were generated. For each dataset, the number of erroneous marker pairs and the running time were collected. The mean and standard deviation for both performance measures were computed, and are summarized in Figure 4.

As Figure 4 illustrates, MERGEMAP is very accurate in detecting the problematic markers and deleting them before merging the individual maps. In most cases, the number of erroneous marker pairs is less than ten, and in a few cases the number of erroneous pairs is equal to zero. When η or γ increases, the problem becomes harder and the quality of the consensus map deteriorates. On the contrary, as m increases the number of erroneous pairs decreases[c].

The running time of MERGEMAP increases as m or η or γ increase, but our software tool is relatively efficient. For the largest dataset with $m = 500$ markers, $\eta = 30$ and $\gamma = 6$, MERGEMAP finishes within 2-3 hours. In contrast, JOINMAP takes several **weeks** to assemble maps with 300 or so markers.

6.2. Comparison with JOINMAP

The objective of this set of experiments is to evaluate the entire process of building consensus maps from "scratch" (i.e., starting from synthetic genotyping data). The synthetic genotyping datasets are generated according to a procedure which is controlled by six parameters. We attempted to model the genotyping process to be as realistic as possible. The parameters are the number K of mapping populations, the number m of markers, the number R of "bad markers" on each mapping populations, the genotyping error rate η and the missing rate γ. The sixth parameter x controls the percentage of the markers shared by two individual maps. The latter is an attempt to model what happens in practice, where the data for individual maps only represent a *subset* of the universe of genetic markers.

The entire procedure to generate a synthetic genotyping dataset can be divided into four steps. In the first step, a "skeleton" map is produced with m markers. The markers on the skeleton map are spaced at a distance of 0.5 centimorgan plus a random distance according to a Poisson process with a mean of 2 centimorgans. The "skeleton" map serves the role of the *true* map.

Following the generation of the skeleton map, in the second step the raw genotyping data for the K mapping populations are then generated sequentially. Here we assume that the mapping populations are all of the DH (double haploid) type, and that each population consists of 100 individuals. The genotypes for the individuals are generated as follows. The genotype at the first marker is generated at random with probability 0.5 of being A and probability 0.5 of being B. The genotype at the next marker depends upon the genotype at the previous marker and the distance between them on the skeleton map. If the distance between the current marker and the previous marker is d centimorgans, then with probability $d/100$, the genotype at the current locus will be the opposite of the one at the previous locus, and with probability $1-(d/100)$ the two genotypes will be the same. Finally, according to the specified error rate and missing rate, the genotype state is flipped to model the introduction of a genotyping error or is simply deleted to model a missing observation.

[c]The only outlier in the figure is the case $m = 300$, $\eta = 20$ and $\gamma = 6$. We examined the raw data, and found that the high mean and standard deviation is due to one single data set, for which our algorithm failed to place one single marker in the right place. This single bad marker contributed 172 erroneous marker pair in total. When averaged over the ten runs, the single bad marker contributed 17 to the average number of erroneous pairs.

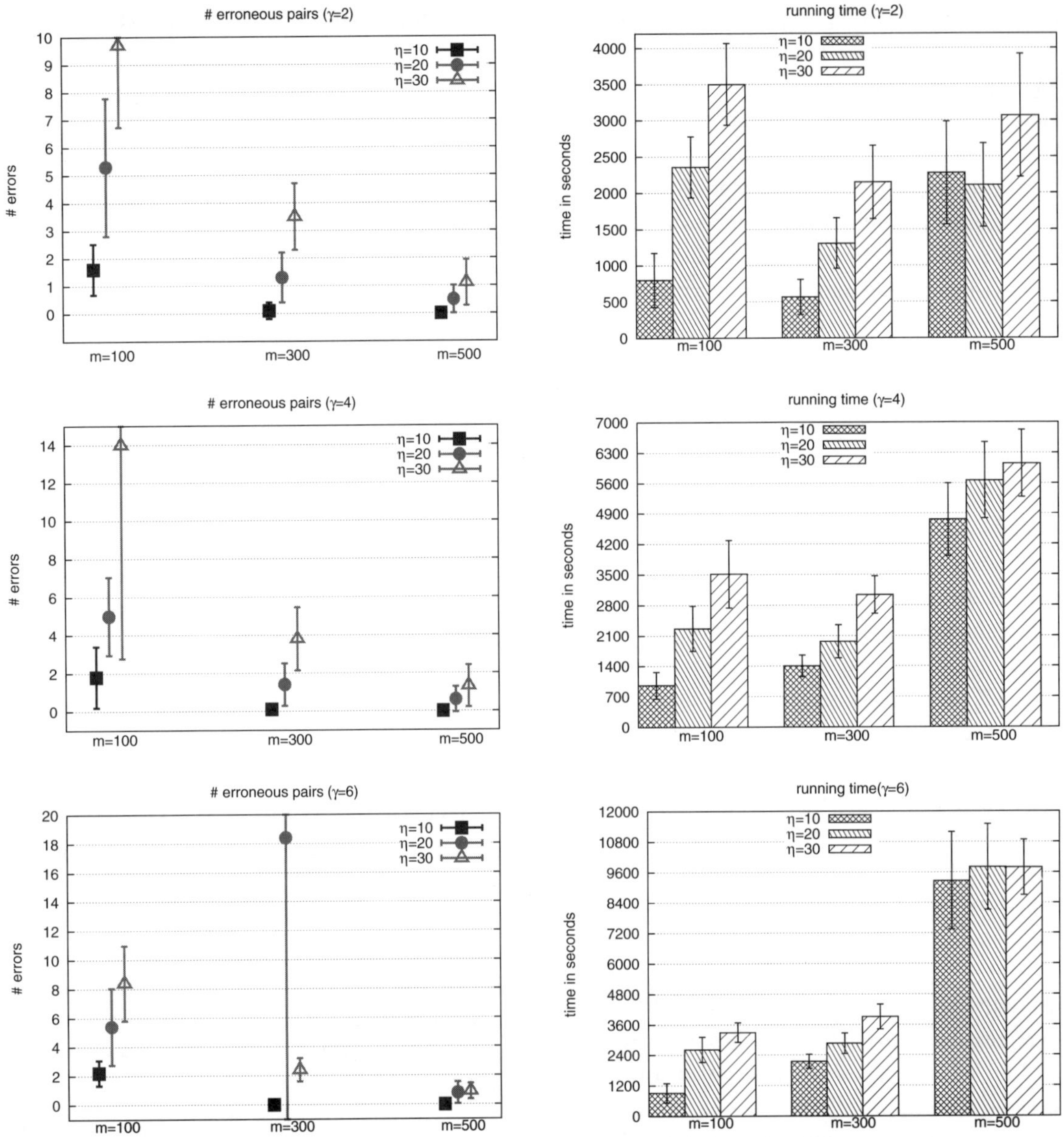

Fig. 4. The number of erroneous marker pairs obtained with MERGEMAP (LEFT) and the average running time (RIGHT) for various choices of m, η and γ. Each point in the figure is an average of the results obtained from ten independent data sets. The standard deviation for the corresponding statistic is represented as the error bar in the above figures.

In the third step, "bad markers" are added to each mapping population. To do so, R markers are first selected at random from each population. The genotypes for those chosen markers across all the 100 individuals are flipped with probability 0.3. Due to the very high error rate introduced for these markers (they are "bad" after all), their positions in the individual genetic maps will be unpredictable. We note that R is small relative to m, and therefore the probability that two individual populations share a common bad marker is very small. When they do, we discard the entire dataset and generate a new one.

The fourth step of generation procedure involves removing a fraction of markers from each individual map. A random subset of $(1 - x)m$ markers is deleted from each mapping population, where x varies from 0.35 to 0.7 in our experiments. As a result, two mapping populations share x^2m markers on average.

For each data set, individual genetic maps are assembled by our tool MSTMAP[6]. The individual maps are then fed into MERGEMAP to build the consensus map. We denote this approach of building the consensus maps as MSTMAP+MERGEMAP.

Here we compare the performance of MSTMAP+MERGEMAP against the software tool JOINMAP. According to our knowledge, JOINMAP is the most popular tool for building consensus map in the community. However, due to the fact that JOINMAP is GUI-based (non-scriptable) and is extremely slow as soon as the number of markers exceeds 150, we were only able to collect results for a few relatively small data sets. As mentioned in the introduction, an alternative approach to the problem of constructing consensus maps is to pool the genotype data for all the individual populations and then apply any existing genetic mapping algorithms by treating the pooled data set as a single population. When pooling individual datasets, a large number of missing observations will have to be introduced. Following this idea, we constructed a consensus map with MSTMAP by first combining the raw mapping data from multiple populations into a pooled dataset. We call this latter approach MSTMAP-C.

We consider two parameter sets, which we thought to be rather realistic. In the first, the parameters are $m = 100, K = 6, x = 0.7, \eta = 0.001, \gamma = 0.001$, and $R = 0$. In the second we changed $R = 2$, while the rest of the parameters were kept identical. For each choice of the parameters, ten random data sets are generated. The number of erroneous marker pairs and the running time is recorded. The results for the two parameters set are presented in Figure 5.

Figure 5-RIGHT shows that MSTMAP+MERGEMAP is orders of magnitude faster than JOINMAP (the y-axis is in log-scale). The difference in running time becomes more apparent when m is large. Also observe that MSTMAP-C can be faster than MSTMAP+MERGEMAP.

Figure 5-LEFT shows that (1) the consensus maps obtained by MSTMAP+MERGEMAP are significantly more accurate than the ones produced by JOINMAP and that (2) MSTMAP-C have comparable accuracy to JOINMAP. We believe that the same conclusions can be derived for larger datasets.

In order to investigate the extend of the advantages brought upon by the tool MERGEMAP we performed an extensive comparison between MSTMAP-C and MSTMAP+MERGEMAP for a variety of parameter settings. For example, Table 1 summarizes the results for $K = 6, x = 0.7$. For this choice of parameters, it is clear that MSTMAP+MERGEMAP outperforms MSTMAP-C for each choice of the parameters. The running time for MSTMAP+MERGEMAP is comparable with those presented in Figure 4 (data not shown), whereas the running time for MSTMAP-C is always very short, within a few minutes regardless of the size of the input. Similar conclusions can be drawn for the cases where $K = 8$, $K = 10$ and $K = 12$ (data not shown due to space restrictions).

7. CONCLUSIONS

We presented a suite of novel algorithms to construct a consensus map from a set of genetic maps given as DAGs. The individual genetic maps are merged into a consensus graph on the basis of shared vertices. Cycles in the consensus graph indicate ordering conflicts among the individual maps, which are resolved according to a parsimonious approach that takes into account two types of errors that may be present in the individual maps, namely, local reshuffles and global displacement. From the set of experimental results reported here, we can conclude that our tool outperforms JOINMAP both in terms of accuracy and running time.

References

1. Jansen J, de Jong AG, van Ooijen JW. Constructing dense genetic linkage maps. *Theor Appl Genet 102* (2001), 1113–1122.
2. Schiex T, Gaspin C. CARTHAGENE: Constructing and joining maximum likelihood genetic maps. In *Proceeding of ISMB* (1997), pp. 258–267.
3. Iwata H, Ninomiya S. AntMap: constructing genetic linkage maps using an ant colony optimization algorithm. *Breeding Science 56* (2006), 371–377.
4. Os HV, Stam P, Visser RGF, Eck HJV. RECORD: a novel method for ordering loci on a genetic linkage map. *Theor Appl Genet 112* (2005), 30–40.
5. Cartwright DA, Troggio M, Velasco R, Gutin A. Genetic mapping in the presence of genotyping errors. *Genetics 174* (2007), 2521–2527.
6. Wu Y, Bhat PR, Close TJ, Lonardi S. Efficient and accurate construction of genetic linkage maps from noisy and missing genotyping. In *Proceeding of WABI* (2007), pp. 395–406.
7. Dib C, Faure S, Fizames C, *et al.* A comprehensive genetic map of the human genome based on 5,264 microsatellites. *Nature 380* (1996), 152–154.

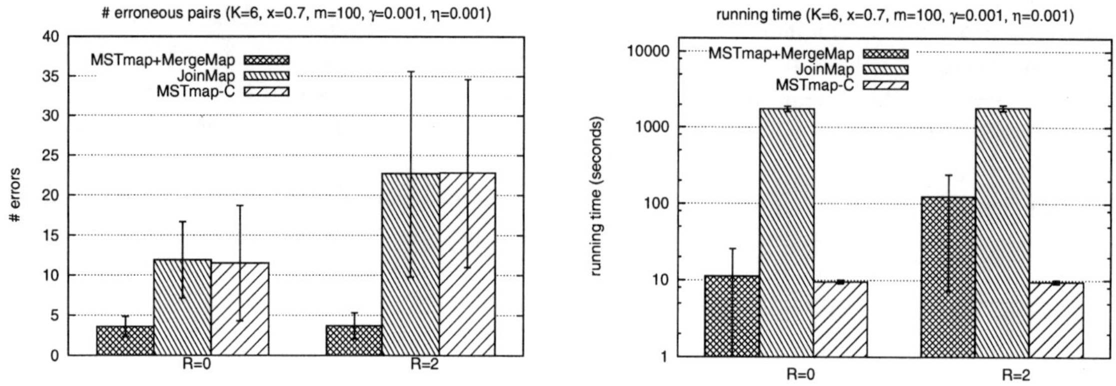

Fig. 5. Comparison between MSTMAP+MERGEMAP, JOINMAP and MSTMAP-C in terms of number of erroneous marker pairs (LEFT) and running time (RIGHT) for $R = 0$ and $R = 2$ respectively. The rest of the parameters are as shown in the title of the figures. Each bar represents an average of ten runs and the error bar indicates the standard deviation.

Table 1. Comparison between MSTMAP+MERGEMAP and MSTMAP-C for $K = 6, x = 0.7$. Each number in the table is the average of the results obtained from ten independent runs.

	# erroneous pairs for MSTMAP+MERGEMAP				# erroneous pairs for MSTMAP-C			
	$\eta = 0.001$	$\eta = 0.005$	$\eta = 0.01$	$\eta = 0.02$	$\eta = 0.001$	$\eta = 0.005$	$\eta = 0.01$	$\eta = 0.02$
	$\gamma = 0.001$	$\gamma = 0.005$	$\gamma = 0.01$	$\gamma = 0.02$	$\gamma = 0.001$	$\gamma = 0.005$	$\gamma = 0.01$	$\gamma = 0.02$
$R = 0$								
$m = 100$	3.6	10.0	16.3	17.9	11.5	15.1	18.0	21.0
$m = 300$	13.7	25.4	34.4	48.6	29.4	29.1	42.1	59.2
$m = 500$	20.9	43.0	56.0	86.9	42.2	56.2	74.3	99.3
$R = 2$								
$m = 100$	3.2	8.4	13.4	18.5	15.3	38.5	32.9	34.0
$m = 300$	11.0	27.6	37.2	55.8	36.9	45.3	48.7	64.9
$m = 500$	19.6	45.0	62.8	81.6	54.1	68.8	84.1	120.1
$R = 4$								
$m = 100$	3.3	12.0	10.6	16.4	24.4	32.1	37.0	44.1
$m = 300$	12.3	23.8	36.2	50.7	39.3	54.6	63.8	69.0
$m = 500$	18.4	46.8	61.2	76.8	59.0	75.2	89.2	120.9
$R = 6$								
$m = 100$	4.1	8.2	10.2	17.7	25.8	24.4	36.4	49.4
$m = 300$	9.6	22.1	31.3	46.4	40.9	52.4	64.6	78.2
$m = 500$	16.2	43.3	56.9	77.6	59.6	73.5	88.9	125.2

8. Ihara N, Takasuga A, Mizoshita K, *et al.* A comprehensive genetic map of the cattle genome based on 3802 microsatellites. *Genome Research 14* (2004), 1987–1998.
9. Dietrich WF, Miller JC, Steen RG, *et al.* A genetic map of the mouse with 4,006 simple sequence length polymorphisms. *Nature Genetics 7* (1994), 220 – 245.
10. Steen RG, Kwitek-Black AE, Glenn C, *et al.* A high-density integrated genetic linkage and radiation hybrid map of the laboratory rat. *Genome Research 9*, 6 (1999).
11. Jackson BN, Aluru S, Schnable PS. Consensus genetic maps: A graph theoretic approach. In *Proceeding of CSB* (2005), pp. 35–43.
12. Beavis WD, Grant D. A linkage map based on information from four f_2 populations of maize (*Zea mays L.*). *Theor Appl Genet 82*, 5 (Oct 1991), 636–644.
13. Stam P. Construction of integrated genetic linkage maps by means of a new computer package: Joinmap. *The Plant Journal 3* (1993), 739–744.
14. Yapa IV, Schneiderb D, Kleinberg J, *et al.* A graph-theoretic approach to comparing and integrating genetic, physical and sequence-based maps. *Genetics 165* (Dec 2003), 2235–2247.
15. Plotkin SA, Shmoys DB, Tardos E. Fast approximation algorithms for fractional packing and covering problems. In *Proceeding of FOCS* (1991), pp. 495–504.
16. Girvan M, Newman MEJ. Community structure in social and biological networks. *PNAS 99* (Jun 2002), 7821–7826.

EFFICIENT HAPLOTYPE INFERENCE FROM PEDIGREES WITH MISSING DATA USING LINEAR SYSTEMS WITH DISJOINT-SET DATA STRUCTURES

Xin Li and Jing Li*

Department of Electrical Engineering and Computer Science, Case Western Reserve University, Cleveland, OH, 44106
**Email: jingli@case.edu*

We study the haplotype inference problem from pedigree data under the zero recombination assumption, which is well supported by real data for tightly linked markers (i.e., *single nucleotide polymorphisms* (SNPs)) over a relatively large chromosome segment. We solve the problem in a rigorous mathematical manner by formulating genotype constraints as a linear system of inheritance variables. We then utilize disjoint-set structures to encode connectivity information among individuals, to detect constraints from genotypes, and to check consistency of constraints. On a tree pedigree without missing data, our algorithm can output a general solution as well as the number of total specific solutions in a nearly linear time $O(mn \cdot \alpha(n))$, where m is the number of loci, n is the number of individuals and α is the inverse Ackermann function[4], which is a further improvement over existing ones[3, 8, 12, 15]. We also extend the idea to looped pedigrees and pedigrees with missing data by considering existing (partial) constraints on inheritance variables. The algorithm has been implemented in C++ and will be incorporated into our PedPhase package[8]. Experimental results show that it can correctly identify all 0-recombinant solutions with great efficiency. Comparisons with other two popular algorithms show that the proposed algorithm achieves 10 to 10^5-fold improvements over a variety of parameter settings. The experimental study also provides empirical evidences on the complexity bounds suggested by theoretical analysis.

1. INTRODUCTION

Experimental data have shown that genetic variation is structured in haplotypes rather than isolated SNPs[14] and haplotypes may provide substantially increased power in detecting gene-disease association. However, the human genome is a diploid and, in practice, haplotype data are not collected directly, especially in large scale sequencing projects mainly due to cost considerations. Hence, efficient and accurate computational methods and computer programs for the inference of haplotypes from genotypes are highly needed.

Recent years have witnessed intensive research on haplotyping methods (see reviews[2, 5, 6, 10, 17]), mainly driven by the HapMap project[14]. For family data, there exist two types of haplotyping methods, statistical methods and combinatorial (or rule-based) methods. And there is a tendency to merge these two types of approaches[9, 16]. In general, the goal of statistical approaches[1] is to find a haplotype assignment for each individual with the maximum likelihood or to output all consistent solutions with their corresponding probabilities. Recently, population haplotype frequencies have been taken into considerations[1] to account for correlations among tightly linked markers (known as *linkage disequilibrium*). A key step in most statistical approaches is to enumerate all possible inheritance patterns and to check the genotype consistency for each of them[1]. Due to the large degrees of freedom, this step usually leads to high time complexity (usually exponential hence computational intractable for large data sets). On the other hand, rule-based algorithms first partially infer haplotypes or inheritance vectors based on genotype constraints, and then search final solutions from the reduced space. Therefore, rule-based algorithms[3, 8, 9, 12, 15] can potentially gain advantage over statistical methods in efficiency. The zero recombinant assumption states that recombination is nonexistent within a pedigree for a sufficiently large number of tightly linked markers. As a realistic assumption, it has been used in both statistical approaches as well as rule-based approaches. Furthermore, a solution to the problem with no recombinant can be served as a subroutine of a general procedure to solve the general haplotype inference problem. Therefore, investigation of efficient algorithms

*Corresponding author.

to obtain all 0-recombinant solutions from a pedigree is of high interests.

For a given pedigree, the goal of the *zero recombinant haplotype configuration* (ZRHC) problem is to identify all possible haplotype assignments with no recombination. An important advance in the development of rule-based algorithms for haplotype inference in pedigrees in general and the ZRHC problem in particular is the introduction of variables to represent uncertainties. The problem can then be discussed and solved with mathematical rigor. Li and Jiang[8] first formulate the problem as a linear system on "*ps*" (a binary indicator of parental source) variables and solve it using Gaussian elimination with a complexity of $O(m^3n^3)$, where m is the number of markers and n is the number of individuals. More recently, Xiao et al.[15] formulate another linear system on "h" (a binary indicator of inheritance relationship) variables, and lower the complexity to $O(mn^2 + n^3 \log^2 n \log\log n)$. For loop-free (tree) pedigrees, Xiao's method can produce a general solution in $O(mn^2 + n^3)$ and a particular solution in $O(mn + n^3)$ time. Here, a particular solution means a specific assignment for each variable which satisfies the constraints, while a general solution is a description of all solutions in a general form where some variables are designated as free (meaning that they are allowed to take any value), and the remaining variables are represented by a linear combination of these free variables. For tree pedigrees, Chan et al.[3] further reduce the complexity of finding a particular solution to a linear time $O(mn)$ by manipulating the constraints on a graph structure. Liu and Jiang[12] also propose an algorithm to produce a particular solution in $O(mn)$ and a general solution in $O(mn^2)$ by further exploring features of their h variable system on a tree pedigree. However, with missing data, it has been shown that ZRHC is NP-hard[11]. Therefore, it seems impossible to incorporate missing data into a pure linear constraint system without enumerations. Li and Jiang[9] propose an integer linear programming algorithm for the minimum recombinant haplotype inference problem, and it can solve ZRHC with missing as a special case. But because it does not use zero recombinant constraints explicitly, it may need to enumerate almost all possible haplotype assignments.

In this paper, we propose an elegant and more efficient algorithm for detecting, recording and consistency checking of constraints on h variables. Notice that it is not necessary to solve the h variable system explicitly, as it was in Ref. 15. Instead, we encode constraints on h variables using disjoint-set forests. By applying an adapted disjoint-set union-find procedure, we can update the disjoint-set structures incrementally upon new constraints, and determine the consistency of the encoded linear system simultaneously. Based on the disjoint-set union-find procedure, the proposed algorithm can produce a general solution in almost linear time ($O(mn \cdot \alpha(n))$ for a tree pedigree, where α is the inverse Ackermann function[4], improved from the best known algorithm with $O(mn^2)$ time complexity[12]. We further extend the algorithm to looped pedigrees and pedigrees with missing data, by utilizing the constraints imposed from existing data. Experimental results show that the algorithm can output all solutions with zero recombinant and it is much more efficient than two popular existing algorithms because of the significant reduction of the enumeration space.

The rest of the paper is organized as follows. In section 2, we introduce the linear system on h variables together with some basic concepts and notations concerning the ZRHC problem. By representing constraints using a linear system, one can formally investigate different strategies to solve the problem in a rigorous manner. And different strategies of manipulating and integrating the constraints will result in different complexities. Our algorithm of detecting and processing constraints from pedigree data is presented in section 3. In both sections, we assume input genotype data has no missing alleles and the ZRHC problem under this case is polynomially solvable. Our algorithm is almost optimal by achieving a nearly linear time complexity on tree pedigrees with complete data. In section 4, we show how to extend the algorithm to cope with missing data and looped pedigrees by effectively reducing the search space before enumerations. The performance of our algorithm and comparisons with other two programs are examined in section 5. We discuss future directions and make concluding remarks in section 6.

2. PRELIMINARIES

A *pedigree graph* indicates the parent-child relationships among an extended family. Figures 1(a) and 2(a) present pedigrees in a conventional manner. The pedigree in Figure 1(a) has a *mating loop*, where an offspring (node 9) is produced by the mating of two relatives (node 6 and 7). A pedigree without mating loops is called a *tree pedigree*. A *nuclear family* only consists of both parents and their children. For any pair of homologous chromosomes from a diploid organism such as human, exactly one is from its father and the other one is from its mother, as illustrated in Figure 1(b). A physical position on a chromosome is called a *locus* and the status of a locus is called an *allele*, represented using an integer ID. We focus on SNP data in this study thus assume there are only two alternative alleles (i.e., bi-allelic data), which turns to be the hardest case for ZRHC[8, 15].

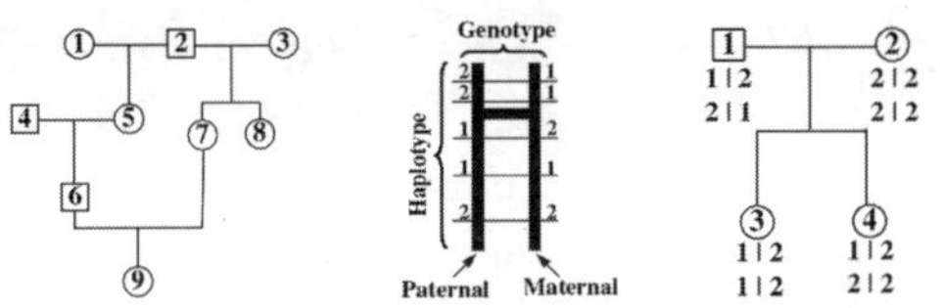

(a) (b) (c)

Fig. 1. (a) A pedigree graph. We use a circle to represent a female, a square to represent a male in a pedigree. (b) A haplotype is composed of all alleles on one chromosome segment. Each allele is an integer value representing the status of a marker at a chromosome locus. (c) A recombination event occurs when a child does not inherit a complete haplotype from its parent. Individual 3 has a paternal haplotype 11 which is not seen in his father. So there must be a crossover between two chromosomes of his father in meiosis, which results in a recombinant haplotype.

At each locus i, a child may inherit either of the paternal or maternal allele of a parent. We use binary variables to indicate parental source (ps) of the two alleles in a child.

Definition 2.1. ps variable $p_i^x \in \{0,1\}$ is defined for each locus i of each individual x. $p_i^x = 0$ if the smaller allele of locus i is of paternal source, $p_i^x = 1$ if it is of maternal source. We technically let $p_i^x = 0$ if locus i is homozygous (two alleles being the same).

Loosely speaking, a *haplotype* consists of all alleles on a chromosome. *Recombination* events or *crossovers* occur when a child inherits a shuffled version of its parent's two haplotypes (see Figure 1(c) for an example). However, for a sufficiently large segment of chromosome with m SNPs, the likelihood of recombination between a parent-child pair is extremely small. For example, a rough estimation of the relationship of genetic distances and physical distances is about 1 Mbps/cM. The average marker interval distance of a 500K SNP chip is about 6 Kbps. Therefore, the probability of seeing a single recombination event from a parent-child pair of 170 SNP markers (~1Mbps) is only ~1%. One can assume a child inherits an entire haplotype segment from a parent for a sufficiently large number of SNPs (*i.e.*, zero recombinant assumption). In such a case, the inheritance behavior between a parent-child pair is unique throughout all m loci, and it is convenient and practically appealing to use a single binary variable (h) to indicate the inheritance behavior between a parent-child pair.

Definition 2.2. Inheritance variable $h^{x_1x_2} \in \{0,1\}$ is defined between a parent x_1 and a child x_2. $h^{x_1x_2} = 0$ if x_2 inherits the paternal haplotype of x_1, $h^{x_1x_2} = 1$ if x_2 inherits the maternal haplotype of x_1.

2.1. Mendelian constraints as a linear system

Mendelian laws of inheritance impose constraints on ps and h variables for each parent-child pair at each locus. These constraints can be represented by a linear relationship of ps and h variables over the group $(Z_2, +)$ (where $0+0=0$, $0+1=1$, $1+1=0$). Table 1 summarizes all cases of constraints at a certain locus i for a parent-child pair. When an individual is homozygous at a certain locus, its ps variable at this locus is determined by definition. When one or both of the parents of an individual are homozygous at a certain locus, this individual's ps variable at this locus is also determined. In both cases, the ps variable is pre-determined. In all the other cases, there is a constraint for each parent-child pair between ps variables and the h variable, as shown in the last three cases in Table 1. The constraints introduced by the zero recombinant assumption is enforced by the single h variable between each parent-child pair. Therefore, the system formed by the sets of constraints collected based on Table 1 consists of all the constraints from data. The satisfiability (or consistency) of this system is equivalent to whether there is a zero recombinant solution.

Table 1. Constraints for a parent-child pair x, y.

genotype		constraint		
x	y		if x is father	if x is mother
1/1	1/2	$p_i^x = 0$	$p_i^y = 0$	$p_i^y = 1$
2/2	1/2	$p_i^x = 0$	$p_i^y = 1$	$p_i^y = 0$
1/2	1/2		$p_i^y = p_i^x + h^{xy}$	$p_i^y = p_i^x + h^{xy} + 1$
1/2	1/1	$p_i^y = 0$	$p_i^y = p_i^x + h^{xy}$	$p_i^y = p_i^x + h^{xy}$
1/2	2/2	$p_i^y = 0$	$p_i^y = p_i^x + h^{xy} + 1$	$p_i^y = p_i^x + h^{xy} + 1$

2.2. Locus graphs

To process constraints, Xiao et al.[15] introduced the concept of locus graphs. We give a brief introduction here for the sake of completeness. A locus graph $L_i(V, E_i)$ is constructed for each locus i to record the constraints on h variables. V consists of all individuals as nodes. There exists an edge in E_i between a parent-child pair only if the *ps* variables of this pair is constrained on the correspondent h variable, i.e., when the parent is heterozygous at locus i (the last three cases in Table 1). Each edge is also labeled by the h variable and the constant associated with the constraint. We refer to this kind of constraints (a linear equation consists of *ps* variables and a h variable) as *edge constraints.* Figure 2(b) shows an example of a locus graph.

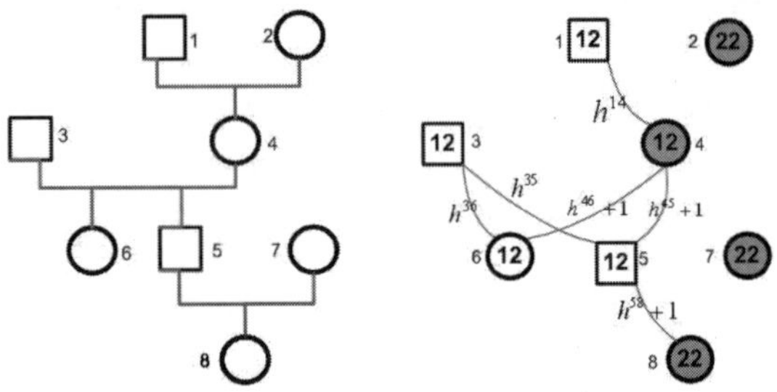

(a) (b)

Fig. 2. (a) A pedigree with 8 members. (b) Given the genotype at a a certain locus i, the correspondent locus graph L_i and h variable constraints. *ps* variables of shaded members (2, 4, 7, 8) are pre-determined. From this locus graph, we can generate two non-redundant h variable constraints, one is a cycle constraint, $h^{35} + h^{36} + h^{45} + h^{46} = 0$ (formed by individual 3, 4, 5, 6), the other is a path constraint, $h^{45} + h^{58} = 0$ (from individual 4 to 8 via 5).

The original idea of Ref. 15 was to integrate edge constraints to construct a new subsystem that only consists of h variables. Their algorithm then solved the subsystem and used h variable solutions to solve *ps* variables. We also record edge constraints on locus graphs. However, instead of explicitly listing and solving the constraints on h variables, we use disjoint-set structures to collect, encode and thus examine the consistency of these constraints, which help us achieve a better time complexity result to obtain a general solution.

2.3. Linear constraints on h variables

There are essentially two types of constraints on h variables in a locus graph L_i: *path constraints* and *cycle constraints.* Notice that the classification of constraints here is more succinct than those in previous work[12, 15] because our method of handling constraints does not require further discrimination of them. According to Table 1, each edge e_{xy} in a locus graph represents an edge constraint in the form $p_i^x + p_i^y = h^{xy} + c_i^{xy}$, where c_i^{xy} is a constant $\in \{0, 1\}$. We use a subscript i for c_i^{xy} because for different loci, the constant between a parent-child pair may be different, which depends on the genotype at that locus as specified in Table 1. For a path $P_{\widetilde{v_s, v_t}}$ from individual s to individual t in locus graph L_i, if we sum up all edge constraints on this path, we have

$$\sum_{e_{xy} \in P_{\widetilde{v_s, v_t}}} (p_i^x + p_i^y) = p_i^s + p_i^t = \sum_{e_{xy} \in P_{\widetilde{v_s, v_t}}} (h^{xy} + c_i^{xy}).$$

If p_i^s and p_i^t are pre-determined constants, we end up with a path constraint on h variables, which is

$$\sum_{e_{xy} \in P_{\widetilde{v_s, v_t}}} h^{xy} = p_i^s + p_i^t + \sum_{e_{xy} \in P_{\widetilde{v_s, v_t}}} c_i^{xy}, \qquad (1)$$

where the right-hand side is a constant. Similarly, for a cycle C in locus graph L_i, which may exist even on a tree pedigree (e.g., when a nuclear family has more than one heterozygous children), we sum up all edge constraints on C,

$$\sum_{e_{xy} \in C} (p_i^x + p_i^y) = 0 = \sum_{e_{xy} \in C} (h^{xy} + c_i^{xy}),$$

and finally have a cycle constraint on h variables

$$\sum_{e_{xy} \in C} h^{xy} = \sum_{e_{xy} \in C} c_i^{xy}.$$

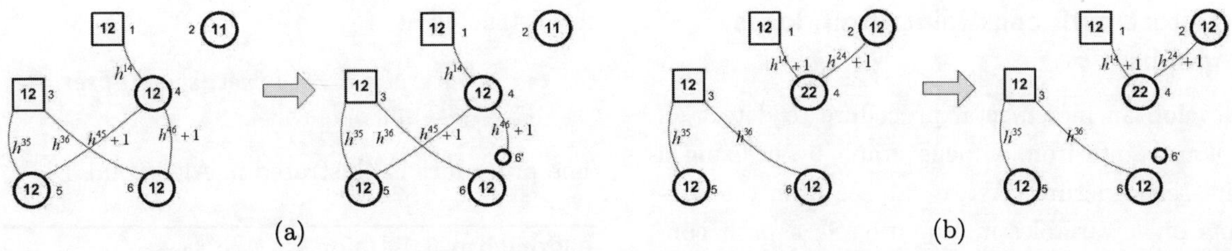

Fig. 3. Node splitting applied to a nuclear family at two loci to remove local cycles. (a) The original locus graph (left), and the locus graph (right) with edges remounted after node 6 was duplicated. (b) A locus graph at another locus before (left) and after (right) node 6 was split. Though no local cycle exists in the locus graph in b, node 6 was also duplicated so that all locus graphs will still have the same number of nodes after splitting.

3. METHODS

By exploiting special features of the constraints on h variables, it is not necessary to explicitly list every path and cycle constraint to check their consistency. We employ disjoint-set structures to detect and to check the consistency of constraints on h variables. For each locus graph L_i, we build a disjoint-set structure D_i to encode its connectivity information. We update the disjoint-set structure incrementally upon processing each edge constraint on a locus graph. Path constraints on a locus graph are detected during this process and will be stored in another disjoint-set structure D. The whole algorithm works on $m+1$ such disjoint-set structures, one D_i for each locus graph L_i and one D for encoding all path constraints. In this section, we assume the inputs are tree pedigrees with complete data. Cycles on a locus graph from a tree pedigree can only be generated within a nuclear family when it has multiple children. We first discuss a node splitting strategy in subsection 3.1 to break all such short cycles, to obtain only path constraints for further processing. Construction of D_i from each locus graph L_i to detect path constraints will be discussed in subsection 3.2. Processing of constraints and consistency check will be discussed in subsection 3.3 and a general solution of h variables will be decoded from the disjoint-set structure D. Solutions of ps variables will then be obtained. The analysis of time complexity and correctness of the algorithm on tree pedigrees will be discussed in subsection 3.4. One of the advantages of the proposed algorithm is that it can be easily extended to the general cases of looped pedigrees and pedigrees with missing data. And we show these extensions in section 4.

3.1. Split nodes to break cycles

In order to simplify the constraint detection, we first transform cycle constraints to path constraints by breaking cycles in locus graphs. There are essentially two kinds of cycles in a locus graph: *global* cycles that are introduced by marriages between relatives and *local* cycles that are introduced by multiple children within one nuclear family (e.g., Figure 2(b)). Only local cycles will exist in a tree pedigree and will be dealt with in this subsection. The treatment of global cycles will be deferred to subsection 4.1 when we discuss the extension to looped pedigrees. We break local cycles for each nuclear family with multiple children by splitting some child nodes and by remounting their edges on each locus graph. More specifically, when a nuclear family has multiple children, any child node v (except an arbitrarily fixed one v_0) and its genotypes will be duplicated to create a new node v' in the same manner across all locus graphs. New *ps* variables will be introduced for these duplicated nodes. For each splitting node v, the edge from its mother (if there is) will be reconnected to node v'. All other edges regarding node v remain untouched. Figure 3 shows an example on how node splitting is performed. By doing so, we technically avoid the treatment of cycle constraints. After the duplication, all new locus graphs (actually *locus trees* now) still have the same set of nodes. Notice that one has to record all local cycle constraints on h variables and constraints that the *ps* variables of duplicated nodes must have the same assignments as those in their original copies. Their constraints can be easily dealt with for local cycles because they only involve local structures (nuclear families). This will be further discussed in the next subsection.

3.2. Detect path constraints from locus graphs

We develop an incremental procedure to detect all path constraints from a locus graph by utilizing a disjoint-set structure. As we can see from the constraints on h variables in Equation 1, a path constraint is specified by the ps variables of its end nodes and summation of the constant parity value c_i^{xy} associated with the edge constraint on each of its edges. Our goal is to detect each non-redundant path on a locus graph with pre-determined end nodes and meanwhile obtain the constant parity summation associated with that path.

To do so, we maintain a disjoint-set structure D_i for each locus graph L_i and update it incrementally. The disjoint-set structure is defined by a pair of values $rep_i[v]$, $offset_i[v]$ for each node v in $V(L_i)$. We use subscript i here to emphasize that the disjoint-set structure D_i is specific for each locus graph. $rep_i[v]$ indicates the node which acts as the representative of the set containing v. And the offset of a node $offset_i[v]$ indicates the summation of the constants associated with the edge constraints on the path from v to its rep_i. Namely, if $rep_i[v] = v_0$, then $offset_i[v] = \sum_{e_{xy} \in P_{\widetilde{v,v_0}}} c_i^{xy}$, where $P_{\widetilde{v,v_0}}$ is the path with end nodes v and v_0, c_i^{xy} is the constant associated with the edge constraint on edge e_{xy} (as specified in the last 3 cases of Table 1). Initially, for every node in V: $rep_i[v] \leftarrow v$, $offset_i[v] \leftarrow 0$. We examine each $e^{xy} \in L_i$ and update D_i by considering the edge constraint $p_i^x + p_i^y = h^{xy} + c_i^{xy}$ represented by e^{xy}. If both $p_i^{rep_i[x]}$ and $p_i^{rep_i[y]}$ are pre-determined, we report a path constraint and record it in D for consistency check (see subsection 3.3). The two sets represented by $rep_i[x]$ and $rep_i[y]$ will always be merged into one because they are connected by an edge e^{xy} and we always let one pre-determined representative be the representative of the new set if there is such one. At the end, any two nodes connected by a path in L_i will be merged into one set and a set in D_i only consists of connected nodes in L_i. By doing so, we can safely detect all path constraints on L_i. Furthermore, the constant associated with a path constraint between two nodes v_s and v_t in the same set can be reconstructed as

$$\sum_{e_{xy} \in P_{\widetilde{v_s,v_t}} \in L_i} c_i^{xy} = offset_i[s] + offset_i[t].$$

The procedure is illustrated in Algorithm 3.1.

Algorithm 3.1 Union$_i(x,\ y,\ c_i^{xy})$

if both $p_i^{rep_i[x]}$ and $p_i^{rep_i[y]}$ are pre-determined **then**
  Report a path constraint P from node $rep_i[x]$ to $rep_i[y]$: $\sum_{e_{xy} \in P} h_i^{xy} = c$, where $c = p_i^{rep_i[x]} + p_i^{rep_i[y]} + offset_i[x] + offset_i[y] + c_i^{xy}$. Encode the constraint in D by applying $Union(rep_i[x], rep_i[y], c)$.
end if
if $p_i^{rep_i[y]}$ is not pre-determined **then**
  $offset_i[rep_i[y]] \leftarrow offset_i[y] + offset_i[x] + c_i^{xy}$
  $rep_i[rep_i[y]] \leftarrow rep_i[x]$
else
  $offset_i[rep_i[x]] \leftarrow offset_i[x] + offset_i[y] + c_i^{xy}$
  $rep_i[rep_i[x]] \leftarrow rep_i[y]$
end if

We also need to capture all constraints that may have not been processed yet in the above procedure due to node splitting. This is easy for a tree pedigree which only possibly has local cycles to split. There are *three* possible types of constraints that need special attentions due to node splitting, *i.e.*, local cycle constraints themselves, ps variables between duplicated nodes and their corresponding splitting nodes, and some path constraints originally existing in the locus graph before splitting, but broken by splitting. We can prove by case analysis that all these constraints can be safely recovered as path constraints on locus graphs after node splitting. We leave the proof in our extended journal version and we illustrate the cases using an example in Figure 4.

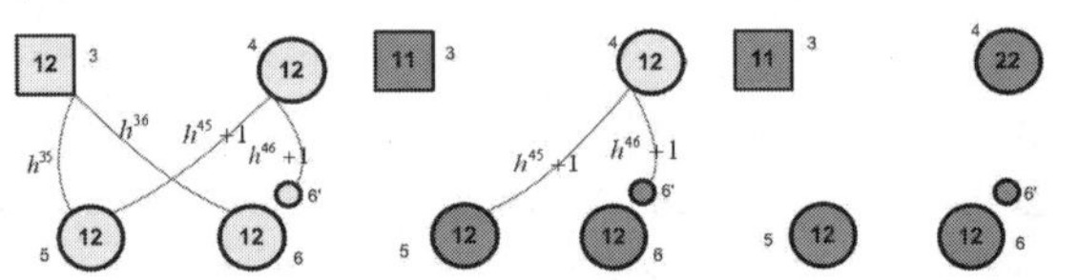

(a) (b) (c)

Fig. 4. This example illustrates all possible patterns of locus graphs of a nuclear family on a tree pedigree. (a) If neither of the parents is homozygous at this locus, then there should be a loop constraint, $h^{36} + h^{35} + h^{45} + h^{46} = c$. Since we split node 6, it is expressed as a path constraint on path $P_{\widetilde{v_6,v_{6'}}}$. Since the locus graph is still connected, no path via this nuclear family will be broken up due to the split of node 6. (b)(c) If one or both of the parents are homozygous at this locus, then both of the children are pre-determined. In this situation, path constraints such as $P_{\widetilde{v_5,v_{6'}}}$ will only take the children as end nodes such that they remain on a consecutive path, unaffected by the split of node 6.

Figure 5 gives an example on how to detect constraints on a locus graph L_i. In the actual implementation of a disjoint-set forest, a node may not directly point to its set representative, we omit the details (see Ref. 4, 13) here for clear demonstration purpose.

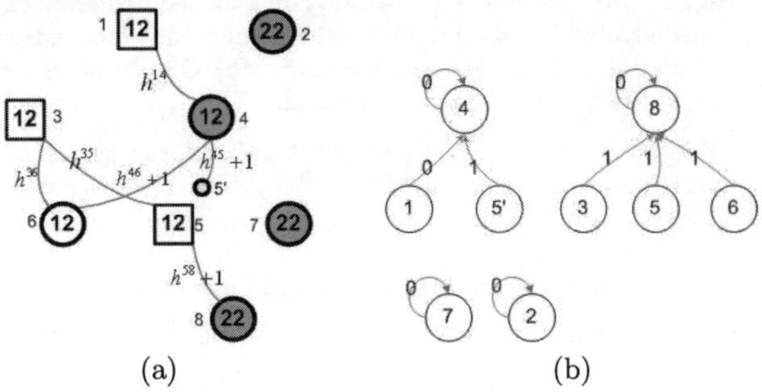

Fig. 5. An example shows the detection of all constraints from a locus graph after node splitting. (a) Locus graph L_i of a pedigree with 8 nodes at a certain locus i. Shaded nodes are pre-determined. (b) The disjoint-set forest formed by adding edges 1-4, 4-5', 3-5, 3-6 and 5-8 of the locus graph L_i in (a). No path constraint has been detected so far. We simply merge the sets containing each pair of nodes. A pointer is annotated with the offset of a node to its representative. If we further process edge 4-6 of L_i. Because both 4 and 6 have a representative with *ps* variable pre-determined, a path between the two representatives (node 4 and 8) will induce a path constraint, which is $\sum_{e_{xy} \in P_{\widetilde{v_4,v_8}}} h^{xy} = 0$.

3.3. Encode path constraints in disjoint-set structure D

Once we detect a path constraint, we encode this constraint also in a disjoint-set structure D. As usual, D is defined by a pair of values $rep[v]$ and $offset[v]$ for each node $v \in V$. $rep[v]$ is a pointer to a node and $offset[v] \in \{0,1\}$ is a constant. We maintain this disjoint-set structure D such that any two nodes k and l in the same set encode a path constraint in the form of $\sum_{e_{xy} \in P_{\widetilde{k,l}}} h^{xy} = offset[k] + offset[l]$. Initially, $rep[v] \leftarrow v$, $offset[v] \leftarrow 0$, for any $v \in V$. When processing a path constraint $\sum_{e_{xy} \in P_{\widetilde{i,j}}} h^{xy} = c$, we check whether the representatives of the two end nodes i and j are the same. *If they are not the same*, which means no constraints on h variables between these two nodes have been discovered so far, we merge the two sets represented by $rep[i]$ and $rep[j]$ as illustrated in Algorithm 3.2. *When $rep[i]$ and $rep[j]$ are the same* (a constraint already exists before seeing the current constraint), we must check the consistency and redundancy between the current constraint and the previous constraint. This can be easily done by comparing the constant c associated with the new constraint and the constant associated with the existing constraint $offset[i] + offset[j]$. If the two constants are the same, the new constraint is redundant and will be dropped; otherwise, inconsistency exists and the program reports no solutions with zero recombinations and terminates. The procedure is summarized in Algorithm 3.2.

Algorithm 3.2 Union(i, j, c)

```
if rep[i] = rep[j] then
  if offset[i] + offset[j]! = c then
    Report inconsistency
  end if
else
  offset[rep[j]] ← offset[j] + offset[i] + c
  rep[rep[j]] ← rep[i]
end if
```

After all path constraints have been processed, the nodes will form several independent sets. A general solution of h variables can be easily decoded from D. More specifically, for each set representative v of D, we define a free binary variable α_v (notice α_v is not the same as *ps* variables). A general solution of h variables can be represented by a linear system of α variables (which are all free) in the form of

$$h^{xy} = \alpha_{rep[x]} + offset[x] + \alpha_{rep[y]} + offset[y]. \quad (2)$$

We can prove that such a solution satisfies all path constraints. We can further argue that there are no other h variable assignments that satisfy all path constraints. We leave the proof in our extended journal version.

Next, let's consider how to compute *ps* variable solutions from h variable solutions. For each node v in D_i, v is connected to its set representative $rep_i[v]$ through a path P on L_i. We have $p_i^v + p_i^{rep_i[v]} = \sum_{e_{xy} \in P \in L_i} (h^{xy} + c_i^{xy}) = \sum_{e_{xy} \in P} h^{xy} + \sum_{e_{xy} \in P} c_i^{xy} = \sum_{e_{xy} \in P} h^{xy} + offset_i[v]$. By plugging in the solution of h variables in Equation 2, we will finally get a general solution for the ZRHC problem,

$$\begin{aligned} p_i^v = p_i^{rep_i[v]} &+ \alpha_{rep[rep_i[v]]} + offset[rep_i[v]] + \\ &\alpha_{rep[v]} + offset[v] + offset_i[v]. \end{aligned} \quad (3)$$

If $p_i^{rep_i[v]}$ is not pre-determined, we have one more degree of freedom in Equation 3.

3.4. Analysis of the algorithm on tree pedigrees with complete data

The overall algorithm is summarized in Algorithm 3.3. We omit the preprocessing steps (such as node splitting, construction of locus graphs) because all those operations can be done in linear time. Here we also state our main result of the algorithm as a theorem. We leave the proof in the extended journal version due to space limitation.

Algorithm 3.3 Process_All_Constraints

```
for i = 1 to m do
    for all edge e_xy ∈ L_i do
        Union_i(x, y, c_i^xy)
    end for
    for all splitting node v do
        if rep_i[v] = rep_i[v'] then
            Union(v, v', offset_i[v] + offset_i[v'])
        end if
    end for
end for
```

Theorem 3.1. *For a tree pedigree with complete data, Algorithm 3.3 correctly outputs a general solution (Equation 2 and 3) and the number of specific solutions (degrees of freedom) for the ZRHC problem if it has a solution, and reports inconsistency otherwise. Its running time is bounded from above by $O(mn\alpha(n))$, where m is the number of loci, n is the number of individual and $\alpha()$ is the inverse Ackermann function*[4].

4. EXTENSION TO GENERAL CASES

4.1. Pedigrees with mating loops

We can further extend the above algorithm to pedigrees with mating loops and pedigrees with missing data. For a looped pedigree, we apply a similar splitting rule to locus graphs as we did for a tree pedigree, except that for a mating between two relatives all their children are duplicated in order to break a global cycle. We use the same method described in section 3.2 and 3.3 to detect all path constraints on each locus graph. However, Theorem 3.1 does not hold anymore in this case because the method does not guarantee the detection of all necessary constraints. The difference lies in the detection of path constraints broken by splitting nodes. All such path constraints can be recovered when breaking a local cycle but may not be recovered when breaking a global cycle. Figure 6 gives such an example on a looped pedigree.

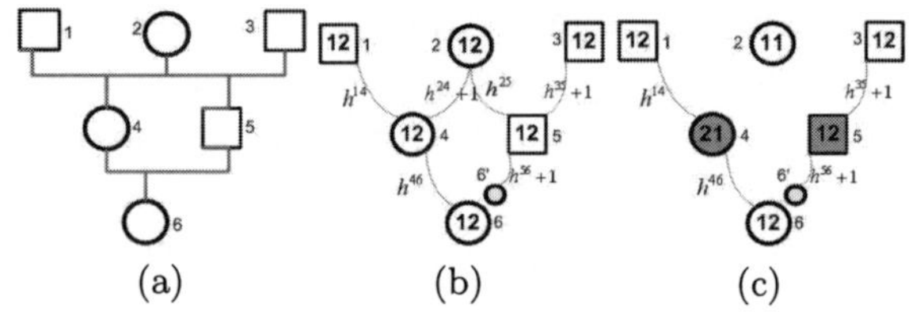

Fig. 6. An example of constraints on a looped pedigree. (a) A pedigree with a mating loop, where node 6 is produced by the mating of two relatives 4 and 5. (b) One locus graph L_i, where there is a path constraint $\sum_{e_{xy}\in P_{\widehat{v_6,v_{6'}}}} h^{xy} = h^{24} + h^{25} + h^{46} + h^{56} = 0$. (c) Another locus graph L_j, where there is a constraint $h^{46} + h^{56} = 0$. Due to the splitting at node 6, this constraint is not on a consecutive path.

Although the set of constraints are not sufficient, we can still obtain all the solutions for a looped pedigree using the following procedure. If there are already inconsistent constraints during consistency check, no solutions with zero recombinant exist. Otherwise, all the h variable solutions obtained based on the general solution (Equation 2) will be examined. If a specific h variable assignment is not consistent with the genotype, we simply drop that assignment. Otherwise, it will result in real haplotype solutions. To check the consistency of an h variable assignment with existing genotypes, we use another disjoint-set structure to encode constraints on alleles. This step is the same for pedigrees with loops and pedigrees with missing data. Essentially for looped pedigrees, we avoid cycle constraints by splitting nodes with the expense that we may miss some constraints. We start to enumerate h variables after processing existing partial constraints. However, as it will be shown in the experiment, the number of all possible h variable assignments from this set of partial constraints is usually very small for a pedigree with complete data, and in most times there is only one solution for pedigrees with 20 or more loci. Therefore, the above extension can efficiently handle looped pedigrees in practice.

4.2. Pedigrees with missing data

For an algorithm to be practically useful, it has to be applicable on real data. Most real data contains missing. One advantage of the proposed algorithm is that it can be easily extended to deal with missing data. Extension of existing work[3, 12, 15] to handle missing is not trivial at all. We take a similar approach as in subsection 4.1 to deal with missing data. Partial constraints on h variables will be collected based on existing genotype data. Solutions of h variables will be obtained based on the set of partial constraints and will be checked for consistency with existing genotype data. More specifically, for a pedigree with missing data, we construct the locus graph L_i for each locus i as usual with node splitting if necessary. The edges in L_i will only be constructed by examining every parent-child pair whose genotypes are complete at this locus i. We apply Algorithm 3.3 to process all edge constraints from such locus graphs. And from the partial constraints on h variables, we get a solution in its general form (Equation 2). The degree of freedom, which is $n_D - 1$ where n_D is the number of independent sets, usually is significantly smaller than the degree of freedom of the original h variables without constraints, which is usually close to $2n$. Therefore, our algorithm has the potential to be significantly faster than those algorithms based on the enumeration of all possible h variables (such as Merlin[1]).

For each specific h variable assignment, the compatibility check with the input genotype data is also examined by utilizing another disjoint-set structure on allele variables. For space limitations, we leave the details in the extended journal version. By doing so, we can efficiently check the consistency between a given h variable assignment and the input genotype data, and generate a set of assignments of alleles that are consistent with the h variable assignment. The total number of h variable assignments is 2^{n_D-1}, and for each assignment, the complexity of genotype consistency check is $O(mn \cdot \alpha(n))$.

5. EXPERIMENTAL RESULTS

We study the performance of our program (denoted as DSS) under different settings (pedigree size, number of loci, missing rate, pattern of missing) and compare its performance with two representative programs Merlin[1] and PedPhase (the integer linear programming ILP algorithm in Ref. 9). Merlin is one of the most widely used statistical packages for linkage analysis and we only use its haplotyping functionality in this comparison. It also uses the zero recombinant assumption. But it examines all possible configurations of inheritance variables and only outputs those compatible ones. PedPhase.ILP is another widely used rule-based algorithm developed by our own group. It can produce all optimal haplotype solutions with minimal recombinants for any pedigree structures with missing data. It can solve the zero recombinant problem as a special case. But because it does not use the zero recombinant assumption explicitly, its efficiency is expected to be inferior to the current algorithm. Under the zero recombinant assumption, all three methods are **exact** algorithms that output all compatible solutions. Our experiments show that their implementations indeed generate the same set of haplotype assignments on same inputs. This again shows that the ZRHC formulation is valid for tightly linked markers, and the set of solutions is the same as the set of solution obtained based on likelihood approaches. Therefore, we only present results on the efficiency comparison.

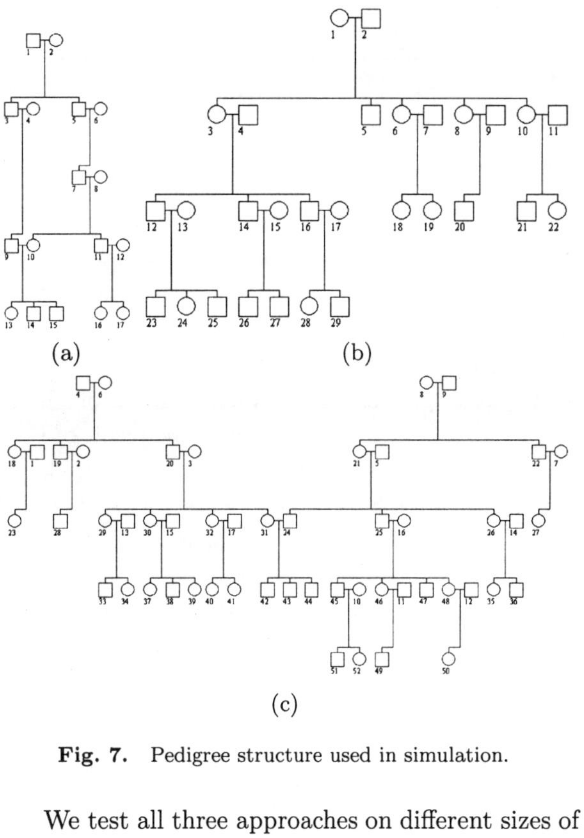

Fig. 7. Pedigree structure used in simulation.

We test all three approaches on different sizes of pedigrees (17, 29, 52, 128), all are real human pedigree structures obtained from literatures. Different number of loci (20, 50, 100, 200), different missing rates (0.05, 0.10, 0.15, 0.20) and different missing patterns are considered. We run Merlin and DSS on a Linux machine with two 3.0GHz Quad-Core Xeon

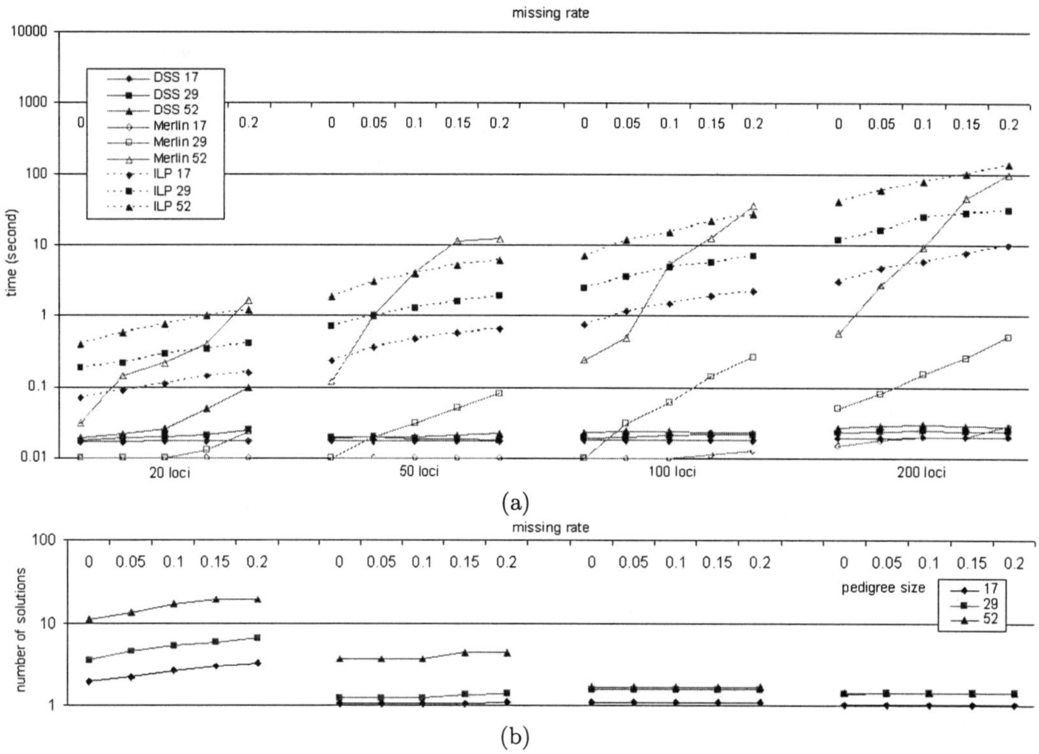

Fig. 8. (a) Comparison of running time (in seconds). (b) Average number of solutions.

5365 processors and 16G memory. PedPhase.ILP only has a Windows version, and it was tested on a much slower Windows machine with a much small memory (Pentium 4 3.2GHz with 2G memory). We measure the time needed for each of the algorithms to output all possible haplotyping solutions of a pedigree. Due to hardware limitations, the result of PedPhase.ILP on pedigree size 128 is not acquired. To generate genotype data that closely resemble real data, we use the Simulated Rheumatoid Arthritis (RA) Data from Genetic Analysis Workshop (GAW) 15. Chromosome 6 of GAW data mimics a 300K SNP chip with an average inter-marker spacing of 9,586 bp. The beginning 20, 50, 100 and 200 loci are truncated to test the three algorithms. Population haplotype frequencies are calculated based on the true haplotype assignments in the simulated data, and are then fed to *SimPed*[7], together with each pedigree structure. *SimPed* will then sample founder haplotypes based on their population frequencies and generate genotype data for each member in a pedigree assuming no recombinations. The three pedigree structures are shown in Figure 7, among which the pedigree with size 17 (Figure 7(a)) is a looped one. The pedigree with size 128 is too large to fit in one page and will be provided on our website.

We designate two ways to generate samples with missing data so as to examine the behavior of the methods with respect to both missing rate and missing pattern variations. We generate the first set of samples by randomly assigning a locus to be missing at a specified missing rate. Second, we make all top generation of a pedigree completely missing for all loci, which is common in real data. For each testing category, we simulate 100 independent data sets and report the average running time. For the random missing case, Figure 8(a) shows the running time of the three programs under different settings, except for the pedigree size 128, for which the running time of Merlin is too large to be juxtaposed with DSS. The result on the pedigree with size 128 is listed in Table 2. The running time of Merlin increases exponentially with the pedigree size, the number of loci and also the missing rate. The running time of PedPhase.ILP (on a slower machine) also has an exponential growth with the increase of the missing rate

Table 2. Comparison of running time (in seconds) between DSS and Merlin on pedigree size 128. The running time of Merlin under some data settings exceeds an hour, and are thus omitted from our measurement.

number of loci	20					50				
missing rate	0.00	0.05	0.10	0.15	0.20	0.00	0.05	0.10	0.15	0.20
DSS	0.0267	0.1539	0.3517	0.4991	0.6540	0.0259	0.0361	0.0368	0.0378	0.0360
Merlin	70	300	600	800	1100	360	800	1000	>1300	—
number of loci	100					200				
missing rate	0.00	0.05	0.10	0.15	0.20	0.00	0.05	0.10	0.15	0.20
DSS	0.0311	0.0426	0.0373	0.0431	0.0461	0.0433	0.0587	0.0518	0.0575	0.0503
Merlin	800	1200	>2400	—	—	—	—	—	—	—

and the number of loci but with a smaller constant compared to Merlin. It also shows a much smaller growth rate with the pedigree size. In contrast, DSS scales smoothly with all parameters (except for the missing rate when the number of loci is 20), and the improvement over Merlin or PedPhase.ILP is from 10 to 10^5 folds for large pedigrees with large number of loci or high rate of missing. In fact, neither Merlin nor PedPhase.ILP can successfully infer haplotypes from the pedigree with size 128 when the number of marker is 200. However, DSS can obtain all solutions within 0.05 second, even for data with 20% missing. This shows that by solving the linear system based on partial constraints from existing data, we significantly reduce the enumeration space of inheritance variables. The experimental results show that when the number of loci is large, the program can still maintain the same linear complexity even for data with 20% missing. But for small number of loci, the running time of DSS increases as missing rate increases (though DSS can finish all the cases within 0.1 second). This is because the number of constraints on h variables is roughly in proportion to the number of loci. So for small number of loci, the remaining degrees of freedom on inheritance variables after solving the linear system could still be high. This number could be partly reflected by the number of all compatible solutions in the end. Figure 8(b) compares the number of h variable solutions in different circumstances. It grows with both the pedigree size and the missing rate, but decreases with the number of loci.

Next, we investigate the performance of all three algorithms on special missing patterns. Figure 9 gives some representative result on the pedigree with size 52, for which all individuals at the top generation (members 4, 6, 8, 9) are missing. For this pedigree, such missing equals a missing rate of ~7.7%. In terms of absolute time, DSS ($0.2 \sim 0.8$ sec) is much better than the other two algorithms ($0.2 \sim 100$ sec). However, the running time is higher than its own running time with a missing rate 10%. The running time of Merlin and PedPhase.ILP on this special data set is between those of missing rate 5% and 10%. DSS is somewhat sensitive to this special missing pattern because when all genotypes of an individual are missing, none of the inheritance variables between her and her parents or children could be determined. A further investigation on this special missing pattern is warranted.

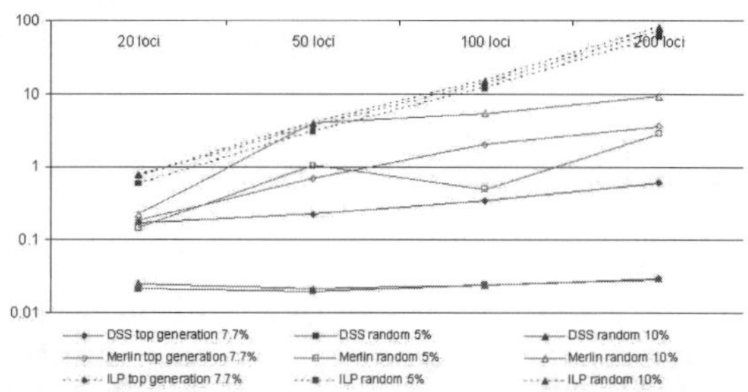

Fig. 9. Comparison of DSS and Merlin on different patterns of missing data.

6. DISCUSSION

We propose an algorithm for haplotype inference from pedigree data without recombinant using disjoint-set structures. The proposed algorithm can output a general solution for a tree pedigree with complete data in time $O(mn\alpha(n))$, which is a further improvement upon existing results. For a general pedigree, or a pedigree with missing data, by using the same framework, our method can significantly reduce degrees of freedom on inheritance variables and thus narrow down the search scope. Experimental results show that the algorithm is efficient in practice for both complete data and missing data, and outperforms two popular algorithms on large data sets. For data with large number of markers, the performance of the algorithm hardly deteriorates as the missing rate increases.

Though several theoretical results of ZRHC were recently reported[3, 12, 15], none of them have been implemented. The empirical examination of the performance of our algorithm offers some evidence for the theoretical bounds on the complexity of such haplotyping approaches based on linear systems.

The performance of our algorithm on pedigrees with missing data depends on the number of constraints the linear system can capture. We observe that the efficiency of this linear system is influenced by variation in missing patterns. So as a possible piece of future work, we can consider a special strategy to handle individuals with all loci missing. Other possible directions are to combine the proposed algorithm with statistical approach to assign a probability likelihood for each of the assignments, and to design algorithms for whole chromosome by calling the current algorithm as a subroutine. Theoretically, it also remains open whether the linear time complexity can be observed for a general pedigree with complete data.

Acknowledgement

This research is supported by National Institutes of Health/National Library of Medicine grant LM008991, and in part by National Institutes of Health/National Center for Research Resources grant RR03655. Support for generation of the GAW15 simulated data was provided from NIH grants 5RO1-HL049609-14, 1R01-AG021917-01A1, the University of Minnesota, and the Minnesota Supercomputing Institute. We would also like to acknowledge GAW grant R01 GM031575.

References

1. Abecasis GR, Wigginton JE. Handling marker-marker linkage disequilibrium pedigree analysis with clustered markers. *Am. J. Hum. Genet* 2005; **77**: 754–767.
2. Bonizzoni P, Vedova GD, Dondi R, Li J. The haplotyping problem: an overview of computational models and solutions. *J Comp Sci Tech* 2003; **18(6)**: 675–88.
3. Chan MY, Chan W, Chin F, Fung S, Kao M. Linear-Time Haplotype Inference on Pedigrees without Recombinations. *Proc. of the 6th Annual Workshop on Algorithms in Bioinformatics (WABI'06)* 2006: 56–67.
4. Cormen TH, Leiserson CE, Rivest RL, Stein C. *Introduction to Algorithms.* 2nd edition, McGraw-Hill Book Company, Boston, MA. 2003: 498–517.
5. Gusfield D. An overview of combinatorial methods for haplotype inference. *Lecture Notes in Computer Science (2983): Computational Methods for SNPs and Haplotype Inference.* 2004: 9–25.
6. Halldórsson BV, Bafna V, Edwards N, Lippert R, Yooseph S, Istrail S. A survey of computational methods for determining haplotypes. *Lecture Notes in Computer Science (2983): Computational Methods for SNPs and Haplotype Inference.* 2004: 26–47.
7. Leal SM, Yan K, Müller-Myhsok B. SimPed: a simulation program to generate haplotype and genotype data for pedigree structures. *Hum Hered* 2005; **60**: 119–122.
8. Li J, Jiang T. Efficient Inference of Haplotypes from Genotype on a Pedigree. *Journal of Bioinformatics and Computational Biology(JBCB)* 2003; **1(1)**: 41–69.
9. Li J, Jiang T. Computing the Minimum Recombinant Haplotype Configuration from Incomplete Genotype Data on a Pedigree by Integer Linear Programming. *Journal of Computational Biology* 2005; **12**: 719–739.
10. Li J, Jiang T. A survey on haplotyping algorithms for tightly linked markers. *Journal of Bioinformatics and Computational Biology* 2008; **6(1)**: 241–259.
11. Liu L, Chen X, Xiao J, Jiang T. Complexity and approximation of the minimum recombination haplotype configuration problem. *In Proc. 16th International Symposium on Algorithms and Computation (ISAAC'05)* 2005: 370–379.
12. Liu L, Jiang T. Linear-Time Reconstruction of Zero-Recombinant Mendelian Inheritance on Pedigrees without Mating Loops. *Proc. of Genome Informatics Workshop (GIW'2007)* 2007: 95–106.
13. Tarjan RE, Leeuwen J. Worst-case analysis of set union algorithms. *Journal of the ACM* 1984; **31(2)**: 245–281.
14. The international HapMap Consortium. A second generation human haplotype map of over 3.1 million SNPs. *Nature* 2007; **449**: 851–61.
15. Xiao J, Liu L, Xia L, Jiang T. Fast Elimination of Redundant Linear Equations and Reconstruction of Recombination-Free mendelian Inheritance on a Pedigree. *Proc. of 18th Annual ACM-SIAM Symoposium on Discrete Algorithms (SODA'07)* 2007: 655–664.
16. Zhang K, Zhao H. A comparison of several methods for haplotype frequency estimation and haplotype reconstruction for tightly linked markers from general pedigrees. *Genetic Epidemiology* 2006; **30(5)**: 423–437.
17. Zhang XS, Wang RS, Wu LY, Chen L. Models and Algorithms for Haplotyping Problem. *Current Bioinformatics* 2006; **1(1)**: 105–114.

Computational Methods

KNOWLEDGE REPRESENTATION AND DATA MINING FOR BIOLOGICAL IMAGING

Wamiq M. Ahmed
Purdue University Cytometry Laboratories, Bindley Bioscience Center
1203 W. State Street, West Lafayette, IN 47907
wahmed@flowcyt.cyto.purdue.edu

Biological and pharmaceutical research relies heavily on microscopically imaging cell populations for understanding their structure and function. Much work has been done on automated analysis of biological images, but image analysis tools are generally focused only on extracting quantitative information for validating a particular hypothesis. Images contain much more information than is normally required for testing individual hypotheses. The lack of symbolic knowledge representation schemes for representing semantic image information and the absence of knowledge mining tools are the biggest obstacles in utilizing the full information content of these images. In this paper we first present a graph-based scheme for integrated representation of semantic biological knowledge contained in cellular images acquired in spatial, spectral, and temporal dimensions. We then present a spatio-temporal knowledge mining framework for extracting non-trivial and previously unknown association rules from image data sets. This mechanism can change the role of biological imaging from a tool used to validate hypotheses to one used for automatically generating new hypotheses. Results for an apoptosis screen are also presented.

1. INTRODUCTION

Microscopic imaging is extensively used to image cell samples in two or three spatial dimensions, a spectral dimension, and a temporal dimension.[1] This leads to five-dimensional image sets, and any combination of these dimensions can be acquired depending on the requirements of the application.[2] The general approach is that biologists first develop a hypothesis and then image biological samples to validate their hypotheses. However, images contain much more information than is needed for analyzing a particular hypothesis. For example, a drug-screening study may use an apoptosis assay and image analysis tools to find out which drug is best for inducing apoptosis in cancer cells. This type of analysis, while very useful for a particular application (apoptosis in cancer cells), is not able to extract all of the information from the imaging data. For example, let us assume there is a link between a cell undergoing apoptosis and induction of apoptosis in its neighboring cells after a certain time because of some underlying biological phenomenon. This information, while present in the images collected during the above mentioned drug screening study, will not be extracted. We believe a data mining approach for analyzing biological data can be extremely useful for harnessing the full potential of the information content of biological images. Such an approach can be used to generate new hypotheses and greatly facilitate biological research. Realization of this goal, however, requires the development of schemes for capturing the semantic content of biological images and development of data mining formalisms for extracting association rules. In order to achieve this goal, we present a graph-based knowledge representation scheme that captures the semantic knowledge contained in multi-dimensional biological images. Then we present a framework for extracting non-trivial, previously unknown association rules in such data. This approach can be used for analyzing large repositories of cellular images and can significantly help in biological discovery.

Association-rule mining for knowledge discovery in databases was proposed by Agrawal et al. and has since been extensively used for finding association rules.[3] Application of these tools to imaging data is hampered by the fact that image data require the extraction and representation of semantic information before data mining algorithms can be applied. Classification and clustering techniques have been previously applied to image data in different domains such as medical imaging and weather monitoring[4], but there has not been any work on association-rule mining on cellular images. The challenge lies in developing powerful knowledge representation schemes to capture the semantic information contained in multi-dimensional images and developing formalisms for mining association rules using these schemes. In this paper we propose a graph-based knowledge representation scheme and a data mining formalism for capturing the semantic image information and for extracting association rules. This approach has the potential to exploit the maximum information content of imaging data for automated biological discovery and can potentially change the role of biological imaging from merely a tool for hypothesis validation to a more powerful tool for generating new hypotheses as well.

2. GRAPH-BASED REPRESENTATION OF SEMANTIC CONTENT

Attribute relational graphs (ARGs) have been used for representing image content for content-based retrieval.[5] The nodes of the ARG represent the objects and the edges represent the relations. In order to develop an integrated representation for multi-dimensional biological images that include two or three spatial dimensions, a spectral, and a temporal dimension we extend the concept of an ARG to a colored attribute relational graph (CARG). A CARG is a special ARG where each node of the ARG contains a color attribute that specifies the spectral band (fluorescence channel) in which this image was acquired. Formally a CARG is a four tuple $G = (V, E, A_v, A_e)$ where V is a set of vertices, E is a set of edges between vertices, A_v is a set of attributes of vertices, and A_e is a set of attributes of edges. The vertices represent the objects in the images and vertex attributes contain attributes of objects such as area, perimeter, texture, and shape descriptors. Edge attributes represent the spatial relations between objects. In our experiments we use four spatial relations that include 'overlap' (o), 'contain' (c), 'near neighbor' (n), and 'far neighbor' (f), and four object attributes that include area, major axis length, minor axis length, and perimeter. CARG captures the image information in three spatial dimensions and the spectral dimension. Information in the temporal dimension is captured using a temporal sequence of CARGs each representing the spatial and spectral information at a time instant. An example of a series of CARGs showing 3 cells imaged in three different fluorescence channels for an apoptosis screen is shown in Figure 1 (a-c). Here white, light gray, and dark gray nodes represent Hoechst, Annexin V fluorescein isothiocyanate (FITC), and propedium iodide (PI) respectively. At time instant 1, Cell 1 is in an early apoptotic state (overlap of white and light gray nodes) whereas it is in a late apoptotic state at time instants 2 and 3. Similarly, Cell 3 is in the live state (white node disjoint from other nodes) at time instant 1 and 2 and in an early apoptotic state at time instant 3.

Most biological events involve spatio-temporal changes in the attributes of biological objects (cells, intracellular compartments) or changes in the spatial relations between different objects.[6,7] The ARG model can be used for representing the information about spatio-temporal events and the spatial relations among them. Each node of the ARG represents an event. Attributes of the nodes include the type of event, participating objects along with their attributes, start time, duration, and the decomposition into simpler events for composite events. For example an apoptosis event may be considered to be made up of sub-events such as 'normal' when cell is alive and 'apoptotic' when the cell undergoes apoptosis. Figure 1(d) shows the representation of four apoptosis events as an ARG.

3. DATA MINING FRAMEWORK

The graph-based knowledge representation scheme proposed in Section 2 provides a data structure for storing image information in terms of the objects and the events happening in the images. Data mining algorithms can then be applied on this symbolic representation. This can help discover interesting patterns in imaging data. Such patterns could be in the form of association rules between different features of biological objects (between roundness and size) or between features of biological objects and different semantic classes of objects (between roundness and mitotic state). These association rules may also have a temporal dependence (between roundness and size after a time interval, or between roundness and cell division after a time interval). In order to capture these patterns we mine six different types of rules as shown in Figure 2. Association rules are generally represented as $(X \rightarrow Y)$ where X is the antecedent and Y is the consequent. The support and confidence values for mined rules are defined as follows.

Support = Number of transactions where both X and Y appear / Number of transactions in the database.

Confidence = Number of transactions where both X and Y appear / Number of transactions where only X appears.

4. EXPERIMENTAL RESULTS AND DISCUSSION

In this section we report the results of applying the association-rule mining algorithms to the images generated by an apoptosis screen. A fluorescent marker (Hoechst) was used for labeling the nuclei whereas Annexin-V-FITC was used to label cells as apoptotic or non-apoptotic. Nuclear features, including area, major axis length, minor axis length, and perimeter, were extracted. Nuclei neighbors were determined using the distance between the centroids of different nuclei. The extracted features were discretized by dividing the range of each feature into 4 ranges as shown in Table 1.

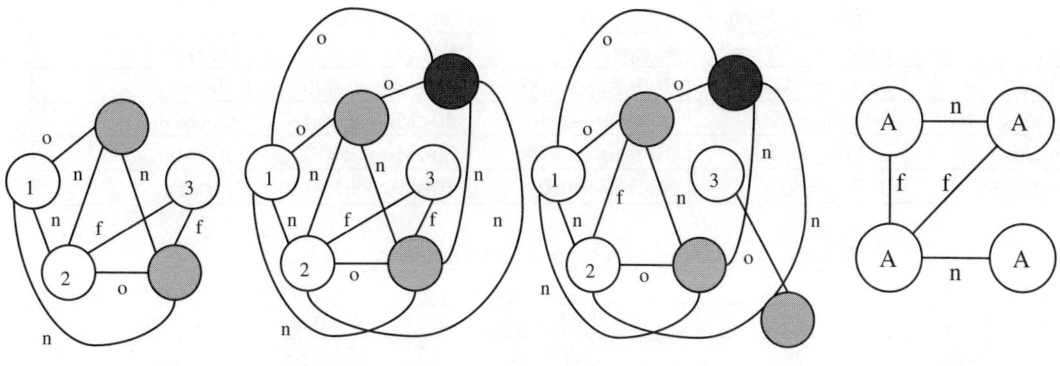

(a) t = 1 (b) t = 2 (c) t = 3 (d)

Fig. 1. (a-c) A sequence of CARGs as an integrated representation of spatial, spectral, and temporal information for three cells. White nodes represent nuclear stain which is used to identify the cells. Light gray nodes represent Annexin-V-FITC and dark gray nodes represent PI (d) Representation of 4 apoptosis events.

Same object spatial rules

1. $Attr_X(A) \rightarrow Attr_Y(A),\ X \neq Y$
2. $Attr_X(A) \rightarrow SemClass_Y(A),\ X \neq Y$
3. $SemClass_X(A) \rightarrow SemClass_Y(A),\ X \neq Y$
4. $SemClass_X(A) \rightarrow Attr_Y(A),\ X \neq Y$

Same object temporal rules

1. $Attr_X(A) \xrightarrow{TempRel} Attr_Y(A),\ X \neq Y$
2. $Attr_X(A) \xrightarrow{TempRel} SemClass_Y(A),\ X \neq Y$
3. $SemClass_X(A) \xrightarrow{TempRel} SemClass_Y(A),\ X \neq Y$
4. $SemClass_X(A) \xrightarrow{TempRel} Attr_Y(A),\ X \neq Y$

Object neighborhood spatial rules

1. $Attr_X(A) \rightarrow Attr_Y(B),\ X \neq Y,\ B \in Neighborhood(A)$
2. $Attr_X(A) \rightarrow SemClass_Y(B),\ X \neq Y,\ B \in Neighborhood(A)$
3. $SemClass_X(A) \rightarrow SemClass_Y(B),\ X \neq Y,\ B \in Neighborhood(A)$
4. $SemClass_X(A) \rightarrow Attr_Y(A),\ X \neq Y,\ B \in Neighborhood(A)$

Object neighborhood temporal rules

1. $Attr_X(A) \xrightarrow{TempRel} Attr_Y(B),\ X \neq Y,\ B \in Neighborhood(A)$
2. $Attr_X(A) \xrightarrow{TempRel} SemClass_Y(B),\ X \neq Y,\ B \in Neighborhood(A)$
3. $SemClass_X(A) \xrightarrow{TempRel} SemClass_Y(B),\ X \neq Y,\ B \in Neighborhood(A)$
4. $SemClass_X(A) \xrightarrow{TempRel} Attr_Y(A),\ X \neq Y,\ B \in Neighborhood(A)$

Spatial event rules

1. $Event_X(A) \rightarrow Attr_Y(A)$
2. $Event_X(A) \rightarrow Event_Y(A)$
3. $Event_X(A) \rightarrow Event_Y(B),\ A,B \in ObjList(X)$
4. $Event_X(A) \rightarrow Event_Y(B),\ B \in Neighborhood(A)$

Temporal event rules

1. $Event_X(A) \xrightarrow{TempRel} Attr_Y(A), TempRel \in \{D,M,O,C,S,E,CO\}$
2. $Event_X(A) \xrightarrow{TempRel} Event_Y(A), TempRel \in \{D,M,O,C,S,E,CO\}$
3. $Event_X(A) \xrightarrow{TempRel} Event_Y(B),\ A,B \in ObjList(X), TempRel \in \{D,M,O,C,S,E,CO\}$
4. $Event_X(A) \xrightarrow{TempRel} Event_Y(B),\ B \in Neighborhood(A), TempRel \in \{D,M,O,C,S,E,CO\}$

Fig. 2. Different types of spatial and temporal rules. $Attr_X(A)$ means attribute X of object A, $SemClass_X(A)$ means semantic class X involving object A, and $Event_X(A)$ means event X involving object A. *TempRel* refers to the temporal relations between different events.[6]

We also use two other features, 'state' and 'nbr,' where state can be either 'live' or 'apoptotic' and 'nbr' can be either 'none,' implying none of the cell's neighbors is in apoptotic state or 'oneplus,' implying one or more of a cell's neighbors are apoptotic. Association-rule mining formalism was then applied to the semantic information extracted from a set of 200 images (100 fields of view x 2 fluorescent channels). Using a minimum support of 0.2 and confidence of 0.6, a total of 90 rules were found. A subset of these rules is shown in Table 2. Some of these rules are obvious such as the dependence of area on the major and minor axis lengths. However the relationship between the apoptotic state of a cell and its features, or that between the apoptotic state of a cell and the state of its neighbors, can be a useful one.

Table 1. Feature ranges used for discretization of features.

Feature	Range1	Range2	Range3	Range4
Area (A)	$0<A_{R1}<=200$	$200<A_{R2}<=400$	$400<A_{R3}<=600$	$600<A_{R4}$
Major axis (Mj)	$0<Maj_{R1}<=20$	$20<Maj_{R2}<=40$	$40<Maj_{R3}<=60$	$60<Maj_{R4}$
Minor axis (Mi)	$0<Min_{R1}<=10$	$10<Min_{R2}<=20$	$20<Min_{R3}<=30$	$30<Min_{R4}$
Perimeter (P)	$0<P_{R1}<=50$	$50<P_{R2}<=100$	$100<P_{R3}<=150$	$150<P_{R4}$

Table 2. Mined rules with support and confidence values.

No.	Antecedent	Consequent	Support	Confidence
1	$Mj = Maj_{R1}$	$Mi = Min_{R2}$	0.447	0.981
2	$Mi = Min_{R2}$	$A = A_{R2}$	0.585	0.778
3	State = live	Nbr = none	0.244	0.854
4	State = apoptotic, $Mj = Maj_{R1}$	$Mi = Min_{R2}$	0.274	0.992
5	State = apoptotic, $A = A_{R2}$	$P = P_{R2}$, $Mi = M_{R2}$	0.432	0.836
6	Nbr = oneplus	State = apoptotic	0.221	0.841
7	$P = P_{R2}$, $Mi = Min_{R2}$	$A = A_{R2}$	0.582	0.819
8	$Mi = Min_{R2}$, Nbr = none, State = apoptotic	$A = A_{R2}$	0.287	0.831
9	$P = P_{R2}$, $A = A_{R2}$, $Mj = Maj_{R2}$	$Mi = Min_{R2}$	0.301	1
10	$A = A_{R2}$, $Mj = Maj_{R1}$	$P = P_{R2}$, $Mi = Min_{R2}$	0.301	1

5. CONCLUSION

Mining association rules from cellular images can be a powerful tool for discovering new biological knowledge in an automated manner. In this paper we have presented a graph-based model for representation of the semantic content of cellular images and a formalism for mining association rules at the level of object features and at the level of biological events. Our experiments did not involve temporal data mining, although our formalism provides for mining temporal association rules. In the future we plan to generate significantly large data sets for applying our data mining formalism on spatial as well as temporal data.

References

1. Ahmed WM, Leavesley SJ, Rajwa B, Ayyaz MN, Ghafoor A, and Robinson, JP. State of the art in information extraction and quantitative analysis for multi-modality biomolecular imaging *Proceedings of the IEEE* 2008; **96, 3**: 512-531.
2. Swedlow JR, Goldberg I, Brauner E, Sorger P K. Informatics and Quantitative Analysis in Biological Imaging. *Science* 2003; **300, 5616**: 100-102.
3. Agrawal R, Imielinski T, Swami A. Mining association rules between sets of items in large databases *Proc.of the ACM SIGMOD Conference on Management of Data* 1993; 207-216.
4. Antonie M, Zaiane OR, Coman A. Application of data mining techniques for medical image classification *Second International ACM SIGKDD Workshop on Multimedia Data Mining* 2001; 94-101.
5. Petrakis EGM, Faloutsos C, Lin K. ImageMap: An image indexing method based on spatial similarity *IEEE Transactions on Knowledge and Data Engineering* 2002; **14, 5**; 979-987.
6. Ahmed WM, Ghafoor A, Robinson JP. Knowledge extraction for biological imaging *IEEE Multimedia* 2007; **14, 4**: 52-62.
7. Ahmed WM, Lenz D, Liu J, Robinson JP. XML-based data model and architecutre for a knowledge-based grid-enabled problem-solving environment for high-throughput biological imaging *IEEE Transactions on Information Technology in Biomedicine* 2008; **12, 2**: 226-240.

FAST MULTISEGMENT ALIGNMENTS FOR TEMPORAL EXPRESSION PROFILES

Adam. A. Smith* and Mark Craven

Departments of Computer Sciences and Biostatistics & Medical Informatics, University of Wisconsin, Madison, Wisconsin 53706, USA

Email: aasmith@cs.wisc.edu, craven@biostat.wisc.edu

We present two heuristics for speeding up a time series alignment algorithm that is related to dynamic time warping (DTW). In previous work, we developed our *multisegment* alignment algorithm to answer similarity queries for toxicogenomic time-series data. Our multisegment algorithm returns more accurate alignments than DTW at the cost of time complexity; the multisegment algorithm is $O(n^5)$ whereas DTW is $O(n^2)$. The first heuristic we present speeds up our algorithm by a constant factor by restricting alignments to a cone shape in alignment space. The second heuristic restricts the alignments considered to those near one returned by a DTW-like method. This heuristic adjusts the time complexity to $O(n^3)$. Importantly, neither heuristic results in a loss in accuracy.

1. INTRODUCTION

Characterizing and comparing temporal gene-expression responses is an important computational task for answering a variety of questions in biological studies. We have previously presented an approach for answering similarity queries about gene-expression time series that is motivated by the task of characterizing the potential toxicity of various chemicals[1]. This approach is designed to handle the plethora of problems that arise in comparing gene expression time series, including sparsity, high-dimensionality, noise in the measurements, and the local distortions that can occur in similar time series. Our experimental evaluation showed that our approach produces more accurate alignments and classifications of gene-expression time series than a handful of alternative approaches, and is robust to relative distortions in time between similar chemical treatments. However, this accuracy comes at the cost of efficiency: the algorithm's time complexity is $O(n^5)$. In this paper, we present two heuristic methods for speeding up our alignment algorithm. We show that these heuristics result in significant speedups without sacrificing the accuracy of the resulting alignments.

The task that we consider is motivated by the need for faster, more cost-efficient protocols for characterizing the potential toxicity of industrial chemicals. The effects of toxic chemicals may often be predicted by how they influence global gene expression over time[2]. By using microarrays, it is possible to measure the expression of thousands of genes simultaneously. It is likely that transcriptional profiles will soon become a standard component of toxicology assessment and government regulation of drugs and other chemicals.

The source we use for toxicology-related gene expression data is the EDGE (Environment, Drugs and Gene Expression) database[2]. EDGE contains expression profiles from mouse liver tissue following exposure to a variety of chemicals and physiological changes, which we refer to as *treatments*. Some of the treatments in EDGE have been assayed as time series. Figure 1-A provides a simplified illustration of the type of data with which we are concerned. The small database in this figure contains time series data for four different treatments, each of which includes measurements for three genes. The true, underlying expression response is not known, but instead the database contains sampled observations which may be noisy. We use the term *observation* to refer to the expression measurements made at a single time point in a treatment.

Figure 1-B then shows the computational task. Given an expression profile as a query, we want to identify the treatment in the database that has the expression profile most similar to the query. In the general case, the query and/or some of the database treatments are time series. In this case, we want to also determine the temporal correspondence between the query and putatively similar treatments in the database. In the toxicology domain, we are

*Corresponding author.

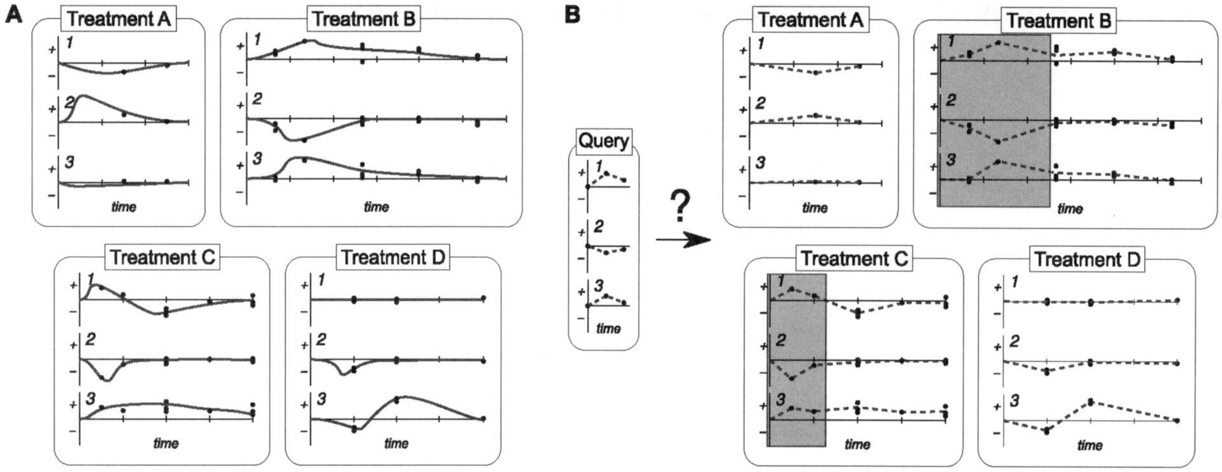

Fig. 1. An example of the similarity-query task for four different treatments with three genes. In Panel A the curves show the actual hidden expression profile for each treatment, even though we must rely on the noisy sampled observations (the points). In Panel B we have reconstructed the profiles at unobserved times, and used them to perform a similarity query. The highlighted areas represent possible good matches.

interested in answering this type of query in order to characterize poorly understood chemicals.

We have developed an approach that is designed to handle several key challenges that this task presents.

- The time series available from toxicological studies are typically sparse, containing measurements from fewer than ten time points.
- Because the time series have been sampled at non-uniform time intervals which vary between treatments, the time points present in a given query may not correspond to measured points in database series.
- Queries may vary in their number of observations or their extent. Some queries may consist of only a single observation, whereas others may contain multiple time points. Some may span only a few hours whereas others include measurements taken over days.
- A given query and its best match in the database may differ in the amplitude, temporal offset, or temporal extent of their responses. For example, the expression profile represented by a query treatment may be similar to a database treatment except that the gene expression responses are attenuated, or occur later, or take place more slowly. Alternatively, the query may be similar to a truncated version of the database series, or vice versa.

Our approach to this task involves first using an interpolation algorithm to reconstruct unobserved expression values from sparse time series, and then using an alignment algorithm to find the highest scoring alignment of the query series to each treatment series in the database. The approach returns the treatment from the database that is most similar to the query, and the calculated alignment between the two series.

Several different methods have been applied to the task of aligning gene-expression time series. Aach and Church[3] were the first to apply the method of *dynamic time warping*[4, 5] to gene expression profiles, and other groups have followed[6, 7]. Dynamic time warping, originally developed for speech recognition problems, is an approach for aligning pairs of time series that uses dynamic programming to find an optimal alignment with respect to a given scoring function. Although DTW has a time complexity of $O(n^2)$, Ratanamahatana and Keogh[8] have shown that using bounding heuristics can effectively make DTW run in $O(n)$. However their method is designed for *global* alignments, which align all of one series to the entirety of the other. Our method, in contrast, does not force the two series to be globally aligned. Instead, it permits a type of *local* alignment in which the end of one series is unaligned. We refer to this case as *shorting* the alignment. This aspect of the approach is motivated by the consideration that one of the series may show more of the temporal response than the other (e.g., one series may not have been measured for as long, or may have responded more quickly).

Bar-Joseph et al.[9] used a warping method that finds a linear mapping between the two time series being aligned. Although it allows local alignments like our method, the linear model does not adequately represent complex alignments. Our method considers alignments that represent a middle ground, in terms of expressiveness, between dynamic time warping and linear warping approaches. Our method is based on a "multisegment" model that warps different regions of the series by different amounts.

Another time-series alignment approach that is somewhat similar to our multisegment method is *correlation optimized warping* (COW)[10]. This method compares time series by dividing them into several roughly equal segments and summing the Pearson's correlations of corresponding segments. Unlike our approach, the COW method assumes that the series are to be globally aligned, without any shorting. Further, the use of correlation can be limiting as COW is unable to distinguish between two series that are proportional to one another.

In previous work, we evaluated our multisegment alignment method in the context of aligning and classifying gene-expression profiles from the EDGE database. This empirical evaluation showed that our multisegment method returned more accurate alignments and classifications than dynamic time warping, two linear alignment methods, and the COW algorithm. The disadvantage of the multisegment method is that its computational complexity is $O(n^5)$ where n is the number of time points in the series being aligned (assuming the two series have the same length). Although the number of *observed* time points is typically small in the series we consider, n is considerably larger because each series is represented by interpolated "pseudo-observations" in addition to the observed time points.

2. TIME SERIES ALIGNMENT METHODS

In this section we discuss two previously developed methods for aligning two series. Figure 2 illustrates the type of alignment problem we consider. The figure shows the types of alignments calculated by dynamic time warping and by our multisegment method[1]. (For simplicity, the figure shows each treatment as consisting of only a single gene.) These alignment paths exist in *alignment space*, where each dimension represents the time of one of the aligned series. A point (x, y) on a path corresponds to a mapping between Series 1 at time x and Series 2 at y. The diagonal represents a special path, in which no warping of time takes place. Here, both alignments short, so that the whole of Series 2 is aligned with only a portion of Series 1.

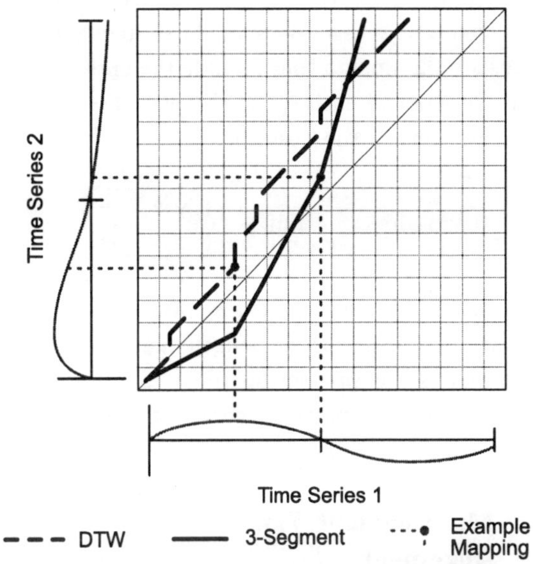

Fig. 2. Aligning two time series in *alignment space* with dynamic time warping and our three-segment model. We refer to this graph as an *alignment matrix*. A point (x, y) corresponds to a mapping between Series 1 at time x and Series 2 at y. The paths thus show the overall alignments chosen by both methods. Notice that the alignments *short* before Series 1 has ended, as there is no evidence that Series 2's expression has begun to increase again at the end.

2.1. Dynamic Time Warping

Dynamic time warping[4, 5] is often used for time series alignment problems. Briefly, this method computes an *alignment matrix* Γ from two series as shown in Figure 2. In our context, the series are a query series q and a database series d. Each element $\gamma(x, y)$ holds the best score aligning q, up to time x, against d, up to time y. The matrix elements are calculated recursively as:

$$\gamma(x, y) = D_E(q_x, d_y) + \min \begin{cases} \gamma(x-1, y) \\ \gamma(x, y-1) \\ \gamma(x-1, y-1) \end{cases} \quad (1)$$

where $D_E(q_x, d_y)$ is the Euclidean distance between points q_x and d_y in the two series. The base element $\gamma(0, 0)$ is just the Euclidean distance at time 0.

Traditional DTW then returns $\gamma(q.r, d.r)$, where $q.r$ and $d.r$ are the rightmost (last) times of the two series, along with the path that resulted in this score. However we are interested in possibly shorting the alignment, finding a local alignment rather than a global one. In this case, allowed alignments are those that explain the entire extent of at least one of the two given time series. We scan the elements of Γ that represent alignments that include the entirety of the query series, the entirety of the database series, or both, and return the best one:

$$\text{bestscore} = \min_{a \leq q.r,\ b \leq d.r} \begin{cases} \frac{\gamma(a, d.r)}{\sqrt{|a|^2 + |d.r|^2}} \\ \frac{\gamma(q.r, b)}{\sqrt{|q.r|^2 + |b|^2}} \end{cases} . \quad (2)$$

The variables a and b represent positions in the two time series. Given series of length m and n, the alignment matrix has mn entries to be calculated. Each of these calculations takes constant time. Thus, if $m \approx n$, the time complexity is $O(n^2)$.

2.2. Multisegment Time Series Alignment

The three-segment path in Figure 2 shows the type of alignment that our multisegment model calculates. In each segment the amplitude and stretching relationships between the two series are somewhat different. We use the term *stretching* to refer to distortions in the rate of some response, and the term *amplitude* to refer to distortions in the magnitude of the response. The total number of segments is specified in advance.

To determine the similarity between a query time series q and a particular database series d, we can calculate how likely it is that q is a somewhat distorted exemplar of the same process that resulted in d. In particular, we can think of a generative process that uses d to generate similar expression profiles. We can then ask how probable q looks under this generative process.

Given this generative process idea, we calculate the probability of a particular alignment of query q given a database series d as follows:

$$P(q|d, s, a) = P_m(m) \prod_{i=1}^{m} P_s(s_i) P_a(a_i) P_e(q_i|d_i, s_i, a_i), \quad (3)$$

where m is the number of segments in the alignment, q_i and d_i refer to the expression measurements for the ith query and database segments respectively, and s_i is the stretching value and a_i is the amplitude value for the ith segment. The location of each segment pair is assumed to be given here. P_m represents a probability distribution over the number of segments in an alignment, up to some maximum number M of allowed segments. P_s represents a probability distribution over possible stretching values for a pair of segments, P_a represents a probability distribution over possible amplitude values, and P_e represents a probability distribution over expression observations in the query series, given the database series and the stretching and amplitude parameters.

We omit the details of this generative model here, but we note that it results in a scoring function that can be used to assess the likelihood of any given alignment. Moreover, we can find optimal scoring alignments under this model using dynamic programming. The core of the dynamic program involves filling in a three-dimensional matrix Γ in which each element $\gamma(i, x, y)$ represents the best score found with i segments that align the query subseries from time 0 to x with the database subseries from time 0 to y. Here, x and y represent time points in the two series.

We define $\gamma(i, x, y)$ with the following recurrence relation:

$$\gamma(i, x, y) = \max_{a<x,\ b<y} \begin{cases} \log P_m(i) + \gamma(i-1, a, b) \\ \quad + \mathsf{score}(a, x, b, y) \\ \quad \textbf{if } x = q.r \textbf{ or } y = d.r \\ \gamma(i-1, a, b) \\ \quad + \mathsf{score}(a, x, b, y) \\ \quad \textbf{otherwise} \end{cases} . \quad (4)$$

Here, $\log P_m(i)$ is a scoring term that considers only the number of segments used in the alignment, and score is the remainder of the scoring function, which takes into account the stretching and amplitude distortions involved in aligning a pair of segments, in addition to how well the observations from the two segments match given these distortions. Again, the indices a and b represent positions in the two time series and $q.r$ and $d.r$ refer to the rightmost (last) time coordinates in the query series and the database series, respectively. The base case is similar. We then find the highest scoring element of Γ that corresponds to a legal shorting:

$$\text{bestscore} = \max_{i,\ a \leq q.r,\ b \leq d.r} \begin{cases} \gamma(i, a, d.r) \\ \gamma(i, q.r, b) \end{cases}, \quad (5)$$

along with its generating path.

Assuming that the two series are of equal length n, Γ is of size $O(n^2)$. For each entry of Γ we must scan through $O(n^2)$ previous segments, and perform $O(n)$ calculations to score a pair of segments. This results in a final time complexity of $O(n^5)$.

2.3. Data

The data we use in our experiments comes from the EDGE toxicology database[2], and can be downloaded from *http://edge.oncology.wisc.edu/*. Our data set consists of 216 unique observations of microarray data, each of which represents the expression values for 1600 different genes. Each of these expression values is calculated by taking the average expression level from four treated animals, divided by the average level measured in four control animals. The data are then converted to a logarithmic scale, so that an expression of 0.0 corresponds to the average basal level observed in the control animals.

Each observation is associated with a treatment and a time point. The treatment refers to the chemical to which the animals were exposed and its dosage. The time point indicates the number of hours elapsed since exposure occurred. Times range from six hours up to 96 hours. The data used in our computational experiments span 11 different treatments, and for each treatment there are observations taken from at least three different time points.

The alignment methods we use all assume that the data is sampled at regular intervals. Because that is not the case here, all our experiments use an interpolation preprocessing step to generate regularly sampled pseudo-observations. Database and query series are interpolated using third order splines in most cases. We use second order splines when there are only two data points.

2.4. Experiments

In order to evaluate our multisegment alignment method on the EDGE data, we assembled query series by randomly sampling a random number of observations of the same treatment but at different times. We then tested the query against a database built from all the remaining observations. In some cases, the expression responses induced by similar treatments may evolve at different rates. To simulate this situation, we temporally distort some query series. For example, one of the distortions doubles all times in the first 48 hours (i.e., it stretches the first part of the series), and then halves all times (plus an offset for the doubling) for the next 24 hours. The other distortions were similar.

We note that this task is only a surrogate for the actual task with which we are concerned: classifying uncharacterized chemicals and aligning them with the most similar treatment in the database. It is a useful surrogate, however, because it is a task in which we know the most similar treatment and the correct alignment of the query to this treatment.

We preprocessed each query and the eleven database treatments using splines to reconstruct pseudo-observations at every four hours. In this experiment, our method returned the database treatment with the highest scoring alignment, as defined by Equation 5. We then measured how accurately we are able to (i) identify the treatment from which each query series was extracted, and (ii) align the query points to their actual time points in the treatment. We refer to the former as *classification accuracy* and the latter as *alignment accuracy*.

In the experiments we report here, our multisegment method is limited to three segments in its alignments. Our previous experiments [1] indicate that the accuracy of the multisegment method remains stable when it is allowed to use more segments.

We considered several other alignment methods as baselines. Dynamic time warping is described in Section 2.1. Linear parametric warping finds the coefficient which, when multiplied by the query times, results in the smallest Euclidean distance between the series. It is similar to the method used by Bar-Joseph et al.[9]. Finally *correlation optimized warping*

(COW)[10] is another segment-based method that divides both series into the same number of segments and then sums the cross correlations of corresponding segments.

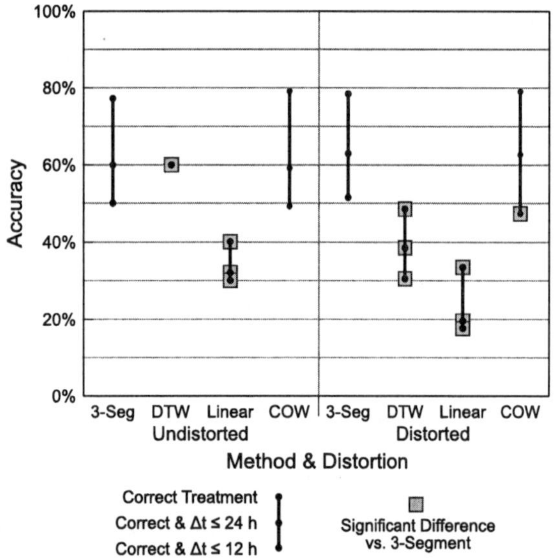

Fig. 3. Classification and alignment accuracies of several methods, including our three-segment model. Each line represents a different method. In each, the top point represents classification accuracy, while the bottom two points add the additional correctness criteria of the average alignment error being less than 24 hours and 12 hours, respectively. Highlighted points are those that are significantly different from the multisegment method ($p \leq 0.05$), by McNemar's χ^2 test.

The results of these experiments are shown in Figure 3. The left half represents those cases in which we did not distort the queries temporally, while in the right half we show the cases in which we did. Each line represents a different method. For each method the top point represents classification accuracy, the middle point represents alignment accuracy by adding the criterion that the average time error in the mapping is less than or equal to 24 hours, and the bottom point shows alignment accuracy where this tolerance is decreased to 12 hours. Highlighted points are those that are significantly different from the multisegment method ($p \leq 0.05$) according to McNemar's χ^2 test. Only COW exhibits accuracies competitive with those of our multisegment method. These estimates of COW's alignment accuracies are optimistic, however, because we have run COW with many settings for its parameters and report only the best results here.

Thus our multisegment method dominates the others in terms of both classification and alignment accuracy, but this comes at the cost of efficiency. Its time complexity of $O(n^5)$ is much greater than the other algorithms. With spline interpolation providing a pseudo-observation every four hours, a typical value for n is on the order of 25. The three-segment method takes about three minutes to do a single alignment. By contrast, the $O(n^2)$ DTW does the calculation in a fraction of a second. We would like our multisegment algorithm to be able to scale to handle queries for large databases of expression time series.

3. THE CONE HEURISTIC

We now describe the first of two new heuristics which address the efficiency limitation of our multisegment algorithm.

The alignment methods we use work by filling in an alignment matrix Γ. One well studied heuristic in similar time-series alignment problems is to restrict the cells of the matrix that are calculated. Several ways of doing this are illustrated in Figure 4. Each panel shows the alignment space when warping one series against another, and the shaded elements indicate the area to which the search is restricted. The so-called *Sakoe-Chiba Band*[4] and *Itakura Parallelogram*[11] are the most commonly used heuristics. The former restricts the search to a constant distance from the diagonal, while the latter allows progressively more warping closer to the middle. However both of these methods implicitly assume that the alignment being sought is a global one, in which there is no shorting of either input series. Here we consider a novel approach which confines the search to a cone starting at the origin and centered on the diagonal, as illustrated in Panel C of the figure.

Formally, we define the cone by a slope $c > 1$. We modify Equation 4 so that:

$$\gamma(i, x, y) = \textit{undefined} \textbf{ if } \frac{x}{y} > c \textbf{ or } \frac{y}{x} > c. \quad (6)$$

With this heuristic, we do not compute undefined values and we do not consider segments anchored in them.

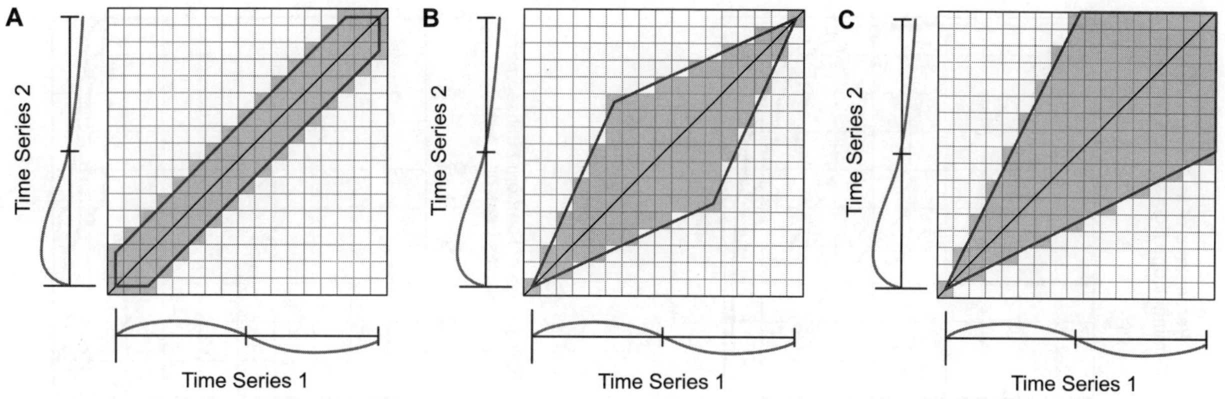

Fig. 4. Restricting the search in alignment space by shape. Each cell represents one element of the alignment matrix, and the shaded areas are those that are actually calculated. Both the Sakoe-Chiba Band (A) and the Itakura Parallelogram (B) are intended for globally aligning whole series to each other. By contrast, our cone (C) is designed for local alignments in which one series might be shorted.

3.1. Theoretical Analysis

Here we do a theoretical analysis of the expected speed-up in restricting the search space to a cone. The primary effect is to reduce the number of segments calculated by a constant factor, so we do not expect to see an improvement in its time complexity of $O(n^5)$. Instead we expect to see a constant speed-up proportional to the relative size of the cone to the alignment matrix. For a square matrix (i.e. where both series are of the same length), the relative size of the cone depends only on the slope of its bounding rays. With a slope of c, this ratio is $\frac{c-1}{c}$. We expect the execution time of the multisegment method with the cone heuristic to take roughly this fraction of the exact calculation's time.

We expect deviation from this value in two cases. First, nonsquare matrices will exhibit smaller ratios, as the cone covers proportionately less of their areas. Second, our calculation assumes that an alignment matrix can be split to an arbitrarily fine granularity. Because the matrix really consists of discrete elements, the ratio of elements covered by a cone will be more than expected for small matrices. For example, in a 5×5 matrix a cone with $c = 2$ will cover 13 cells, for a ratio of $\frac{13}{25}$ rather than $\frac{1}{2}$.

3.2. Experiments

Here we evaluate restricting the search in alignment space to a cone, in order to assess (i) its speedup relative to our original multisegment method, and (ii) its ability to find high-scoring alignments and produce accurate time-series classifications.

For the experiments in this section, we again use the data described in Section 2.3 and the same set of queries we used in Section 2.4.

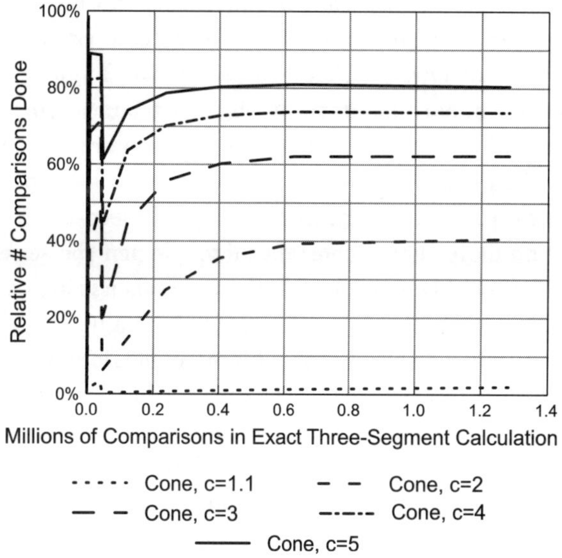

Fig. 5. Relative speed of the cone heuristic method. Time is measured by the number of comparisons of a point in each series.

First, we determine how much faster the cone method makes our calculations. We measure time in terms of *point comparisons*, or comparisons of a single time point in one series with a single time point in another. This is a good surrogate for calculation time

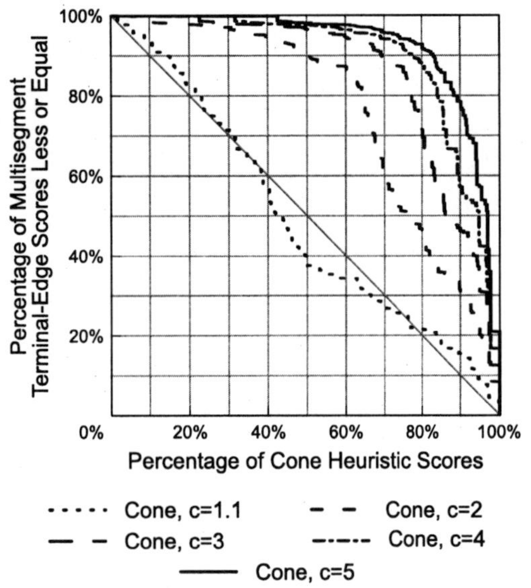

Fig. 6. Comparison of the cone heuristic method scores to the scores on the terminal edges of the alignment matrix of the exact multisegment calculation. These are the best alignments found for each legal shorting of the alignment.

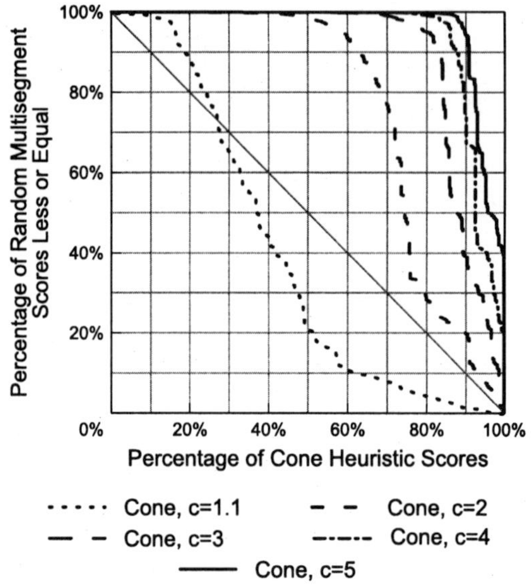

Fig. 7. Comparison of the cone heuristic method scores to the scores of 1000 randomly sampled multisegment alignments.

needed. Dynamic time warping performs $O(n^2)$ of these calculations, while our method performs $O(n)$ for each of $O(n^4)$ pairs of segments compared, for a total of $O(n^5)$. Figure 5 graphs the number of point comparisons done by the exact multisegment method versus the fraction done by the heuristic cone method, for the undistorted experiment we ran in Section 2.4. Although we obtain good speed-up, it is by no more than a constant factor. When the series are of roughly equal length, the time taken is in good agreement with our earlier calculated value of $\frac{c-1}{c}$. We predicted earlier that nonsquare matrices, which have less area covered by the cones, would run faster. These account for the dips visible in the figure. Additionally as predicted, smaller matrices tend to have larger values on this graph, because of the way cones divide the cells discretely.

Next we assess the alignments found using the cone heuristic, when we align each query to the database observations derived from the same treatment. We compare the score of each such alignment to a standard set of other alignments of the same query and database series. The first standard set we use consists of possible alignments found when doing the exact calculation. Recall that by Equation 5, the multisegment method chooses the best score from among all the possible shortings. We use all these scores—the best found for each shorting—as the standard set. Because the scores are drawn from the terminal edges of the alignment matrix, we refer to this set as the terminal-edge scores. We illustrate the comparison in Figure 6. Here we graph the percentage of heuristic-based scores that are better or equal to a percentage of the terminal-edge scores. For example, 80% of the cone-based scores are better or equal to at least roughly 70% of the edge scores when $c = 3$. If each query's score were drawn from the same distribution as its standard set, we would expect the curves to roughly coincide with the graphed diagonal. Thus with $c = 1.1$, the alignment found will likely not be better than picking one of the edge alignments at random. However with larger cones ($c \geq 2$), there is a good chance that the heuristic will lead to an alignment that scores well.

Figure 7 shows a similar comparison, but this time we compare against the scores from 1000 randomly sampled three-segment alignments as our standard set. These alignments are determined by randomly picking three contiguous segments in alignment space from origin to a terminal edge, and then picking the best amplitude coefficient for each segment by least squares. As before, when $c = 1.1$

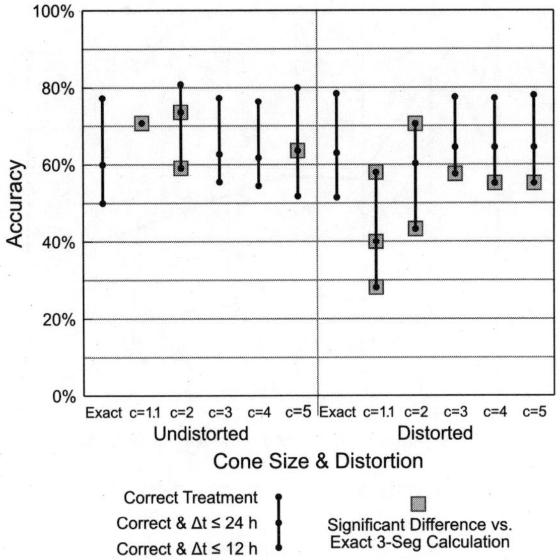

Fig. 8. Classification and alignment accuracies for the cone heuristic method with varying values of the slope parameter. Also shown for reference are the accuracy values for the exact three-segment method. Highlighted points are significantly different ($p \leq 0.05$) from the exact method by McNemar's χ^2 test.

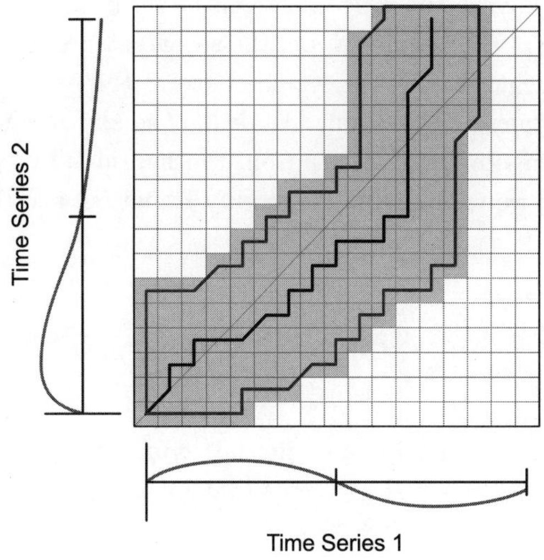

Fig. 9. Alignment space diagram of the hybrid-DTW heuristic for the multisegment method. We first find an alignment path using our hybrid-DTW method, and then we restrict the multisegment search to elements within a spread s of this path. Here s is two.

the alignment does not appear to be much better than random. With wider cones, however, the resulting alignments have better scores than most random alignments. Thus we conclude that, given a large enough cone, the segments that are not calculated are often not part of the optimal alignment. The alignment that is found by the cone heuristic will likely be comparable to the best alignment found by the exact method.

Finally we perform the classification/alignment task as in Section 2.4, except using the cone heuristic with the multisegment method. The results are shown in Figure 8. Except for the most narrow cones considered ($c = 1.1$), there is not a loss in accuracy due to finding suboptimal alignments. There may be some benefit to accuracy in using a moderate cone, with $c = 2$ or $c = 3$. If so, this is because such cones preclude the multisegment method from using extreme alignments, such as mapping the beginning of one series to the end of the other or using too great a slope in alignment space.

4. THE HYBRID-DTW HEURISTIC

Now we consider an alternative method for speeding up our multisegment alignment method. Here we restrict the search space by doing a first pass with a DTW-like method, and then considering only multisegment alignments that are close to the DTW path in alignment space. This method is illustrated in Figure 9.

We refer to the first pass method as *hybrid-DTW*, because it combines the dynamic programming of dynamic time warping with a scoring function similar to that used in our multisegment algorithm. The scoring function we use for it is:

$$D(q_x, d_y) = \begin{array}{c} D_E^2(q_x, \alpha d_y) + D_E^2(\frac{1}{\alpha} q_x, d_y) \\ -D_E^2(q_x, \mu_{DB,y}) - D_E^2(\mu_{DB,x}, d_y) \end{array} \qquad (7)$$

where D_E is the Euclidean distance, α is a value chosen by least squares to minimize the first two terms, and $\mu_{DB,x}$ is the average value in the database for time x. Unlike classic DTW, any element $\gamma(x, y)$ can either add to or subtract from the final score. In DTW, the final score is a normalized sum, so it has a strong bias to reduce the number of elements on its alignment path. Our hybrid-DTW is able to avoid this bias.

Given the alignment path returned by the hybrid-DTW calculation, the second step of our approach restricts the search space of the exact multisegment calculation. We define the *spread* s to be the maximum distance from the hybrid-DTW path that we will search. We modify Equation 4 so that:

$$\gamma(i,x,y) = \begin{array}{l} \textit{undefined} \\ \textbf{if } |x - x_h| > s \textbf{ or } |y - y_h| > s \end{array} \tag{8}$$

for all points (x_h, y_h) in the hybrid path. Thus we only consider segments that both begin and end within s of the hybrid path.

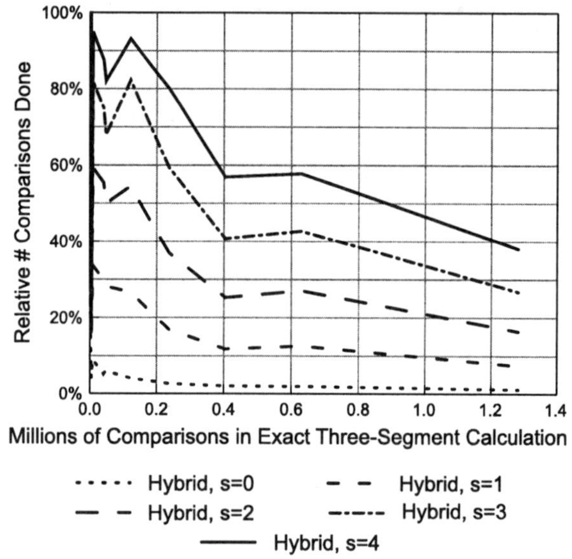

Fig. 10. Relative speed of the multisegment method with and without the hybrid-DTW heuristic. Again, time is measured by the number of comparisons of a point in each series.

4.1. Theoretical Analysis

Like classic DTW, the time complexity of hybrid-DTW is $O(mn)$, where m and n are the lengths of the series being aligned. The maximum length of the path it returns is $m + n$. This gives us a maximum bound on the number of segments calculated for the multisegment method of:

$$\sum_{i=1}^{m+n} (i-1)(2s+1) = \frac{1}{2}((m+n)^2 - (m+n))(2s+1), \tag{9}$$

where s is the spread. Assuming that $s \ll m \approx n$, the number of segments considered is $O(n^2)$. We multiply this value by the $O(n)$ time required to calculate the score of a segment, and obtain a total time complexity of $O(n^3)$.

4.2. Experiments

As with the cone heuristic, we assess the hybrid-DTW heuristic by considering (i) its speedup relative to the original multisegment method, and (ii) the quality of the alignments it finds. Recall that we have interpolated pseudo-observations at four-hour intervals. Thus we evaluate these criteria with s ranging from zero elements (zero hours) up to four elements (16 hours).

Speedup is shown in Figure 10, which is again measured in terms of point comparisons. With $s = 0$ or $s = 1$ we obtain speedups of an order of magnitude. We again see the dips corresponding to non-square matrices. The best speed-ups occur for the largest matrix sizes. In contrast to the cone heuristic, the ratio of comparisons done still appears to be decreasing for the largest values. This is what we would expect with a better time complexity.

Next we consider the resulting alignment scores, and as before we compare them to the scores of both terminal-edge alignments and random alignments. The results of these score comparisons are shown in Figures 11 and 12. The hybrid-DTW heuristic method does well here, even when s is zero. Most of the scores found using this heuristic are better than or equal to the edge and random scores for the same query.

Finally, Figure 13 shows the classification and alignment accuracies for the exact multisegment calculation and the calculation using the hybrid-DTW first pass. There is no significant difference in accuracy when using the heuristic versus doing the exact multisegment calculation. For completeness, the figure also shows the accuracies when using the hybrid-DTW method by itself (i.e. not as a first pass for the multisegment method). Although it is more ro-

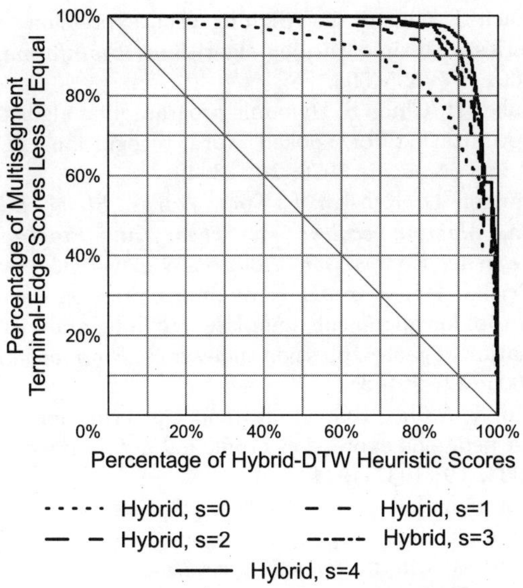

Fig. 11. Comparison of the hybrid-DTW heuristic scores to the scores on the terminal edges of the alignment matrix of the exact multisegment method.

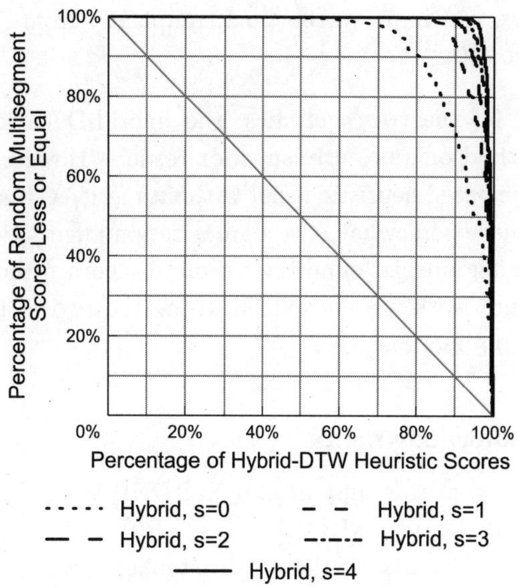

Fig. 12. Comparison of the hybrid-DTW heuristic scores to the scores of 1000 randomly sampled multisegment alignments.

bust to distortion than ordinary DTW and seems to align well, its accuracy—especially classification accuracy—is still significantly worse than that of the multisegment method.

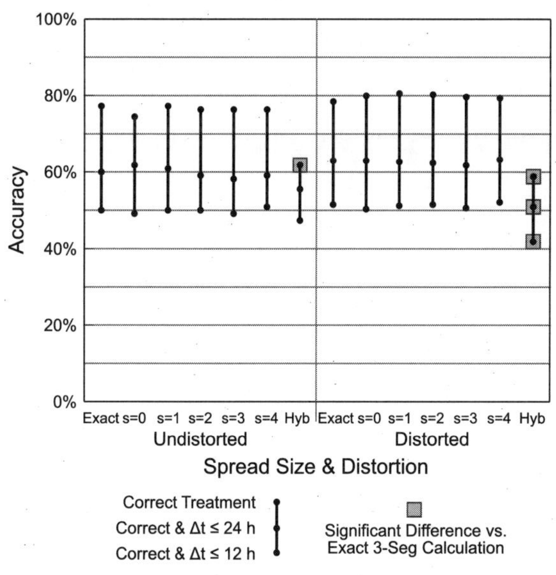

Fig. 13. Classification and alignment accuracies for the hybrid-DTW heuristic method with varying values of the spread parameter. Also shown for reference are the accuracy values for the exact three-segment method and the hybrid-DTW method on its own. As before, highlighted points are significantly different ($p \leq 0.05$) from the exact method by McNemar's χ^2 test.

Taken together, these results imply that the hybrid-DTW's alignment paths are a good approximation to those found by the multisegment method. Using spread values of zero or one has the potential to speed up the calculation greatly while not significantly harming the accuracy provided.

5. DISCUSSION

In a previous investigation[1] we showed that our multisegment alignment method is more accurate than other methods, both in terms of classification and alignment accuracy. In this study, we have presented two heuristics that can be used to speed up its calculation.

- By restricting the search space in the warping matrix to a cone, we may speed up the calculation by a constant factor. The cone may also serve as a useful bias, preventing alignments that are shorted too much when there is reason to believe that they should be excluded. The cone shape is analogous to both the Sakoe-Chiba Band and the Itakura Parallelogram, but it allows non-global alignments.
- By restricting the search space to alignments that are near those found by a modified DTW method,

we can improve the time complexity from $O(n^5)$ to $O(n^3)$.

Of the two heuristics, the hybrid-DTW based method offers clearly superior results. However the cone based heuristic is not without merit, as it seems to have somewhat of a regularization effect, biasing the alignments found toward more accurate ones. In future work we will explore using the two methods in conjunction.

Acknowledgments

This work was supported by NIH/NIEHS grant R01 ES012752, and NIH/NLM grant R01 LM07050. We would also like to thank Christopher Bradfield and Aaron Vollrath of the EDGE project.

References

1. Smith AA, Vollrath A, Bradfield C, Craven M. Similarity queries for temporal toxicogenomic expression profiles. *PLoS Computational Biology* 2008; In press.
2. Hayes K, Vollrath A, Zastrow G, McMillan B, Craven M, Jovanovich S, Walisser J, Rank D, Penn S, Reddy J, Thomas R, Bradfield C. EDGE: A centralized resource for the comparison, analysis and distribution of toxicogenomic information. *Molecular Pharmacology* 2005; **67**: 1360–1368.
3. Aach J, Church G. Aligning gene expression time series with time warping algorithms. *Bioinformatics* 2001; **17**: 495–508.
4. Sakoe H, Chiba S. Dynamic programming algorithm optimization for spoken word recognition. *IEEE ASSP Magazine* 1978; **26**: 43–49.
5. Sankoff D, Kruskal J. *Time Warps, String Edits, and Macromolecules: The Theory and Practice of Sequence Comparison*. Addison-Wesley 1983.
6. Criel J, Tsiporkova E. Gene time expression warper: A tool for alignment, template matching and visualization of gene expression time series. *Bioinformatics* 2006; **22**: 251–252.
7. Liu X, Müller HG. Modes and clustering for time-warped gene expression profile data. *Bioinformatics* 2003; **19**: 1937–1944.
8. Ratanamahatana C, Keogh EJ. Three myths about dynamic time warping data mining. In: *Proceedings of SIAM International Conference on Data Mining*. SIAM, 506–510.
9. Bar-Joseph Z, Gerber G, Gifford D, Jaakkola T, Simon I. Continuous representations of time-series expression data. *Journal of Computational Biology* 2003; **10**: 341–356.
10. Nielsen NV, Carstensen JM, Smedsgaard J. Aligning of single and multiple wavelength chromatographic profiles for chemometric data analysis using correlation optimised warping. *Journal of Chromatography A* 1998: 17–35.
11. Itakura F. Minimum prediction residual principle applied to speech recognition. *IEEE Transactions on Acoustics, Speech, and Signal Processing* 1975; **23**: 67–72.

GRAPH WAVELET ALIGNMENT KERNELS FOR DRUG VIRTUAL SCREENING

Aaron Smalter[1], Jun Huan[1], and Gerald Lushington[2]
[1]*Department of Electrical Engineering and Computer Science*
[2] *Molecular Graphics and Modeling Laboratory*
University of Kansas
Email :{*asmalter, jhuan, glushington*}*@ku.edu*

In this paper we introduce a novel graph classification algorithm and demonstrate its efficacy in drug design. In our method, we use graphs to model chemical structures and apply a wavelet analysis of graphs to create features capturing graph local topology. We design a novel graph kernel function to utilize the created feature to build predictive models for chemicals. We call the new graph kernel a graph *wavelet-alignment kernel.*

We have evaluated the efficacy of the wavelet-alignment kernel using a set of chemical structure-activity prediction benchmarks. Our results indicate that the use of the kernel function yields performance profiles comparable to, and sometimes exceeding that of the existing state-of-the-art chemical classification approaches. In addition, our results also show that the use of wavelet functions significantly decreases the computational costs for graph kernel computation with more than 10 fold speed up.

1. INTRODUCTION

The fast accumulation of data describing chemical compound structures and biological activity calls for the development of efficient informatics tools. *Cheminformatics* is a rapidly emerging research discipline that employs a wide array of statistical, data mining, and machine learning techniques with the goal of establishing robust relationships between chemical structures and their biological properties. Cheminformatics hence is an important component on the application side of applying informatics approach to life science problems. It has a broad range of applications in chemistry and biology; arguably the most commonly known roles are in the area of drug discovery where cheminformatics tools play a central role in the analysis and interpretation of structure-activity data collected by various means of modern high throughput screening technology. Traditionally the analysis of large chemical structure-activity databases was done only within pharmaceutical companies and up until recently the academic community has had only limited access to such databases. This situation, however, has changed dramatically in very recent years.

In 2002, the National Cancer Institute created the Initiative for Chemical Genetics (ICG) with the goal of offering to the academic research community a large database of chemicals with their roles in cancer research [18]. Two years later, the National Health Institute (NIH) launched a Molecular Libraries Initiative (MLI) that included the formation of the national Molecular Library Screening Centers Network (MLSCN). MLSCN is a consortium of 10 high-throughput screening centers for screening large chemical libraries [1]. Collectively, ICG and MLSCN aim to offer to the academic research community the results of testing about a million compounds against hundreds of biological targets. To organize this data and to provide public access to the results, the PubChem database and the Chembank database have been developed as the central repository for chemical structure-activity data. These databases currently contain more than 18 million chemical compound records, more than 1000 bioassay results, and links from chemicals to bioassay description, literature, references, and assay data for each entry.

These publicly-available large-scale chemical compound databases have offered tremendous opportunities for designing highly efficient *in silico* drug design method. Many machine learning and data mining algorithms have been applied to study the structure-activity relationship of chemicals. For example, Xue *et al.* reported promising results of applying five different machine learning algorithms: logistic regression, C4.5 decision tree, k-nearest neighbor, probabilistic neural network, and support vector machines to predicting the toxicity of chemicals against an organism of Tetrahymena pyriformis [21].

Advanced techniques, such as random forest and MARS (Multivariate Adaptive Regression Splines) have also been applied to cheminformatics applications [15, 17].

Recently Support Vector Machines (SVM) have gained popularity in drug design. Support vector machines work by constructing a hyperplane in a high dimensional feature space. Two key insights of SVM are the utilization of kernel functions (i.e. inner product of two points in a Hilbert Space) to transform a non-linear classification to a linear classification and the utilization of a large margin classifier to separate points with different class labels. Large margin classifiers have low chance of over fitting and works efficiently in high dimensional feature spaces.

Support vector machines have been widely utilized in cheminformatics study. Traditional way of applying SVM to cheminformatics is to first create a set of features (or *descriptors* in many quantitative structure-properity relationship studies) and then use SVM to train a predictive model [6, 16]. Recently using graphs to model chemical structures and using data mining approach to obtain high quality, task-relevant features gain popularity in cheminformatics [16]. In this paper, we report a novel application graph wavelet analysis in creating high quality localized structure features for cheminformatics. Specifically, in our method, we model a chemical as a graph where a *node* represents an *atom* and an *edge* represents an *chemical bond* in the chemical. We leverage wavelet functions applied to graph-structured data in order to construct a graph kernel function. The wavelet functions are used to condense neighborhood information about an atom into a single feature of that atom, rather than features spread over it's neighboring atoms. By doing so, we extract (local) features with various topological scales about chemical structures and use these wavelet features to compute an alignment of two chemical graphs.

We have applied our graph kernel methods to several chemical structure-activity benchmarks. Our results indicate that our approaches yields performance profiles at least competitive with, and sometimes exceeding that of current state-of-the-art approaches. In addition, the identification of highly discriminative patterns for chemical activity classification provides evidence that our methods can make generalizations about chemical function given molecular structure. More over, our results also show that the use of wavelet functions significantly decreases the computational costs for graph kernel computation.

The rest of the paper is organized in the following way. Section 2 presents an overview of related work on quantitative chemical structure-property relationship study. In Section 3, we present background information about graph representation of chemical structures, graph database mining, and graph kernel function. Section 4 discusses the algorithmic details of our work, and in Section 5 we examine an empirical study of the proposed algorithm using several chemical structure benchmarks. We conclude with a short discussion of the pros and cons of our proposed methods.

2. RELATED WORK

A *target property* of the chemical compound is a measurable quantity of the compound. There are two categories of target properties: continuous (e.g., binding affinities to a protein) and discrete target properties (e.g. active compounds vs. inactive compounds).

The relationship between a chemical compound and its target property is typically investigated through a quantitative structure-property relationship (*QSPR*) [a]. Abstractly, any QSPR method may be generally defined as a function that maps a chemical space to a property space in the form of

$$P = \hat{k}(D) \tag{1}$$

where D is a chemical structure, P is a property, and the function $\hat{k}$ is an estimated mapping from a chemical space to a property space.

Different QSPR methodologies can be understood in terms of the types of target property values (continuous or discrete), types of features, and algorithms that map descriptors to target properties. Many classification methods has been applied to build QSPR models and recent ones include Decision Trees, Classification based on association [2], and

[a]Such study also known as a quantitative structure-activity relationship (QSAR) but *property* refers to a broader range of applications than activity.

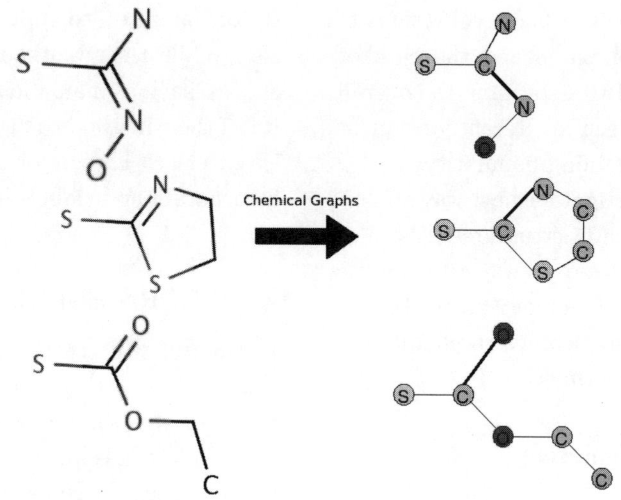

Fig. 1. Left: three sample chemical structures. Right: Graph representations of the three chemical structures.

Random Forest among many others. In our subsequent investigation, we focus on graph representation of chemical structures, graph wavelet analysis, and graph kernel methods that work well in high and even infinite dimensional feature space with low chance of over-fitting.

3. BACKGROUND

Before we proceed to algorithmic details, we present some general background regarding a computational analysis of chemical structure-property relationship which includes (i) a graph representation of chemical structures, (ii) graph kernel functions, and (iii) graph wavelet analysis.

3.1. Chemical Structure and Graph Modeling of Chemical Structures

Chemical compounds are organic molecules that are easily modeled by a graph representation. In our representation, we use *nodes* in a graph to model *atoms* in a chemical structure and *edges* in the graph to model chemical *bonds* in the chemical structure. In our representation, nodes are labeled with the atom element type, and edges are labeled with the bond type (single, double, and aromatic bond). The edges in the graph are undirected, since there is no directionality associated with chemical bonds. Figure 1 shows three sample chemical structures and their graph representation. Since we always use graphs to model chemical structures, in the following discussion, we make little distinction about graphs and chemicals, nodes and atoms, and edges and chemical bonds.

3.2. Graph Kernel Functions

The term *kernel function* in our context refers to an operation of computing the inner product between two objects (e.g. graphs) in a feature space, thus avoiding the explicit computation of coordinates in that feature space. Depends on the dimensionality of the feature space, we divide the current graph kernel function into two groups.

The first group works in a finite dimensional feature space [16]. Algorithms in the group first compute a set of features and performs subsequent classification in this feature space. Many existing application of machine learning algorithms to cheminformatics problems follow into this category.

The second group works in an infinite dimensional feature space. Example of this group include kernels that work on paths [12], on cyclic graphs [10]. The kernel computation in infinite dimensional feature space is usually challenging. To ease the prohibitive computational cost, Kashima et al[12] developed a Markov model to randomly generate walks of a labeled graph. The random walks are created using a transition probability matrix combined with a walk termination probability. These collections of random walks are then compared and the number of shared sequences is used to determine the overall similarity between two molecules. Recently frequent pattern based kernels are gaining popularity.

In this paper, we investigate a new way to create features of chemical graph structures. We also present an efficient computational way to compute such kernel called wavelet-alignment graph kernel. Our experimental study has demonstrated the efficiency and efficacy of our method.

3.3. Graph Wavelets Analysis

Wavelet functions are commonly used as a means for decomposing and representing a function or signal as its constituent parts, across various resolutions or scales. Wavelets are usually applied to numerically valued data such as communication signals or mathematical functions, as well as to some regularly structured numeric data such as matrices and images. Graphs, however, are arbitrarily structured and may represent innumerable relationships and topologies between data elements. Recent work has established the successful application of wavelet functions to graphs for multi-resolution analysis. Two examples of wavelet functions, the Haar and the mexican hat, are depicted in Figure 2.

Crovella et al. [4] have developed a multi-scale method for network traffic data analysis. For this application, they are attempting to determine the scale at which certain traffic phenomena occur. They represent traffic networks as graphs labeled with some measurement such as bytes carried per unit time. In their method, they use the *hop* distance between vertices in a graph, defined as the length of the shortest path between them, and apply a weighted average function to compute the difference between the average of measurements close to a vertex and measurements that are far, up to a certain distance. This process produces a new measurement for a specific vertex that captures and condenses information about the vertex neighborhood. Figure 3 shows a diagram of wavelet function weights overlayed on a chemical structure.

Maggioni et al. [13] demonstrate a general-purpose biorthogonal wavelet for graph analysis. In their method, they use the dyadic powers of an diffusion operator to induce a multiresolution analysis. While their method applies to a large class of spaces, such as manifolds and graphs, the applicability of their method to attributed chemical structures is not clear. The major technical difficulty is how to incorporate node labels in a multiresolution analysis.

4. ALGORITHM DESIGN

In the following sections we outline the algorithms that drive our experimental method. In short, we measure the similarity of graph structures whose nodes and edges have been labeled with various features. These features represent different kinds of chemical structure information including atoms and chemical bonds types among others. To compute the similarity of two graphs, the nodes of one graph are aligned with the nodes of the second graph, such that the total overall similarity is maximized with respect to all possible alignments. Vertex similarity is measured by comparing vertex descriptors, and is computed recursively so that when comparing two nodes, we also compare the immediate neighbors of those nodes, the neighbors of immediate neighbors, and so on so forth.

4.1. Graph Alignment Kernel

An *alignment* of two graphs G and G' (assuming $|V[G] \leq |V[G']|$) is a 1-1 mapping $\pi : V[G] \rightarrow V[G']$. Given an alignment π, we define the similarity between two graphs, as measured by a kernel function k_A, below:

$$k_A(G, G') := \max_\pi \sum_{v \in V[G]} k_n(v, \pi(v)) + \sum_{u,v} k_e((u, v), (\pi(u), \pi(v))) \quad (2)$$

The function k_n is a kernel function to measure the similarity of nodes and the function k_e is a kernel function to measure the similarity of edges. Intuitively in Equation 2 we use an additive model to compute the similarity between two graphs by computing the sum of the similarity of nodes and the similarity of edges. The maximal similarity among all possible alignments is defined as the similarity between two graphs.

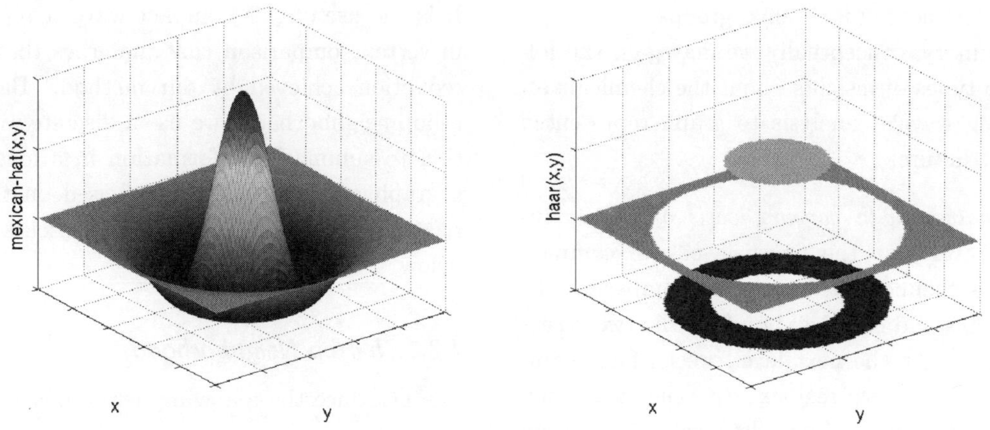

Fig. 2. Two examples of wavelet functions in 3 dimensions, the mexican hat on the left, and the Haar on the right.

4.2. Simplified Graph Alignment Kernel

A direct computation of the graph alignment kernel is computationally intensive and is unlikely scalable to large graphs. With no surprise, the graph alignment kernel computation is no easier than the subgraph isomorphism problem, a known NP-hard problem [b]. To derive efficient algorithm scalable to large graphs, we simplify the graph kernel function with the following formula:

$$k_M(G, G') = \max_{\pi} \sum_{v \in V[G]} k_a(f(v), f(\pi(v))) \qquad (3)$$

Where $\pi : V[G] \rightarrow V[G']$ denotes an alignment of graph G and G'. $f(v)$ is a set of features associated with a node that not only include node features but also include information about topology of the graph where v belongs to.

By equation 3, we are trying to compute a maximal weighted bipartite graph, which has an efficient solution known as the Hungarian algorithm. The complexities of the algorithm is $O(|V[G]|^3)$. See [7] for further details.

Below we provide an efficient method, based on graph wavelet analysis, to crate features to capture the topological structure of a graph.

4.3. Graph Wavelet Analysis

Originally proposed to analyze time series signals, wavelet analysis transforms a series of signals to a set of summaries with different scale. Two of the key insights of wavelet analysis of signals are (i) using localized basis functions and (ii) analysis with different scales. Wavelet analysis offers efficient tools to decompose and represent a function with arbitrary shape [5, 8]. Since invented, wavelet analysis has quickly gained popularity in a wide range of applications outside time series data, such as image analysis and geography data analysis. In all these applications, the level of detail, or *scale* is considered as an important factor in data comparison and compression. We show two examples of wavelet functions in a 3D space in Figure 2.

Our Intuition. With wavelet analysis as applied to graph represented chemical structure, for each atom, we may collect features about the atom and its local environment with different scales. For example, we may collect information about the average charge of an atom and atoms surrounding the atom and assign the average value as a feature to the atom. We may also collect information about whether an atom belongs to a nearby functional group, whether the surrounding atoms of a particular atom belong to a nearby functional group, and the local topology of

[b]Formally, we need to show a reduction from the graph alignment kernel to the subgraph isomorphism problem. The details of such reduction are omitted due to their loose connection to the main theme of the current paper, which is advanced data mining approach as applied to cheminformatics applications

an atom to its nearby functional groups.

In summary, conceptually we may gain the following two types of insights about the chemicals after applying wavelet analysis to graph represented chemical structure:

- *Analysis with varying levels of scale.* Intuitively, at the finest level, we compare two chemical structures by comparing the atoms and chemical bonds in the two structures. At the next level, we perform comparison of two regions (e.g. chemical functional groups) of two chemicals. At an even coarser level, small regions may be grouped into larger ones (e.g. pharmacophore), and we compare two chemicals by comparing the large regions and the connections among large regions.
- *Non-local connection.* In a chemical structure, two atoms that are not directly connected by a chemical bond may still have some kind of interaction. Therefore when comparing two graphs and their vertices, we cannot depend only on the *local environment* immediately surrounding an atom, but rather must consider both local and non-local environment.

Though conceptually appealing, current wavelet analysis is often limited to numerical data with regularly structures such as matrices and images. Graphs, however, are arbitrarily structured and may represent innumerable relationships and topologies between data elements. In order to define a reasonable graph wavelet functions, we have introduced the following two important concepts:

- h-hop neighborhood
- Discrete wavelet functions

The former, h-hop neighborhood, is essentially used to project graphs from a high dimensional space with arbitrary topology into a Euclidean space suitable for operation with wavelets. The h-hop measure defines a distance metric between vertices that is based on the shortest path between them. The discrete wavelet function then operates on a graph projection in the h-hop Euclidean space to compactly represent the information about the local topology of a graph. It is the use of this compact wavelet representation in vertex comparison that underlies the complexity reduction achieved by our method. Based on the h-hop neighborhood, we use a discrete wavelet function to summarize information in a local region of a graph and create features based on the summarization. These two concepts are discussed in detail below.

4.3.1. *h-Hop Neighborhood*

We introduce the following definitions.

Definition 4.1. Given a node v in a graph G the **h-hope neighborhood** of v, denoted by $N_h(v)$, is the set of nodes that are (according to the shortest path) exactly h hops away from v.

For example if $h = 0$, we have $N_0(v) = v$ and if $h = 1$, we have $N_1(v) = \{u|(u, v) \in E[G]\}$.

We use f_v denotes the feature vector associated with a node v in a graph G. $|f|$ is the feature vector length (number of features in the feature vector). The average feature measurement, denoted by $\overline{f}_j(v)$ for nodes in $N_j(v)$ is

$$\overline{f}_j(v) = \frac{1}{|N_j(v)|} \sum_{u \in N_j(v)} f_u \tag{4}$$

Example 4.1. The left part of the Figure 3 shows a chemical graph. Given a node v in the graph G, we label the shortest distance of nodes to v in the G. In this case $N_0(v) = v$ and $N_1(v) = \{t, u\}$. If our feature vector contains a single feature of atomic number, $\overline{f}_1(v)$ is the average atomic number of atoms that are at most 1-hop away from v. In our case, since $N_1(v) = \{t, u\}$ and $\{t, u\}$ are both carbon with atomic number equal to eight, then $\overline{f}_1(v)$ is equal to eight as well.

4.3.2. *Discrete Wavelet Functions*

In order to adapt a wavelet function to discrete structure such as graphs, we convert a wavelet function $\psi(x)$ to apply to the h-hop neighborhood. Towards that end, we scale a wavelet function $\psi(x)$ (such as the Haar, or Mexican Hat) to have support on the domain $[0, 1)$, with integral 0, and partition the function into $h+1$ intervals. We then compute the aver-

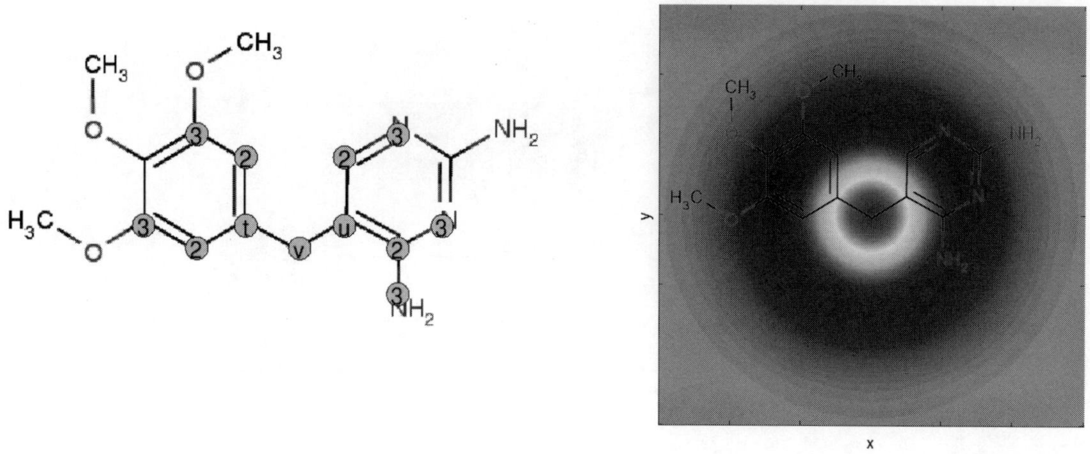

Fig. 3. Left: chemical graph centered on vertex v, with adjacent vertices t and u. Vertices more than one hop away are labeled with the hop number, up to hop distance three. Right: Superposition of a wavelet function on the chemical graph. Note here we can see the intensity of the wavelet function corresponds to the hop distance from the central vertex. Also this represents an idealized case where the hop distance between vertices corresponds roughly to their spatial distance. Unlabeled vertices correspond to carbon (C); hydrogens are shown without explicit bonds (edges).

age, $\psi_{j,h}$, as the average of $\psi(x)$ over the jth interval, $0 \le j \le h$ as below.

$$\psi_{j,h} \equiv \frac{1}{h+1} \int_{j/(h+1)}^{(j+1)/(h+1)} \psi(x)dx \quad (5)$$

With neighborhood and discrete wavelet functions, we are ready to apply a wavelet analysis to graphs. We call our analysis *wavelet measurements*, denoted by $\Gamma_h(v)$, for a node v in a graph G at scale up to $h > 0$.

$$\Gamma_h(v) = C_{h,v} * \sum_{j=0}^{h} \psi_{j,h} * \overline{f}_j(v) \quad (6)$$

where $C_{h,v}$ is a normalization factor with $C(h,v) = (\sum_{j=0}^{h} \frac{\psi_{j,h}^2}{|N_k(v)|})^{-1/2}$

We define $\Gamma^h(v)$ as the sequence of wavelet measurements as applied to a node v with scale value up to h. That is $\Gamma^h(v) = \{\Gamma_1(v), \Gamma_2(v), \ldots, \Gamma_h(v)\}$. We call $\Gamma^h(v)$ the *wavelet measurement vector* of node v. Finally we plug the wavelet measurement vector into the alignment kernel with the following formula.

$$k_\Gamma(G, G') = \max_\pi \sum_{v \in V[G]} k_a(\Gamma^h(v), \Gamma^h(\pi(v))) \quad (7)$$

where $k_a(\Gamma^h(v), \Gamma^h(\pi(v))$ is a kernel function defined on vectors. Two popular choices are linear kernel and radius based function kernel.

Example 4.2. The right part of Figure 3 shows a chemical graph overlayed with a wavelet function centered on a specific vertex. We can see how the wavelet is most intense at the central vertex, hop distance of zero, corresponding to a strongly positive region of the wavelet function. As the hop distance increases the wavelet function becomes strongly negative, as we can see roughly at hop distances of one and two. At hop distance greater than two, the wavelet function returns to zero intensity, indicating negligible contribution from vertices at this distance.

5. EXPERIMENTAL STUDY

We have conducted classification experiments on five different biological activity data sets, and measured support vector machine (SVM) classifier prediction accuracy for several different feature generation methods. We describe the data sets and classification methods in more detail in the following subsections, along with the associated results. Figure 4 gives a graphical overview of the process.

We performed all of our experiments on a desktop computer with a 3Ghz Pertium 4 processor and

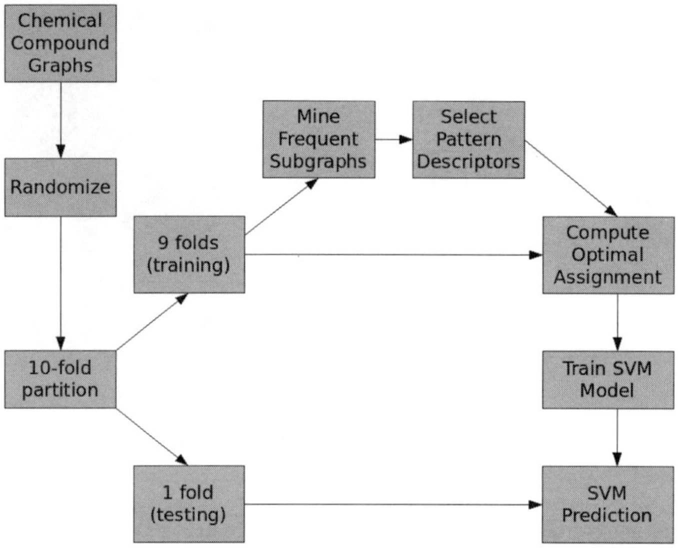

Fig. 4. Experimental workflow for a single cross-validation trial.

1 GB of RAM.

5.1. Data Sets

We have selected five data sets representing typical chemical benchmarks in drug design to evaluate our classifier performance. The Predictive Toxicology Challenge data set, discussed by Helma et al[9], contains a set of chemical compounds classified according to their toxicity in male rats (PTC-MR), female rats (PTC-FR), male mice (PTC-MM), and female mice (PTC-FM). The Human Intestinal Absorption (HIA) data set (Wessel et al.[19]) contains chemical compounds classified by intestinal absorption activity. The remaining data set (MD) is from Patterson et al[14], and was used to validate certain molecule descriptors. Various statistics for these data sets can be found in Table 1.

All of these data sets exist natively as binary classification problems, therefore in the case of the MD and HIA data sets, some preprocessing is required to transform them into regression and multi-class problems. For regression, this is a straightforward process of using the compound activity directly as the regression target. In the case of multi-class problems the transformation is not as direct. We chose to use a histogram of compound activity values to visualize which areas of the activity space are more dense, allowing natural and intuitive placement of class separation thresholds.

5.2. Methods

We evaluated the performance of the SVM classifier trained with different methods. The first two methods (WA-linear, WA-RBF) are both computed using our wavelet-alignment kernel, but use different functions for computing atom-atom similarity; we tested both a linear and RBF function here. In our experimental study, we experimentally evaluated different hop distance threshold and fixed $h = 3$ in all experimental study.

The method optimal alignment (OA) consists of the similarity values computed by the optimal assignment kernel, as proposed by Frölich et al[7]. There are several reasons that we consider OA as the current-state-of-the-art graph based chemical structure classification method. First, the OA method is developed specifically for chemical graph classification. Second the OA method contains a large library to compute different features for chemical structures. Third, the OA method has developed a sophisticated kernel function to compute the similarity between two chemical structures. Our experimental study shows that with the wavelet analysis we obtain performance profiles comparable to, and sometimes ex-

Table 1. Data set and class statistics.

Dataset	# Graphs	Class	Labels	Count
HIA	86	regression	0 - 100	86
		binary	0	39
			1	47
		multi-class	1	21
			2	18
			3	21
			4	26
MD	310	regression	0 - 7000	310
		binary	0	162
			1	148
		multi-class	1	46
			2	32
			3	37
			4	35
PTC-MR	344	binary	0	192
			1	152
PTC-MM	336	binary	0	207
			1	129
PTC-FR	351	binary	0	230
			1	121
PTC-FM	349	binary	0	206
			1	143

ceeding that of the existing state-of-the-art chemical classification approaches. In addition, we achieve a significant computational time reduction by using the wavelet analysis. The details of the experimental study are shown below.

In our experiments, we used the support vector machine (SVM) classifier in order to generate activity predictions. We used the LibSVM classifier implemented by Chang et al[3] as included in the Weka data-mining software package by Witten et al. [20]. The SVM parameters were fixed across all methods, and we use a linear kernel. For (binary) classification we used nu-SVC for multi-class classification with nu = 0.5. We used the Haar wavelet function in our WA experiments. Classifier performance was averaged over a 10-fold cross-validation set.

We developed and tested most of our algorithms under the MATLAB programming environment. The OA software was provided by [7] as part of their JOELib software, a computational chemistry library implemented in java. [11]

5.3. Results

Below we report our experimental study of the wavelet-alignment kernel with two focuses: (i) classification accuracy and (ii) computational efficiency.

5.3.1. *Classification Accuracy*

Table 2 reports the average and standard deviation of the prediction results over 10 trials. For classification problems, results are in prediction accuracy, and for regression problems they are in mean squared error (MSE) per sample. From the table, we observe that for the HIA data set, WA-RBF kernel significantly outperforms OA for both binary and multi-class classification. For MD data set, OA does best for both classification sets, but WA-linear is best for regression. For the PTC binary data, the WA-linear method outperforms the others in 3 of the 4 sets.

5.3.2. *Computational Efficiency*

In Table 3, we document the kernel computation time for both OA and WA methods using 6 different data sets. The runtime advantage of our WA algorithm over OA is clear, showing improved computation efficiency by factors of over 10 fold for the WA-linear kernel and over 5 fold for the WA-RBF kernel.

Figure 5 shows the kernel computation time across a range of dataset sizes, with chemical compounds sampled from the HIA data set. Using simple random sampling with replacement, we create data sets sized from 50 to 500. We did not try to run OA

Table 2. Prediction results of cross-validation experiments, averaged 10 randomized trials, with standard deviation in parentheses. For regression data sets labeled with real values, result is mean squared error (lower is better); for classification the result is prediction accuracy (higher is better). The best result for each data and label set is marked with an asterisk.

Dataset	Labels	OA	WA-RBF	WA-linear
HIA	real	979.82(32.48)*	989.72(33.60)	989.31(24.62)
	binary	51.86(3.73)	61.39(2.77)*	57.67(3.54)
	multi-class	29.30(2.23)	39.06(0.63)*	29.76(5.73)
MD	real	3436395(1280)	3436214(1209)*	3440415(1510)
	binary	67.16(0.86)*	52.51(3.34)	65.41(0.42)
	multi-class	39.54(1.65)*	33.35(3.83)	33.93(1.87)
PTC-FM	binary	58.56(1.53)*	51.46(3.45)	55.81(1.31)
PTC-FR	binary	58.57(2.11)	52.87(2.65)	59.31(1.95)*
PTC-MM	binary	58.23(1.25)	52.36(0.93)	58.91(2.078)*
PTC-MR	binary	51.51(1.20)	52.38(3.48)	52.09(2.61)*

Table 3. Running time for the computation of OA, WA-linear, and WA-RBF) kernels, in seconds. Speedup is computed as the ratio between the OA processing time and that of WA.

Dataset	Kernel	Time	Speedup
HIA	OA	75.87	-
	WA-RBF	13.76	5.51
	WA-linear	4.91	15.45
MD	OA	350.58	-
	WA-RBF	50.85	6.89
	WA-linear	26.82	13.07
PTC-FM	OA	633.13	-
	WA-RBF	103.95	6.09
	WA-linear	44.87	14.11
PTC-FR	OA	665.95	-
	WA-RBF	116.89	5.68
	WA-linear	54.64	12.17
PTC-MM	OA	550.41	-
	WA-RBF	99.39	5.53
	WA-linear	47.51	11.57
PTC-MR	OA	586.12	-
	WA-RBF	101.68	5.80
	WA-linear	45.93	12.73

on even larger data set since the experimental results clearly demonstrate the efficiency of the WA kernel already.

What these run time results do not demonstrate is the *even greater* computational efficiency afforded by our WA algorithm when operating on general, non-chemical graph data. As noted at the end of section four, chemical graphs have some restrictions on their general structure. Specifically, the number of atom neighbors is bound by a small constant (4 or so). Since the OA computation time is much more dependent on the number of neighbors, we can see that WA is even more advantageous in these circumstances. Unfortunately, since the OA software is designed as part of the JOELib chemoinformatics library specifically for use with chemical graphs, it will not accept generalized graphs as input, and hence we could not empirically demonstrate this aspect of our algorithm.

6. CONCLUSIONS

Graph structures are a powerful and expressive representation for chemical compounds. In this paper we present a new method *wavelet-assignment*, for computing the similarity of chemical compounds, based on the use of an optimal assignment graph

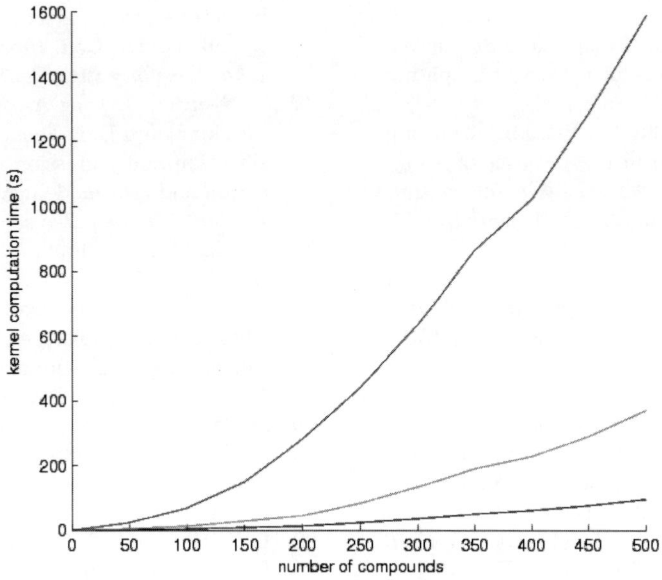

Fig. 5. Comparison of computation times between OA (top line), WA-RBF (middle line) and WA-linear (bottom line) kernels.

kernel function augmented with pattern and wavelet based descriptors. Our experimental study demonstrates that our wavelet-based method deliver an improved classification model, along with an order of magnitude speedup in kernel computation time. For high-volume, real world data sets, this algorithm is able to handle a much greater number of graph objects, demonstrating it's potential for processing both chemical and non-chemical data in large amounts. In our present study, we only used limited number of atom features. In the future, we plan to involve domain experts to evaluate the performance of our algorithms, including the prediction accuracy and the capability for identifying important features in diverse chemical structure data sets.

Acknowledgments

This work has been supported by the Kansas IDeA Network for Biomedical Research Excellence (NIH/NCRR award #P20 RR016475), the KU Center of Excellence for Chemical Methodology and Library Development (NIH/NIGM award #P50 GM069663), and NIH grant #R01 GM868665.

References

1. CP Austin, LS Brady, TR Insel, and FS Collins. Nih molec-
ular libraries initiative. *Science*, 306(5699):1138–9, 2004.
2. Yiming Ma Bing Liu, Wynne Hsu. Integrating classification and association rule mining. In *Proceedings of the Fourth International Conference on Knowledge Discovery and Data Mining*, 1998.
3. C. Chang and C. Lin. Libsvm: a library for support vector machines, 2001. Software available at http://www.csie.ntu.edu.tw/ cjlin/libsvm.
4. M. Crovella and E. Kolaczyk. Graph wavelets for spatial traffic analysis. *Infocom*, 3:1848–1857, 2003.
5. Antonios Deligiannakis and Nick Roussopoulos. Extended wavelets for multiple measures. In *Proceedings of the 2003 ACM SIGMOD international conference on Management of data*, 2003.
6. M. Deshpande, M. Kuramochi, and G. Karypis. Frequent sub-structure-based approaches for classifying chemical compounds. *IEEE Transactions on Knowledge and Data Engineering*, 2005.
7. Fröohlich, J. Wegner, F. Sieker, and A. Zell. Kernel functions for attributed molecular graphs - a new similarity-based approach to adme prediction in classification. QSAR & Combinatorial Science, 2006.
8. M. Garofalakis and Amit Kumar. Wavelet synopses for general error metrics. *ACM Transactions on Database Systems (TODS)*, 30(4):888–928, 2005.
9. C. Helma, R. King, and S. Kramer. The predictive toxicology challenge 2000-2001. *Bioinformatics*,

17(1):107–108, 2001.
10. Tamas Horvath, Thomas Gartner, and Stefan Wrobel. Cyclic pattern kernels for predictive graph mining. *SIGKDD*, 2004.
11. Jun Huan, Wei Wang, and Jan Prins. Efficient mining of frequent subgraph in the presence of isomorphism. In *Proceedings of the 3rd IEEE International Conference on Data Mining (ICDM)*, pages 549–552, 2003.
12. H. Kashima, K. Tsuda, and A. Inokuchi. Marginalized kernels between labeled graphs. In *Proc. of the Twentieth Int. Conf. on Machine Learning (ICML)*, 2003.
13. M. Maggioni, J. Bremer Jr, R. Coifman, and A. Szlam. Biorthogonal diffusion wavelets for multiscale representations on manifolds and graphs. In *Proc. SPIE Wavelet XI*, volume 5914, 2005.
14. D. Patterson, R. Cramer, A. Ferguson, R. Clark, and L. Weinberger. *Neighbourhood Behaviour: A Useful Concept for Validation of "Molecular Diversity" Descriptors*, 39:3049–3059, 1996.
15. Put R, Xu QS, Massart DL, and Vander Heyden Y. Multivariate adaptive regression splines (mars) in chromatographic quantitative structure-retention relationship studies. *J Chromatogr A.*, 1055(1-2), 2004.
16. Aaron Smalter, Jun Huan, and Gerald Lushington. Structure-based pattern mining for chemical compound classification. *Proceedings of the 6th Asia Pacific Bioinformatics Conference*, 2008.
17. V. Svetnik, C. Tong A. Liaw, J. C. Culberson, R. P. Sheridan, and B. P. Feuston. Random forest: A classification and regression tool for compound classification and qsar modeling. *Journal of chemical information and computer sciences*, 43, 2003.
18. Nicola Tolliday, Paul A. Clemons, Paul Ferraiolo, Angela N. Koehler, Timothy A. Lewis, Xiaohua Li, Stuart L. Schreiber, Daniela S. Gerhard, and Scott Eliasof. Small molecules, big players: the national cancer institute's initiative for chemical genetics. *Cancer Research*, 66:8935–42, 2006.
19. M. Wessel, P. Jurs, J. Tolan, and S. Muskal. Prediction of human intestinal absorption of drug compounds from molecular structure. *J. Chem. Inf. Comput. Sci.*, 38(4):726–735, 1998.
20. I. Witten and E. Frank. Morgan Kaufmann, San Francisco, CA, 2nd edition edition, 2005.
21. Y. Xue, H. Li, C. Y. Ung, C. W. Yap, and Y. Z. Chen. Classification of a diverse set of tetrahymena pyriformis toxicity chemical compounds from molecular descriptors by statistical learning methods. *Chem. Res. Toxicol.*, 19 (8), 2006.

AUTHOR INDEX